SpringerWienNewYork

CISM COURSES AND LECTURES

The series presents lecture notes, monographs, edited works and proceedings in the field of Mechanics, Engineering, Computer Science and Applied Mathematics.
Purpose of the series is to make known in the international scientific and technical community results obtained in some of the activities organized by CISM, the International Centre for Mechanical Sciences.

INTERNATIONAL CENTRE FOR MECHANICAL SCIENCES

COURSES AND LECTURES - No. 480

APPLIED MICROMECHANICS OF POROUS MATERIALS

EDITED BY

LUC DORMIEUX
ECOLE NATIONALE DES PONTS ET CHAUSSEES,
CHAMPS SUR MARNE, FRANCE

FRANZ-JOSEF ULM
MASSACHUSETTS INSTITUTE OF TECHNOLOGY,
CAMBRIDGE MA, U.S.A.

SpringerWienNewYork

The publication of this volume was co-sponsored and co-financed by the UNESCO Venice Office - Regional Bureau for Science in Europe (ROSTE) and its content corresponds to a CISM Advanced Course supported by the same UNESCO Regional Bureau.

This volume contains 80 illustrations

Printed in Italy
SPIN 11500322

In order to make this volume available as economically and as rapidly as possible the authors' typescripts have been reproduced in their original forms. This method unfortunately has its typographical limitations but it is hoped that they in no way distract the reader.

ISBN-10 3-211-26362-4 SpringerWienNewYork
ISBN-13 978-3-211-26362-4 SpringerWienNewYork

PREFACE

All natural composite materials (soils, rocks, woods, hard and soft tissues, etc.) and many engineered composites (concrete, bioengineered tissues, etc.) are multiphase and multiscale material systems. The multiphase composition of such materials is permanently evolving over various scales of time and length, creating in the course of this process the most heterogeneous class of materials in existence, with heterogeneities that manifest themselves from the nanoscale to the macroscale. The most prominent heterogeneity of such natural composite materials is the porosity, ie. the space left in between the different solid phases at various scales, ranging from interlayer spaces in between minerals filled by a few water molecules, to the macropore space in between microstructural units of the material in the micrometer to millimeter range. This porosity is the key to understanding and prediction of macroscopic material behavior, ranging from diffusive or advective transport properties to stiffness, strength and deformation behavior.

The specific nature of the mechanical behavior of multiphase porous materials was early on recognized in the groundbreaking works of M.A. Biot and K. Terzaghi, who developed the macroscopic basis of what is now known as 'Poromechanics'. Ever since, poromechanics has entered a large number of engineering applications ranging from civil and environmental engineering to petroleum engineering and more recently biomechanical engineering. In the 1970's, a breakthrough was achieved with pioneering works that relate macroscopic laws to microstructural properties. Furthermore, as new experimental techniques such as nanoindentation, now provide an unprecedented access to micromechanical properties and morphologies of materials, it becomes possible to trace these features from the nanoscale to the macroscale of day-to-day engineering applications, and predict transport properties, stiffness, strength and deformation behaviors within a consistent framework of 'Micromechanics of Porous Media'. The focus of this course which took place in July 2004, was to review fundamentals and applications of this rapidly emerging discipline of Applied Mechanics and Engineering Science.

This book assembles the lecture notes on 'Applied micromechanics of porous materials'. It is composed of three parts: (I) Transport properties of porous media; (II) Microporomechanics; (III) Materials Applications. Part I and II introduces the two fundamental homogenization theories of micromechanics of porous media, namely asymptotic expansion techniques and linear and nonlinear mean-field theories based on the concept of a representative elementary volume (Part II). The first is illustrated for the upscaling of fluid mass transport phenomena through the pore space that involve both advection and diffusion, and allows for a rigorous derivation of the permeability and tortuosity tensor. Linear and nonlinear mean-field theories are most effective for upscaling of the elastic and inelastic solid response of porous materials, which is illustrated for cracked porous media and for plastic deformation of saturated porous materials. Finally, the combination of microporomechanics theory with advanced experimental micromechanical techniques (incl. Nanoindentation, Atomic Force Microscopy and Environmental Scanning Electron Microscopy) is illustrated in Part III of these lecture notes, and is applied to the multiscale investigation of the poroelastic properties of cement-based materials, shales and bones.

From the onset, the course was designed as a well-balanced blend of theory and hands-on application of microporomechanics to a large range of porous materials in the linear and nonlinear regimes. Intended for a first course in 'micromechanics of porous media' for graduate students, researchers and engineers working at the forefront of Engineering Mechanics and Materials Science, we trust that these lecture notes be a source of imagination.

Luc Dormieux
Franz-Josef Ulm

CONTENTS

Part I

Transport properties of porous media

Transport in Porous Media: Upscaling by Multiscale Asymptotic Expansions

Jean-Louis Auriault*

* Laboratoire Sols, Solides, structures (3S), UJF, INPG, CNRS UMR 5521, Grenoble, France

Abstract Transport in porous media is investigated by upscaling the pore scale behaviour. We use the technique of multiscale asymptotic expansions which seems to be the most efficient method to obtaining macroscopic equivalent behaviours. Different transport phenomena are addressed: fluid flow through a saturated porous medium (Darcy's law), diphasic flow (coupled Darcy's laws), solute transport (diffusion, advection, dispersion) and fluid flow through deformable porous media (consolidation).

1 Physical Motivation

Heterogeneous media with a large number of heterogeneities cannot be described by considering each of the heterogeneities, that would yield to intractable boundary value problems. The well known method is to replace, *if possible*, the heterogeneous media by an homogeneous one, the description of which is valid at a very large scale (the macroscopic scale) with respect to the heterogeneity scale. As a continuous equivalent description, the derived macroscopic behaviour should be intrinsic to the medium and to the excitation, i.e. it should be independent of the macroscopic boundary conditions.
There are two main ways of deriving this macroscopic description. The first one is a directly macroscopic approach, which is often associated with experiments and is called the phenomenological approach. Many physical laws have been first derived by this kind of approaches. It is the case of Darcy's law, Darcy (1856) and of Biot's laws Biot (1962). The second kind of continuous approach allows to derive the macroscopic behaviour from the local description. This is an upscaling technique. Upscaling techniques allow the derivation of an equivalent macroscopic continuous description from the description of a Representative Elementary Volume (REV). The equivalent description is called the homogenized description. Different techniques are available which address random as well as periodic heterogeneous media. Among all, the most popular are the statistic modelling (Kröner, 1972), the self-consistent method, see Zaoui (1987), the volume averaging method, see, e.g., Nigmatulin (1981) and the homogenization for periodic structures (Sanchez-Palencia, 1974; Keller, 1977; Bensoussan et al., 1978).
When an upscaling technique is in use, the macroscopic behaviour is derived from the description at the heterogeneity scale that describes the physical process over a representative elementary volume (REV). The existence of a such a volume is required for any

continuous macroscopic representation of the physical system, and, as a consequence, is required for applying any upscaling technique. By definition, the REV is i) sufficiently large for representing the heterogeneity scale[1], and ii) small compared to the macroscopic volume. As a consequence, a condition of separation of scales is required. This fundamental condition can be expressed as:

$$\frac{l}{L} = \varepsilon << 1, \tag{1.1}$$

where l and L are the characteristic lengths at the REV scale and at the macroscopic scale, respectively.
This definition intuitively conjures up a geometrical separation of scales, whereas this fundamental condition must also be verified regarding the excitation (i.e., the physical process). For instance, consider the propagation of a wave in an heterogeneous medium. The wavelength actually constitutes a third characteristic length. Intuitively, we see that a continuous approach for describing this physical process will be possible only if the heterogeneity scale is small compared to the wavelength; a wavelength of the order of the heterogeneity length scale would lead to wave-trapping effects, which could not be completely described by an equivalent continuous behaviour at the macroscopic scale. For fluid flow in porous media as the medium is excited by a pressure gradient, the characteristic length of the excitation is thus related to the pressure gradient.
Therefore, the fundamental condition of separation of scales is expressed as $l/L << 1$, where L is the macroscopic characteristic length and is either geometrical or related to the excitation: the existence of the REV and, as a consequence, the condition of separation of scales, are not only constrained to geometrical considerations but also related to the excitation (i.e., the physical process). The analysis should focus on what we call the physical system, that consists of both the medium and the excitation.

In the lecture, we use the *method of homogenization for periodic structures*, which is also called the method of multiscale asymptotic expansions to investigate transport in porous media. This method has been introduced by Sanchez-Palencia (1974, 1980), Keller (1977), and Bensoussan et al. (1978). More recently, a more physical methodology based on dimensionless analysis has been introduced by Auriault (1991). This approach highlights the conditions under which homogenization can be applied. The presentation of this methodology is the purpose of the present paper. In this technique, a systematic use is made of the separation of scale parameter which, although small, is not generally null ($\varepsilon \neq 0$). That procures some advantages to the technique: (1) avoiding prerequisites at the macroscopic scale: the macroscopic equivalent description is obtained from the heterogeneity scale description plus the condition of separation of scales, only; (2) modelling finite size macroscopic samples and phenomena with finite macroscopic characteristic lengths ($\varepsilon \neq 0$); (3) modelling macroscopically nonhomogeneous media or phenomena;

[1]The REV size is typically equal or less than times the heterogeneity size, Cherel et al. (1988); Anguy et al. (1994). Therefore, a macroscopic equivalent behaviour cannot be deduced from an arbitrary boundary value problem concerning a single such REV: the result will be strongly dependent on the boundary conditions in use.

(4) modelling problems with several separations of scales by introducing several separation of scales parameter; (5) modelling several simultaneous phenomena; (6) determining whether the system "medium+phenomena" is homogenizable or not, i.e., whether or not a continuous equivalent macroscopic description exists; and (7) providing the domains of validity of the macroscopic models.
Evidently, real porous media are rarely periodic. However, there exists similarity in the behaviour of periodic and random media, on the condition that a separation of scales is present. Consider the right-angled parallelepipedal REV Ω_{REV} of a random medium. Construct a period, Ω, by introducing three successive plane symmetries with respect to three nonparallel faces of Ω_{REV}. The periodic medium of period Ω, and the random medium of REV Ω_{REV}, possess similar structure of their macroscopic description (although an eventual anisotropy could be modified by the plane symmetries). Therefore, we assume in the following that the heterogeneous medium is periodic, without lost of generality. In many applications, it was shown that this gives reasonable results (Quintard and Whitaker, 1993) for effective parameters.

In part 2, we rapidly present the method of multiscale asymptotic expansions. Then we successively address flow in saturated rigid porous media in part 3, diphasic flow in part 4, solute transport in part 5 and flow in deformable porous media in part 6. All these investigations show two separated scales, only.

2 Multiscale Asymptotic Expansion Technique

A periodic medium of period Ω and a separation of scales, $l/L = \varepsilon \ll 1$ are assumed. Such a period of a porous medium is shown in Fig.1. Let $\mathbf{X}$ be the physical space

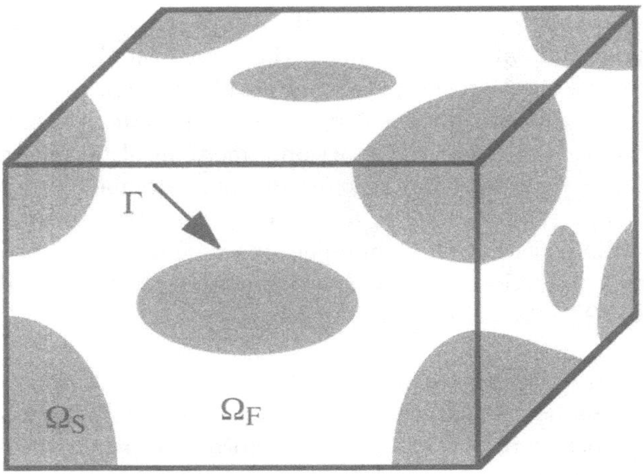

Figure 1. Period of the porous medium

variable of the medium. From the two characteristic lengths l and L, two dimensionless space variables can be defined:

$$\begin{aligned} \mathbf{y} &= \mathbf{X}/l \quad \text{is the dimensionless "microscopic" space variable,} \\ \mathbf{x} &= \mathbf{X}/L \quad \text{is the dimensionless "macroscopic" space variable.} \end{aligned}$$

As a consequence of the separation of scales, $\mathbf{y}$ and $\mathbf{x}$ are two separated variables. Any space dependent quantity φ required to describe the physical process can be described, *a priori*, as a function of these two separated variables:

$$\varphi = \varphi(\mathbf{y}, \mathbf{x}). \tag{2.1}$$

Obviously, these space variables are related by $\mathbf{x} = \varepsilon \mathbf{y}$. The macroscopic equivalent model is obtained from the description at the heterogeneity scale by (Auriault, 1991):
1. assuming the medium to be periodic, without loss of generality since the separation of scales condition is fulfilled. That implies that the different φ's are periodic with respect to space variable $\mathbf{y}$. Note that the eventual dependence of φ with respect to $\mathbf{x}$ shows the eventual macroscopic inhomogeneity of the medium,
2. writing the local description in a dimensionless form, by using l or L as the characteristic length (the choice of l or L is a technical point of no effect on the final result) and by using characteristic values φ_c for the different quantities φ entering the local description: $\varphi = \varphi_c \varphi^*$, where an asterisk shows a dimensionless quantity,
3. deriving dimensionless numbers from the governing equations and evaluating these

dimensionless numbers with respect to the scale ratio ε. A dimensionless number Q is said to be $\mathcal{O}(\varepsilon^p)$ if

$$\varepsilon^{p+1} << Q << \varepsilon^{p-1}. \tag{2.2}$$

For the sake of simplicity, we assume here that all dimensionless numbers can be evaluated by integer powers of ε. Once all dimensionless numbers have been estimated, the governing equations can then be written as :

$$\sum_{q=0}^{\infty} \varepsilon^q \, A^{*(q)} = 0, \tag{2.3}$$

where the q's are integers. The $A^{*(q)}$'s are generally some dimensionless operators. Although the value of ε is fixed in each point of the medium for a given upscaling problem, the corresponding macroscopic equivalent model at stake, if it exists, is valid for all values of ε that respect the above evaluations of the dimensionless numbers. Thus parameter ε can be considered as varying in some range.

4. looking for the dimensionless unknown fields in the form of asymptotic expansions in powers of ε. As a consequence of both the separation of scales and the periodicity, all physical quantities can be looked for in the form of asymptotic expansions in powers of ε (Bensoussan et al., 1978):

$$\varphi^*(\mathbf{y},\mathbf{x}) = \varphi^{(0)}(\mathbf{y},\mathbf{x}) + \varepsilon^1\varphi^{(1)}(\mathbf{y},\mathbf{x}) + \varepsilon^2\varphi^{(2)}(\mathbf{y},\mathbf{x}) + ... \tag{2.4}$$

where the $\varphi^{(i)}$'s are $\mathbf{y}$ periodic and $\varphi^{(0)}$ is $\mathcal{O}(1)$,

5. solving the successive local boundary-value problems that are obtained after introducing these expansions in the local dimensionless description, and equating like powers of ε,

6. obtaining the macroscopic equivalent model from compatibility conditions which are the sufficient and necessary conditions for the existence of solutions to the successive local boundary-value problems. At some given order of ε, a local balance equation will yield an equation of the form

$$\nabla_y \,.\, \mathbf{\Phi}^{(i+1)} + \nabla_x \,.\, \mathbf{\Phi}^{(i)} = F^*, \tag{2.5}$$

where $\mathbf{\Phi}^{(i+1)}$ and $\mathbf{\Phi}^{(i)}$ are $\mathbf{y}$-periodic tensors and F^* is a source term. For simplicity, we assume here that (2.5) is valid over all Ω^* in the distribution sense. Other situations are possible, see, e.g., balance (3.22) below. This latter equation will lead to the macroscopic description. It actually expresses the balance at the $\mathbf{y}$ scale of the periodic quantity $\mathbf{\Phi}^{(i+1)}$, in presence of the source term $S^* = F^* - \nabla_x.\mathbf{\Phi}^{(i)}$. Integrating it over the period gives what is called a "compatibility condition". This compatibility condition must be checked (Fredholm alternative), otherwise the original equation has no solution and the problem is not homogenizable. It is written as

$$< \nabla_x \,.\, \mathbf{\Phi}^{(i)} >_{\Omega^*} - < F^* >_{\Omega^*} = 0, \tag{2.6}$$

where $< . >_{\Omega^*}$ denotes the volume average with respect to space variable $\mathbf{y}$ over the period and is defined by

$$< . >_{\Omega^*} = \frac{1}{\mid \Omega^* \mid} \int_{\Omega_y^*} . \, \mathrm{dV}^*, \tag{2.7}$$

and

7. checking if the aforementioned process yields results that verify the estimations of dimensionless numbers in point 3: if not, e.g. if $A^{*(0)} = 0$, then the system (medium + phenomena) under consideration is not homogenizable: there exists no equivalent macroscopic description. Of course, an averaging process is possible in such a case, but it yields pseudo-effective coefficients that depend on the boundary conditions of the macroscopic boundary value problem and on L, thus leading to scale effects. The macroscopic mathematical model loses its legitimacy. The homogeneised solution is valid in the bulk macroscopic volume. Near the macroscopic boundary, the periodicity is broken and boundary layers are introduced to match the homogeneised solution to the macroscopic boundary conditions. The thickness of these boundary layers is generally of approximately one period size for diffusion or elastic problems, Sanchez-Palencia (1987). Their investigation yields the homogeneised boundary condition to be used, see Ene and Sanchez-Palencia (1975) for an example. Boundary layers are not studied in the present lecture.

3 Darcy's Law

For a sufficiently low pore Reynolds number Re, the steady state flow of incompressible Newtonian fluids through a Galilean porous media with connected pores is described by the celebrated Darcy's law. The velocity $\mathbf{v}$ is linearly related to the gradient of pressure

$$v_i = -\frac{K_{ij}}{\mu}\frac{\partial p}{\partial X_j}, \tag{3.1}$$

where $\mathbf{K}$ is the permeability tensor and μ is the viscosity. Let us recover Darcy's law by upscaling the pore scale behaviour, see Ene and Sanchez-Palencia (1975), Sanchez-Palencia (1980).

We consider a periodic rigid porous matrix of period Ω saturated by an incompressible fluid. The pores are connected; they occupy Ω_F and the pore surface is denoted Γ, see Figure 1. The macroscopic characteristic length is L and the characteristic pore size is l. Characteristic lengths L and l are generally difficult to estimate *a priori*. We just assume for the moment a separation of scales

$$\frac{l}{L} = \varepsilon \ll 1. \tag{3.2}$$

Characteristic lengths L and l will be made precise later.

3.1 Pore Scale Description of the Flow

The flow is slow, laminar, permanent and isothermal: at the pore scale, the Reynolds number and the transient Reynolds number are assumed as small, $Re_l \leq \mathcal{O}(\varepsilon)$, $Rt_l \leq \mathcal{O}(\varepsilon)$ (Subscript l shows that the dimensionless numbers are calculated by using l as the characteristic length). Therefore, the fluid velocity $\boldsymbol{v}$ and the pressure p verify

$$\mu\frac{\partial^2 v_i}{\partial X_j \partial X_j} - \frac{\partial p}{\partial X_i} = 0, \tag{3.3}$$

$$\frac{\partial v_i}{\partial X_i} = 0 \qquad \text{in } \Omega_F, \tag{3.4}$$

$$v_i = 0 \qquad \text{on } \Gamma. \tag{3.5}$$

The above set is put in dimensionless form by introducing the dimensionless space variable $y_i = X_i/l$ and adequate characteristic quantities, v_c, p_c and μ_c : $\mathbf{v} = v_c\mathbf{v}^*$, $p = p_c p^*$, $\mu = \mu_c\mu^*$. The above set introduces the dimensionless number

$$\mathcal{Q}_l = \frac{|\frac{\partial p}{\partial X_i}|}{|\mu\frac{\partial^2 v_i}{\partial X_j \partial X_j}|} = \frac{l p_c}{\mu_c v_c}. \tag{3.6}$$

Number $\mathcal{Q}_l$ can be evaluated in function of ε by noticing that the flow at the pore level is caused by a macroscopic gradient of pressure which equilibrates the viscous force

$$\frac{p_c}{L} = \mathcal{O}(\frac{\mu_c v_c}{l^2}). \tag{3.7}$$

We deduce that $\mathcal{Q}_l = \mathcal{O}(\varepsilon^{-1})$. The dimensionless form of the set (3.3-3.5) is the following, with $\mathcal{Q}_l = \varepsilon^{-1}$

$$\varepsilon\mu^* \frac{\partial^2 v_i^*}{\partial y_j \partial y_j} - \frac{\partial p^*}{\partial y_i} = 0, \tag{3.8}$$

$$\frac{\partial v_i^*}{\partial y_i} = 0 \qquad \text{in } \Omega_F^*, \tag{3.9}$$

$$v_i^* = 0 \qquad \text{on } \Gamma^*. \tag{3.10}$$

3.2 Upscaling

By following the multiple scale expansion technique (Sanchez-Palencia, 1980), the velocity $\mathbf{v}^*$ and the pressure fluctuation p^* are looked for in the form of asymptotic expansions of powers of ε

$$\mathbf{v}^* = \mathbf{v}^{(0)}(\mathbf{x},\mathbf{y}) + \varepsilon\, \mathbf{v}^{(1)}(\mathbf{x},\mathbf{y}) + \varepsilon^2\, \mathbf{v}^{(2)}(\mathbf{x},\mathbf{y}) + \ldots \tag{3.11}$$

$$p^* = p^{(0)}(\mathbf{x},\mathbf{y}) + \varepsilon\, p^{(1)}(\mathbf{x},\mathbf{y}) + \varepsilon^2\, p^{(2)}(\mathbf{x},\mathbf{y}) + \ldots \tag{3.12}$$

Introducing these expansions in (3.8-3.10) gives by identification of the like powers of ε successive boundary value problems to be investigated. The first order approximation of the pressure verifies

$$\frac{\partial p^{(0)}}{\partial y_i} = 0, \qquad p^{(0)} = p^{(0)}(\mathbf{x}). \tag{3.13}$$

The first order approximation of the velocity $\mathbf{v}^{(0)}$ and the second order approximation of the pressure $p^{(1)}$ are determined by the following set

$$\mu^* \frac{\partial^2 v_i^{(0)}}{\partial y_j \partial y_j} - J_i - \frac{\partial p^{(1)}}{\partial y_i} = 0, \tag{3.14}$$

$$\frac{\partial v_i^{(0)}}{\partial y_i} = 0 \qquad \text{in } \Omega_F^*, \tag{3.15}$$

$$v_i^{(0)} = 0 \qquad \text{on } \Gamma^*, \tag{3.16}$$

where $\mathbf{v}^{(0)}$ and $p^{(1)}$ are Ω^*-periodic. Vector $J_i = \partial p^{(0)}/\partial x_i$ is the macroscopic pressure gradient that causes the flow. To investigate this boundary value problem, it is convenient to introduce the space $\mathcal{V}$ of periodic divergence free vectors, defined on Ω_F^*, null on Γ^*, with the scalar product:

$$(\mathbf{u}, \mathbf{v})_{\mathcal{V}} = \int_{\Omega_F^*} \mu^* \frac{\partial u_i}{\partial y_j} \frac{\partial v_i}{\partial y_j} \, \mathrm{dV}^*, \tag{3.17}$$

which satisfies the following symmetry

$$(\mathbf{u}, \mathbf{v})_{\mathcal{V}} = (\mathbf{v}, \mathbf{u})_{\mathcal{V}}. \tag{3.18}$$

Now, multiply the two members of (3.14) by $\mathbf{u} \in \mathcal{V}$ and integrate over Ω_F^*. By using integration by parts, the divergence theorem, the periodicity and the boundary condition on Γ^*, we obtain

$$\forall\ \mathbf{u} \in \mathcal{V}, \qquad (\mathbf{u}, \mathbf{v}^{(0)})_{\mathcal{V}} = -(\mathbf{u}, \mathbf{J})_{\mathrm{L}^2}, \tag{3.19}$$

It gives $\mathbf{v}^{(0)}$ as a linear vectorial function of $\mathbf{J}$

$$v_i^{(0)} = -\lambda_{ij}^* J_j, \tag{3.20}$$

where $\boldsymbol{\lambda}^*(\mathbf{y})$ is the pore scale conductivity tensor field. When the viscosity μ is a constant, this relation can be put in the form

$$v_i^{(0)} = -\frac{k_{ij}^*}{\mu^*} J_j, \tag{3.21}$$

where $\mathbf{k}^*(\mathbf{y})$ is the pore scale permeability tensor field.

Then the volume balance of $\mathbf{v}^{(1)}$

$$\frac{\partial v_i^{(1)}}{\partial y_i} + \frac{\partial v_i^{(0)}}{\partial x_i} = 0, \tag{3.22}$$

gives by averaging a compatibility condition which stands for the macroscopic volume balance and introduces the Darcy's law

$$\frac{\partial \langle v_i^{(0)} \rangle}{\partial x_i} = 0, \qquad \langle v_i^{(0)} \rangle = -\Lambda_{ij}^* J_j = -\frac{K_{ij}^*}{\mu^*} J_j, \qquad \mathbf{K}^* = \langle \mathbf{k}^* \rangle = \frac{1}{|\Omega^*|} \int_{\Omega_F^*} \mathbf{k}^* \, \mathrm{dV}^*, \tag{3.23}$$

where $\boldsymbol{\Lambda}^*$ and $\mathbf{K}^*$ are the macroscopic dimensionless conductivity and permeability tensors, respectively.

The vector $\mathbf{k}_{.j}^*$ is solution of the following variational form

$$\forall\ \mathbf{u} \in \mathcal{V}, \qquad (\mathbf{u}, \mathbf{k}_{.j}^*)_{\mathcal{V}} = \mu^* (\mathbf{u}, \mathbf{I}_{.j})_{\mathrm{L}^2}, \tag{3.24}$$

where $\mathbf{I}$ is the identity matrix.

3.3 Macroscopic Flow Model

By returning to dimensional quantities, relations (3.23) become

$$\frac{\partial \langle v_i \rangle}{\partial X_i} = \mathcal{O}(\varepsilon \frac{\partial \langle v_i \rangle}{\partial X_i}), \qquad \langle v_i \rangle = -\frac{K_{ij}}{\mu} \frac{\partial p}{\partial X_j} + \mathcal{O}(\varepsilon \langle v_i \rangle), \qquad \mathbf{K} = l^2 \mathbf{K}^*. \tag{3.25}$$

Remark that the macroscopic model is defined within a relative error $\mathcal{O}(\varepsilon)$. By using formulation (3.24), it is possible to show that tensor $\mathbf{K}$ is positive and symmetrical, (Sanchez-Palencia, 1980), see below.

Volume averaged velocity is a flux. It is important to check the physical meaning of the macroscopic quantities that are introduced by the upscaling process. Quantity p in (3.25) does not pose problem: it is directly defined as the first approximation of the local pore pressure, which first approximation is a constant over the period. Therefore, quantity p in (3.25) is a pressure. On an other hand $\langle \mathbf{v} \rangle$, which is a volume average can be shown to be a Darcy velocity, i.e. a flux. This is the consequence of the solenoidal character of $\mathbf{v}^{(0)}$. Let start from the identity :

$$\frac{\partial}{\partial y_k}(v_k^{(0)} y_i) = \frac{\partial v_k^{(0)}}{\partial y_k} y_i + v_k^{(0)} I_{ik}.$$

After integrating over Ω_F^*, using the divergence theorem and the adherence condition on Γ^*, we obtain:

$$\langle v_i^{(0)} \rangle = |\Omega^*|^{-1} \int_{\delta\Omega_F^* \cap \delta\Omega^*} v_k^{(0)} y_i N_k \, \mathrm{dS}^*,$$

where $\delta\Omega_F^*$ and $\delta\Omega^*$ are the boundaries of Ω_F^* and Ω^*, respectively, and N is the unit outward normal to Ω_F^*. Let l_i^* be the dimensionless length of the period along direction y_i and let Σ_i^* be the cross-section of the period at $y_i = l_i^*$, see Figure 2. Surface $\Sigma_{F_i}^*$ is the fluid part of Σ_i^* . Because of the Ω^*-periodicity of $v_k^{(0)} y_i$ in direction $y_j, j \neq i$, and because it cancels out at $y_i = 0$, we are left with :

$$< v_i^{(0)} >= |\Omega^*|^{-1} \int_{\Sigma_{F_i}^*} v_i^{(0)} l_i^* \, \mathrm{dS}^* = |\Sigma_i^*|^{-1} \int_{\Sigma_{F_i}^*} v_i^{(0)} \mathrm{dS}^*,$$

(no summation on i). Therefore, $< \mathbf{v}^{(0)} >$ is a flux and $< \mathbf{v} >$ too at its first order of approximation.

Tensors $\mathbf{\Lambda}$ and $\mathbf{K}$ are symmetrical. Consider formulation (3.19) with on the one hand $\mathbf{v}^{(0)} = \boldsymbol{\lambda}^*_{.q}$, $\mathbf{u} = \boldsymbol{\lambda}^*_{.p}$ and on the other hand $\mathbf{v}^{(0)} = \boldsymbol{\lambda}^*_{.p}$, $\mathbf{u} = \boldsymbol{\lambda}^*_{.q}$. Due to the symmetry of the scalar product, we obtain

$$\int_{\Omega_F^*} \mu^* \frac{\partial \lambda_{ip}}{\partial y_j} \frac{\partial \lambda_{iq}^*}{\partial y_j} \, \mathrm{dV}^* = -\int_{\Omega_F^*} \lambda_{pq}^* \mathrm{dV}^* = -\int_{\Omega_F^*} \lambda_{qp}^* \mathrm{dV}^*.$$

Tensor $\mathbf{\Lambda}$ is symmetrical, and tensor $\mathbf{K}$ too.

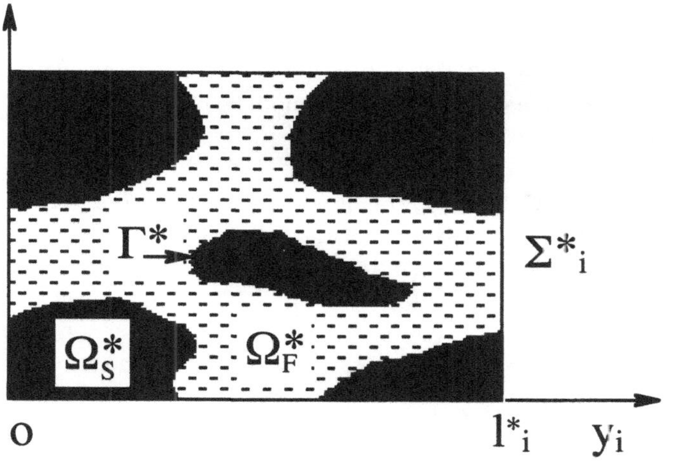

Figure 2. Period Ω^* of the porous medium, two-dimensional case

Tensors $\mathbf{\Lambda}$ and $\mathbf{K}$ are positive. Consider now (3.19) with $\mathbf{u} = \mathbf{v}^{(0)}$

$$|\Omega^*|^{-1} \int_{\Omega_F^*} \mu^* \frac{\partial v_i^{(0)}}{\partial y_j} \frac{\partial v_i^{(0)}}{\partial y_j} \, \mathrm{dV}^* = -|\Omega^*|^{-1} \int_{\Omega_F^*} v_i^{(0)} \frac{\partial p^{(0)}}{\partial x_i} \, \mathrm{dV}^* = \frac{\partial p^{(0)}}{\partial x_i} \Lambda_{ij}^* \frac{\partial p^{(0)}}{\partial x_j}.$$

The left hand member is positive. Then, the right hand member is positive and tensor $\mathbf{\Lambda}$ is positive, as well as tensor $\mathbf{K}$.

Evaluation of l and L. It is possible to *a posteriori* obtain an evaluation of l. It generally depends on the flow direction $\mathbf{n}$. From the third relation in (3.25) we have

$$l_{\mathbf{n}} \approx \sqrt{K_{ij} n_i n_j}, \tag{3.26}$$

which in the isotropic case reduces to

$$l \approx \sqrt{K}. \tag{3.27}$$

The macroscopic characteristic length L is either a characteristic size of the macroscopic boundary value problem either a length related to the flow itself. This later one can be *a posteriori* evaluated from the macroscopic pressure field

$$L \approx \frac{p}{|\nabla p|}. \tag{3.28}$$

A consequence of these estimations is that L generally depends on $\mathbf{X}$ if the porous medium is not macroscopically homogeneous. It results in a dependence of ε on $\mathbf{X}$, $\varepsilon = \varepsilon(\mathbf{X})$.

3.4 Domain of validity of the model

For permanent and quasi-static slow flow, physical considerations showed that $\mathcal{Q}_l = \mathcal{O}(\varepsilon^{-1})$, which is the domain of validity of the model (3.25). Let us have a look to other estimations of $\mathcal{Q}_l$.

Smaller estimations of $\mathcal{Q}_l$. Let $\mathcal{Q}_l$ be decreased to $\mathcal{Q}_l = \varepsilon^q$, $q \geq 0$, and let us explore the case $\mathcal{Q}_l = 1$. The dimensionless form of the set (3.3-3.5) is now in the form

$$\mu^* \frac{\partial^2 v_i^*}{\partial y_j \partial y_j} - \frac{\partial p^*}{\partial y_i} = 0. \tag{3.29}$$

$$\frac{\partial v_i^*}{\partial y_i} = 0 \qquad \text{in } \Omega_F^*, \tag{3.30}$$

$$v_i^* = 0 \qquad \text{on } \Gamma^*. \tag{3.31}$$

The first orders approximation of the velocity $\mathbf{v}^{(0)}$ and of the pressure $p^{(0)}$ are determined by the following set

$$\mu^* \frac{\partial^2 v_i^{(0)}}{\partial y_j \partial y_j} - \frac{\partial p^{(0)}}{\partial y_i} = 0, \tag{3.32}$$

$$\frac{\partial v_i^{(0)}}{\partial y_i} = 0 \qquad \text{in } \Omega_F^*, \tag{3.33}$$

$$v_i^{(0)} = 0 \qquad \text{on } \Gamma^*, \tag{3.34}$$

where $\mathbf{v}^{(0)}$ and $p^{(0)}$ are Ω^*-periodic. This system is similar to the system (3.14-3.16) after replacing $p^{(1)}$ by $p^{(0)}$ and letting $\mathbf{J} = \mathbf{0}$. Therefore, we obtain $\mathbf{v}^{(0)} = 0$, which is in a contradiction with $|\mathbf{v}^{(0)}| = \mathcal{O}(1)$. Estimation $\mathcal{Q}_l = 1$ does not correspond to a homogenizable situation. We also obtain $p^{(0)} = p^{(0)}(\mathbf{x})$. The next order yields a boundary value problem for $\mathbf{v}^{(1)}$ and $p^{(1)}$. It is easy to check that this problem is similar to system (3.14-3.16) after replacing $\mathbf{v}^{(0)}$ by $\mathbf{v}^{(1)}$. We then recover Darcy's flow

$$v_i^{(1)} = -\frac{k_{ij}^*}{\mu^*} J_j,$$

The homogenization process shifts $\mathcal{Q}_l = \mathcal{O}(1)$ to $\mathcal{Q}_l = \mathcal{O}(\varepsilon^{-1})$. Estimation $\mathcal{Q}_l = \varepsilon^q$, $q \geq 0$ yields $\mathbf{v}^{(i)} = 0$, $i \leq q+1$, which is also a non-homogenizable situation. All these estimations are of no physical meaning.

Higher estimations of $\mathcal{Q}_l$. If $\mathcal{Q}_l$ is increased by increasing the gradient of pressure, $\mathcal{Q}_l = \varepsilon^{-q}$, $q \geq 2$, we obtain also a non-homogenizable situation when assuming Stokes equation to be valid. In the case $\mathcal{Q}_l = \varepsilon^{-2}$, we have at the local scale

$$\varepsilon^2 \mu^* \frac{\partial^2 v_i^*}{\partial y_j \partial y_j} - \frac{\partial p^*}{\partial y_i} = 0,$$

$$\frac{\partial v_i^*}{\partial y_i} = 0 \qquad \text{in } \Omega_F^*,$$

$$v_i^* = 0 \qquad \text{on } \Gamma^*.$$

At the lower order we recover

$$\frac{\partial p^{(0)}}{\partial y_i} = 0, \qquad p^{(0)} = p^{(0)}(\mathbf{x}).$$

However, the second order gives

$$\frac{\partial p^{(0)}}{\partial x_i} + \frac{\partial p^{(1)}}{\partial y_i} = 0.$$

The periodicity of $p^{(1)}$ yields $p^{(0)}$ = constant, which is not admissible since $\partial p^{(0)}/\partial x_i$ should be $\mathcal{O}(1)$. We deduce that the estimation $\mathcal{Q}_l = \varepsilon^{-2}$ corresponds to a non-homogenizable situation. However, continuing the homogenization process gives the Darcy's flow at the next order. The homogenization process shifts $\mathcal{Q}_l = \mathcal{O}(\varepsilon^{-2})$ to $\mathcal{Q}_l = \mathcal{O}(\varepsilon^{-1})$. It is easy to show that estimations $\mathcal{Q}_l = \varepsilon^{-q}$, $q \geq 2$ are non-homogenizable.

Concluding remarks. The above results are resumed in Table 1. Obviously, the adopted pore scale behaviour (3.3-3.5) is not convenient to describe other numerous flows in porous media. When the gradient of pressure is increased, inertial non-linearities appear. Deviation to Darcy's law in this case was investigated theoretically by Wodie and Levy (1991) and Mei and Auriault (1991), experimentally in Rasoloarijaona and Auriault (1994), Skjetne and Auriault (1999a), and numerically by Firdaouss et al. (1997). For small Reynolds numbers the obtained flow law is cubic with respect of the seepage velocity. For compressible fluids, the local balance equation (3.4) is modified. Darcy's law remains valid, but the macroscopic volume balance becomes non-linear, Auriault et al. (1990). in the case of low pressure gas flow, wall-slip appears on the pore surface, that

Table 1. Stokes' flow through rigid porous media.

$\mathcal{Q}_l = \mathcal{O}(\varepsilon^q)$, $q \geq 0$	$\mathcal{Q}_l = \mathcal{O}(\varepsilon^{-1})$	$\mathcal{Q}_l = \mathcal{O}(\varepsilon^{-q})$, $q \geq 2$
Non-homogenizable	Darcy's law	Non-homogenizable

yields Klinkenberg's law, Skjetne and Auriault (1999b), Chastanet et al. (2004). Transient flows were also investigated by homogenization in Levy (1979), Auriault (1980) and Burridge and Keller (1981), in which the phenomenological law previously introduced by Biot (1941) is recovered. Let us mention also non-Newtonian fluid flow in Auriault et al. (2002a) and flow in non-Galilean porous matrix, Auriault et al. (2000), Auriault et al. (2002b). All these studies concern single porosity porous media. Double or multiple porosity media introduce more than two separated scales, that generally strongly modifies the macroscopic behaviour, see e.g. Auriault and Boutin (1992, 1993).

4 Immiscible Two-phase Flow in Porous Media

The aim of this part is to investigate the governing equations that describe the flow of two immiscible fluids through a Galilean rigid porous medium (Auriault and Sanchez-Palencia, 1986; Auriault, 1987). The two fluids are shown by subscripts 1 and 2, respectively. To simplify the analysis we assume that

- A separation of scales is present, $l/L = \varepsilon \ll 1$, see the previous part for the definition of the two lengths l and L.

- The porous medium is rigid and Galilean and the two fluids are viscous Newtonian and incompressible.

- The two viscosities are of the same order of magnitude with respect to ε.

- As in the previous part, flows are slow (inertia is negligible), isothermal and time variations are neglected in Navier-Stokes equations.

- Each volume Ω_1 and Ω_2 occupied by each fluid is connected. In such flows, the interface Γ_{12} is generally moving. We assume this movement to be slower than the flow itself: if $\mathbf{n}$ is a unit normal to Γ_{12}, the fluid velocity on this surface verifies $\mathbf{v}.\mathbf{n} = \mathcal{O}(\varepsilon|\mathbf{v}_1|) = \mathcal{O}(\varepsilon|\mathbf{v}_2|)$. This assumption is coherent with equation (4.12) below. Integrating (4.12) over Ω^*_α, applying the divergence theorem and considering the periodicity of $\mathbf{v}^{(0)}_\alpha$ yields

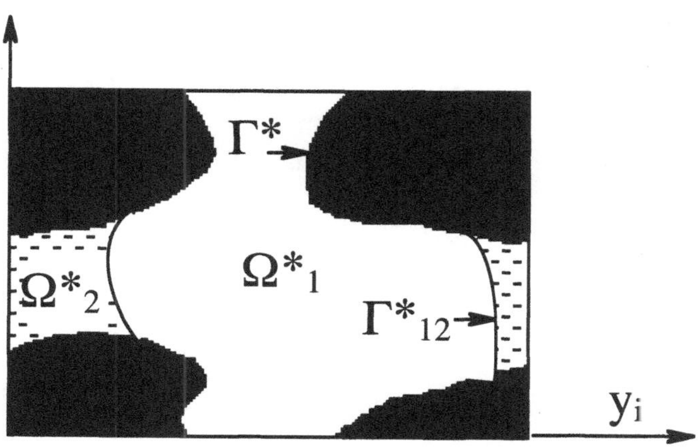

Figure 3. Period Ω^* of the non-saturated porous medium, cross section

$$\int_{\Gamma^*_{12}} \mathbf{v}^{(0)}_\alpha.\mathbf{n}\ \mathrm{dS}^* = 0.$$

- The Weber number $\mathcal{W}$ is $\mathcal{O}(1)$: the capillary pressure p_{ca} is of the same order of magnitude as the pressure in the two fluids. Laplace equation is written

$$p_{ca} = \sigma(\frac{1}{R_1} + \frac{1}{R_2}), \tag{4.1}$$

where σ is the surface tension assumed here as a constant and the two principal radii of curvature R_1 and R_2 are of the order of magnitude of the characteristic length l of the pores. Therefore, we have

$$\mathcal{W} = p_c l \sigma^{-1} = \mathcal{O}(1). \tag{4.2}$$

- There is no mass transfer between the two fluids.

4.1 Pore Scale Description.

By following the same route as for the one fluid flow, it can be shown that

$$\mathcal{Q}_{1_l} = \mathcal{O}(\mathcal{Q}_{2_l}) = \mathcal{O}(\varepsilon^{-1}). \tag{4.3}$$

Therefore, with the above assumptions, the dimensionless local description is in the form

$$\varepsilon\mu_\alpha^* \frac{\partial^2 v_{\alpha i}^*}{\partial y_j \partial y_j} - \frac{\partial p_\alpha^*}{\partial y_i} = 0, \tag{4.4}$$

$$\frac{\partial v_{\alpha i}^*}{\partial y_i} = 0 \qquad \text{in } \Omega_\alpha^*, \tag{4.5}$$

$$v_{\alpha i}^* = 0 \qquad \text{on } \Gamma^*, \tag{4.6}$$

where $\alpha = 1, 2$ and on Γ_{12}^*

$$(\sigma_{2_{ij}}^* - \sigma_{1_{ij}}^*) n_j = p_{ca}^* n_i, \tag{4.7}$$

where the $\sigma_{\alpha_{ij}}^* = -p_\alpha^* I_{ij} + \varepsilon\mu^*(\frac{\partial v_{\alpha_i}^*}{\partial y_j} + \frac{\partial v_{\alpha_j}^*}{\partial y_i})$ are the stress components, and

$$v_{1_i}^* = v_{2_i}^*, \qquad v_{\alpha_i}^* n_i = \mathcal{O}(\varepsilon |\mathbf{v}_\alpha^*|). \tag{4.8}$$

4.2 Upscaling.

After introducing *ad hoc* asymptotic expansions in the above description, and identifying like powers of ε, we obtain from (4.4) at the first order

$$\frac{\partial p_\alpha^{(0)}}{\partial y_i} = 0, \qquad p_\alpha^{(0)} = p_\alpha^{(0)}(\mathbf{x}), \tag{4.9}$$

and from (4.7)

$$p_{ca}^{(0)} = p_2^{(0)} - p_1^{(0)} = p_{ca}^{(0)}(\mathbf{x}). \tag{4.10}$$

The first approximation of the capillary pressure is a constant over the period, as well as the first approximation of the mean curvature of the surface Γ_{12}^*.

At the following order, we obtain a boundary value problem for $\mathbf{v}_\alpha^{(0)}$ and $p_\alpha^{(1)}$, $\alpha = 1, 2$

$$\mu_\alpha^* \frac{\partial^2 v_{\alpha i}^{(0)}}{\partial y_j \partial y_j} - \frac{\partial p_\alpha^{(1)}}{\partial y_i} - \frac{\partial p_\alpha^{(0)}}{\partial x_i} = 0, \tag{4.11}$$

$$\frac{\partial v_{\alpha i}^{(0)}}{\partial y_i} = 0 \qquad \text{in } \Omega_\alpha^*, \tag{4.12}$$

$$v_{\alpha i}^{(0)} = 0 \qquad \text{on } \Gamma^*, \tag{4.13}$$

and on Γ_{12}^*

$$(\sigma_{2_{ij}}^{(1)} - \sigma_{1_{ij}}^{(1)}) n_j = p_{ca}^{(1)} n_i, \tag{4.14}$$

$$v_{1_i}^{(0)} = v_{2_i}^{(0)}, \qquad v_{\alpha_i}^{(0)} n_i = 0, \tag{4.15}$$

and where $\mathbf{v}_\alpha^{(0)}$ and $p_\alpha^{(1)}$ are Ω^*-periodic. To investigate the above boundary value problem, let us introduce the space $\mathcal{V}_1$ of Ω^*-periodic vectors, defined over $\Omega_F^* = \Omega_1^* + \Omega_2^*$, with zero divergence, zero valued on Γ^*, continuous on Γ_{12}^*, with zero flux through Γ_{12}^*, and with the scalar product

$$(\mathbf{u}, \mathbf{v})_{\mathcal{V}_1} = \int_{\Omega_F^*} \mu_\alpha^* \frac{\partial u_i}{\partial y_j} \frac{\partial v_i}{\partial y_j} \, \mathrm{dV}^*. \tag{4.16}$$

Multiply the two members of equation (4.11) by $\mathbf{u} \in \mathcal{V}_1$ and integrate over Ω_F^*. Since this equation can be written

$$\frac{\partial \sigma_{\alpha_{ij}}^{(1)}}{\partial y_j} = -\frac{\partial p_\alpha^{(0)}}{\partial x_i},$$

we have in Ω_1^*, after using the periodicity, the adherence condition on Γ^*, and the incompressibility at the first order

$$\int_{\Omega_1^*} u_i \frac{\partial \sigma_{1_{ij}}^{(1)}}{\partial y_j} \, \mathrm{dV}^* = -\int_{\Omega_1^*} u_i \frac{\partial p_1^{(0)}}{\partial x_i} \, \mathrm{dV}^* = \int_{\Omega_1^*} \left(\frac{\partial u_i \sigma_{1_{ij}}^{(1)}}{\partial y_j} - \sigma_{1_{ij}}^{(1)} \frac{\partial u_i}{\partial y_j}\right) \mathrm{dV}^*$$

$$= \int_{\Gamma_{12}^*} u_i \sigma_{1_{ij}}^{(1)} N_{1_j} \, \mathrm{dS}^* - \int_{\Omega_1^*} \mu_1^* \frac{\partial u_i}{\partial y_j} \frac{\partial v_{1_i}^{(0)}}{\partial y_j} \, \mathrm{dV}^*,$$

where $\mathbf{N}_1$ is the outward unit normal to Ω_1^*. In the same way, we obtain in Ω_2^*

$$\int_{\Omega_2^*} u_i \frac{\partial \sigma_{2_{ij}}^{(1)}}{\partial y_j} \, \mathrm{dV}^* = -\int_{\Gamma_{12}^*} u_i \sigma_{2_{ij}}^{(1)} N_{1_j} \mathrm{dS}^* - \int_{\Omega_2^*} \mu_2^* \frac{\partial u_i}{\partial y_j} \frac{\partial v_{2_i}^{(0)}}{\partial y_j} \, \mathrm{dV}^*.$$

Finally, adding member to member the two last equalities, and noticing that, because of $(4.15)_2$

$$\int_{\Gamma_{12}^*} u_i (\sigma_{1_{ij}}^{(1)} - \sigma_{2_{ij}}^{(1)}) N_{1_j} \mathrm{dS}^* = -\int_{\Gamma_{12}^*} u_i p_{ca}^{(1)} N_{1_i} \mathrm{dS}^* = 0,$$

give the equivalent variational form

$$\forall\, \mathbf{u} \in \mathcal{V}_1, \qquad \int_{\Omega_F^*} \mu_\alpha^* \frac{\partial u_i}{\partial y_j} \frac{\partial v_{\alpha i}^{(0)}}{\partial y_j} \, \mathrm{dV}^* = \int_{\Omega_F^*} u_i \frac{\partial p_\alpha^{(0)}}{\partial x_i} \, \mathrm{dV}^*. \tag{4.17}$$

This formulation ensures the uniqueness of $\mathbf{v}^{(0)}$. The problem being linear, let us introduce particular solutions. Let

$$\boldsymbol{\lambda}_1^{*j}(\mathbf{y}) = \begin{cases} \boldsymbol{\lambda}_{11.j}^* & \text{in } \Omega_1^* \\ \boldsymbol{\lambda}_{12.j}^* & \text{in } \Omega_2^* \end{cases}$$

be the solution for

$$\frac{\partial p_1^{(0)}}{\partial x_i} = \delta_{ij}, \qquad \frac{\partial p_2^{(0)}}{\partial x_i} = 0,$$

and

$$\boldsymbol{\lambda}_2^{*j}(\mathbf{y}) = \begin{cases} \boldsymbol{\lambda}_{21.j}^* & \text{in } \Omega_1^* \\ \boldsymbol{\lambda}_{22.j}^* & \text{in } \Omega_2^* \end{cases}$$

be the solution for

$$\frac{\partial p_1^{(0)}}{\partial x_i} = 0, \qquad \frac{\partial p_2^{(0)}}{\partial x_i} = \delta_{ij}.$$

Then the general solution is in the form

$$v_{1_i}^{(0)} = -\lambda_{11_{ij}}^* \frac{\partial p_1^{(0)}}{\partial x_j} - \lambda_{12_{ij}}^* \frac{\partial p_2^{(0)}}{\partial x_j}, \tag{4.18}$$

$$v_{2_i}^{(0)} = -\lambda_{21_{ij}}^* \frac{\partial p_1^{(0)}}{\partial x_j} - \lambda_{22_{ij}}^* \frac{\partial p_2^{(0)}}{\partial x_j}, \tag{4.19}$$

The volume balance gives at the second order

$$\frac{\partial v_{\alpha i}^{(1)}}{\partial y_i} + \frac{\partial v_{\alpha i}^{(0)}}{\partial x_i} = 0. \tag{4.20}$$

Each equation for $\alpha = 1, 2$, yields a compatibility equation which is obtained by volume averaging

$$\frac{\partial \langle v_{1_i}^{(0)} \rangle}{\partial x_i} = -|\Omega^*|^{-1} \int_{\Gamma_{12}^*} v_{1_i}^{(1)} N_{1_i} \mathrm{dS}^* = \phi \frac{\partial S_1}{\partial t^*}, \tag{4.21}$$

$$\frac{\partial \langle v_{2_i}^{(0)} \rangle}{\partial x_i} = -|\Omega^*|^{-1} \int_{\Gamma_{12}^*} v_{2_i}^{(1)} N_{2_i} \mathrm{dS}^* = \phi \frac{\partial S_2}{\partial t^*}, \tag{4.22}$$

where ϕ is the porosity and S_α is the saturation of fluid α, with $S_1 + S_2 = 1$. The macroscopic volume balances (4.21) and (4.22) yield a generalized flow law in the form

$$\langle v_{1_i}^{(0)} \rangle = -\Lambda_{11_{ij}}^* \frac{\partial p_1^{(0)}}{\partial x_j} - \Lambda_{12_{ij}}^* \frac{\partial p_2^{(0)}}{\partial x_j}, \tag{4.23}$$

$$\langle v_{2_i}^{(0)} \rangle = -\Lambda_{21_{ij}}^* \frac{\partial p_1^{(0)}}{\partial x_j} - \Lambda_{22_{ij}}^* \frac{\partial p_2^{(0)}}{\partial x_j}, \tag{4.24}$$

where

$$\mathbf{\Lambda}_{\alpha\beta}^* = \langle \boldsymbol{\lambda}_{\alpha\beta}^* \rangle = \frac{1}{|\Omega^*|} \int_{\Omega_\alpha^*} \boldsymbol{\lambda}_{\alpha\beta}^* \, \mathrm{dV}^*.$$

4.3 Macroscopic Two-phase Flow Model

By returning to dimensional quantities, relations (4.23) and (4.24) become

$$\frac{\partial \langle v_{\alpha i} \rangle}{\partial X_i} = -\phi \frac{\partial S_\alpha}{\partial t} + \mathcal{O}(\varepsilon \frac{\partial \langle v_{\alpha i} \rangle}{\partial X_i}), \quad \langle v_{\alpha i} \rangle = -\frac{\Lambda_{\alpha\beta_{ij}}}{\mu} \frac{\partial p_\beta}{\partial X_j} + \mathcal{O}(\varepsilon \langle v_{\alpha i} \rangle), \tag{4.25}$$

where

$$\mathbf{\Lambda}_{\alpha\beta} = \frac{l^2}{\mu_c} \mathbf{\Lambda}_{\alpha\beta}^*, \qquad \alpha = 1, 2, \tag{4.26}$$

are conductivity tensors. The macroscopic problem is closed by giving the relation between the capillary pressure p_{ca} and one of the saturations, e.g., S_1.

Remark that, on the contrary to one fluid flow, permeabilities cannot be introduced unless in the very restrictive case where $\mu_1 = \mu_2 =$ constant. Volume average velocities $\mathbf{v}_\alpha^{(0)}$ are shown to be fluxes by following the same route as for the one fluid flow. It is also possible to show (Auriault, 1987) that the conductivities verifies the symmetries

$$\Lambda_{\alpha\alpha_{ij}} = \Lambda_{\alpha\alpha_{ji}}, \qquad \Lambda_{\alpha\beta_{ij}} = \Lambda_{\beta\alpha_{ji}}. \tag{4.27}$$

In practical applications, the coupling terms are often neglected: $\Lambda_{\alpha\beta_{ij}} = 0$ when $\alpha \neq \beta$, see Zarcone and Lenormand (1994) for an experimental checking. That can be obtained from the above analysis by letting negligible the interface between the two fluids: $|\Gamma_{12}| \approx 0$.

5 Diffusion/Convection in Porous Media

We consider again a rigid Galilean porous matrix with connected pores. The pores are saturated by a viscous Newtonian incompressible fluid which contains a solute at low concentration. The solute is transported in the pores by diffusion and advection. Temperature is constant. We neglect adsorption on the pore surface Γ and absorption in the solid Ω_S.
It is convenient to introduce (Auriault and Adler, 1995), characteristic times associated with advection and diffusion, respectively: $T_l^{adv} = l/v_c$, $T_l^{diff} = l^2/D_c$ at scale l and $T_L^{adv} = L/v_c$, $T_L^{diff} = L^2/D_c$ at scale L, where D_c and v_c are characteristic values of the molecular diffusion tensor and the fluid velocity, respectively. Their ratio is equal to the Péclet number $\mathcal{P}e$, which is also the ratio of the advection term to the diffusion term in equation (5.1) below

$$\mathcal{P}e_l = \frac{T_l^{diff}}{T_l^{adv}}, \qquad \mathcal{P}e_L = \frac{T_L^{diff}}{T_L^{adv}}.$$

The whole study can be done by comparing $\mathcal{P}e$ to the ratio ε between the microscopic and the macroscopic space scales of the medium. Four regimes can be distinguished. When the local Péclet number $\mathcal{P}e_l = v_c l/D_c = T_l^{diff}/T_l^{adv}$ is of order ε^2 (macroscopic Péclet number $\mathcal{P}e_L = v_c L/D_c = T_L^{diff}/T_L^{adv} = \mathcal{O}(\varepsilon)$), diffusion is predominant at the large and small scales; convection does not appear at the macroscopic scale at the first order of approximation, and one obtains a macroscopic diffusion equation with a macroscopic diffusion tensor. When the local Péclet number $\mathcal{P}e_l$ is of order ε, $\mathcal{P}e_L = \mathcal{O}(1)$, diffusion is still predominant at the small scale, but convection becomes of the same order of magnitude at the large scale. When the local Péclet number $\mathcal{P}e_l$ is equal to 1, $\mathcal{P}e_L = \mathcal{O}(\varepsilon^{-1})$, the situation is more complex; convection and diffusion are equivalent at the small spatial scale, but convection is predominant for short times and large spatial scale. It is only at the second order of approximation that diffusion appears at the large scale in the form of dispersion. Finally, when the local Péclet number $\mathcal{P}e_l$ is assumed to be of the order of ε^{-1}, $\mathcal{P}e_L = \mathcal{O}(\varepsilon^{-2})$, convection is predominant everywhere and the concentration is constant along streamlines at the macroscopic scale. Hence, the concentration distribution depends mostly upon the concentration imposed at the external boundaries of the macroscopic volume. Because of this dependency, the problem cannot be homogenized anymore.

5.1 Formulation of the Local Problem

The physical processes of molecular diffusion and convection of the solute can be described by the following mass balance equation

$$\frac{\partial c}{\partial t} + \frac{\partial}{\partial X_i}(-D_{ij}\frac{\partial c}{\partial X_i} + v_i c) = 0 \qquad \text{in } \Omega_F, \tag{5.1}$$

$$D_{ij}\frac{\partial c}{\partial X_i}N_i = 0 \qquad \text{on } \Gamma, \tag{5.2}$$

where c is the concentration (mass of pollutant per unit volume of fluid), $\mathbf{D}$ is the molecular diffusion tensor which is generally an isotropic tensor, t is the time variable, $\mathbf{N}$ is the unit vector normal to Γ and $\mathbf{v}$ is the fluid velocity. Velocity $\mathbf{v}$ is investigated in the previous part 3 and we will make use of the analysis herein in the present study.

We aim at describing diffusion and advection in the porous medium at the macroscopic scale. Therefore, we use T_L^{diff} as the characteristic time when diffusion is predominant or equivalent to advection at macroscale to make dimensionless the above set of equations

$$t_c = T_L^{diff} = \frac{L^2}{D_c} \text{ when } \mathcal{P}e_l = \mathcal{O}(\varepsilon^2) \text{ or } \mathcal{P}e_l = \mathcal{O}(\varepsilon), \tag{5.3}$$

and we use $t_c = T_L^{adv}$ when advection is predominant

$$t_c = T_L^{adv} = \frac{L}{v_c} \text{ when } \mathcal{P}e_l = \mathcal{O}(1) \text{ or } \mathcal{P}e_l = \mathcal{O}(\varepsilon^{-1}). \tag{5.4}$$

The dimensionless form of the local description is as follows (l is the characteristic length in use)

$$\mathcal{P}_l\frac{\partial c^*}{\partial t^*} + \frac{\partial}{\partial y_i}(-D^*_{ij}\frac{\partial c^*}{\partial y_j} + \mathcal{P}e_l v^*_i c^*) = 0 \qquad \text{in } \Omega^*_F, \tag{5.5}$$

$$D^*_{ij}\frac{\partial c^*}{\partial y_j}N_i = 0 \qquad \text{on } \Gamma^*. \tag{5.6}$$

It introduces two dimensionless numbers

$$\mathcal{P}_l = \frac{|\frac{\partial c}{\partial t}|}{|\frac{\partial}{\partial X_i}(D_{ij}\frac{\partial c}{\partial X_j})|} = \frac{l^2}{D_c t_c}, \tag{5.7}$$

and the Péclet number is defined by

$$\mathcal{P}e_l = \frac{|\frac{\partial}{X_i}(v_i c)|}{|\frac{\partial}{\partial X_i}(D_{ij}\frac{\partial c}{\partial X_j})|} = \frac{v_c l}{D_c}. \tag{5.8}$$

In the following, we apply the homogenization process to the four above mentioned different cases of interest: predominant diffusion at macroscale, $t_c = T_L^{diff}$, $\mathcal{P}e_l = \varepsilon^2$, $\mathcal{P}_l = \varepsilon^2$; diffusion and advection equivalent at macroscale, $t_c = T_L^{diff} = T_L^{adv}$, $\mathcal{P}e_l = \varepsilon$, $\mathcal{P}_l = \varepsilon^2$; predominant advection at macroscale, $t_c = T_L^{adv}$, $\mathcal{P}e_l = 1$, $\mathcal{P}_l = \varepsilon$; strong advection, $t_c = T_L^{adv}$, $\mathcal{P}e_l = \varepsilon^{-1}$, $\mathcal{P}_l = 1$. We look for c^* in the form of an asymptotic expansion

$$c^* = c^{(0)}(\mathbf{x},\mathbf{y},t^*) + \varepsilon c^{(1)}(\mathbf{x},\mathbf{y},t^*) + \varepsilon^2 c^{(2)}(\mathbf{x},\mathbf{y},t^*) + \cdots, \tag{5.9}$$

where the components $c^{(i)}(\mathbf{x},\mathbf{y},t^*)$ are $\mathbf{y}$-periodic of period Ω^*, and where space variable $\mathbf{x}$ is expressed by $\mathbf{x} = \varepsilon\mathbf{y}$.

5.2 Predominant Diffusion at Macroscale

The observation time is now

$$t_c = T_L^{diff} = \varepsilon T_L^{adv},$$

which yields

$$\mathcal{P}e_l = \varepsilon^2, \qquad \mathcal{P}_l = \varepsilon^2.$$

The normalized equations (5.5) and (5.6) take the following form

$$\varepsilon^2 \frac{\partial c^*}{\partial t^*} + \frac{\partial}{\partial y_i}(-D^*_{ij}\frac{\partial c^*}{\partial y_j} + \varepsilon^2 v_i^* c^*) = 0 \qquad \text{in } \Omega_F^*, \tag{5.10}$$

$$D^*_{ij}\frac{\partial c^*}{\partial y_j} N_i = 0 \qquad \text{on } \Gamma^*. \tag{5.11}$$

Upscaling. After introducing expansion (5.9), we obtain at the lower order of ε

$$\frac{\partial}{\partial y_i}(D^*_{ij}\frac{\partial c^{(0)}}{\partial y_j}) = 0 \qquad \text{in } \Omega_F^*, \tag{5.12}$$

$$D^*_{ij}\frac{\partial c^{(0)}}{\partial y_j} N_i = 0 \qquad \text{on } \Gamma^*, \tag{5.13}$$

where $c^{(0)}$ is a periodic function of $\mathbf{y}$. Let us introduce the space $\mathcal{C}$ of periodic functions defined in Ω_F^*, of zero average over Ω_F^* (an extraneous condition just introduced to insure the Hilbertian character of the space), with the scalar product

$$(\alpha,\beta)_{\mathcal{C}} = \int_{\Omega_F^*} D^*_{ij}\frac{\partial \alpha}{\partial y_i}\frac{\partial \beta}{\partial y_j}\,\mathrm{dV}^*.$$

Multiplying the two sides of (5.12) by $\alpha \in \mathcal{C}$, and integrating over Ω_F^* yield after some analysis to the following weak formulation

$$\forall \alpha \in \mathcal{C}, \qquad \int_{\Omega_F^*} D^*_{ij}\frac{\partial \alpha}{\partial y_i}\frac{\partial c^{(0)}}{\partial y_j}\,\mathrm{dV}^* = 0. \tag{5.14}$$

This formulation insures the existence and uniqueness of $c^{(0)} \in \mathcal{C}$. That yields to the trivial solution of (5.16)-(5.17)

$$c^{(0)} = c^{(0)}(\mathbf{x}, t^*). \tag{5.15}$$

The first order concentration is a constant over the period Ω^*. After taking into account this result, the second order term $c^{(1)}$ is the solution of the set

$$\frac{\partial}{\partial y_i}(D^*_{ij}(\frac{\partial c^{(0)}}{\partial x_j} + \frac{\partial c^{(1)}}{\partial y_j})) = 0 \qquad \text{in } \Omega^*_F, \tag{5.16}$$

$$D^*_{ij}(\frac{\partial c^{(0)}}{\partial x_j} + \frac{\partial c^{(1)}}{\partial y_j})N_i = 0 \qquad \text{on } \Gamma^*, \tag{5.17}$$

where $c^{(1)}$ is a periodic function of $\mathbf{y}$. Multiplying the two sides of (5.16) by $\alpha \in \mathcal{C}$, and integrating over Ω^*_F yield to the following weak formulation

$$\forall \alpha \in \mathcal{C}, \qquad \int_{\Omega^*_F} D^*_{ij}\frac{\partial \alpha}{\partial y_i}(\frac{\partial c^{(0)}}{\partial x_j} + \frac{\partial c^{(1)}}{\partial y_j})\ \mathrm{dV}^* = 0. \tag{5.18}$$

This formulation insures the existence and uniqueness of $c^{(1)} \in \mathcal{C}$, that gives the solution of the linear problem (5.16)-(5.17) in the form

$$c^{(1)} = \chi^{*+}_i \frac{\partial c^{(0)}}{\partial x_i} + \bar{c}^{(1)}(\mathbf{x}, t^*), \tag{5.19}$$

where $\boldsymbol{\chi}^{*+}$ is a periodic vector, of zero average which component χ^{*-}_k verifies

$$\forall \alpha \in \mathcal{C}, \qquad \int_{\Omega^*_F} D^*_{ij}\frac{\partial \alpha}{\partial y_i}(I_{jk} + \frac{\partial \chi^{*+}_k}{\partial y_j})\ \mathrm{dV}^* = 0. \tag{5.20}$$

Function $\bar{c}^{(1)}$ is an arbitrary function of $\mathbf{x}$ and t^* which is introduced due to the extraneous condition $< \boldsymbol{\chi}^{*+} >_{\Omega^*_\Gamma} = 0$. Finally, the boundary value problem for $c^{(2)}$ reduces to, after introducing the above result for $c^{(0)}$ (the term $\partial/\partial y_i(v^{(0)}c^{(0)})$ cancels out)

$$\frac{\partial c^{(0)}}{\partial t^*} - \frac{\partial}{\partial y_i}(D^*_{ij}(\frac{\partial c^{(1)}}{\partial x_j} + \frac{\partial c^{(2)}}{\partial y_j})) - \frac{\partial}{\partial x_i}(D^*_{ij}(\frac{\partial c^{(0)}}{\partial x_j} + \frac{\partial c^{(1)}}{\partial y_j})) = 0 \qquad \text{in } \Omega^*_F, \tag{5.21}$$

$$D^*_{ij}(\frac{\partial c^{(1)}}{\partial x_j} + \frac{\partial c^{(2)}}{\partial y_j})N_i = 0 \qquad \text{on } \Gamma^*, \tag{5.22}$$

where $c^{(2)}$ is $\mathbf{y}$-periodic. On the contrary to the boundary value problems for $c^{(0)}$ and $c^{(1)}$, the boundary value problem for $c^{(2)}$ yields a compatibility condition which is obtain by integrating (5.21) over Ω^*

$$\phi\frac{\partial c^{(0)}}{\partial t^*} - \frac{\partial}{\partial x_i}(D^{*+}_{ij}\frac{\partial c^{(0)}}{\partial x_j}) = 0, \tag{5.23}$$

where ϕ is the porosity and where the effective dimensionless diffusion tensor $\mathbf{D}^{*+}$ is defined by

$$D_{ij}^{*+} = \frac{1}{|\Omega^*|} \int_{\Omega_F^*} D_{ik}^*(I_{jk} + \frac{\partial \chi_j^{*+}}{\partial y_k}) \, \mathrm{dV}^*. \tag{5.24}$$

Macroscopic diffusion model. By returning to physical quantities, we obtain

$$\phi \frac{\partial c}{\partial t} - \frac{\partial}{\partial X_i}(D_{ij}^+ \frac{\partial c}{\partial X_j}) = \mathcal{O}(\varepsilon \phi \frac{\partial c}{\partial t}), \qquad \mathbf{D}^+ = D_c \, \mathbf{D}^{*+}, \tag{5.25}$$

which relative accuracy is $\mathcal{O}(\varepsilon)$. Tensor $\mathbf{D}^+$ is a purely diffusive effective tensor. By following a similar route as for investigating the properties of the permeability $\mathbf{K}$ in part 3, it can be shown from formulation (5.18) that tensor $\mathbf{D}^+$ is positive and symmetrical.

5.3 Diffusion and Advection Equivalent at Macroscale

The observation time is now

$$t_c = T_L^{diff} = T_L^{adv},$$

which gives the following dimensionless numbers

$$\mathcal{P}e_l = \varepsilon, \qquad \mathcal{P}_l = \varepsilon^2.$$

The normalized equations (5.5) and (5.6) take now the following form

$$\varepsilon^2 \frac{\partial c^*}{\partial t^*} + \frac{\partial}{\partial y_i}(-D_{ij}^* \frac{\partial c^*}{\partial y_j} + \varepsilon v_i^* c^*) = 0 \qquad \text{in } \Omega_F^*, \tag{5.26}$$

$$D_{ij}^* \frac{\partial c^*}{\partial y_j} N_i = 0 \qquad \text{on } \Gamma^*. \tag{5.27}$$

Upscaling. After introducing expansion (5.9) in the above set, it easy to show that the boundary value problems for $c^{(0)}$ and $c^{(1)}$ are unchanged. Therefore results (5.15) and (5.19) are still valid

$$c^{(0)} = c^{(0)}(\mathbf{x}, t^*), \qquad c^{(1)} = \chi_i^{*+} \frac{\partial c^{(0)}}{\partial x_i} + \bar{c}^{(1)}(\mathbf{x}, t^*). \tag{5.28}$$

However, the boundary value problem for $c^{(2)}$ is changed into

$$\frac{\partial c^{(0)}}{\partial t^*} - \frac{\partial}{\partial y_i}(D_{ij}^*(\frac{\partial c^{(1)}}{\partial x_j} + \frac{\partial c^{(2)}}{\partial y_j})) - \frac{\partial}{\partial x_i}(D_{ij}^*(\frac{\partial c^{(0)}}{\partial x_j} + \frac{\partial c^{(1)}}{\partial y_j}))$$

$$+ \frac{\partial}{\partial y_i}(c^{(0)} v_i^{(1)} + c^{(1)} v_i^{(0)}) + \frac{\partial}{\partial x_i}(c^{(0)} v_i^{(0)}) = 0 \quad \text{in } \Omega_F^*, \tag{5.29}$$

$$D^*_{ij}(\frac{\partial c^{(1)}}{\partial x_j}+\frac{\partial c^{(2)}}{\partial y_j})N_i=0 \qquad \text{on } \Gamma^*, \tag{5.30}$$

where $c^{(2)}$ is $\mathbf{y}$-periodic. The compatibility condition is now in the form

$$\phi\frac{\partial c^{(0)}}{\partial t^*}-\frac{\partial}{\partial x_i}(D^{*+}_{ij}\frac{\partial c^{(0)}}{\partial x_j}-<v^{(0)}>_i c^{(0)})=0, \tag{5.31}$$

where $\mathbf{D}^{*+}$ is the effective diffusion tensor already defined in the previous case. The macroscopic velocity $<\mathbf{v}^{(0)}>$ was defined in part 3.

Macroscopic diffusion-advection model. After coming back to dimensional quantities, we have

$$\phi\frac{\partial c}{\partial t}-\frac{\partial}{\partial X_i}(D^+_{ij}\frac{\partial c}{\partial X_j}-<v>_i c)=\mathcal{O}(\varepsilon\phi\frac{\partial c}{\partial t}), \qquad \mathbf{D}^+=D_c\,\mathbf{D}^{*+}, \tag{5.32}$$

which relative accuracy is $\mathcal{O}(\varepsilon)$. Tensor $\mathbf{D}^+$ is a purely diffusive tensor: it is not modified by the advection.

5.4 Predominant Advection at Macroscale

The observation time is now

$$t_c=T_L^{adv}=\varepsilon T_L^{diff},$$

and we have

$$\mathcal{P}e_l=1, \qquad \mathcal{P}_l=\varepsilon.$$

The normalized equations (5.5) and (5.6) take the following form

$$\varepsilon\frac{\partial c^*}{\partial t^*}+\frac{\partial}{\partial y_i}(-D^*_{ij}\frac{\partial c^*}{\partial y_j}+v^*_i c^*)=0 \qquad \text{in } \Omega^*_F, \tag{5.33}$$

$$D^*_{ij}\frac{\partial c^*}{\partial y_j}N_i=0 \qquad \text{on } \Gamma^*. \tag{5.34}$$

Upscaling. The boundary value problem for $c^{(0)}$ is now in the form

$$\frac{\partial}{\partial y_i}(D^*_{ij}\frac{\partial c^{(0)}}{\partial y_j}-v^{(0)}_i c^{(0)})=0 \qquad \text{in } \Omega^*_F, \tag{5.35}$$

$$D^*_{ij}\frac{\partial c^{(0)}}{\partial y_j}N_i=0 \qquad \text{on } \Gamma^*, \tag{5.36}$$

where $c^{(0)}$ is a periodic function of $\mathbf{y}$. The associated weak formulation is similar to (5.14)

$$\forall \alpha \in \mathcal{C}, \qquad \int_{\Omega_F^*} D_{ij}^* \frac{\partial \alpha}{\partial y_i} \frac{\partial c^{(0)}}{\partial y_j} \, \mathrm{dV}^* = 0. \tag{5.37}$$

Therefore results (5.15) is still valid

$$c^{(0)} = c^{(0)}(\mathbf{x}, t^*). \tag{5.38}$$

The boundary value problem for $c^{(1)}$ is now very different

$$\frac{\partial c^{(0)}}{\partial t^*} - \frac{\partial}{\partial y_i}(D_{ij}^*(\frac{\partial c^{(0)}}{\partial x_j} + \frac{\partial c^{(1)}}{\partial y_j}))$$

$$+\frac{\partial}{\partial y_i}(c^{(0)} v_i^{(1)} + c^{(1)} v_i^{(0)}) + \frac{\partial}{\partial x_i}(c^{(0)} v_i^{(0)}) = 0 \quad \text{in } \Omega_F^*, \tag{5.39}$$

$$D_{ij}^*(\frac{\partial c^{(0)}}{\partial x_j} + \frac{\partial c^{(1)}}{\partial y_j}) N_i = 0 \qquad \text{on } \Gamma^*, \tag{5.40}$$

where $c^{(1)}$ is $\mathbf{y}$-periodic. The above boundary value problem shows a compatibility condition for the existence of $c^{(1)}$, which is obtained as usual by volume averaging equation (5.39) over Ω^*

$$\phi \frac{\partial c^{(0)}}{\partial t^*} + \frac{\partial}{\partial x_i}(< v^{(0)} >_i c^{(0)}) = 0. \tag{5.41}$$

As expected the macroscopic behaviour shows only advection at the first order of approximation. To obtain the first corrector, where diffusion is present, we first evaluate the time derivative from (5.41) and we introduce it back into equation (5.39). It is easily obtained

$$\frac{\partial}{\partial y_i}(D_{ij}^*(\frac{\partial c^{(0)}}{\partial x_j} + \frac{\partial c^{(1)}}{\partial y_j})) - v_i^{(0)} \frac{\partial c^{(1)}}{\partial y_i} =$$

$$(v_i^{(0)} - \phi^{-1} < v^{(0)} >_i) \frac{\partial c^{(0)}}{\partial x_i} \quad \text{in } \Omega_F^*. \tag{5.42}$$

The solution of boundary value problem (5.42)-(5.40) is investigated in space $\mathcal{C}$, and we use the linearity of the problem to put it in the form

$$c^{(1)} = \chi_i^{*++} \frac{\partial c^{(0)}}{\partial x_i} + \bar{c}^{(1)}(\mathbf{x}, t^*), \tag{5.43}$$

where $\boldsymbol{\chi}^{*++}$ is a periodic vector, of zero average which component χ_k^{*++} verifies

$$\frac{\partial}{\partial y_i}(D^*_{ij}(I_{jk} + \frac{\partial \chi_k^{*++}}{\partial y_j})) - v_i^{(0)}\frac{\partial \chi_k^{*++}}{\partial y_i} = v_k^{(0)} - \phi^{-1} < v^{(0)} >_k \quad \text{in } \Omega_F^*, \tag{5.44}$$

$$D^*_{ij}(I_{jk} + \frac{\partial \chi_k^{*++}}{\partial y_j})N_i = 0 \quad \text{on } \Gamma^*. \tag{5.45}$$

It must be noticed that $\boldsymbol{\chi}^{*++} \neq \boldsymbol{\chi}^{**}$, since now $\boldsymbol{\chi}^{*++}$ depends on the velocity field and through it on the macroscopic gradient of pressure of components $\partial p^{(0)}/\partial x_i$. It is interesting to see that the above set is exactly the set obtained by Brenner (1980), with the method of moments, equation (52) where $\mathbf{B} = -\boldsymbol{\chi}^{*++}$.
To obtain the macroscopic equation verified by the corrector $c^{(1)}$, we look for the system which gives $c^{(2)}$

$$\frac{\partial c^{(1)}}{\partial t^*} - \frac{\partial}{\partial y_i}(D^*_{ij}(\frac{\partial c^{(1)}}{\partial x_j} + \frac{\partial c^{(2)}}{\partial y_j})) - \frac{\partial}{\partial x_i}(D^*_{ij}(\frac{\partial c^{(0)}}{\partial x_j} + \frac{\partial c^{(1)}}{\partial y_j}))$$

$$+\frac{\partial}{\partial y_i}(c^{(0)}v_i^{(2)} + c^{(1)}v_i^{(1)} + c^{(2)}v_i^{(0)}) + \frac{\partial}{\partial x_i}(c^{(0)}v_i^{(1)} + c^{(1)}v_i^{(0)}) = 0 \quad \text{in } \Omega_F^*, \tag{5.46}$$

$$D^*_{ij}(\frac{\partial c^{(1)}}{\partial x_j} + \frac{\partial c^{(2)}}{\partial y_j})N_i = 0 \quad \text{on } \Gamma^*, \tag{5.47}$$

where $c^{(2)}$ is $\mathbf{y}$-periodic. Equation (5.46) is volume averaged, that gives

$$\frac{\partial < c^{(1)} >}{\partial t^*} - \frac{\partial}{\partial x_i}(< D^*_{ij}(I_{jk} + \frac{\partial \chi_k^{*++}}{\partial y_j}) > \frac{\partial c^{(0)}}{\partial x_k}$$

$$- < v^{(1)} >_i c^{(0)} - < c^{(1)}v^{(0)} >_i) = 0, \tag{5.48}$$

which represents the condition for the existence of $c^{(2)}$. This equation can be slightly modified by the introduction of equations (3.20) and (5.43)

$$\frac{\partial < c^{(1)} >}{\partial t^*} - \frac{\partial}{\partial x_i}(D_{ij}^{*++}\frac{\partial c^{(0)}}{\partial x_j} - < v^{(1)} >_i c^{(0)} - \bar{c}^{(1)} < v^{(0)} >_i) = 0, \tag{5.49}$$

where $\mathbf{D}^{*++}$ is the effective dispersion tensor

$$D_{ij}^{*++} = \frac{1}{|\Omega^*|}\int_{\Omega_F^*} (D^*_{ik}(I_{jk} + \frac{\partial \chi_j^{*++}}{\partial y_k}) + \lambda^*_{ik}\chi_j^{*++}\frac{\partial p^{(0)}}{\partial x_k}) \, \mathrm{dV}^*. \tag{5.50}$$

Equation (5.49) represents the behaviour of the first corrector.

Macroscopic dispersion equation. Up to the first order in ε, the macroscopic concentration and velocity are written

$$< c^* > \approx < c^{(0)} > + \varepsilon < c^{(1)} > = \phi c^{(0)} + \varepsilon \phi \bar{c}^{(1)} \tag{5.51}$$

$$< \mathbf{v}^* > \approx < \mathbf{v}^{(0)} > + \varepsilon < \mathbf{v}^{(1)} > \tag{5.52}$$

By adding member to member equation (5.41) and equation (5.49) multiplied by ε, we obtain

$$\frac{\partial < c^* >}{\partial t^*} - \frac{\partial}{\partial x_i}(\varepsilon D_{ij}^{*++} \frac{\partial < c^* >}{\partial x_j} - < v^* >_i \frac{< c^* >}{\phi}) = \mathcal{O}(\varepsilon^2), \tag{5.53}$$

where we see that the approximation of the model is of order ε^2. We notice also that the diffusion term is small compared to the advection term, $Pe_L = \varepsilon^{-1}$. Let us come back to physical quantities. We have

$$\frac{\partial < c > t_c}{\partial t \quad c_c} - \frac{\partial}{\partial X_i} L(\varepsilon \frac{L}{D_c c_c} D_{ij}^{++} \frac{\partial < c >}{\partial X_j} - \frac{1}{v_c c_c} < v >_i \frac{< c >}{\phi}) = \mathcal{O}(\varepsilon^2). \tag{5.54}$$

After recalling that $t_c = T_L^{adv} = L/v_c$ and $v_c L/D_c = Pe_L = \varepsilon^{-1}$, we finally have

$$\frac{\partial < c >}{\partial t} - \frac{\partial}{\partial X_i}(D_{ij}^{++} \frac{\partial < c >}{\partial X_j} - < v >_i \frac{< c >}{\phi}) = \mathcal{O}(\varepsilon^2 \frac{\partial < c >}{\partial t}). \tag{5.55}$$

As shown in Auriault and Adler (1995), the dispersion tensor $\mathbf{D}^{++}$ is positive but generally non-symmetric. Its experimental measure poses some difficulty, due to the length L of the sample to be used, (Auriault and Lewandowska, 1997), see below.
An other question concerns the physical meaning of the macroscopic quantity $< \mathbf{v} >$, which includes the first corrector $< \mathbf{v}^{(1)} >$. From (3.22), $\mathbf{v}^{(1)}$ is not divergence free. Therefore $< \mathbf{v}^{(1)} >$ is not generally a flux (see above the reasoning for $\mathbf{v}^{(0)}$). As a consequence $< \mathbf{v} >$ itself is not a flux in all cases. More, due to $< \mathbf{v}^{(1)} >$ behaviour, it does not generally verify Darcy's law. This problem is addressed in a forthcoming paper (Auriault et al., 2004).

5.5 Very strong advection

The observation time is now

$$t_c = T_L^{adv} = \varepsilon^2 T_L^{diff},$$

from which we obtain

$$\mathcal{P}e_l = \varepsilon^{-1}, \qquad \mathcal{P}_l = 1.$$

The normalized equations (5.5) and (5.6) take now the form

$$\frac{\partial c^*}{\partial t^*} + \frac{\partial}{\partial y_i}(-D_{ij}^* \frac{\partial c^*}{\partial y_j} + \varepsilon^{-1} v_i^* c^*) = 0 \qquad \text{in } \Omega_F^*, \tag{5.56}$$

$$D_{ij}^* \frac{\partial c^*}{\partial y_j} N_i = 0 \qquad \text{on } \Gamma^*. \tag{5.57}$$

Upscaling. The previous expansions are easily performed again. At the lowest order we have

$$\frac{\partial}{\partial y_i}(c^{(0)}v_i^{(0)}) = 0 \qquad \text{in } \Omega_F^*, \tag{5.58}$$

which is written equivalently

$$v_i^{(0)}\frac{\partial c^{(0)}}{\partial y_i} = 0. \tag{5.59}$$

On Γ^* we have again

$$D_{ij}^*\frac{\partial c^{(0)}}{\partial y_j}N_i = 0. \tag{5.60}$$

Equation (5.59) shows that $c^{(0)}$ does not depend upon the curvilinear abscissa along the streamlines: the concentration in the bulk of the porous medium depends directly upon its value on the external macroscopic boundary of the medium. Since $c^{(0)}$ directly depends upon these macroscopic boundary conditions, the problem cannot be homogenized anymore. This is emphasized after considering the boundary value problem at the next order

$$\frac{\partial c^{(0)}}{\partial t^*} + \frac{\partial}{\partial y_i}(-D_{ij}^*\frac{\partial c^{(0)}}{\partial y_j} + c^{(0)}v_i^{(1)} + c^{(1)}v_i^{(0)}) = 0,$$

which by integration yields the compatibility condition

$$\frac{\partial < c^{(0)} >}{\partial t^*} = 0.$$

This result is in a contradiction with $\mathcal{P}_l = \mathcal{O}(1)$.

5.6 Conclusion

The different macroscopic models for solute transport in porous media and their domains of validity are regrouped in Table 2. By decreasing the Péclet number, there exists a continuous passage from the dispersion model to the diffusion-advection model and then to the diffusion model.

Experimental measure of dispersion coefficient poses some difficulty, due to the length L of the sample to be used, (Auriault and Lewandowska, 1997). The above macroscopic models are valid in the bulk material where the separation of scales is verified. That introduces limitations of two different kinds to model validity. Firstly, the condition of separation of scales (1.1) should be verified by both the sample geometry

$$\varepsilon_g = \frac{l}{L} \ll 1, \tag{5.61}$$

where L is the characteristic length of the sample, and by the phenomenons under investigation. In the case of solute transport this can be written, for example in a column

Table 2. Macroscopic models for solute transport in porous media.

$\mathcal{P}e_l$	$\mathcal{P}_l$	Observation time t_c	Diffusion/dispersion tensor	Model
$\leq \varepsilon^2$	ε^2	$T_L^{diff} = L^2/D_c$	Diffusion: $\mathbf{D}^+$	Diffusion
ε	ε^2	$T_L^{diff} = L^2/D_c = T_L^{adv}$	Diffusion: $\mathbf{D}^+$	Diffusion-advection
1	ε	$T_L^{adv} = L/v_c$	Dispersion: $\mathbf{D}^{++}$	Advection $\gg$ Diffusion Dispersion
$\geq \varepsilon$	1	$T_L^{adv} = L/v_c$		Non homogenizable

test

$$\varepsilon_{ph} = \frac{l}{L_{ph}} \ll 1, \qquad L_{ph} \approx |\frac{\partial}{\partial X}(\frac{c}{c_0})|^{-1}, \tag{5.62}$$

where c_0 is the constant solute concentration at the inlet of the column. Such limitations are analysed for such column tests in Auriault and Lewandowska (1997). Secondly, near the macroscopic boundaries the separation of scales is not verified. Thus, the complete macroscopic solution needs the introduction of boundary layers to match the homogenized solution to the boundary values, (Bensoussan et al., 1978). Of course, the determination of effective coefficients from experiments will be valid if the thickness of these boundary layers is small compared to the sample size or if the measures are performed outside these boundary layers (Lewandowska and Auriault, 1998). It is admitted that in elliptic problems such as purely diffusive problems or elastic problems, the thickness of the boundary layers is approximately l, (Sanchez-Palencia, 1987), with $l \approx 5$ to 10 grain size, (Bear, 1972), (Anguy et al., 1994). Thus the limitation to the model validity in absence of advection reduces to the above conditions $\varepsilon_g \ll 1$ and $\varepsilon_{ph} \ll 1$. When advection is present at macroscale, the boundary layer thickness is stretched from l to h. Let us evaluate h. The characteristic time T_l^{diff} for a solute particle to diffuse through a distance l is

$$T_l^{diff} = \frac{l^2}{D_c}.$$

When advection is present, the solute particle is also convected with a velocity v_c. During the time T_l^{diff}, the solute particle is shifted on a distance

$$h = v_c T_l^{diff} = \frac{l^2 v_c}{D_c}.$$

Therefore the boundary layer thickness in presence of advection becomes $l + h$. Let us compare $h + l$ to l

$$\frac{h+l}{l} = 1 + \frac{v_c l}{D_c} = 1 + Pe_l.$$

The boundary layer thickness increases with the local Péclet number, which necessitates increasing sample length, specially in case of dispersion.

6 Deformable Saturated Porous Media.

The originality of saturated deformable porous media stands in the presence at pore scale of two very different materials: a porous matrix which material is a solid and a fluid which occupies the pores. This local strong heterogeneity yields non-classical macroscopic behaviours. Two different classes of macroscopic phenomenological behaviours can be pointed out:

- Monophasic classical macroscopic behaviours with a single displacement field, where the viscosity of the fluid is considered or not (see, e.g., Tsytovich and Zaretskij (1969); Gassmann (1951)).

- Diphasic macroscopic behaviours with two displacement fields, one for the solid and one for the fluid, which where introduced by Biot (1941, 1955, 1962).

The above mentioned investigations were conducted directly at the macroscopic scale by using experimental and phenomenological approaches. In the present lecture we upscale the pore scale description by applying the method of asymptotic expansions described in part 2, which was used for the problem under consideration by, e.g., Auriault and Sanchez-Palencia (1977), Sanchez-Palencia (1980), Chapt. 8, Burridge and Keller (1981) and Boutin and Auriault (1990). For the sake of simplicity, the material of the porous matrix is assumed elastic and the deformations are small, whereas the fluid is viscous Newtonian and incompressible. We also limit the study to quasi-statics: inertia is neglected.
In the first subsection, we investigate the macroscopic behaviour of an empty porous matrix, which will appear as a basic problem for the saturated case. Then the saturated porous media is addressed. Three different macroscopic behaviours are obtained in function of a contrast of property dimensionless number: diphasic behaviour, monophasic elastic behaviour and monophasic viscoelastic behaviour.

6.1 Empty Porous Matrix

We investigate the behaviour of a periodic Galilean porous matrix with empty pores (no stress on Γ, see figure 1). The stress $\boldsymbol{\sigma}_S$ and the displacement $\mathbf{u}_S$ of the matrix verify the following equations where gravity is not considered without lost of generality:

$$\frac{\partial \sigma_{Sij}}{\partial X_j} = 0, \tag{6.1}$$

$$\sigma_{Sij} = a_{ijkh}\ e_{Xkh}(\mathbf{u}_S) \qquad \text{in } \Omega_S, \tag{6.2}$$

$$\sigma_{Sij} N_j = 0 \qquad \text{on } \Gamma. \tag{6.3}$$

In these equations $\mathbf{a}$ is the elastic tensor, with possible discontinuities where the classical continuities of displacement and normal stress are valid. Tensor $\mathbf{a}$ verifies the symmetries

$$a_{ijkh} = a_{khij} = a_{jikh} = a_{ijhk},$$

and the ellipticity condition

$$a_{ijkh}\ e_{ij}\ e_{kh} \geq \gamma\ e_{ij}\ e_{ij}, \qquad \gamma > 0$$

Tensor $\mathbf{e}$ is the small deformation

$$e_{ij} = \frac{1}{2}\left(\frac{\partial u_i}{\partial X_j} + \frac{\partial u_j}{\partial X_i}\right).$$

Finally, relation (6.3) stands for the no-stress condition on the pore surface.

Macroscopic behaviour of the empty matrix Let us upscale the above system. We first normalize it by adopting the macroscopic point of view (L is the characteristic length to make dimensionless the local description). The dimensionless set takes the form, with $\mathbf{a} = a_c\ \mathbf{a}^*$ and $\mathbf{u}_S = u_c\ \mathbf{u}_S^*$

$$\frac{\partial \sigma^*_{Sij}}{\partial x_j} = 0, \tag{6.4}$$

$$\sigma^*_{Sij} = a^*_{ijkh}\ e^*_{xkh}(\mathbf{u}^*_S) \qquad \text{in } \Omega^*_S, \tag{6.5}$$

$$\sigma^*_{Sij} N_j = 0 \qquad \text{on } \Gamma^*. \tag{6.6}$$

We look for $\mathbf{u}^*_S$ in the form

$$\mathbf{u}^*_S = \mathbf{u}^{(0)}(\mathbf{x},\mathbf{y}) + \varepsilon \mathbf{u}^{(1)}(\mathbf{x},\mathbf{y}) + \varepsilon^2 \mathbf{u}^{(2)}(\mathbf{x},\mathbf{y}) + \cdots,$$

with $\mathbf{y} = \mathbf{x}/\varepsilon$. After introducing the above expansion in the dimensionless set, we obtain at the lowest order a boundary value problem for $\mathbf{u}^{(0)}$:

$$\frac{\partial}{\partial y_j}(a^*_{ijkh}\ e_{ykh}(\mathbf{u}^{(0)})) = 0 \qquad \text{in } \Omega^*_S, \tag{6.7}$$

$$a^*_{ijkh}\ e_{ykh}(\mathbf{u}^{(0)}) N_j = 0 \qquad \text{on } \Gamma^*, \tag{6.8}$$

where $\mathbf{u}^{(0)}$ is $\mathbf{y}$-periodic. To investigate this boundary value problem, let us introduce the space $\mathcal{E}$ of vectors $\mathbf{u}$, that are defined on Ω^*_S, periodic, of zero average over Ω^*_S :

$$\int_{\Omega^*_S} \mathbf{u}\ \mathrm{dV}^* = 0, \tag{6.9}$$

with the scalar product

$$(\mathbf{u},\alpha)_{\mathcal{E}} = \int_{\Omega^*_S} e_{yij}(\mathbf{u})\ e_{yij}(\alpha)\ \mathrm{dV}^*. \tag{6.10}$$

Multiply the two members of (6.7)) by $\alpha \in \mathcal{E}$ and integrate over Ω_S^*. After integrating by parts and using the divergence theorem, we obtain

$$0 = \int_{\Omega_S^*} \frac{\partial \sigma_{Sij}^*(\mathbf{u}^{(0)})}{\partial y_j} \alpha_i \, \mathrm{dV}^* = \int_{\delta\Omega_S^*} \sigma_{Sij}^*(\mathbf{u}^{(0)}) \alpha_i N_j \, \mathrm{dV}^* - \int_{\Omega_S^*} \sigma_{Sij}^*(\mathbf{u}^{(0)}) \frac{\partial \alpha_i}{\partial y_j} \, \mathrm{dV}^*.$$

The surface integral cancels out on Γ^* because of (6.8) and on the rest of $\delta\Omega_S^*$ by periodicity. Finally, the symmetry of σ_S^* yields

$$\sigma_{Sij}^* \frac{\partial \alpha_i}{\partial y_j} = \sigma_{Sji}^* \frac{\partial \alpha_j}{\partial y_i} = \sigma_{Sij}^* \frac{\partial \alpha_j}{\partial y_i} = \frac{1}{2} \sigma_{Sij}^* \left(\frac{\partial \alpha_i}{\partial y_j} + \frac{\partial \alpha_j}{\partial y_i} \right),$$

which gives the following variational form

$$\forall \alpha \in \mathcal{E}, \qquad \int_{\Omega_S^*} a_{ijkh}^* \, e_{ykh}(\mathbf{u}^{(0)}) \, e_{yij}(\alpha) \, \mathrm{dV}^* = 0. \tag{6.11}$$

This formulation is equivalent to the system (6.7-6.8) with $\mathbf{u}^{(0)}$ of zero average. Relation (6.9) is an added extraneous condition, just introduced for (6.10) to be a scalar product. In definitive, the solution of (6.7-6.8) is the solution of (6.11) to an added arbitrary vector function of $\mathbf{x}$, only.

Lax-Milgram lemma, (Necas, 1967), yields the existence and uniqueness of $\mathbf{u}^{(0)}$ in (6.11), and obviously we have $\mathbf{u}^{(0)} = 0$. Thus, the solution of (6.7-6.8) is in the form

$$\mathbf{u}^{(0)} = \mathbf{u}^{(0)}(\mathbf{x}).$$

At the following order, we obtain the boundary value problem for $\mathbf{u}^{(1)}$:

$$\frac{\partial}{\partial y_j} (a_{ijkh}^* \; (e_{ykh}(\mathbf{u}^{(1)})) + e_{xkh}(\mathbf{u}^{(0)}))) = 0 \qquad \text{in } \Omega_S^*, \tag{6.12}$$

$$a_{ijkh}^* \; (e_{ykh}(\mathbf{u}^{(1)})) + e_{xkh}(\mathbf{u}^{(0)})) N_j = 0 \qquad \text{on } \Gamma^*, \tag{6.13}$$

where $\mathbf{u}^{(1)}$ is Ω^*-periodic in $\mathbf{y}$. The weak equivalent formulation (to an added arbitrary vector $\overline{\mathbf{u}}^{(1)}(\mathbf{x})$) is obtained as above by multiplying the two members of (6.12) by $\boldsymbol{\alpha} \in \mathcal{E}$ and integrating over Ω_S^*. The formulation is in the form

$$\forall \boldsymbol{\alpha} \in \mathcal{E}, \qquad \int_{\Omega_S^*} a_{ijkh}^* (e_{ykh}(\mathbf{u}^{(1)})) + e_{xkh}(\mathbf{u}^{(0)})) \, e_{yij}(\boldsymbol{\alpha}) \, \mathrm{dV}^* = 0. \tag{6.14}$$

Again, the properties of the elastic tensor $\mathbf{a}$ insures the existence and uniqueness of the solution of (6.14). The solution is a linear vectorial function of $\mathbf{e}_x(\mathbf{u}^{(0)})$. If the vector $\boldsymbol{\xi}^{*lm}$ is the solution for $e_{xij}(\mathbf{u}^{(0)}) = \delta_{il}\delta_{jm}$, l and m fixed, the solution of (6.12-6.13) is in the form

$$u_i^{(1)} = \xi_i^{*lm} e_{xlm}(\mathbf{u}^{(0)}) + \overline{u}_i^{(1)}(\mathbf{x}),$$

where $\overline{\mathbf{u}}^{(1)}(\mathbf{x})$ is an arbitrary vector and $\boldsymbol{\xi}^*(\mathbf{y})$ belongs to $\mathcal{E}$

$$\int_{\Omega_S^*} \boldsymbol{\xi}^* \ \mathrm{dV}^* = 0.$$

The boundary value problem for $\mathbf{u}^{(2)}$ is obtained at the following order

$$\frac{\partial \sigma_{Sij}^{(1)}}{\partial y_j} + \frac{\partial \sigma_{Sij}^{(0)}}{\partial x_j} = 0 \qquad \text{in } \Omega_S^*, \tag{6.15}$$

$$\sigma_{Sij}^{(1)} N_j = 0 \qquad \text{on } \Gamma^*, \tag{6.16}$$

where $\boldsymbol{\sigma}_S^{(0)}$ and $\boldsymbol{\sigma}_S^{(1)}$ are the zero and the one order term in the expansion of $\boldsymbol{\sigma}_S$ in powers of ε. Equation (6.15) is the momentum balance of the periodic stress $\boldsymbol{\sigma}_S^{(1)}$, which verifies the condition (6.16) on Γ^*, with the source term $-\dfrac{\partial \sigma_{Sij}^{(0)}}{\partial x_j}$. Therefore there exists a compatibility condition (existence of $\mathbf{u}^{(2)}$) which is obtained by volume averaging (6.15). By using (6.16) and the periodicity of $\boldsymbol{\sigma}_S^{(1)}$, we obtain

$$< \frac{\partial \sigma_{Sij}^{(0)}}{\partial x_j} > = \frac{\partial < \sigma_{Sij}^{(0)} >}{\partial x_j} = \frac{\partial < \sigma_{Sij}^{T(0)} >}{\partial x_j} =$$

$$-|\Omega^*|^{-1} \int_{\Omega_S^*} \frac{\partial \sigma_{Sij}^{(1)}}{\partial y_j} \ \mathrm{dV}^* = -|\Omega^*|^{-1} \int_{\delta\Omega_S^*} \sigma_{Sij}^{(1)} N_j \ \mathrm{dS}^* = 0.$$

Let us define the total stress $\boldsymbol{\sigma}^{*T}$ by

$$\boldsymbol{\sigma}^{*T} = \begin{cases} \boldsymbol{\sigma}_S^* & \text{in } \Omega_S^* \\ 0 & \text{in } \Omega_F^*. \end{cases}$$

Introducing the definition of $\boldsymbol{\sigma}_S^{(0)}$ yields

$$< \sigma_{ij}^{T(0)} > = < a_{ijkh}^* \ (e_{ykh}(\mathbf{u}^{(1)})) + e_{xkh}(\mathbf{u}^{(0)})) >$$

$$= < a_{ijkh}^* \ (e_{ykh}(\boldsymbol{\xi}^{*lm}) \ e_{xlm}(\mathbf{u}^{(0)}) + e_{xkh}(\mathbf{u}^{(0)})) >$$

$$= c_{ijkh}^* \ e_{xkh}(\mathbf{u}^{(0)}),$$

with the effective elastic tensor $\mathbf{c}^*$ defined by

$$c_{ijkh}^* = < a_{ijkh}^* + a_{ijlm}^* \ e_{ylm}(\boldsymbol{\xi}^{*kh}) > .$$

Finally, after coming back to dimensional quantities, the macroscopic behaviour is at the first order of approximation

$$\frac{\partial < \sigma_{ij}^T >}{\partial x_j} = \mathcal{O}(\varepsilon \frac{\partial < \sigma_{ij}^T >}{\partial x_j}), \tag{6.17}$$

$$\sigma_{ij}^T = c_{ijkh} \ e_{xkh}(\mathbf{u}_S) + \mathcal{O}(\varepsilon \ \sigma_{ij}^T), \qquad \mathbf{c} = a_c \mathbf{c}^*. \tag{6.18}$$

Investigation of the macroscopic behaviour The structure of the system (6.17-6.18) is similar to the structure of the system (6.1-6.2). However, the physical meaning of the macroscopic quantities and the properties of the effective elastic tensor should be investigated.

Physical meaning of $< \boldsymbol{\sigma}^{T(0)} >$ *and* $\boldsymbol{u}^{(0)}$. Tensor $< \boldsymbol{\sigma}^{T(0)} >$ is the volume average of a stress. By following a similar route as for the Darcy velocity, let us show that it is also a surface average, i.e. a stress. Consider the identity

$$\frac{\partial(\sigma_{Sij}^{(0)} y_k)}{\partial y_j} = \frac{\partial \sigma_{Sij}^{(0)}}{\partial y_j} y_k + \sigma_{Sik}^{(0)}.$$

Integrate over Ω_S^*. From (6.8), the divergence of $\boldsymbol{\sigma}_S^{(0)}$ cancels out and the divergence theorem together with the condition on Γ^* (cf. Figure 2) yield

$$< \sigma_{Sik}^{(0)} >= |\Omega^*|^{-1} \int_{\delta\Omega_S^* \cap \delta\Omega^*} \sigma_{Sij}^{(0)} y_k N_j \text{ dS}^*.$$

The right hand member can be seen as a stress. Let the period defined by $0 \leq y_i \leq l_i^*, i = 1, 2, 3$, where l_i^* represents the period length along axis y_i. The integrals on $y_j = 0$ and $y_j = l_j^*, j \neq k$ cancel out by periodicity, as well as the integrals on $y_k = 0$. By taking $\boldsymbol{\sigma}_S^{(0)} = 0$ in Ω_F^* and by considering that for $y_k = l_k^*$, $N_j = \delta_{jk}$, we are left with

$$\int_{\delta\Omega_S^* \cap \delta\Omega^*} \sigma_{Sij}^{(0)} y_k N_j \text{ dS}^* = l_k^* \int_{y=l_k^*} \sigma_{Sij}^{(0)} \delta_{jk} \text{ dS}^*,$$

$$< \sigma_{Sik}^{(0)} >= l_k^* |\Omega^*|^{-1} \int_{y=l_k^*} \sigma_{Sij}^{(0)} \text{ dS}^* = |\Sigma_k^*|^{-1} \int_{y=l_k^*} \sigma_{Sij}^{(0)} \text{ dS}^*,$$

no summation on k and where Σ_k^* is the cross-section $y_k = l_k$ of Ω^*. The right end member represents a surface average: it is a stress. Remark that the results comes from the solenoidal character of $\boldsymbol{\sigma}_S^{(0)}$:

$$\frac{\partial \sigma_{Sij}^{(0)}}{\partial y_j} = 0.$$

By contrast, the volume average of the second approximation $\boldsymbol{\sigma}_S^{(1)}$ of the stress is generally not a stress since we have

$$\frac{\partial \sigma_{Sij}^{(1)}}{\partial y_j} = -\frac{\partial \sigma_{Sij}^{(0)}}{\partial x_j} \neq 0.$$

The physical meaning of $\mathbf{u}^{(0)}$ does not pose any problem since it is the same physical quantity at both pore and macroscopic scales: $\mathbf{u}^{(0)}$ is a function of $\mathbf{x}$, only.

Tensor $\mathbf{c}$ *is an elastic tensor.* Firstly, tensor $\mathbf{c}$ shows the symmetries of an elastic tensor. On the one hand we have by construction

$$c_{ijkh} = c_{jikh} = c_{ijhk}.$$

On the other hand, $c_{ijkh} = c_{khij}$. To see this, consider formulation (6.14) where we put successively

$$e_{xij}(\mathbf{u}^{(0)}) = \delta_{ip}\delta_{jq} \qquad \text{and} \qquad \boldsymbol{\alpha} = \boldsymbol{\xi}^{*rs},$$

and

$$e_{xij}(\mathbf{u}^{(0)}) = \delta_{ir}\delta_{jr} \qquad \text{and} \qquad \boldsymbol{\alpha} = \boldsymbol{\xi}^{*pq}.$$

The symmetries of the scalar product and tensor $\mathbf{a}^*$ yield

$$\int_{\Omega_S^*} a^*_{ijkh}\, e_{yij}(\boldsymbol{\xi}^{*pq})\, e_{ykh}(\boldsymbol{\xi}^{*rs})\, \mathrm{dV}^* = -\int_{\Omega_S^*} a^*_{pqkh}\, e_{ykh}(\boldsymbol{\xi}^{*rs})\, \mathrm{dV}^* = -\int_{\Omega_S^*} a^*_{rskh}\, e_{ykh}(\boldsymbol{\xi}^{*pq})\, \mathrm{dV}^*.$$

The last equality demonstrates the symmetry of $\mathbf{c}^*$ (see the definition of $\mathbf{c}^*$).
Secondly, tensor $\mathbf{c}^*$ verifies an ellipticity condition as for $\mathbf{a}^*$

$$c^*_{ijkh}\, e_{xij}\, e_{xkh} \geq \beta\, e_{xij}\, e_{xij}, \qquad \beta > 0. \tag{6.19}$$

The left hand member of this equality is written at the first order

$$c^*_{ijkh}\, e_{xij}(\mathbf{u}^{(0)})\, e_{xkh}(\mathbf{u}^{(0)}) = <\sigma^{(0)}_{Sij}>\; e_{xij}(\mathbf{u}^{(0)})$$

$$= |\Omega^*|^{-1} e_{xij}(\mathbf{u}^{(0)}) \int_{\Omega_S^*} a^*_{ijkh}\, (e_{ykh}(\mathbf{u}^{(1)})) + e_{xkh}(\mathbf{u}^{(0)}))\, \mathrm{dV}^*.$$

Formulation (6.14) with $\boldsymbol{\alpha} = \mathbf{u}^{(1)}$ is in the form

$$\int_{\Omega_S^*} a^*_{ijkh}\, (e_{ykh}(\mathbf{u}^{(1)})) + e_{xkh}(\mathbf{u}^{(0)}))\, e_{yij}(\mathbf{u}^{(1)})\, \mathrm{dV}^* = 0.$$

By combining member to member the two last equalities, we obtain

$$c^*_{ijkh}\, e_{xij}(\mathbf{u}^{(0)})\, e_{xkh}(\mathbf{u}^{(0)}) =$$

$$= |\Omega^*|^{-1} \int_{\Omega_S^*} a^*_{ijkh}\, (e_{yij}(\mathbf{u}^{(1)})) + e_{xij}(\mathbf{u}^{(0)}))\, (e_{ykh}(\mathbf{u}^{(1)})) + e_{xkh}(\mathbf{u}^{(0)}))\, \mathrm{dV}^*, \tag{6.20}$$

and considering the ellipticity property of tensor $\mathbf{a}$ yields

$$c^*_{ijkh}\, e_{xij}(\mathbf{u}^{(0)})\, e_{xkh}(\mathbf{u}^{(0)}) \geq$$

$$\gamma\, |\Omega^*|^{-1} \int_{\Omega_S^*} (e_{yij}(\mathbf{u}^{(1)})) + e_{xij}(\mathbf{u}^{(0)}))\, (e_{yij}(\mathbf{u}^{(1)})) + e_{xij}(\mathbf{u}^{(0)}))\, \mathrm{dV}^*.$$

The right hand member cannot cancel out. Otherwise the periodicity of $\mathbf{u}^{(1)}$ would yield $e_{xij}(\mathbf{u}^{(0)}) = 0$, which is not acceptable. Therefore, the above inequality can be put in the form (6.19).

Energetic consistency. Let us show that the volume average of the pore scale energy density equals the macroscopic energy density. That comes directly from (6.20) by noting that this relation can be written in the form

$$<\sigma^{(0)}_{Sij}>\, e_{xij}(\mathbf{u}^{(0)}) = |\Omega^*|^{-1} \int_{\Omega_S^*} \sigma^{(0)}_{Sij}\, e^{(0)}_{ij}\, \mathrm{dV}^*.$$

6.2 Saturated Deformable Porous Matrix

We look for the macroscopic behaviour of a saturated porous matrix which contains a filtrating fluid when the movement is quasi-static: velocities are small and accelerations are negligible. The temperature is a constant. Both solid and fluid parts are connected. However, this condition for the fluid is not necessary when the macroscopic behaviour is monophasic.

For the sake of simplicity, we adopt the hypotheses that were previously advanced for the constituents:

- the material of the porous matrix is linear elastic and deformations are small. The matrix is Galilean and periodic.
- the fluid is viscous Newtonian and incompressible. The viscosity is a constant.

Pore scale description and estimations For the solid, the pore scale behaviour is as above

$$\frac{\partial \sigma_{Sij}}{\partial X_j} = 0, \tag{6.21}$$

$$\sigma_{Sij} = a_{ijkh}\ e_{Xkh}(\mathbf{u}_S) \qquad \text{in } \Omega_S. \tag{6.22}$$

The fluid verifies Stokes equation

$$\frac{\partial \sigma_{Fij}}{\partial X_j} = 0, \tag{6.23}$$

$$\sigma_{Fij} = 2\mu D_{ij} - pI_{ij}, \tag{6.24}$$

and the incompressibility condition

$$\frac{\partial v_i}{\partial X_i} = 0 \qquad \text{in } \Omega_F. \tag{6.25}$$

On the interface Γ, the velocity and the normal stress are continuous

$$\sigma_{Sij}N_j = \sigma_{Fij}N_j, \tag{6.26}$$

$$v_S i = v_F i \qquad \text{on } \Gamma. \tag{6.27}$$

We assume in the following that the macroscopic stresses and the displacements in the solid and the fluid are of similar order of magnitude, respectively

$$\boldsymbol{\sigma}_S^{macro} = \mathcal{O}(\boldsymbol{\sigma}_F^{macro}), \qquad \mathbf{u}_S = \mathcal{O}(\mathbf{u}_F).$$

With these assumptions, we are left with only one dimensionless number to be estimated

$$\mathcal{Q} = |\nabla p| / |\mu \Delta \mathbf{v}|.$$

We choose L to make dimensionless the system (6.21-6.27). When the matrix is deformable, the above estimations enable a new form of $\mathcal{Q}$

$$\mathcal{Q}_L = \frac{p_c}{L}\frac{L^2}{\mu_c v_c} = \frac{a_c u_c}{L}\frac{L\, t_c}{\mu_c u_c} = \frac{a_c t_c}{\mu_c},$$

which now appears as a contrast of properties number. Time t_c is the characteristic time of the excitation. In part 3 where the matrix was rigid, we have demonstrated that the only homogenizable situation corresponded to estimation

$$\mathcal{Q}_l = \mathcal{O}(\varepsilon^{-1}), \qquad \mathcal{Q}_L = \mathcal{O}(\varepsilon^{-2}),$$

which yields the Darcy's law. Due to the deformable matrix, different estimations are now possible which yield to different homogenizable situations. The three cases of interest are the following

a. $\mathcal{Q}_L = \mathcal{O}(\varepsilon^{-2})$ yields a diphasic macroscopic behaviour.

b. $\mathcal{Q}_L = \mathcal{O}(\varepsilon^{-1})$ gives an elastic monophasic macroscopic behaviour.

c. $\mathcal{Q}_L = \mathcal{O}(1)$ corresponds to a viscoelastic monophasic macroscopic behaviour.

We successively address these three cases.

Diphasic macroscopic behaviour: Biot's model The system (6.21-6.27) is made dimensionless with $\mathbf{X} = L\mathbf{x}$, $\mathbf{u}_S = u_c \mathbf{u}_S^*$, $t = t_c t^*$, $\mathbf{v} = \frac{u_c}{t_c}\mathbf{v}^*$ and $\mathbf{a} = a_c \mathbf{a}^*$, with $\mathcal{Q}_L = \mathcal{O}(\varepsilon^{-2})$

$$\frac{\partial}{\partial x_j}(a^*_{ijkh}\; e^*_{kh}(\mathbf{u}^*_S)) = 0, \tag{6.28}$$

$$\sigma^*_{Sij} = a^*_{ijkh}\; e^*_{kh}(\mathbf{u}^*_S) \qquad \text{in } \Omega^*_S, \tag{6.29}$$

$$\varepsilon^2 \mu^* \frac{\partial^2 v^*_i}{\partial x_j \partial x_j} - \frac{\partial p^*}{\partial x_i} = 0, \tag{6.30}$$

$$\sigma^*_{Fij} = 2\mu^* \varepsilon^2 D^*_{ij} - p^* I_{ij}, \qquad D^*_{ij} = \frac{1}{2}\left(\frac{\partial v^*_i}{\partial x_j} + \frac{\partial v^*_j}{\partial x_i}\right) \qquad \text{in } \Omega^*_F, \tag{6.31}$$

$$\sigma^*_{Fij} N_j = \sigma^*_{Fij} N_j, \tag{6.32}$$

$$v^*_{Fi} = v^*_{Fi} \qquad \text{on } \Gamma^*. \tag{6.33}$$

We look for the unknowns $\mathbf{u}_S^*$, $\mathbf{v}^*$ et p^* in the form

$$\mathbf{u}_S^*(\mathbf{x},\mathbf{y},t^*) = \mathbf{u}^{(0)}(\mathbf{x},\mathbf{y},t^*) + \varepsilon\mathbf{u}^{(1)}(\mathbf{x},\mathbf{y},t^*) + \varepsilon^2\mathbf{u}^{(2)}(\mathbf{x},\mathbf{y},t^*) + \cdots,$$

$$\mathbf{v}^*(\mathbf{x},\mathbf{y},t^*) = \mathbf{v}^{(0)}(\mathbf{x},\mathbf{y},t^*) + \varepsilon\mathbf{v}^{(1)}(\mathbf{x},\mathbf{y},t^*) + \varepsilon^2\mathbf{v}^{(2)}(\mathbf{x},\mathbf{y},t^*) + \cdots,$$

$$p^*(\mathbf{x},\mathbf{y},t^*) = p^{(0)}(\mathbf{x},\mathbf{y},t^*) + \varepsilon p^{(1)}(\mathbf{x},\mathbf{y},t^*) + \varepsilon^2 p^{(2)}(\mathbf{x},\mathbf{y},t^*) + \cdots, \tag{6.34}$$

with $\mathbf{y} = \varepsilon^{-1}\mathbf{x}$ and where $\mathbf{u}^{(i)}$, $\mathbf{v}^{(i)}$ and $p^{(i)}$ are Ω^*-periodic in $\mathbf{y}$. Note that we have introduced a possible time dependence which could be due to fluid volume change.

At the lowest order, we obtain for $\mathbf{u}^{(0)}$ a closed boundary value problem already encountered when investigating the empty matrix

$$\frac{\partial}{\partial y_j}(a_{ijkh}^* \, e_{ykh}(\mathbf{u}^{(0)})) = 0 \qquad \text{in } \Omega_S^*,$$

$$a_{ijkh}^* \, e_{ykh}(\mathbf{u}^{(0)})N_j = 0 \qquad \text{on } \Gamma^*,$$

where $\mathbf{u}^{(0)}$ is Ω^*-periodic in $\mathbf{y}$. Therefore we have

$$\mathbf{u}^{(0)} = \mathbf{u}^{(0)}(\mathbf{x},t^*).$$

Similarly, at the lowest order, Stokes equation yields

$$\frac{\partial p^{(0)}}{\partial y_i} = 0,$$

and we retrieve

$$p^{(0)} = p^{(0)}(\mathbf{x},t^*).$$

At the following order, Stokes equation and the adherence condition give a boundary value problem for $\mathbf{v}^{(0)}$

$$\mu^* \frac{\partial^2 v_i^{(0)}}{\partial y_j \partial y_j} - \frac{\partial p^{(1)}}{\partial y_i} - \frac{\partial p^{(0)}}{\partial x_i} = 0,$$

$$\frac{\partial v_i^{(0)}}{\partial y_i} = 0 \qquad \text{in } \Omega_F^*,$$

$$v_i^{(0)} = \frac{\partial u_i^{(0)}}{\partial t^*} \qquad \text{on } \Gamma^*,$$

with $\mathbf{v}^{(0)}$ and $p^{(1)}$ Ω^*-periodic in $\mathbf{y}$. Let us introduce the relative velocity to the solid $w_i^* = v_i^{(0)} - \partial u_i^{(0)}/\partial t^*$: we recover the boundary value problem for the Darcy's flow in a rigid matrix, after replacing $\mathbf{v}^{(0)}$ by $\mathbf{w}^*$

$$\mu^* \frac{\partial^2 w_i^*}{\partial y_j \partial y_j} - \frac{\partial p^{(1)}}{\partial y_i} - \frac{\partial p^{(0)}}{\partial x_i} = 0,$$

$$\frac{\partial w_i^*}{\partial y_i} = 0 \qquad \text{in } \Omega_F^*,$$
$$w_i^* = 0 \qquad \text{on } \Gamma^*,$$

$\mathbf{w}^*$ and $p^{(1)}$ Ω^*-periodic in $\mathbf{y}$.
Therefore, we have

$$w_i^* = -\frac{k_{ij}^*}{\mu^*}\frac{\partial p^{(0)}}{\partial x_j},$$

and by averaging the incompressibility condition at the second order, we obtain a Darcy's law relatively to the moving matrix

$$< \mathbf{w}^* >=< \mathbf{v}^{(0)} > -\phi\frac{\partial \mathbf{u}^{(0)}}{\partial t^*} = -\frac{\mathbf{K}^*}{\mu^*}\nabla_x p^{(0)}, \qquad \mathbf{K}^* =< \mathbf{k}^* > . \tag{6.35}$$

Tensor $\mathbf{K}^*$ is the permeability tensor of the matrix considered as rigid.

Let us come back to the matrix. The following order yields a boundary value problem for $\mathbf{u}^{(1)}$:

$$\frac{\partial}{\partial y_j}(a_{ijkh}^* \; (e_{ykh}(\mathbf{u}^{(1)})) + e_{xkh}(\mathbf{u}^{(0)}))) = 0 \qquad \text{in } \Omega_S^*,$$

$$a_{ijkh}^* \; (e_{ykh}(\mathbf{u}^{(1)})) + e_{xkh}(\mathbf{u}^{(0)}))N_j = -p^{(0)}N_i \qquad \text{on } \Gamma^*,$$

where $\mathbf{u}^{(1)}$ is Ω^*-periodic in $\mathbf{y}$. This system with $p^{(0)} = 0$ is similar to the system investigated for the empty porous matrix. However, the unknown $\mathbf{u}^{(1)}$ is now a linear vectorial function of $\mathbf{e}_x(\mathbf{u}^{(0)})$ and of $p^{(0)}$, to an added arbitrary vector $\overline{\mathbf{u}}^{(1)}$ which is a function of $\mathbf{x}$ and t^*

$$u_i^{(1)} = \xi_i^{*lm} e_{xlm}(\mathbf{u}^{(0)}) - \eta_i^* p^{(0)} + \overline{u}_i^{(1)}(\mathbf{x}, t^*),$$

where $\boldsymbol{\xi}^*(\mathbf{y})$ is the third order tensor already defined and $\boldsymbol{\eta}^*(\mathbf{y})$ is a new vector field which is obtain with $\mathbf{e}_x(\mathbf{u}^{(0)}) = 0$ and $p^{(0)} = -1$.

Consider now the momentum balances of the solid and the fluid at order ε^0, together with the normal stress continuity condition on Γ^* at order ε :

$$\frac{\partial \sigma_{Sij}^{(1)}}{\partial y_j} + \frac{\partial \sigma_{Sij}^{(0)}}{\partial x_j} = 0 \qquad \text{in } \Omega_S^*,$$

$$\frac{\partial \sigma_{Fij}^{(1)}}{\partial y_j} - \frac{\partial p^{(0)}}{\partial x_i} = 0 \qquad \text{in } \Omega_F^*,$$

$$\sigma_{Sij}^{(1)} N_j = \sigma_{Fij}^{(1)} N_j \qquad \text{on } \Gamma^*.$$

Stresses $\boldsymbol{\sigma}_S^{(1)}$ and $\boldsymbol{\sigma}_F^{(1)}$ are Ω^*-periodic. These balances yield a compatibility condition by integration on their respective domain of validity Ω_S^* and Ω_F^*. Let us now define the total stress $\boldsymbol{\sigma}^{*T}$ by

$$\boldsymbol{\sigma}^T = \begin{cases} \boldsymbol{\sigma}_S^* & \text{in } \Omega_S^*, \\ \boldsymbol{\sigma}_F^* & \text{in } \Omega_F^*. \end{cases}$$

We have

$$< \frac{\partial \sigma_{ij}^{T(0)}}{\partial x_j} >= 0, \tag{6.36}$$

$$\sigma_{ij}^{T(0)} = c^*_{ijkh}\ e_{xkh}(\mathbf{u}^{(0)}) - \alpha^*_{ij}\ p^{(0)}, \tag{6.37}$$

where

$$c^*_{ijkh} =< a^*_{ijkh} + a^*_{ijlm}\ e_{ylm}(\boldsymbol{\xi}^{*kh}) >$$

is the dimensionless effective elastic tensor above introduced by the macroscopic behaviour of the empty porous matrix, and

$$\alpha^*_{ij} = \phi I_{ij} + < a^*_{ijlm}\ e_{ylm}(\boldsymbol{\eta}^*) >$$

is a new elastic tensor.

A second compatibility condition is obtained from the fluid volume balance at order ε^0

$$\frac{\partial v_i^{(0)}}{\partial x_i} + \frac{\partial v_i^{(1)}}{\partial y_i} = 0.$$

Integrating over Ω^*_F, and using the displacement continuity condition on Γ^* at order ε and the periodicity of $\mathbf{v}^{(1)}$ give

$$\frac{\partial(< v_i^{(0)} > -\phi \frac{\partial u_i^{(0)}}{\partial t^*})}{\partial x_i} = -\gamma^*_{ij}\ \frac{\partial e_{xij}(\mathbf{u}^{(0)})}{\partial t^*} - \beta^*\ \frac{\partial p^{(0)}}{\partial t^*}, \tag{6.38}$$

which introduces two new elastic coefficients

$$\gamma^*_{ij} = \phi I_{ij} - < \xi^{*ij}_{p,p} >,$$

$$\beta^* =< \eta^*_{p,p} > .$$

Finally, the macroscopic behaviour is given by (6.36-6.37-6.38) and (6.35). The dimensional behaviour is given by

$$\frac{\partial < \sigma^T_{ij} >}{\partial X_j} = \mathcal{O}(\varepsilon \frac{\partial < \sigma^T_{ij} >}{\partial X_j}), \tag{6.39}$$

$$< \sigma^T_{ij} >= c_{ijkh}\ e_{Xkh}(\mathbf{u}_S) - \alpha_{ij}\ p + \mathcal{O}(\varepsilon < \sigma^T_{ij} >), \tag{6.40}$$

$$\frac{\partial(< v_i > -\phi \frac{\partial u_{Si}}{\partial t})}{\partial X_i} = -\gamma_{ij}\ \frac{\partial e_{Xij}(\mathbf{u}_S)}{\partial t} - \beta\ \frac{\partial p}{\partial t} + \mathcal{O}(\varepsilon\beta\ \frac{\partial p}{\partial t}), \tag{6.41}$$

$$< v_i > -\phi \frac{\partial u_{Si}}{\partial t} = -\frac{K_{ij}}{\mu} \frac{\partial p}{\partial X_j} + \mathcal{O}(\varepsilon(< v_i > -\phi \frac{\partial u_{Si}}{\partial t})), \tag{6.42}$$

$$\mathbf{c} = a_c \mathbf{c}^*, \qquad \boldsymbol{\alpha} = \boldsymbol{\alpha}^*, \qquad \boldsymbol{\gamma} = \boldsymbol{\gamma}^*, \qquad \beta = \frac{\beta^*}{a_c},$$

which represent seven scalar equations for seven unknowns u_{Si}, $< v_i >$ and p. This system is similar to Biot's system. The porous saturated medium is described by two displacement fields for the solid and the liquid, respectively. The phenomenon under consideration is called consolidation. In practise, one tries to reduce the number of unknowns, that will be shown below.

Properties of the diphasic macroscopic behaviour We first address the physical meaning of macroscopic quantities and the symmetry and positiveness of effective coefficients. Then we study particular cases of homogeneity and isotropy. We also show that the number of unknowns can be reduced and finally we give some information about the case of slightly compressible fluids.

Properties of the macroscopic quantities and of the effective coefficients

- The macroscopic quantities that are present in the macroscopic behaviour were previously analysed, with the exception of $< \boldsymbol{\sigma}^T >$, for which the reasoning about $< \boldsymbol{\sigma}_S >$ can be done again in the same form. Then, $< \boldsymbol{\sigma}^T >$ and $< \mathbf{v} >$ are fluxes and they represent a stress and a Darcy's velocity, respectively. Tensors $\mathbf{c}$ et $\mathbf{K}$ were also studied. Consider tensors $\boldsymbol{\alpha}$, $\boldsymbol{\gamma}$ and β.
- The coupling of (6.40) and (6.41) is symmetric, $\boldsymbol{\alpha} = \boldsymbol{\gamma}$.
  The property comes from the symmetry of the scalar product in space $\mathcal{E}$ and from the variational form which gives $\mathbf{u}^{(1)}$

$$\forall \boldsymbol{\alpha} \in \mathcal{E}, \qquad \int_{\Omega_S^*} a^*_{ijkh}(e_{ykh}(\mathbf{u}^{(1)})) + e_{xkh}(\mathbf{u}^{(0)}))\ e_{yij}(\boldsymbol{\alpha})\ \mathrm{dV}^* = -\int_{\Gamma^*} p^{(0)} \alpha_i N_i \mathrm{dS}^*,$$

where $\mathbf{N}$ is the outward unit normal to Ω_S^*. Consider this form with on the one hand $\mathbf{u}^{(1)} = \boldsymbol{\eta}^*$ and $\boldsymbol{\alpha} = \boldsymbol{\xi}^{*lm}$ and on the other hand $\mathbf{u}^{(1)} = \boldsymbol{\xi}^{*lm}$ and $\boldsymbol{\alpha} = \boldsymbol{\eta}^*$. The symmetry of the scalar product yields

$$\int_{\Omega_S^*} a^*_{ijkh}\ e_{ykh}(\boldsymbol{\xi}^{*lm})\ \ e_{yij}(\boldsymbol{\eta}^*)\ \mathrm{dV}^* = \int_{\Gamma^*} \xi_i^{*lm} N_i \mathrm{dS}^* = -\int_{\Omega_S^*} a^*_{ijlm}\ e_{yij}(\boldsymbol{\eta}^*)\ \mathrm{dV}^*.$$

The periodicity of vector $\boldsymbol{\xi}^{*lm}$ and the divergence theorem then give

$$\int_{\Gamma^*} \xi_i^{*lm} N_i \mathrm{dS}^* = \int_{\delta\Omega_S^*} \xi_i^{*lm} N_i \mathrm{dS}^* = \int_{\Omega_S^*} \xi_{i,i}^{*lm} \mathrm{dV}^* = -\int_{\Omega_S^*} a^*_{ijlm}\ e_{yij}(\boldsymbol{\eta}^*)\ \mathrm{dV}^*.$$

After considering the definitions of $\boldsymbol{\alpha}^*$ and $\boldsymbol{\gamma}^*$, the last equality gives

$$\boldsymbol{\alpha}^* = \boldsymbol{\gamma}^*.$$

- Tensor $\boldsymbol{\alpha}^*$ is symmetric. This comes directly from the definition of $\boldsymbol{\alpha}^*$ and the properties of tensor a*

$$\alpha^*_{ij} = \phi I_{ij} + < a^*_{ijlm}\ e_{ylm}(\boldsymbol{\eta}^*) > .$$

Tensor $\boldsymbol{\alpha}^*$ is symmetric and real; therefore it can be made diagonal.

- Coefficient β^* is positive, $\beta^* > 0$. Consider again the variational form for $\mathbf{u}^{(1)}$ with now $\mathbf{u}^{(1)} = \boldsymbol{\alpha}^* = \boldsymbol{\eta}^*$

$$\int_{\Omega^*_S} a^*_{ijkh}\ e_{yij}(\boldsymbol{\eta}^*)\ \ e_{ykh}(\boldsymbol{\eta}^*)\ \mathrm{dV}^* = \int_{\Gamma^*} \eta^*_i N_i \mathrm{dS}^* = \int_{\Omega^*_S} \eta^*_{i,i}\ \mathrm{dV}^*, \tag{6.43}$$

where the periodicity of $\boldsymbol{\eta}^*$ was used. The ellipticity property of tensor **a** yields the positivity of the left hand member of these equalities, that gives the result from the definition of β.

Particular cases.

- Assume the material of the porous matrix is homogeneous. Tensor $\mathbf{a}^*$ is now independent of $\mathbf{y}$. The expression of $\mathbf{c}^*$ becomes

$$c^*_{ijkh} =< a^*_{ijkh} + a^*_{ijlm}\ e_{ylm}(\boldsymbol{\xi}^{*kh}) >= (1-\phi)a^*_{ijkh} + a^*_{ijlm} < e_{ylm}(\boldsymbol{\xi}^{*kh}) > . \tag{6.44}$$

Let us introduce the inverse $\mathbf{d}^*$ of $\mathbf{a}^*$

$$a^*_{ijkh}\ d^*_{\alpha\beta kh} = \frac{1}{2}(I_{\alpha i}I_{\beta j} + I_{\alpha j}I_{\beta i}).$$

We obtain from (6.44)

$$< e_{ylm}(\boldsymbol{\xi}^{*kh}) >= (c^*_{ijkh} - (1-\phi)\ a^*_{ijkh})\ d^*_{ijlm},$$

which yields

$$< \xi^{*kh}_{i,i} >= (c^*_{ijkh} - (1-\phi)\ a^*_{ijkh})\ d^*_{ijtt} = c^*_{ijkh}\ d^*_{ijtt} - (1-\phi)\ I_{kh}.$$

We deduce that $\boldsymbol{\alpha}^*$ and $\boldsymbol{\gamma}^*$ are in the form

$$\alpha^*_{ij} = \gamma^*_{ij} = \phi I_{ij} - < \xi^{*ij}_{l,l} >= I_{ij} - c^*_{khij}\ d^*_{khtt}.$$

On an other hand, we have from the definition of $\boldsymbol{\alpha}^*$

$$e_{ylm}(\boldsymbol{\eta}^*) = (\alpha^*_{ij} - \phi I_{ij})\ d^*_{ijlm} = ((1-\phi)I_{ij} - c^*_{khij}\ d^*_{khtt})\ d^*_{ijlm},$$

and

$$\beta^* =< \eta^*_{l,l} >= (1-\phi)d^*_{iill} - c^*_{khij}\ d^*_{khtt}\ d^*_{ijll}.$$

In some applications, the material of the porous matrix is very slightly compressible. For such cases, the approximation of incompressibility is often adopted, particularly in soil mechanics

$$d^*_{ijll} = 0,$$

which yields

$$\alpha^*_{ij} = I_{ij}, \qquad \beta^* = 0.$$

We recover the classical behaviour in use in soil mechanics.

- Assume the matrix material is homogeneous and isotropic and the matrix is macroscopically isotropic. We have

$$a^*_{ijtr} = \lambda^*_1 I_{ij}\ I_{tr} + \mu^*_1(I_{it}I_{jr} + I_{ir}I_{jt}),$$
$$c^*_{ijtr} = \lambda^* I_{ij}\ I_{tr} + \mu^*(I_{it}I_{jr} + I_{ir}I_{jt}),$$

where λ^*_1, μ^*_1 and λ^*, μ^* are the Lamé coefficients of the matrix material and of the matrix at the macroscopic scale, respectively. The expression of $\boldsymbol{\alpha}^*$ simplifies to

$$\alpha^*_{ij} = I_{ij} - \frac{c^*_{ijll}}{3\lambda^*_1 + 2\mu^*_1} = (1 - \frac{3\lambda^* + 2\mu^*}{3\lambda^*_1 + 2\mu^*_1})I_{ij} = \alpha^* I_{ij}.$$

Since we have

$$\frac{3\lambda^* + 2\mu^*}{3\lambda^*_1 + 2\mu^*_1}) \geq 0,$$

we obviously get

$$\alpha^* \leq 1$$

More,

$$\alpha^* = \phi + (\lambda^*_1 + \frac{2\mu^*_1}{3}) < \eta^*_{i,i} >,$$

and (6.43) yields $\alpha^* > \phi$ and

$$\phi \leq \alpha^* \leq 1.$$

Therefore, we have

$$0 \leq \frac{3\lambda^* + 2\mu^*}{3\lambda^*_1 + 2\mu^*_1} \leq 1 - \phi.$$

Finally we get

$$\beta^* = \frac{3}{3\lambda^*_1 + 2\mu^*_1}((1 - \phi) - \frac{3\lambda^* + 2\mu^*}{3\lambda^*_1 + 2\mu^*_1}),$$

and we retrieve that an incompressible material is characterized by $\alpha^* = 1$ et $\beta^* = 0$.

Equations of the diphasic consolidation: Biot's model. By taking into account $\boldsymbol{\alpha} = \boldsymbol{\gamma}$, the system that describes the macroscopic behaviour of the diphasic consolidation is in the form (for simplicity, the relative errors s $\mathcal{O}(\varepsilon)$ are omitted)

$$\frac{\partial}{\partial X_j}(c_{ijkh}\ e_{Xkh}(\mathbf{u}_S) - \alpha_{ij}\ p) = 0, \tag{6.45}$$

$$< v_i > -\phi\frac{\partial u_{Si}}{\partial t} = -\frac{K_{ij}}{\mu}\frac{\partial p}{\partial X_j}, \tag{6.46}$$

$$\frac{\partial(< v_i > -\phi\frac{\partial u_{Si}}{\partial t})}{\partial X_i} = -\alpha_{ij}\ \frac{\partial e_{Xij}(\mathbf{u}_S)}{\partial t} - \beta\ \frac{\partial p}{\partial t}, \tag{6.47}$$

In practice, it is of interest to reduce the number of unknowns. That can be done in two different ways.

- By eliminating $< \mathbf{v} >$ between the two last equations (6.46) and (6.47). We obtain

$$\frac{\partial}{\partial X_j}(c_{ijkh}\, \mathbf{e}_{Xkh}(\mathbf{u}_S) - \alpha_{ij}\, p) = 0,$$

$$\frac{\partial}{\partial X_i}(\frac{K_{ij}}{\mu}\frac{\partial p}{\partial X_j}) = \alpha_{ij}\frac{\partial e_{Xij}(\mathbf{u}_S)}{\partial t} - \beta\frac{\partial p}{\partial t},$$

which represents the classical system in use in soil mechanics: four scalar equations for four unknowns u_{Si} and p.

- By rendering symmetric the system. Let us integrate with respect to time the volume balance (6.47) and let $< \mathbf{v} >= \phi\dot{\mathbf{u}}_F$ where $\mathbf{u}_F$ is the average displacement of the fluid over Ω_F. We obtain

$$p = -\alpha_{ij}\beta^{-1}e_{Xij}(\mathbf{u}_S) - \phi\beta^{-1}\frac{\partial}{\partial X_i}(u_{Fi} - u_{Si}) + f(\mathbf{X}).$$

As generally p, u_{Si} and u_{Fi} are increments from a given initial state, function $f(\mathbf{X})$ which is independent of time can be considered as null. Introduce the above value of p into the the Darcy's law (6.46)

$$\frac{\partial}{\partial X_i}(\phi p) = -\frac{\partial}{\partial X_i}(\phi\alpha_{lm}\beta^{-1}e_{Xlm}(\mathbf{u}_S) + \phi^2\beta^{-1}\frac{\partial}{\partial X_l}(u_{Fl} - u_{Sl}))$$

$$= -\phi^2\mu K_{ij}^{-1}(\frac{\partial u_{Fj}}{\partial t} - \frac{\partial u_{Sj}}{\partial t}).$$

We recover the formulation obtained by Biot (Biot, 1955) by a phenomenological approach. With the Biot's notation, we have

$$\frac{\partial}{\partial X_j}(Q_{lm}\, e_{Xlm}(\mathbf{u}_S) + R\theta) = b_{ij}(\frac{\partial u_{Fj}}{\partial t} - \frac{\partial u_{Sj}}{\partial t}), \qquad (6.48)$$

$$\phi p = -Q_{ij}e_{Xij}(\mathbf{u}_S) - R\theta,$$

with

$$Q_{ij} = \phi\beta^{-1}(\alpha_{ij} - \phi I_{ij}), \quad R = \phi^2\beta^{-1}, \quad b_{ij} = \phi^2\mu K_{ij}^{-1}, \quad \theta = \frac{\partial u_{Fi}}{\partial X_i}.$$

Let us introduce the matrix partial stress by

$$< \sigma_{Sij} >=< \sigma_{ij}^T > +\phi p I_{ij}.$$

Subtracting member to member equation (6.48) from equation (6.45) gives

$$\frac{\partial}{\partial X_j} < \sigma_{Sij} >= -b_{ij}(\frac{\partial u_{Fj}}{\partial t} - \frac{\partial u_{Sj}}{\partial t}), \qquad (6.49)$$

with

$$< \sigma_{Sij} >= \pi_{ijkh}e_{Xkh}(\mathbf{u}_S) + Q_{ij}\theta,$$

$$\pi_{ijkh} = c_{ijkh} + \beta^{-1}(\alpha_{ij} - \phi I_{ij})(\alpha_{kh} - \phi I_{kh}).$$

Equations (6.48) and (6.49) were introduced by Biot (Biot, 1955). Their advantage is to focus on the symmetry of the solid-fluid coupling, with the tensor $\mathbf{Q}$ present in both equations. Their generalization to acoustics is very well adapted to the study of wave propagations.

Effective stress

Let us introduce the effective stress

$$< \sigma^e_{ij} >=< \sigma^T_{ij} > +pI_{ij}.$$

As the first order of approximation $p^{(0)}$ of p is a constant over a period, the above defined effective stress is similar to its definition in soil mechanics.

Compressible fluid

When the fluid is strongly compressible, the above results are easily generalized (Auriault et al., 1990). When the fluid is slightly compressible, the linearity of the macroscopic behaviour can be preserved. The dimensionless equation of state of the fluid (temperature is assumed as a constant) can be written in the form

$$\varepsilon \frac{\partial p^*}{\partial t^*} = -K^*_\omega \frac{\partial v^*_i}{\partial x_i},$$

where K^*_ω is the rigidity of the fluid. After introducing the asymptotic expansion for the velocity, we obtain as above at the first order

$$\frac{\partial v_i^{(0)}}{\partial y_i} = 0.$$

Therefore Darcy's law remains valid. The compressibility appears at the following order

$$\frac{\partial p^{(0)}}{\partial t^*} = -K^*_\omega \left(\frac{\partial v_i^{(0)}}{\partial x_i} + \frac{\partial v_i^{(1)}}{\partial y_i}\right),$$

which yields to the macroscopic volume balance (the relative error $\mathcal{O}(\varepsilon)$ is not shown)

$$\frac{\partial(< v_i > -\phi\frac{\partial u_{Si}}{\partial t})}{\partial X_i} = -\alpha_{ij}\, \frac{\partial e_{Xij}(\mathbf{u}_S)}{\partial t} - \beta\, \frac{\partial p}{\partial t},$$

where coefficient β is now defined by

$$\beta =< \eta_{l,l} > + \frac{\phi}{K_\omega}.$$

Monophasic elastic macroscopic behaviour The dimensionless number $\mathcal{Q}_L$ is now decreased to

$$\mathcal{Q}_L = \mathcal{O}(\varepsilon^{-1}),$$

which can be done by, e.g., increasing the viscosity. The dimensionless pore scale behaviour is in the form

$$\frac{\partial}{\partial x_j}(a^*_{ijkh}\ e^*_{kh}(\mathbf{u}^*_S)) = 0, \tag{6.50}$$

$$\sigma^*_{Sij} = a^*_{ijkh}\ e^*_{kh}(\mathbf{u}^*_S) \qquad \text{in } \Omega^*_S, \tag{6.51}$$

$$\varepsilon\mu^*\frac{\partial^2 v^*_i}{\partial x_j \partial x_j} - \frac{\partial p^*}{\partial x_i} = 0, \tag{6.52}$$

$$\sigma^*_{Fij} = 2\mu^*\varepsilon D^*_{ij} - p^* I_{ij} \qquad \text{in } \Omega^*_F, \tag{6.53}$$

$$\sigma^*_{Fij}N_j = \sigma^*_{Fij}N_j, \tag{6.54}$$

$$u^*_{Fi} = u^*_{Fi} \qquad \text{on } \Gamma^*. \tag{6.55}$$

At the lower order the boundary value problem for the solid displacement $\mathbf{u}^{(0)}$ is unchanged

$$\mathbf{u}^{(0)} = \mathbf{u}^{(0)}(\mathbf{x}, t^*)$$

The first order boundary value problem for the fluid is for $\mathbf{v}^{(0)}$ and $p^{(0)}$ is in the form

$$\mu^*\frac{\partial^2 v^{(0)}_i}{\partial y_j \partial y_j} - \frac{\partial p^{(0)}}{\partial y_i} = 0,$$

$$\frac{\partial v^{(0)}_i}{\partial y_i} = 0 \qquad \text{in } \Omega^*_F,$$

$$v^{(0)}_i = \frac{\partial u^{(0)}_i}{\partial t^*} \qquad \text{on } \Gamma^*,$$

where $\mathbf{v}^{(0)}$ and $p^{(0)}$ are Ω^*-periodic in $\mathbf{y}$. The solution is trivial

$$v^{(0)}_i = \frac{\partial u^{(0)}_i(\mathbf{x}, t^*)}{\partial t^*}, \qquad p^{(0)} = p^{(0)}(\mathbf{x}, t^*).$$

We see that at the first order of magnitude, due to the increased viscosity, there is no relative displacement of the fluid to the solid matrix.
The equations for $\mathbf{u}^{(1)}$ are unchanged. Therefore, we have

$$u^{(1)}_i = \xi^{*lm}_i e_{xlm}(\mathbf{u}^{(0)}) - \eta^*_i p^{(0)} + \overline{u}^{(1)}_i(\mathbf{x}, t^*).$$

The bulk macroscopic momentum balance is also unchanged

$$< \frac{\partial \sigma_{ij}^{T(0)}}{\partial x_j} >= 0,$$

$$\sigma_{ij}^{T(0)} = c_{ijkh}^* \; e_{xkh}(\mathbf{u}^{(0)}) - \alpha_{ij}^* \; p^{(0)}.$$

By contrast, the volume balance (6.47) is drastically modified. Because of

$$< v_i^{(0)} >= \phi \; \frac{\partial u_i^{(0)}}{\partial t^*},$$

it becomes

$$0 = \alpha_{ij}^* \; \frac{\partial e_{xij}(\mathbf{u}_S^{(0)})}{\partial t^*} + \beta^* \; \frac{\partial p^{(0)}}{\partial t^*},$$

and we have the equality $\boldsymbol{\alpha}^* = \boldsymbol{\gamma}^*$, which remains valid in the present case. Integrating the above equation with respect to time and noting that the constant of integration can be again taken as null since $p^{(0)}$ and $\mathbf{e}_x(\mathbf{u}^{(0)})$ are increments, we obtain

$$p^{(0)} = -\beta^{*-1} \alpha_{ij}^* \; e_{xij}(\mathbf{u}^{(0)}).$$

Therefore, the macroscopic dimensional description is in the form

$$\frac{\partial < \sigma_{ij}^T >}{\partial X_j} = \mathcal{O}(\varepsilon \frac{\partial < \sigma_{ij}^T >}{\partial X_j}),$$

$$< \sigma_{ij}^T >= c_{ijkh}^+ \; e_{Xkh}(\mathbf{u}_S) + \mathcal{O}(\varepsilon < \sigma_{ij}^T >),$$

with

$$c_{ijkh}^+ = c_{ijkh} + \beta^{-1} \alpha_{ij} \; \alpha_{kh}.$$

From the definition of tensor $\mathbf{c}^+$, it is easy to check that the macroscopic behaviour is elastic. We recover the Gassmann model.

Monophasic viscoelastic macroscopic behaviour The viscous stress and the pressure are of the same order of magnitude at the macroscopic scale

$$\mathcal{Q}_L = O(1),$$

and the dimensionless pore scale description is now

$$\frac{\partial}{\partial x_j}(a_{ijkh}^* \; e_{kh}^*(\mathbf{u}_S^*)) = 0, \tag{6.56}$$

$$\sigma_{Sij}^* = a_{ijkh}^* \; e_{kh}^*(\mathbf{u}_S^*) \qquad \text{in } \Omega_S^*, \tag{6.57}$$

$$\mu^* \frac{\partial^2 v_i^*}{\partial x_j \partial x_j} - \frac{\partial p^*}{\partial x_i} = 0, \tag{6.58}$$

$$\sigma^*_{Fij} = 2\mu^* D^*_{ij} - p^* I_{ij} \qquad \text{in } \Omega^*_F, \tag{6.59}$$

$$\sigma^*_{Sij} N_j = \sigma^*_{Fij} N_j, \tag{6.60}$$

$$v^*_{Si} = v^*_{Fi} \qquad \text{on } \Gamma^*. \tag{6.61}$$

Let $\mathbf{u}^*$ be the displacement defined by $\mathbf{u}^*_S$ in Ω^*_S and by $\mathbf{u}^*_F$ in Ω^*_F. It is convenient to consider displacements and pressure at constant frequency ω (ω is small since the movement is quasi-static). All quantities φ are looked for in the form $\varphi e^{i\omega t}$, $i^2 = -1$. Therefore, we get

$$\mathbf{v}^* = i\omega^* \mathbf{u}^*_F.$$

The first order term $\mathbf{u}^{(0)}$ of the displacement is the solution of the following system

$$\frac{\partial}{\partial y_j}(a^*_{ijkh}\ e_{ykh}(\mathbf{u}^{(0)})) = 0 \qquad \text{in } \Omega^*_S,$$

$$\mu^* \frac{\partial^2 (i\omega^* u_i^{(0)})}{\partial y_j \partial y_j} = 0,$$

$$\frac{\partial u_i^{(0)}}{\partial y_i} = 0 \qquad \text{in } \Omega^*_F,$$

$$a^*_{ijkh}\ e_{ykh}(\mathbf{u}^{(0)}) N_j = 2\mu^* e_{yij}(i\omega^* \mathbf{u}^{(0)}) N_j,$$

$$u^{(0)}_{Si} = u^{(0)}_{Fi} \qquad \text{on } \Gamma^*,$$

where $\mathbf{u}^{(0)}$ is Ω^*-periodic in $\mathbf{y}$. It is easy to check that

$$\mathbf{u}^{(0)} = \mathbf{u}^{(0)}(\mathbf{x}, \omega^*).$$

We next obtain a boundary value problem for $\mathbf{u}^{(1)}$ and $p^{(0)}$ in the form

$$\frac{\partial}{\partial y_j}(a^*_{ijkh}\ (e_{ykh}(\mathbf{u}^{(1)})) + e_{xkh}(\mathbf{u}^{(0)}))) = 0 \qquad \text{in } \Omega^*_S,$$

$$\mu^* \frac{\partial^2 (i\omega^* u_i^{(1)})}{\partial y_j \partial y_j} - \frac{\partial p^{(0)}}{\partial y_i} = 0,$$

$$\frac{\partial u_i^{(0)}}{\partial x_i} + \frac{\partial u_i^{(1)}}{\partial y_i} = 0 \qquad \text{in } \Omega^*_F,$$

$$a^*_{ijkh}\ (e_{ykh}(\mathbf{u}^{(1)})) + e_{xkh}(\mathbf{u}^{(0)})) N_j =$$

$$(-p^{(0)}I_{ij} + 2\mu^* i\omega^*(e_{yij}(\mathbf{u}^{(1)})) + e_{xij}(\mathbf{u}^{(0)}))N_j,$$

$$u_{Si}^{(1)} = u_{Fi}^{(1)} \qquad \text{on } \Gamma^*,$$

where $\mathbf{u}^{(1)}$ and $p^{(0)}$ are Ω^*-periodic. Due to the linearity of the problem, the solution can be put in the form

$$u_i^{(1)} = \xi_i^{*++lm} e_{xlm}(\mathbf{u}^{(0)}) + \overline{u}_i^{(1)}(\mathbf{x}, \omega^*),$$

$$p^{(0)} = \zeta^{*++lm} e_{xlm}(\mathbf{u}^{(0)}).$$

Vector $\overline{\mathbf{u}}^{(1)}(\mathbf{x}, \omega^*)$ is arbitrary and independent of $\mathbf{y}$. Tensors $\boldsymbol{\xi}^{*++}$ and $\boldsymbol{\zeta}^{*++}$ are complex valued and they depend on $\mathbf{y}$ and ω^*. Remark that no additive constant appear in the expression for $p^{(0)}$. This is the consequence of the continuity of the normal stress on Γ^*. Finally, a compatibility condition is obtained at the following order which, as usual, stands for the macroscopic behaviour. It takes a similar structure as in the previous case. In dimensional form, we get

$$\frac{\partial < \sigma_{ij}^T >}{\partial X_j} = \mathcal{O}(\varepsilon \frac{\partial < \sigma_{ij}^T >}{\partial X_j}),$$

$$< \sigma_{ij}^T >= c_{ijkh}^{++}\ e_{xkh}(\mathbf{u}_S) + \mathcal{O}(\varepsilon < \sigma_{ij}^T >),$$

$$\mathbf{c}^{++} = a_c \mathbf{c}^{*++}, \qquad c_{ijkh}^{*++} = |\Omega^*|^{-1} \int_{\Omega_S^*} (a_{ijkh}^* + a_{ijlm}^* e_{ylm}(\boldsymbol{\xi}^{*++kh}))\ \mathrm{dV}^*$$

$$+|\Omega^*|^{-1} \int_{\Omega_F^*} (-\zeta_{ij}^{*++}\ I_{kh} + 2i\omega^*\mu^* e_{yij}(\boldsymbol{\xi}^{*++kh}) + 2i\omega^*\mu^* I_{ij} I_{kh})\ \mathrm{dV}^*.$$

However, now tensor $\mathbf{c}^{++}$ is complex valued and it depends on ω. For a transient

Table 3. Macroscopic models for deformable saturated porous media.

$\mathcal{Q}_L = \mathcal{O}(\varepsilon^{-2})$	$\mathcal{Q}_L = \mathcal{O}(\varepsilon^{-1})$	$\mathcal{Q}_L = \mathcal{O}(1)$
Diphasic	Elastic monophasic	Viscoelastic monophasic
Biot model	Gassmann model	Tsytovich and Zaretskij model
	$\longrightarrow$ ‖	$\longleftarrow$

movement, the behaviour is in the form

$$< \sigma_{ij}^T >= \mathcal{F}_{ijkh}^{-1}(\mathbf{c}^{++})\ *\ e_{Xkh}(\mathbf{u}_S) + \mathcal{O}(\varepsilon < \sigma_{ij}^T >),$$

where $\mathcal{F}^{-1}_{ijkh}(\mathbf{c}^{++})$ is the inverse Fourier transform of $\mathbf{c}^{++}$ and $*$ stands for the convolution product. The macroscopic behaviour is linear viscoelastic. We recover the Tsytovich and Zaretskij model.

Concluding remarks. The different models for deformable saturated porous media are regrouped in table 3. There is a continuous passage from the Biot diphasic model to the Gassmann elastic monophasic model. This passage is obtained by decreasing $\mathcal{Q}_L$. That can be done by decreasing t_c, which yields the behaviour of a saturated deformable porous medium under a "rapid" loading (Auriault and Sanchez-Palencia, 1977). A second continuous passage is from the viscoelastic monophasic model to the Gassmann elastic monophasic model, which is obtained by increasing $\mathcal{Q}_L$. These two passages are the only ones possible.

Bibliography

Y. Anguy, R. Eherlich, C.M. Prince, V. Riggert and D. Bernard. The sample support problem for permeability assessment in reservoir sandstones. In J.M. Yarus and R.L. Chambers, editors, *Stochastic Modeling and Geostatistics Principle Methods and Case Studies.* A.A.P.G. Computer Application in Geology, 3: 37–53, 1994.

J.-L. Auriault and E. Sanchez-Palencia. Etude du comportement macroscopique d'un milieu poreux saturé déformable. *Journal de Mécanique*, 16 (4):575–603, 1977 (in French).

J.-L. Auriault. Dynamic behaviour of a porous medium saturated by a Newtonian fluid. *Int. J. Engng. Sc.*, 18:775–785, 1980.

J.-L. Auriault and E. Sanchez-Palencia. Remarques sur la loi de Darcy pour les écoulements biphasiques en milieux poreux. *Journal of Theoretical and Applied Mechanics*, Special issue "Modélisation assymptotique d'écoulements de fluides":141–156, 1986 (in French).

J.-L. Auriault. Non-saturated deformable porous media: quasi-statics. *Transport in Porous Media*, 2(1):45–64, 1987.

J.-L. Auriault, T. Strzelecki, J. Bauer and S. He. Porous deformable media saturated by a very compressible fluid: quasi-statics. *Eur. J. Mech. A/Solids*, 9(4):373–392, 1990.

J.-L. Auriault. Heterogeneous medium. Is an equivalent macroscopic description possible? *Int. J. Eng. Sci.*, 29(7):785–795, 1991.

J.-L. Auriault and C. Boutin. Deformable porous media with double porosity. Quasi-statics: I Coupling effects. *Transport in Porous Media*, 7:63–82, 1992.

J.-L. Auriault and C. Boutin. Deformable porous media with double porosity. Quasi-statics: II Memory effects. *Transport in Porous Media*, 10:153–169, 1993.

J.-L. Auriault and P. Adler. Taylor dispersion in porous media: Analysis by multiple scale expansions. *Advances in Water Resources*, 18 (4):217–226, 1995.

J.-L. Auriault and J. Lewandowska. On the validity of diffusion/dispersion tests in soils. *Engineering Transactions*, 45 (3-4):395–417, 1997.

J.-L. Auriault, C. Geindreau and P. Royer. Filtration law in rotating porous media. *C. R. Acad. Sci. Paris, IIb*, 328:779–784, 2000.

J.-L. Auriault, P. Royer and C. Geindreau. Filtration law for power-law fluids in anisotropic porous media. *Int. J. Engng. Sc.*, 40(10):1151–1163, 2002a.

J.-L. Auriault, C. Geindreau and P. Royer. Coriolis effects on filtration law in rotating porous media. *Transport in Porous Media*, 48:315–330, 2002b.

J.-L. Auriault, C. Geindreau and C. Boutin. Filtration law in porous media with poor separation of scales. *Sub judice*, 2004.

J. Bear. *Dynamics of Fluids in Porous Media*, 1972. American Elsevier, New-York.

A. Bensoussan, J.L. Lions and G. Papanicolaou. *Asymptotic Analysis for Periodic Structures*, 1978. North-Holland, Amsterdam.

M.A. Biot. General theory of three-dimensional consolidation. *J. Appl. Physics*, 12: 155–164, 1941.

M.A. Biot. Theory of elasticity and consolidation for a porous anisotropic solid. *J. Appl. Physics*, 26:182–185, 1955.

M.A. Biot. Generalized theory of acoustic propagation in porous dissipative media. *J. Acoust. Soc. Am.*, 34(9):1254–1264, 1962.

C. Boutin and J.-L. Auriault. Dynamic behaviour of porous media saturated by a viscoelastic fluid. Application to bituminous concretes. *Int. J. Engng. Sc.*, 28 (11): 1157–1181, 1990.

R. Burridge and J.B. Keller. Poroelasticity equations derived from microstructure. *J. Acoust. Soc. Am.*, 70:1140–1146, 1981.

H. Brenner. Dispersion resulting from flow through spatially periodic porous media. *Phil. Trans. R. Soc. Lond.*, A287:81–133, 1980.

J. Chastanet, P. Royer and J.-L. Auriault. Does Klinkenberg law survive upscaling? *MRC*, 31(3):277–286, 2004.

L. Cherel, G. Bonnet and J.-L. Auriault. Locally periodic medium and homogeneization of random media. *Archives of Mechanics*, 40(5-6):529–542, 1988.

H. Darcy. *Les Fontaines Publiques de la Ville de Dijon*, 1856. Dalmont, Paris.

H.I. Ene and E. Sanchez-Palencia. Equations et phénomène de surface pour l'écoulement dans un modèle de milieu poreux. *Journal de Mécanique*, 14 (1):73–108, 1975 (in French).

M. Firdaouss, J.-L. Guermond and P. Le Quéré. Non linear corrections to Darcy's law at low Reynolds numbers. *J. Fluid Mech.*, 353:331–350, 1997.

F. Gassmann. Über die elastizität poröser medien. *Vierteljahrsschrift d. Naturf. Ges. Zürich*, 96:1–23, 1951 (in German).

J.B. Keller. Effective behaviour of heterogeneous media. In U. landman, editor, *Statistical Mechanics and Statistical Methods in Theory and Application.* Plenum, New York, 631–644, 1977.

E. Kröner. *Statistical Continuum Mechanics*, 1972. Springer Verlag, Wien.

T. Levy. Propagation of waves in a fluid saturated porous elastic solid. *Int. J. Eng. Sci.*, 17:1005–1014, 1979.

J. Lewandowska and J.-L. Auriault. Scale separation in diffusion/dispersion tests in porous media. In *Proceedings of the Biot Conference on Poromechanics*, pages 599–604, Louvain-la-Neuve 14-16 September 1998, Balkema 1998.

C.C. Mei and J.-L. Auriault. The effect of weak inertia on flow through a porous medium. *J. Fluid Mech.*, 222:647–663, 1991.

J. Necas. *Les Méthodes Mirectes en Théorie des Equations Elliptiques*, 1967. Masson, Paris.

R.I. Nigmatulin. Three-dimensional averaging in the mechanics of heterogeneous media. *Fluid Mechanics*, 10(4):72–107, 1981.

M. Quintard and S. Whitaker. Transport in ordered and disordered porous media: volume-averaged equations, closure problems, and comparison with experiments. *Chem. Eng. Sci.*, 48(14):2537–2564, 1993.

M. Rasoloarijaona and J.-L. Auriault. Nonlinear seepage flow through a rigid porous medium. *Eur. J. Mech. B/Fluids*, 13:177–195, 1994.

E. Sanchez-Palencia. Comportement local et macroscopique d'un type de milieux physiques hétérogènes. *Int. J. Eng. Sci.*, 12:331–351, 1974 (in French).

E. Sanchez-Palencia. *Nonhomogeneous Media and Vibration Theory*, lecture notes in Physics no. 127, 1980. Springer, Berlin.

E. Sanchez-Palencia. Boundary layers and edge effects in composite. In E. Sanchez-palencia and A. Zaoui, editors, *Homogenization Technique for Composite Media.* Lecture Notes in Physics 272, Springer, 1987.

E. Skjetne and J.-L. Auriault. New insights on steady, nonlinear flow in porous media. *Eur. J. Mech. B/Fluids*, 18(1):131–145, 1999a.

E. Skjetne and J.-L. Auriault. Homogenization of wall-slip gas flow. *Transport in porous Media*, 36(3):293–306, 1999b.

N.A. Tsytovich and Y.K. Zaretskij. The development of the theory of soil consolidation in the USSR 1917-1967. *Geotechnique*, 19(3):357–375, 1969.

J.-C. Wodie and T. Levy. Correction non linéaire de la loi de Darcy. *C. R. Acad. Sci. Paris, II*, 312:157–161, 1991.

A. Zaoui. Approximate statistical modelling and applications. In Lecture Notes in Physics 272, *Homogenization Techniques for Composite Media.* Springer Verlag, Berlin, 338–397, 1987.

C. Zarcone and R. Lenormand. Détermination expérimentale du couplage visqueux dans les écoulements diphasiques en milieux poreux. *C. R. Acad. Sci. Paris, II*, 318: 1429–1435, 1994.

Micromechanical Analysis of Saturated Porous Media with Local Mass Exchange Phenomenon

Dariusz Łydżba[1]

[1] Institute of Geotechnics and Hydrotechnics, Wrocław University of Technology, Wrocław, Poland

Abstract. The note presents some results concerning mechanical and transport properties of saturated porous media. The attention is paid for phenomenon of local mass exchange due to a sorption process as well as its effect on transport and mechanical properties of porous medium. The results are obtained through a micro-macro passage, i.e. by a systematic use of the asymptotic expansion technique. The gas flow through a rigid skeleton with simultaneous sorption process is investigated first. The analysis is then extended for the case of a gas flow through a deformable porous skeleton. It is pointed out that the macroscopic description for the both cases studied exhibits a strong memory effect. A mathematical form of the memory function is presented for a particular microstructure. The reciprocal relations between the overall material constants, involved in the description derived, end the paper.

1 Introduction

Porous media belong to a class of heterogeneous materials for which a mathematical description, useful for engineering calculation, is classically proposed within a context of the continuum mechanics. This means that these materials, from engineering point of view, are treated as macroscopically homogeneous ones. In order to emphasize, however, their heterogeneous structure and at the same time a possibility of defining for them an *equivalent homogeneous medium* they are also marked as the micro-heterogeneous ones.

The aim of the micro-macro approach is to formulate a macroscopic description of a process considered based on knowledge of a mathematical description of the process at the heterogeneities scale. The last description takes into account the distances of the order of pore diameter whereas that for an equivalent medium does not. Roughly speaking, the microscopic description can be treated, in some sense, as the exact one whereas the macroscopic – as some approximation, useful for engineering calculations. Both descriptions are, however, formulated within a concept of the classical physics, so both are of a macroscopic nature.

This note presents some results concerning mechanical and transport properties of saturated porous media. The attention is paid for phenomenon of local mass exchange due to a sorption process as well as its effect on transport and mechanical properties of porous medium. The macroscopic description of the phenomenon studied is obtained by a systematic use of the asymptotic expansion technique. The chapter is written in a following sequence. In the subsequent section some fundamentals of an asymptotic expansion technique are presented. Then, using this technique, the gas flow through a sorbing rigid skeleton is investigated. The analysis is then extended for a case of a deformable porous skeleton. Some final conclusions end the chapter.

Throughout the chapter an index notation with summation convention over repeated indices is used. The sign convention of the continuum mechanics (tension as positive) is applied. The geometrical linearity hypothesis is assumed, i.e. small strains and small displacements are postulated.

2 Fundamentals of Asymptotic Expansion Technique

The first central assumption of the asymptotic expansion technique is the condition that the medium exhibits a periodic structure, i.e. it is a periodicity-generated from a single element, the so-called base cell or a unit cell. The micro-macro passage is realized, within this approach, by considering a sequence of boundary value problems and corresponding to them solutions. Each of the problems corresponds to an actual value of the scale parameter $\varepsilon>0$. This parameter is a measure of a relative length of the unit cell to the length of the entire medium, i.e. if l represents a characteristic length of the unit cell and L represents a characteristic length of the medium considered then $\varepsilon = l/L$ (see Figure 1). The entire medium occupies a domain of a fixed size. So, a particular boundary value problem differs from the others by a number of the unit cells of which the

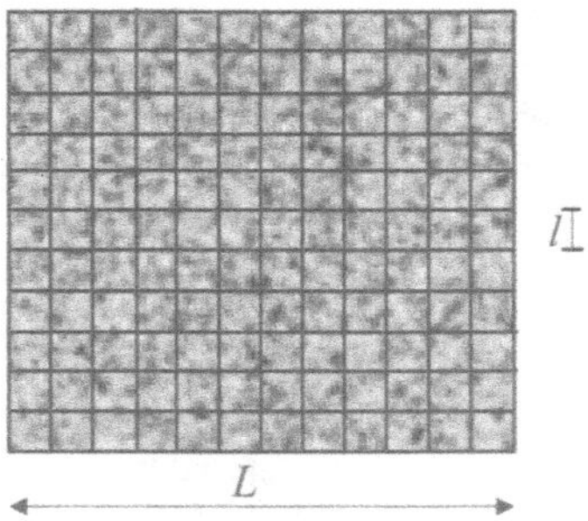

Figure 1. A schematic view at the periodic structure.

medium is composed. If a periodic structure consists of N^3 unit cells then the medium characterizes $\varepsilon = 1/N$. A successive increasing of the number of unit cells in the medium is realized by an appropriate decreasing of the unit cell size. The unit cell is subjected to consecutive homothetic transformations.

Let the equation:

$$A^{\varepsilon}\left(u^{\varepsilon}\right) = 0 \tag{2.1}$$

represents the microscopic description of the phenomenon studied. The symbols A^{ε} and u^{ε} denote a differential operator and a physical field being looked for, respectively. Both depend on an actual value of the parameter ε. The macroscopic variable is identified as the limit of u^{ε}, at $\varepsilon \to 0$, i.e.:

$$u(x) = \lim_{\varepsilon \to 0} u^{\varepsilon}(x) \tag{2.2}$$

whereas the macroscopic description - as the differential equation that this limit satisfies. So, the macroscopic description represents a mathematical description of the medium composed of an infinite number of the unit cells.

The second central assumption of the asymptotic expansion technique is the postulate that the parameterized field being considered exhibits an asymptotic character with respect to the scale parameter ε, i.e.:

$$u^{\varepsilon}(x) = u^{(0)}(x,y) + \varepsilon u^{(1)}(x,y) + \varepsilon^2 u^{(2)}(x,y) \;+ \ldots + \quad \text{where} \quad y = \frac{x}{\varepsilon} \tag{2.3}$$

Furthermore, each term $u^{(i)}(x, y)$ of the asymptotic development is assumed to be Y-periodic in y, i.e. it takes equal values on the opposite site of the unit cell:

$$u^{(i)}(x, y+Y) = u^{(i)}(x,y) \tag{2.4}$$

The arguments x and y are treated as the independent space variables, so the spatial derivative operator is modified as:

$$\frac{d}{dx_i} = \frac{\partial}{\partial x_i} + \varepsilon^{-1}\frac{\partial}{\partial y_i} \tag{2.5}$$

The method consists in incorporating the expansion (2.3) into the microscopic description (2.1). Collecting terms based on their order one gets a sequence of equations for consecutive terms of the asymptotic expansion. The periodicity condition of $u^{(i)}(x, y)$ enables to solve the problem with respect to y-variable, i.e. within the unit cell. In consequence, this process results in the sought macroscopic description of the issue being considered.

It has to be marked that the result obtained using the asymptotic expansion technique is, from the mathematics point of view, formal. It is obtained by postulating the asymptotic character of the parameterized solution of the microscopic description which is not a priori guaranteed. So the result, to be mathematically rigorous, has to be supplemented by a proof that:

$$\lim_{\varepsilon \to 0} u^{\varepsilon}(x) = u^{(0)}(x) \tag{2.6}$$

In most problems, a mathematical proof of the above convergence is available using, for instance, the so-called *energy method* or *the two-scale convergence method* (see for more details the appendix in Hornung (1997)).

An objection which rises as to the applicability of the asymptotic expansion technique to any porous media is the assumption that the medium poses a periodic structure. It appears, however, that independently of whether the medium is random or periodic, the form of the macroscopic equivalent description remains unchanged (Sab (1992), Sab (1994)).

Normalization. The base cell is subjected, during the micro-macro passage, to a homothetic transformation. This transformation does not affect the values of dimensionless microstructure parameters, as for instance: a volume fraction, a relative distances between the components, a shape of the grains. The homothetic transformation affects however the values of microstructure parameters with metric units, as for instance: a specific surface of porous medium or diameters of pores. This becomes obvious when one considers an effect of the homothetic transformation on a value of the specific surface measure of a porous structure (Figure 2.).

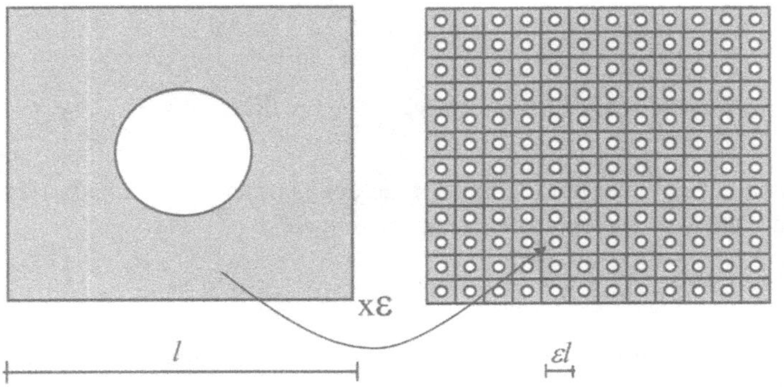

Figure 2. Variation of a specific surface value due to a unit cell homothetic transformation.

According to the definition, a specific surface is determined as:

$$\zeta = \frac{S_p}{V} \tag{2.7}$$

where: ζ denotes a value of the specific surface, S_p is a measure of the pores lateral surface and V denotes a measure of the porous medium volume.
When the domain is filled up to the volume by unit cells of the characteristic length εl then:

$$\zeta^\varepsilon = \frac{S_p^\varepsilon}{V^\varepsilon} = \frac{\varepsilon^2 S_p^1}{\varepsilon^3 V^1} = \varepsilon^{-1}\zeta^1 \xrightarrow[\varepsilon \to 0]{} \infty \tag{2.8}$$

The result presented above shows that the micro-macro procedure described above could lead to incorrect results since the homothetic transformation used during the micro-macro passage could drastically affect the virgin properties of porous medium analyzed. In order to avoid this inconsistency, one has to normalize the equations of the microscopic description before applying the asymptotic expansion. An example given below clarifies this operation.

Consider a flow of an incompressible Newtonian liquid through a long pipe with a small diameter δ. Neglecting flow disturbances at the ends of the pipe, one can assume that the flow within the pipe is a laminar (see Landau and Lifszyc (1994)). An average velocity of the flowing liquid is proportional to applied gradient of liquid pressure, i.e.:

$$\langle v \rangle \propto -\frac{\delta^2}{\mu}\,\mathrm{grad}p \tag{2.9}$$

where: $\langle v \rangle$ - the average value of a liquid velocity and μ – a liquid viscosity.
For the pipe of a diameter equated to $\varepsilon\delta$ (ε<<1) the average velocity drastically decreases:

$$\langle v \rangle \propto -\frac{\varepsilon^2\delta^2}{\mu}\text{grad}p \tag{2.10}$$

This is due to the fact that, at a fixed value of the liquid viscosity, a pore diameter decrease induces a proportional increase of the pipe resistance to liquid flow. The same goes when one considers the flow process through an arbitrary porous medium. So, in order to compensate, due to the unit cell homothetic transformation, an *artificial* increase of the porous medium resistance to liquid flow one has to proportionally decrease a value of the liquid viscosity. Comparing Eqs. (2.9) and (2.10), it is easily deduced that the viscosity should be scaled down as: $\mu \rightarrow \varepsilon^2\mu$. As a consequence, the constitutive equation of the Newtonian liquid has to be normalized, before applying the asymptotic expansion, as follows:

$$\sigma_{ij} = -p\delta_{ij} + \varepsilon^2\mu\left(v_{i,j} + v_{j,i}\right) \tag{2.11}$$

where: v_i – a component of the liquid velocity vector, p – a liquid pressure, σ_{ij} – a component of the liquid stress tensor.

The exhaustive presentation of the asymptotic expansion method, mathematical details as well as many examples of its application, the reader can find in Bensoussan, Lions and Papanicolau (1978), Sanchez-Palencia (1980) and Hornung (1997).

3 Transport Properties of Porous Media with Sorption Phenomenon: Rigid Skeleton Case

The consequence of the contact of two volumetric phases is often a sorption phenomenon. The character of this phenomenon depends on the structure and surface type of the solid body. Two extreme cases can be distinguished (Zarzycki and Chacuk (1993)):

- a solid body is characterized by a flat, nonporous surface; the process takes place on the surface of the solid phase and it consists in the formation of autonomous sorption phase;
- micro-porous structure with the pore sizes comparable to the gas molecule diameters is characteristic of a solid body; gas sorption occurs in the whole volume of the solid body and the process is called as the absorption.

The study is focused on a system of bituminous coal and carbon dioxide. The sorption process occurring in this multiphase medium is of a very complex character. Lasoń (1988) distinguished three ranges of pores in the fabric of bituminous coal, which determine different mechanisms of gas storing:

- the area of network sorption: the radii of capillaries range from 0.3 to 0.5 nm and the absorption which occurs inside them resembles a phenomenon of dissolution;
- the area of micro-pores: the capillaries with radii up to 1.5 nm, in which the phenomenon of filling up to volume is observed;
- the are of meso- and macro-pores which comprises capillaries of greater radii, where single- and multilayered adsorption takes place as well as a gas in a free state is present.

Only a small part of the gas exists in a free state. The main part of the carbon dioxide is constrained in the coal matrix. Contingently upon the value of a free gas pore pressure, the phases may or may not be in a thermodynamic equilibrium. When it is a lack of a thermodynamic equilibrium in the system then phenomena of the gas sorption or a gas desorption occur there. The constrained gas and the free gas act each other through the internal surface of coal.

During the study an internal geometry of the porous medium considered is simplified. It is assumed that the medium is composed only of the solid part and the network of macro-pores. The macro-pores are saturated with gas in a free state. The solid part comprises, however, the porous matrix with the capillaries of smaller sizes where the gas occurs as the constrained phase. We do not distinguish if the sorbed gas occurs in the porous matrix as a gas phase or as a liquid one. It is postulated only that the transport process of the constrained gas is governed by the Fick molecular diffusion law. The equilibrium criterion is assumed in a form of a sorption isotherm equation, i.e.:

$$C - F(p) = 0 \tag{3.1}$$

where: C denotes an amount (concentration) of the gas constrained in the coal matrix, p denotes the value of the free gas pore pressure and $F(p)$ – a sorption isotherm.

If the value of a free gas pore pressure does not satisfy Eq. (3.1) it indicates that the system is unbalanced. The phenomenon of a local mass exchange between the phases occurs in the system.

The gas flow in the macro-pores is assumed to be governed by the Navier-Stokes equations of a barotropic liquid.

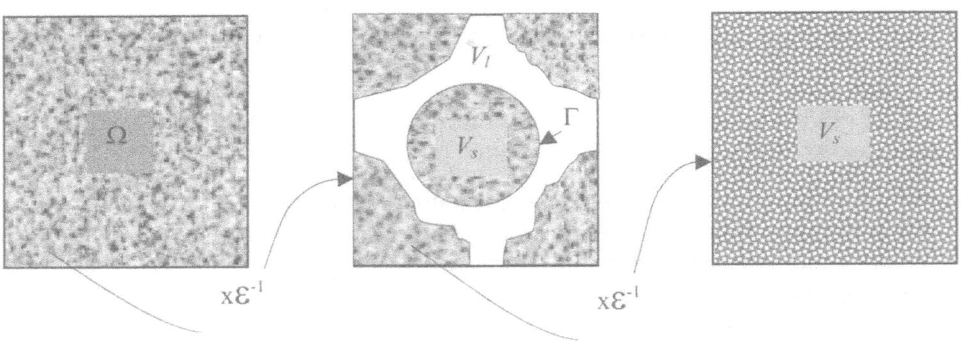

Figure 3. Schematic view of porous coal scales.

3.1 Local Description

With the above assumptions, the micro-scale description can be presented as:
constrained gas in micro-porous skeleton, i.e.
– a mass conservation equation:

$$\frac{\partial C}{\partial t} - DC_{,ii} = 0 \qquad \text{in } V_s \tag{3.2}$$

free gas in macro-pores: i.e.:
- balance equations (small Reynolds number is postulated):

$$\sigma^l_{ij,i} = 0 \qquad \text{in } V_l \tag{3.3}$$

- constitutive equations of a barotropic Newtonian liquid:

$$\sigma^l_{ij} = -p\delta_{ij} + \lambda v_{k,k}\delta_{ij} + \mu\left(v_{i,j} + v_{j,i}\right) \qquad \text{in } V_l \tag{3.4}$$

- a mass conservation equation:

$$\frac{\partial \rho}{\partial t} + \left(\rho v_i\right)_{,i} = 0 \qquad \text{in } V_l \tag{3.5}$$

- the ideal gas law for isothermal processes:

$$\rho = \frac{\rho_a}{p_a} p \qquad \text{in } V_l \tag{3.6}$$

The above system of equations is completed by the boundary conditions on the phase separation surface Γ, i.e.:
- a continuity of the mass flux:

$$\left[\rho v_i + DC_{,i}\right] N_i = 0 \qquad \text{on } \Gamma \tag{3.7}$$

- an adhesion condition:

$$v_i t_i = 0 \qquad \text{on } \Gamma \tag{3.8}$$

- a sorption isotherm equation:

$$C = F(p) \qquad \text{on } \Gamma \tag{3.9}$$

where: v_i– free gas velocity vector, ρ – free gas density, λ and μ–coefficients of viscosity, σ^l_{ij} – stress tensor of the free gas, D–diffusion coefficient, ρ_a –value of a gas density at the atmospheric pressure, p_a–value of the atmospheric pressure, V_s – volume occupied by the micro-porous skeleton within the unit cell, V_l – volume occupied by the macro-pores within the unit cell, N_i and t_i– components of a normal and a tangent vector to the common surface Γ, respectively.

In addition, the thermodynamic equilibrium between the phases is assumed at the initial instant.

3.2 Normalization

For the Newtonian liquid, as has been pointed out previously, the normalization is equivalent to a scaling down the viscosity coefficient. For a barotropic fluid (two viscosity coefficients) this process corresponds to the following substitution: $\mu \rightarrow \varepsilon^2\mu$ and $\lambda \rightarrow \varepsilon^2\lambda$ (see Eq. (2.11)).

In order to normalize the diffusion process one has to distinguish the two different cases, i.e.:

- case 1°: microstructure of the porous medium considered and its diffusion property bring about that the diffusion process is much slower than the filtration one;
- case 2°: microstructure of the porous medium considered and its diffusion property cause that the diffusion process is comparable fast as the filtration one.

The two cases distinguished above clearly show that the system: bituminous coal-carbon dioxide is characterized by the two characteristic times, i.e.: T_f and T_D, of the fluid filtration and fluid diffusion, respectively. The first case distinguished implies $T_D/T_f = O(\varepsilon^{-1})$ whereas the second one - $T_D/T_f = O(1)$. During the micro-macro passage the unit cell of a medium considered is subjected to a homothetic transformation. This operation can not, of course, affect the proportion between the two characteristic times distinguished. Using the Fick law, the characteristic diffusion time can be estimated as:

$$V_D^\varepsilon C = O\left(D\frac{C}{\varepsilon\, l}\right) \Rightarrow T_D^\varepsilon = O\left(\frac{\varepsilon\, l}{V_D^\varepsilon}\right) = O\left(\frac{\varepsilon^2 l^2}{D}\right) \tag{3.10}$$

The characteristic filtration time is given by:

$$T_f^\varepsilon = O\left(\frac{\varepsilon\, l}{V_f^\varepsilon}\right) \tag{3.11}$$

where: l is a characteristic length of the unit cell before its homothetic transformation, V_D^ε and V_f^ε denote mean values of diffusion and filtration velocities corresponding to the actual length of the base cell equated to εl, respectively.
Using Eq. (3.10) and Eq. (3.11), one gets:

$$\frac{T_D^\varepsilon}{T_f^\varepsilon} = O\left(V_f^\varepsilon l \frac{\varepsilon}{D}\right) \tag{3.12}$$

The consequence of the viscosities' scaling ($\mu \to \varepsilon^2\mu$ and $\lambda \to \varepsilon^2\lambda$) is an ε - independence of the filtration velocity V_f^ε. The constant characteristic times' proportion is preserved by scaling, similarly as viscosities, the diffusion coefficient. For the two cases, one gets:
case 1.

$$\frac{T_D^\varepsilon}{T_f^\varepsilon} = O\left(\varepsilon^{-1}\right) \Rightarrow \frac{T_D^\varepsilon}{T_f^\varepsilon} = O\left(V_f^\varepsilon l \frac{\varepsilon}{D}\right) = O\left(\varepsilon^{-1}\right) \Rightarrow D \to \varepsilon^2 D \tag{3.13}$$

case 2.

$$\frac{T_D^\varepsilon}{T_f^\varepsilon} = O(1) \Rightarrow \frac{T_D^\varepsilon}{T_f^\varepsilon} = O\left(V_f^\varepsilon l \frac{\varepsilon}{D}\right) = O(1) \Rightarrow D \to \varepsilon D \tag{3.14}$$

The above scaling and the viscosities' scaling, affect only the equations (3.2), (3.4) and (3.7) of the microscopic description. These equations can be presented as:

$$\frac{\partial C}{\partial t} - \varepsilon^m DC_{,ii} = 0 \tag{3.15}$$

$$\sigma^l_{ij} = -p\delta_{ij} + \varepsilon^2 \lambda v_{k,k}\delta_{ij} + \varepsilon^2 \mu\left(v_{i,j} + v_{j,i}\right) \tag{3.16}$$

$$\left[\rho v_i + \varepsilon^m DC_{,i}\right] N_i = 0 \tag{3.17}$$

where $m=2$ corresponds to the case 1° whereas $m=1$ - to the case 2°.

3.3 Asymptotic Expansions and Averaging Process

Introducing the asymptotic expansions of: v, p, ρ, and C and modifying the spatial derivative operator according to (2.5) one gets:

$$\varepsilon^0 \frac{\partial C^{(0)}}{\partial t} - \varepsilon^{m-2} D \frac{\partial}{\partial y_i}\left(\frac{\partial C^{(0)}}{\partial y_i}\right) +$$

$$-\varepsilon^{m-1} D \left\{ \frac{\partial}{\partial y_i}\left(\frac{\partial C^{(0)}}{\partial x_i} + \frac{\partial C^{(1)}}{\partial y_i}\right) + \frac{\partial}{\partial x_i}\left(\frac{\partial C^{(1)}}{\partial y_i}\right)\right\} + \ldots = 0 \quad \text{in } V_s \tag{3.18}$$

$$-\varepsilon^{-1}\frac{\partial p^{(0)}}{\partial y_i} +$$

$$+\varepsilon^0 \left[-\frac{\partial p^{(0)}}{\partial x_i} - \frac{\partial p^{(1)}}{\partial y_i} + \mu \frac{\partial}{\partial y_j}\left(\frac{\partial v_i^{(0)}}{\partial y_j}\right) + (\lambda + \mu)\frac{\partial}{\partial y_i}\left(\frac{\partial v_j^{(0)}}{\partial y_j}\right)\right] + \ldots = 0 \quad \text{in } V_l \tag{3.19}$$

$$\varepsilon^{-1}\frac{\partial\left(\rho^{(0)} v_i^{(0)}\right)}{\partial y_i} + \varepsilon^0 \left\{ \frac{\partial \rho^{(0)}}{\partial t} + \frac{\partial\left(\rho^{(0)} v_i^{(0)}\right)}{\partial x_i} + \frac{\partial\left(\rho^{(0)} v_i^{(1)}\right)}{\partial y_i} \frac{\partial\left(\rho^{(1)} v_i^{(0)}\right)}{\partial y_i}\right\} + \ldots = 0 \quad \text{in } V_l \tag{3.20}$$

$$\varepsilon^0\left(\rho^{(0)} - \frac{\rho_a}{p_a} p^{(0)}\right) + \varepsilon^1\left(\rho^{(1)} - \frac{\rho_a}{p_a} p^{(1)}\right) + \ldots = 0 \quad \text{in } V_l \tag{3.21}$$

$$\left[\varepsilon^0 \rho^{(0)} v_i^{(0)} + \varepsilon^1 \left\{\rho^{(0)} v_i^{(1)} + \rho^{(1)} v_i^{(0)}\right\}\right] N_i +$$

$$+\left[\varepsilon^{m-1} D \frac{\partial C^{(0)}}{\partial y_i} + \varepsilon^m \left(D\frac{\partial C^{(0)}}{\partial x_i} + D\frac{\partial C^{(1)}}{\partial y_i}\right)\right] N_i = 0 \quad \text{on } \Gamma \tag{3.22}$$

$$\varepsilon^0 v_i^{(0)} t_i + \varepsilon^1 v_i^{(1)} t_i + \ldots = 0 \qquad \text{on } \Gamma \tag{3.23}$$

$$C^{(0)} = F\left(p^{(0)}\right) \qquad \text{on } \Gamma \tag{3.24}$$

Collecting terms based on their order one gets set of local boundary-value problems for the consecutive terms of the asymptotic expansions. For the case 1°, i.e. $m=2$, they are:
I local problem: the equations (3.19) and (3.21), i.e.:

$$O\left(\varepsilon^{-1}\right): -\frac{\partial p^{(0)}}{\partial y_i} = 0 \qquad \text{in } V_l \tag{3.25}$$

$$O\left(\varepsilon^{0}\right): \rho^{(0)} - \frac{\rho_a}{p_a} p^{(0)} = 0 \qquad \text{in } V_l \tag{3.26}$$

II local problem: the equations (3.19), (3.20), (3.22) and (3.23), i.e.:

$$O\left(\varepsilon^{0}\right): -\frac{\partial p^{(0)}}{\partial x_i} - \frac{\partial p^{(1)}}{\partial y_i} + \mu \frac{\partial}{\partial y_j}\left(\frac{\partial v_i^{(0)}}{\partial y_j}\right) + (\lambda + \mu)\frac{\partial}{\partial y_i}\left(\frac{\partial v_j^{(0)}}{\partial y_j}\right) = 0 \qquad \text{in } V_l \tag{3.27}$$

$$O\left(\varepsilon^{-1}\right): \frac{\partial\left(\rho^{(0)} v_i^{(0)}\right)}{\partial y_i} = 0 \qquad \text{in } V_l \tag{3.28}$$

$$O\left(\varepsilon^{0}\right): \rho^{(0)} v_i^{(0)} N_i = 0 \qquad \text{on } \Gamma \tag{3.29}$$

$$O\left(\varepsilon^{0}\right): v_i^{(0)} t_i = 0 \qquad \text{on } \Gamma \tag{3.30}$$

III local problem: the equations (3.18) and (3.24), i.e.:

$$O\left(\varepsilon^{0}\right): \frac{\partial C^{(0)}}{\partial t} - D\frac{\partial}{\partial y_i}\left(\frac{\partial C^{(0)}}{\partial y_i}\right) = 0 \qquad \text{in } V_s \tag{3.31}$$

$$O\left(\varepsilon^{0}\right): C^{(0)} = F\left(p^{(0)}\right) \qquad \text{on } \Gamma \tag{3.32}$$

IV local problem: the equations (3.20) and (3.22), i.e.:

$$O\left(\varepsilon^{0}\right): \frac{\partial \rho^{(0)}}{\partial t} + \frac{\partial\left(\rho^{(0)} v_i^{(0)}\right)}{\partial x_i} + \frac{\partial\left(\rho^{(0)} v_i^{(1)}\right)}{\partial y_i} + \frac{\partial\left(\rho^{(1)} v_i^{(0)}\right)}{\partial y_i} = 0 \qquad \text{in } V_l \tag{3.33}$$

$$O\left(\varepsilon^{1}\right): \left\{\rho^{(0)} v_i^{(1)} + \rho^{(1)} v_i^{(0)} + D\frac{\partial C^{(0)}}{\partial y_i}\right\} N_i = 0 \qquad \text{on } \Gamma \tag{3.34}$$

This is a hierarchy that can be solved in order, first finding the first term of the appropriate asymptotic expansion, then using it to find the second term, and on up to the desired order of approximation. Each term of the asymptotic expansions is Y-periodic. This property will be used to solve each of the local problems distinguished above. In a following this is not any more explicitly recalled.

The I local problem implies:

$$p^{(0)} = p^{(0)}(x,t) \tag{3.35}$$

$$\rho^{(0)} = \frac{\rho_a}{p_a} p^{(0)}(x,t) \tag{3.36}$$

The II local problem represents the classical problem of incompressible fluid flow through a rigid porous medium. The term $p^{(0)}$ is now considered as a known function of x. The fields: $v_i^{(0)}$ and $p^{(1)}$ depends linearly on the macroscopic gradient $\mathbf{grad}_x p^{(0)}$, therefore:

$$v_i^{(0)} = -k_{ij}(y)\frac{\partial p^{(0)}}{\partial x_j} \quad \text{and} \quad p^{(1)} = \omega_i(y)\frac{\partial p^{(0)}}{\partial x_i} + \overline{p_1(x)} \qquad \text{in } V_l \tag{3.37}$$

where : $\overline{p_1(x)}$ is an arbitrary function of x-argument whereas the fields: $k_{ij}(y)$ and $\omega_i(y)$ are the solutions of the following boundary value problem, i.e.:

$$\delta_{ij} - \frac{\partial \omega_j}{\partial y_i} + \mu \frac{\partial}{\partial y_l}\left(\frac{\partial k_{ij}}{\partial y_l}\right) = 0 \qquad \text{in } V_l \tag{3.38}$$

$$\frac{\partial k_{ij}}{\partial y_i} = 0 \qquad \text{in } V_l \tag{3.39}$$

$$k_{ij} = 0 \qquad \text{on } \Gamma \tag{3.40}$$

The volume average of $v_i^{(0)}$ results in the well known Darcy law of the filtration process, i.e.:

$$\left\langle v_i^{(0)} \right\rangle = -K_{ij}\frac{\partial p^{(0)}}{\partial x_j} \tag{3.41}$$

where symbol $\langle\ \rangle$ represents a volume average operator.

The III local problem is considered by applying the substitution:

$$U(x,y,t) = C^{(0)}(x,y,t) - F\left(p^{(0)}\right) \tag{3.42}$$

and transforming the equations (3.31) and (3.32) to the Laplace transform space, i.e.:

$$s\mathcal{L}(U) - \frac{\partial}{\partial y_i}\left(D\frac{\partial \mathcal{L}(U)}{\partial y_i}\right) = -\mathcal{L}\left(\frac{\partial F\left(p^{(0)}\right)}{\partial t}\right) \qquad \text{in } V_s \tag{3.43}$$

$$\mathcal{L}(U) = 0 \qquad \text{on } \Gamma \tag{3.44}$$

where: s is the Laplace complex variable and symbol $\mathcal{L}$ denotes the Laplace transformation. The system (3.43)-(3.44) is linear, hence:

$$\mathcal{L}(U) = -\mathcal{L}(G(y,t))\,\mathcal{L}\left(\frac{\partial F\left(p^{(0)}\right)}{\partial t}\right) \tag{3.45}$$

The function $G(y,t)$ is a solution of the following local problem, i.e.:

$$s\mathcal{L}(G) - \frac{\partial}{\partial y_i}\left(D\frac{\partial \mathcal{L}(G)}{\partial y_i}\right) = 1 \qquad \text{in } V_s \tag{3.46}$$

$$\mathcal{L}(G) = 0 \qquad \text{on } \Gamma \tag{3.47}$$

By virtue of the convolution theorem, one can write:

$$U(x,y,t) = -\int_0^t \frac{\partial F\left(p^{(0)}\right)}{\partial \tau} G(y,t-\tau)\, d\tau \qquad \text{in } V_s \tag{3.48}$$

Using once more the substitution (3.42), the final form of the solution of the III local problem is obtained:

$$C^{(0)}(x,y,t) = F\left(p^{(0)}\right) - \int_0^t \frac{\partial F\left(p^{(0)}\right)}{\partial \tau} G(y,t-\tau)\,d\tau \tag{3.49}$$

Volume average of the above equation results in:

$$\left\langle C^{(0)}(x,t)\right\rangle = (1-n)F\left(p^{(0)}\right) - \int_0^t \frac{\partial F\left(p^{(0)}\right)}{\partial \tau}\left\langle G(t-\tau)\right\rangle d\tau \tag{3.50}$$

where: n is a porosity of the medium, i.e. a volume of the macro-pores in a unit volume of the system.

The equation (3.50) shows that the macroscopic measure of a concentration of the gas constrained in the micro-porous skeleton depends on a history of the first time derivative of the sorption isotherm. The function $\left\langle G(t)\right\rangle$ is called as a memory function.

The last macroscopic equation is a mass conservation law for a free gas in macro-pores. It is a direct consequence of the volume average of the IV local problem. The following form is obtained:

$$n\frac{\partial \rho^{(0)}}{\partial t} + \frac{\partial}{\partial x_i}\left[\rho^{(0)}\left\langle v_i^{(0)}\right\rangle\right] + \frac{\partial \left\langle C^{(0)}\right\rangle}{\partial t} = 0 \tag{3.51}$$

The last term in the above equation represents a source term due to the diffusion-sorption process taking place in the coal fabric.

3.4 Macroscopic Description

The governing set of the macroscopic description of the gas flow through sorbing rigid skeleton is constituted by the following equations:

- the Darcy law of the free gas filtration process:

$$\langle v_i \rangle = -K_{ij} \frac{\partial p}{\partial x_j} \tag{3.52}$$

- the mass conservation law for the free gas :

$$n\frac{\partial \rho}{\partial t} + \frac{\partial \rho \langle v_i \rangle}{\partial x_i} + \frac{\partial \langle C \rangle}{\partial t} = 0 \tag{3.53}$$

- the constitutive relation of the average concentration of the gas sorbed by the coal matrix:

$$\langle C \rangle = (1-n) F(p) - \int_0^t \frac{\partial F(p(\tau))}{\partial \tau} \langle G(t-\tau) \rangle \, d\tau \tag{3.54}$$

The above system of governing equations has to be completed by the sorption isotherm function $F(p)$. For the system of the bituminous coal-carbon dioxide, the Langmuir's sorption isotherm is usually adopted, i.e.:

$$F(p) = \frac{B_1 p / p_a}{B_2 + p / p_a} \tag{3.55}$$

where: B_1 and B_2 are material constants.

When the diffusion process is a comparable fast as the filtration process, i.e. the case 2° distinguished in the section 3.2, then the above governing set of equations still holds true. The equation (3.54), however, simplifies to a form corresponding to $\langle G(t) \rangle = 0$ (see Lydzba and Auriault (1996)).

3.5 Particular Microstructure: Form of the Memory Function

A particular microstructure of the porous medium is assumed in order to present an analytical form of the memory function. It is postulated, in a following, that the porous material is composed of uniform micro-porous grains. Each grain is of a spherical shape. The spatial structure of the grain packing is sketched in Figure 4.

The memory function is defined by the system (3.46)-(3.47) stated in a Laplace variable s domain. The corresponding system in the actual time t domain can be presented as:

$$\frac{\partial G(y,t)}{\partial t} - D\frac{\partial}{\partial y_i}\left(\frac{\partial G(y,t)}{\partial y_i}\right) = 0 \qquad \text{in } V_s,\ t>0 \tag{3.56}$$

$$G(y,t) = 0 \qquad \text{on } \Gamma,\ t>0 \tag{3.57}$$

$$G(y,t) = 1 \qquad \text{in } V_s, \text{ for } t=0 \tag{3.58}$$

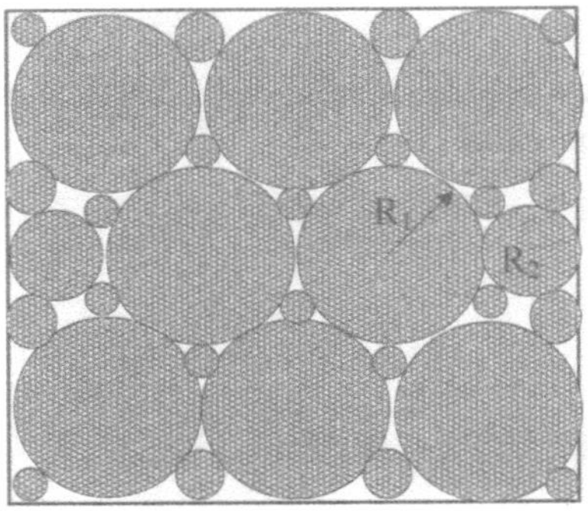

Figure 4. Schematic view of the microstructure analyzed.

The above system can be solved separately for each grain. Using spherical coordinates and substituting $M(r,t)=G(r,t)\cdot r$, the above system transforms to:

$$\frac{1}{D}\frac{\partial M(r,t)}{\partial t}-\frac{\partial^2 M(r,t)}{\partial r^2}=0 \qquad r\in(0,R),\ t>0 \tag{3.59}$$

$$M(r,t)=0 \qquad \text{at } r=0 \text{ and } r=R,\ t>0 \tag{3.60}$$

$$M(r,t)=r \qquad r\in(0,R),\ \text{for } t=0 \tag{3.61}$$

Solution of the above initial-boundary value problem has been found using the method of separation of variables (details of the method can be found in Özişik (1980)). The final form is presented below, i.e.:

$$M(r,t)=-2R\sum_{m=1}^{\infty}\frac{\cos(m\pi)}{m\pi}\times\sin\left(\frac{m\pi}{R}r\right)\times e^{-D(m\pi/R)^2 t} \tag{3.62}$$

Taking into account the substitution used, one gets immediately the form of the function $G(r,t)$:

$$G(r,t)=-\frac{2R}{r}\sum_{m=1}^{\infty}\frac{\cos(m\pi)}{m\pi}\times\sin\left(\frac{m\pi}{R}r\right)\times e^{-D(m\pi/R)^2 t} \tag{3.63}$$

Volume average over the grain domain results in:

$$\langle G(t)\rangle^R=6\sum_{m=1}^{\infty}\frac{1}{(m\pi)^2}\times e^{-D(m\pi/R)^2 t} \tag{3.64}$$

where the symbol $\langle\ \rangle^R$ denotes an averaging operator over the domain occupied by a grain of the radius R. For the medium composed of N families of grains, each family characterizes appropriate radius of grain, the memory function can be expressed as:

$$\langle G(t) \rangle = \sum_{i=1}^{N} c_i \langle G(t) \rangle^{R_i} \tag{3.65}$$

The parameter c_i denotes a volume fraction of the grains, with the radius R_i, in a porous medium. The final form of the memory function is presented below:

$$\langle G(t) \rangle = 6 \sum_{i=1}^{N} \left\{ c_i \sum_{m=1}^{\infty} \frac{1}{(m\pi)^2} \times e^{-D(m\pi / R_i)^2 t} \right\} \tag{3.66}$$

A quantitative presentation of the memory function is given in Figure 5. The plots in this figure have been estimated using the first two thousand terms of the infinite sum of the memory function. It has to be marked, however, that the sum of the memory function converges very quickly with time, so much less terms can be used. A typical value of the diffusion coefficient has been used in the calculations, i.e.: D=10^{-11} [m^2/s]. The curves in Figure 5 clearly show the influence of the grain radius on the memory function of the microstructure analyzed.

The form of the memory function obtained is of no use, from practical point of view. Figure 6 presents an approximation of the memory function by use only of the two terms. The following approximation has been applied:

$$\langle G(t) \rangle^R = b_1 \times e^{-t/T_1} + (1 - b_1) \times e^{-t/T_2} \tag{3.67}$$

Dots in the Figure 6 represent values calculated according to the formula (3.64) whereas a solid line – according to the approximation. The following values of the constants have been determined, i.e.: b_1=0.296, T_1=610 [s] and T_2= 8820 [s]. These values correspond to the microstructure composed of the uniform grains of the radius R=1 [mm].

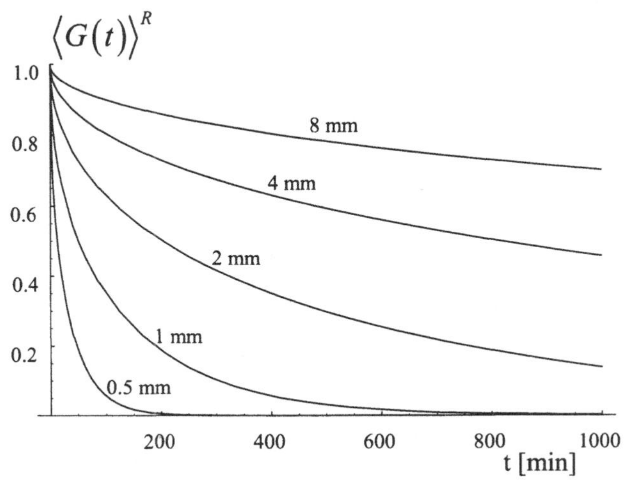

Figure 5. Memory function evolution curves for different grain radii.

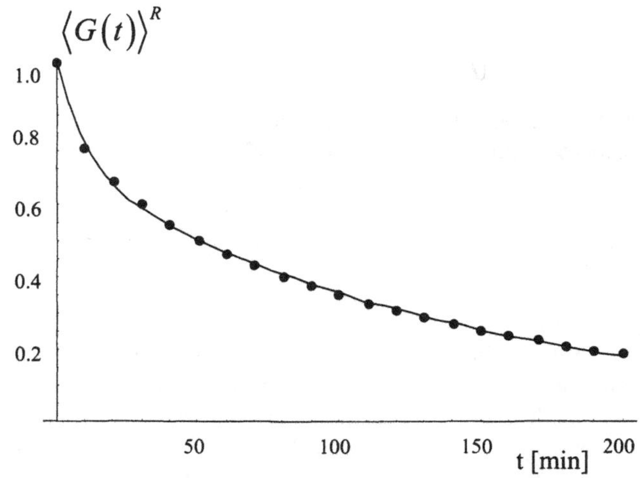

Figure 6. An approximation of the memory function.

The comparison presented in Figure 6 clearly shows that the approximation proposed really satisfactory describes the analytical solution. Visible discrepancies occur only at the first seconds of the process which is not clearly seen in Figure 6. In order to verify usefulness of this approximation another test has been performed. For some history of the sorption isotherm, values of the gas concentration have been determined based on the analytical solution and the approximation. The sorption isotherm history applied as well as a final comparison is shown in Figure 7. Once more a satisfactory fit has been obtained.

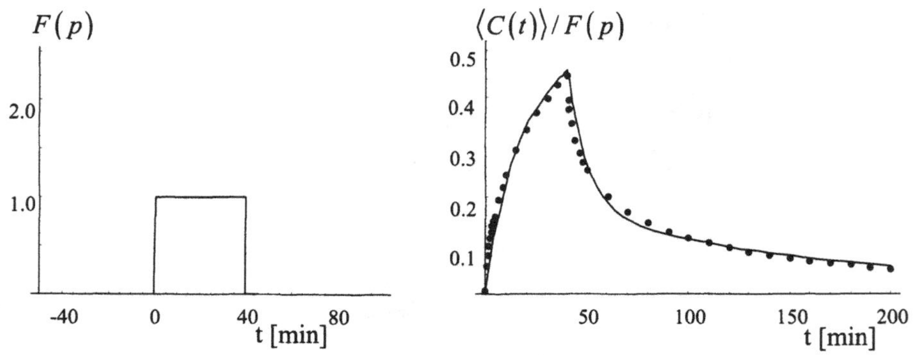

Figure 7. Sorption isotherm history and evolution of a constrained gas concentration.

4 Mechanical Properties of Porous Media with Sorption Phenomenon

The medium considered is the same as in the section 3, i.e. the bituminous coal saturated with carbon dioxide. As before, it is assumed that the medium is composed of a micro-porous coal skeleton and the network of macro-pores. The macro-pores are filled by a gas in a free state. The micro-porous skeleton is a mixture of a solid phase and of the gas constrained in the micro-pores. The transport process of the sorbed gas is assumed to be governed by the Fick molecular diffusion law. The flow of a free gas through macro-pores is postulated to be described by the Navier-Stokes equations of a barotropic fluid. In contrast to the analysis performed in the section 3, this time the micro-porous skeleton is treated as a deformable one. It is postulated that a mechanical behavior of the skeleton with the gas constrained in micro-pores can be modeled by the equations of uncoupled theory of diffuso-elasticity (for details on the theory of diffuso-elasticity as well as the theory of thermo-diffuso-elasticity the reader is referred to Nowacki (1986) and Gałka, Telega and Wojnar (1994)). The uncoupled theory means that the gas constrained in micro-pores acts on a solid stresses whereas the stress state of the solid does not affect the sorption equilibrium criterion, i.e. the sorption isotherm function is independent on the solid stress.

4.1 Local Description

With the above assumptions, the micro-scale description is given by:

micro-porous skeleton with constrained gas:

– balance equations:

$$\sigma^s_{ij,i} = 0 \qquad \text{in } V_s \tag{4.1}$$

– constitutive equations of uncoupled diffuso-elasticity theory

$$\sigma^s_{ij} = a_{ijkh} e_{kh}(u) - \gamma_c C \delta_{ij} \qquad \text{in } V_s \quad \text{where } e_{kh}(u) = \frac{1}{2}\left(u_{k,h} + u_{h,k}\right) \tag{4.2}$$

– mass conservation equation for the constrained gas:

$$\frac{\partial C}{\partial t} - DC_{,ii} = 0 \qquad \text{in } V_s \tag{4.3}$$

free gas in macro-pores:

– balance equations:

$$\sigma^l_{ij,i} = 0 \qquad \text{in } V_l \tag{4.4}$$

– constitutive equations of barotropic Newtonian liquid:

$$\sigma^l_{ij} = -p\delta_{ij} + \lambda v_{k,k}\delta_{ij} + \mu\left(v_{i,j} + v_{j,i}\right) \qquad \text{in } V_l \tag{4.5}$$

– mass conservation equation:

$$\frac{\partial \rho}{\partial t} + \left(\rho v_i\right)_{,i} = 0 \qquad \text{in } V_l \tag{4.6}$$

– the ideal gas law for isothermal processes:

$$\rho = \frac{\rho_a}{p_a} p \qquad \text{in } V_l \tag{4.7}$$

The above equations use the same symbols as that applied in the section 3. In addition, the following notation is introduced, i.e.: u – the skeleton displacement vector, σ_{ij}^s - the skeleton stress tensor, γ_c – a coefficient of a sorption swelling and δ_{ij} – the Kronecker symbol.

The above system of equations is completed by the boundary conditions on the phase separation surface Γ, i.e.:

– continuity of the mass flux:

$$\left[\rho\left(v_i - \dot{u}_i\right) + DC_{,i}\right] N_i = 0 \qquad \text{on } \Gamma \tag{4.8}$$

– sorption isotherm equation:

$$C = F(p) \qquad \text{on } \Gamma \tag{4.9}$$

– an adhesion condition:

$$\left(v_i - \dot{u}_i\right) t_i = 0 \qquad \text{on } \Gamma \tag{4.10}$$

– continuity of the stress vector:

$$\left[\sigma_{ij}^s - \sigma_{ij}^l\right] N_i = 0 \qquad \text{on } \Gamma \tag{4.11}$$

where: $\dot{u}$ represents the skeleton velocity vector.

The thermodynamic equilibrium between the phases is assumed at the initial instant.

4.2 Asymptotic Expansions and Averaging Process

The equations of the local description are normalized and scaled, exactly in the same manner, as it has been done in the section 3. The same two cases are distinguished, i.e.: case 1° corresponding to T_D/T_f=O(ε^{-1}) and case 2° corresponding to T_D/T_f=O(1). It appears also that the macroscopic description for the case 2° represents a simplified version of that for the case 1° (Lydzba (2002)). In a following, the case 1° is only considered.

Introducing the asymptotic expansions of: v, p, ρ, u and C and modifying the spatial derivative operator according to (2.5) one gets:

$$\begin{aligned}
&\varepsilon^{-2}\frac{\partial}{\partial y_i}\left[a_{ijkh}e_{kh}^y\left(u^{(0)}\right)\right] + \varepsilon^{-1}\left\{\frac{\partial}{\partial y_i}\left[a_{ijkh}e_{kh}^x\left(u^{(0)}\right) + a_{ijkh}e_{kh}^y\left(u^{(1)}\right) - \gamma_c C^{(0)}\delta_{ij}\right] + \frac{\partial}{\partial x_i}\left[a_{ijkh}e_{kh}^y\left(u^{(0)}\right)\right]\right\} + \\
&+\varepsilon^{0}\left\{\frac{\partial}{\partial x_i}\left[a_{ijkh}e_{kh}^x\left(u^{(0)}\right) + a_{ijkh}e_{kh}^y\left(u^{(1)}\right) - \gamma_c C^{(0)}\delta_{ij}\right]\right\} + \\
&+\varepsilon^{0}\left\{\frac{\partial}{\partial y_i}\left[a_{ijkh}e_{kh}^x\left(u^{(1)}\right) + a_{ijkh}e_{kh}^y\left(u^{(2)}\right) - \gamma_c C^{(1)}\delta_{ij}\right]\right\} + \ldots = 0 \qquad \text{in } V_s
\end{aligned} \tag{4.12}$$

$$\varepsilon^0\left[\frac{\partial C^{(0)}}{\partial t}-D\frac{\partial}{\partial y_i}\left(\frac{\partial C^{(0)}}{\partial y_i}\right)\right]-\varepsilon^1 D\left\{\frac{\partial}{\partial y_i}\left(\frac{\partial C^{(0)}}{\partial x_i}+\frac{\partial C^{(1)}}{\partial y_i}\right)+\frac{\partial}{\partial x_i}\left(\frac{\partial C^{(0)}}{\partial y_i}\right)\right\}+\ldots=0 \text{ in } V_s \quad (4.13)$$

$$-\varepsilon^{-1}\frac{\partial p^{(0)}}{\partial y_i}+$$

$$+\varepsilon^0\left[-\frac{\partial p^{(0)}}{\partial x_i}-\frac{\partial p^{(1)}}{\partial y_i}+\mu\frac{\partial}{\partial y_j}\left(\frac{\partial v_i^{(0)}}{\partial y_j}\right)+(\lambda+\mu)\frac{\partial}{\partial y_i}\left(\frac{\partial v_j^{(0)}}{\partial y_j}\right)\right]+\ldots=0 \qquad \text{in } V_l \qquad (4.14)$$

$$\varepsilon^{-1}\frac{\partial\left(\rho^{(0)}v_i^{(0)}\right)}{\partial y_i}+\varepsilon^0\left[\frac{\partial \rho^{(0)}}{\partial t}+\frac{\partial\left(\rho^{(0)}v_i^{(0)}\right)}{\partial x_i}+\frac{\partial\left(\rho^{(0)}v_i^{(1)}\right)}{\partial y_i}+\frac{\partial\left(\rho^{(1)}v_i^{(0)}\right)}{\partial y_i}\right]+\ldots=0 \qquad \text{in } V_l, \qquad (4.15)$$

$$\varepsilon^0\left(\rho^{(0)}-\frac{\rho_a}{p_a}p^{(0)}\right)+\varepsilon^1\left(\rho^{(1)}-\frac{\rho_a}{p_a}p^{(1)}\right)+\ldots=0 \qquad \text{in } V_l \qquad (4.16)$$

$$\left[\varepsilon^0\rho^{(0)}\left(v_i^{(0)}-\dot{u}_i^{(0)}\right)+\varepsilon^1\left\{\rho^{(0)}\left(v_i^{(1)}-\dot{u}_i^{(1)}\right)+\rho^{(1)}\left(v_i^{(0)}-\dot{u}_i^{(0)}\right)+D\frac{\partial C^{(0)}}{\partial y_i}\right\}\right]N_i+\ldots=0 \quad \text{on } \Gamma \; (4.17)$$

$$\varepsilon^0\left(C^{(0)}-F\left(p^{(0)}\right)\right)+\ldots=0 \qquad \text{on } \Gamma \qquad (4.18)$$

$$\varepsilon^0\left(v_i^{(0)}-\dot{u}_i^{(0)}\right)t_i+\varepsilon^1\left(v_i^{(1)}-\dot{u}_i^{(1)}\right)t_i+\ldots=0 \qquad \text{on } \Gamma \qquad (4.19)$$

$$\varepsilon^{-1}\left[a_{ijkh}e^y_{kh}\left(u^{(0)}\right)\right]N_i+\varepsilon^0\left[a_{ijkh}e^x_{kh}\left(u^{(0)}\right)+a_{ijkh}e^y_{kh}\left(u^{(1)}\right)-\gamma_c C^{(0)}\delta_{ij}+p^{(0)}\delta_{ij}\right]N_i+$$

$$\text{on } \Gamma \, (4.20)$$

$$+\varepsilon^1\left[a_{ijkh}e^x_{kh}\left(u^{(1)}\right)+a_{ijkh}e^y_{kh}\left(u^{(2)}\right)-\gamma_c C^{(1)}\delta_{ij}+p^{(1)}\delta_{ij}-\lambda\frac{\partial v_k^{(0)}}{\partial y_k}\delta_{ij}-\mu\left(\frac{\partial v_i^{(0)}}{\partial y_j}+\frac{\partial v_j^{(0)}}{\partial y_i}\right)\right]\times N_i=0$$

The following simplified notation has been used above:

$$e^x_{ij}(u)=\frac{1}{2}\left(\frac{\partial u_i}{\partial x_j}+\frac{\partial u_j}{\partial x_i}\right) \quad \text{and} \quad e^y_{ij}=\frac{1}{2}\left(\frac{\partial u_i}{\partial y_j}+\frac{\partial u_j}{\partial y_i}\right)$$

Collecting terms based on their orders one gets set of local boundary-value problems for the consecutive terms of the asymptotic expansions.

I local problem: the equations (4.12) and (4.20), i.e.:

$$O(\varepsilon^{-2}):\ \frac{\partial}{\partial y_i}\left[a_{ijkh}e^{y}_{kh}\left(u^{(0)}\right)\right]=0 \qquad \text{in } V_s \tag{4.21}$$

$$O(\varepsilon^{-1}):\ \left[a_{ijkh}e^{y}_{kh}\left(u^{(0)}\right)\right]N_i=0 \qquad \text{on } \Gamma \tag{4.22}$$

II local problem: the equations (4.14) and (4.16), i.e.:

$$O(\varepsilon^{-1}):\ \frac{\partial p^{(0)}}{\partial y_i}=0 \qquad \text{in } V_l \tag{4.23}$$

$$O(\varepsilon^{0}):\ \rho^{(0)}-\frac{\rho_a}{p_a}p^{(0)}=0 \quad \text{in } V_l \tag{4.24}$$

III local problem: the equations (4.14), (4.15), (4.17) and (4.19), i.e.:

$$O(\varepsilon^{0}):\ -\frac{\partial p^{(0)}}{\partial x_i}-\frac{\partial p^{(1)}}{\partial y_i}+\mu\frac{\partial}{\partial y_j}\left(\frac{\partial v_i^{(0)}}{\partial y_j}\right)+(\lambda+\mu)\frac{\partial}{\partial y_i}\left(\frac{\partial v_j^{(0)}}{\partial y_j}\right)=0 \qquad \text{in } V_l \tag{4.25}$$

$$O(\varepsilon^{-1}):\ \frac{\partial\left(\rho^{(0)}v_i^{(0)}\right)}{\partial y_i}=0 \qquad \text{in } V_l \tag{4.26}$$

$$O(\varepsilon^{0}):\ \rho^{(0)}\left(v_i^{(0)}-\dot{u}_i^{(0)}\right)N_i=0 \quad \text{on } \Gamma \tag{4.27}$$

$$O(\varepsilon^{0}):\ \left(v_i^{(0)}-\dot{u}_i^{(0)}\right)t_i=0 \qquad \text{on } \Gamma \tag{4.28}$$

IV local problem: the equations (4.13), (4.18), i.e.:

$$O(\varepsilon^{0}):\ \frac{\partial C^{(0)}}{\partial t}-D\frac{\partial}{\partial y_i}\left(\frac{\partial C^{(0)}}{\partial y_i}\right)=0 \qquad \text{in } V_s \tag{4.29}$$

$$O(\varepsilon^{0}):\ C^{(0)}-F\left(p^{(0)}\right)=0 \ \text{ on } \Gamma \tag{4.30}$$

V local problem: the equations (4.12), (4.20), i.e.:

$$O(\varepsilon^{-1}):\ \frac{\partial}{\partial y_i}\left[a_{ijkh}e^{x}_{kh}\left(u^{(0)}\right)+a_{ijkh}e^{y}_{kh}\left(u^{(1)}\right)-\gamma_c C^{(0)}\delta_{ij}\right]+\frac{\partial}{\partial x_i}\left[a_{ijkh}e^{y}_{kh}\left(u^{(0)}\right)\right]=0 \quad \text{in } V_s \tag{4.31}$$

$$O(\varepsilon^{0}):\ \left[a_{ijkh}e^{x}_{kh}\left(u^{(0)}\right)+a_{ijkh}e^{y}_{kh}\left(u^{(1)}\right)-\gamma_c C^{(0)}\delta_{ij}+p^{(0)}\delta_{ij}\right]N_i=0 \qquad \text{on } \Gamma \tag{4.32}$$

VI local problem: the equations (4.12) and (4.20), i.e.:

$$O(\varepsilon^0):\ \frac{\partial}{\partial x_i}\left[a_{ijkh}e^x_{kh}\left(u^{(0)}\right)+a_{ijkh}e^y_{kh}\left(u^{(1)}\right)-\gamma_c C^{(0)}\delta_{ij}\right]+$$

$$+\frac{\partial}{\partial y_i}\left[a_{ijkh}e^x_{kh}\left(u^{(1)}\right)+a_{ijkh}e^y_{kh}\left(u^{(2)}\right)-\gamma_c C^{(1)}\delta_{ij}\right]=0 \qquad \text{in } V_s \qquad (4.33)$$

$$O(\varepsilon^1):\ \left[a_{ijkh}e^x_{kh}\left(u^{(1)}\right)+a_{ijkh}e^y_{kh}\left(u^{(2)}\right)-\gamma_c\, C^{(1)}\,\delta_{ij}\right]\times N_i+$$

$$+\left[p^{(1)}\delta_{ij}-\lambda\frac{\partial v^{(0)}_k}{\partial y_k}\delta_{ij}-\mu\left(\frac{\partial v^{(0)}_i}{\partial y_j}+\frac{\partial v^{(0)}_j}{\partial y_i}\right)\right]\times N_i=0 \qquad \text{on } \Gamma \qquad (4.34)$$

VII local problem: the equations (4.15) and (4.17), i.e.:

$$O(\varepsilon^0):\ \frac{\partial\rho^{(0)}}{\partial t}+\frac{\partial\left(\rho^{(0)}v^{(0)}_i\right)}{\partial x_i}+\frac{\partial\left(\rho^{(0)}v^{(1)}_i\right)}{\partial y_i}+\frac{\partial\left(\rho^{(1)}v^{(0)}_i\right)}{\partial y_i}=0 \qquad \text{in } V_l \qquad (4.35)$$

$$O(\varepsilon^1):\ \left\{\rho^{(0)}\left(v^{(1)}_i-\dot{u}^{(1)}_i\right)+\rho^{(1)}\left(v^{(0)}_i-\dot{u}^{(0)}_i\right)+D\frac{\partial C^{(0)}}{\partial y_i}\right\}N_i=0 \qquad \text{on } \Gamma \qquad (4.36)$$

The first local problem implies:

$$u^{(0)}=u^{(0)}(x,t) \qquad (4.37)$$

The next three local problems are of the same or similar form to those considered in the section 3, therefore their final solutions are only presented, i.e.:

$$p^{(0)}=p^{(0)}(x,t) \qquad (4.38)$$

$$\rho^{(0)}=\rho^{(0)}(x,t) \qquad (4.39)$$

$$\left\langle v^{(0)}_i\right\rangle=-K_{ij}\frac{\partial p^{(0)}}{\partial x_j} \qquad (4.40)$$

$$C^{(0)}(x,y,t)=F\left(p^{(0)}\right)-\int_0^t\frac{\partial F\left(p^{(0)}\right)}{\partial t}G(y,t-\tau)\,d\tau \qquad (4.41)$$

$$\left\langle C^{(0)}(x,t)\right\rangle=(1-n)F\left(p^{(0)}\right)-\int_0^t\frac{\partial F\left(p^{(0)}\right)}{\partial t}\left\langle G(y,t-\tau)\right\rangle d\tau \qquad (4.42)$$

where: n – a volume of the macro-pores in the unit volume of the medium.

Taking into account the relation (4.41), the V local problem can be rewritten as:

$$\frac{\partial}{\partial y_i}\left[a_{ijkh}e^x_{kh}\left(u^{(0)}\right)+a_{ijkh}e^y_{kh}\left(u^{(1)}\right)-\gamma_c\delta_{ij}\left\{F\left(p^{(0)}\right)-\int_0^t\frac{\partial F\left(p^{(0)}\right)}{\partial\tau}G(y,t-\tau)\,d\tau\right\}\right]=0 \quad \text{in } V_s \quad (4.43)$$

$$\left[a_{ijkh} e^{x}_{kh}\left(u^{(0)}\right) + a_{ijkh} e^{y}_{kh}\left(u^{(1)}\right) - \gamma_c F\left(p^{(0)}\right)\delta_{ij} + p^{(0)}\delta_{ij} \right] N_i = 0 \qquad \text{on } \Gamma \ (4.44)$$

The above system transformed to the Laplace transform space would be linear. Therefore, the solution of the V local problem can be formulated as:

$$u_i^{(1)}(x,y,t) = \xi_i^{jk}(y) e^{x}_{jk}\left(u^{(0)}\right) - \eta_i(y) p^{(0)} + \eta_i \gamma_c F\left(p^{(0)}\right) + \\ +\gamma_c \int_0^t \frac{\partial F\left(p^{(0)}\right)}{\partial \tau} w_i(y, t-\tau) d\tau + \overline{u}_i(x) \qquad \text{in } V_s \ (4.45)$$

The field $\overline{u}_i(x)$ is undetermined x – dependent function. The vectorial field $w_i(y,t)$ is a solution, in the Laplace transform space, of the following local problem:

$$\frac{\partial}{\partial y_i}\left[a_{ijlm} e_{lmy}\left(\mathcal{L}\left[w(y,t)\right] \right) + \mathcal{L}\left[G(y,t)\right]\delta_{ij} \right] = 0 \qquad \text{in } V_s \qquad (4.46)$$

$$a_{ijlm} e_{lmy}\left(\mathcal{L}\left[w(y,t)\right] \right) N_i = 0 \qquad \text{on } \Gamma \qquad (4.47)$$

The fields $\xi_i^{lm}(y)$, $\eta_i(y)$ are solutions of the local problems presented below, i.e.:

- the field $\xi_i^{lm}(y)$:

$$\frac{\partial}{\partial y_i}\left[a_{ijlm} + a_{ijkh} e^{y}_{kh}\left(\xi^{lm}\right) \right] = 0 \qquad \text{in } V_s \qquad (4.48)$$

$$\left[a_{ijlm} + a_{ijkh} e^{y}_{kh}\left(\xi^{lm}\right) \right] N_i = 0 \qquad \text{on } \Gamma \qquad (4.49)$$

- the field $\eta_i(y)$:

$$\frac{\partial}{\partial y_i}\left[a_{ijkh} e^{y}_{kh}(\eta) \right] = 0 \qquad \text{in } V_s \qquad (4.50)$$

$$\left[a_{ijkh} e^{y}_{kh}(\eta) + \delta_{ij} \right] N_i = 0 \qquad \text{on } \Gamma \qquad (4.51)$$

Volume average of the VI local problem results, after some algebraic transformation, in the macroscopic balance equations for the medium considered. The final form can presented as:

$$\frac{\partial \left\langle \sigma_{ij}^{T} \right\rangle}{\partial x_i} = 0 \qquad (4.52)$$

$$\left\langle \sigma_{ij}^{T} \right\rangle = A_{ijkh} e^{x}_{kh}(u) - \alpha_{ij} p^{(0)} - \gamma_{ij} F(p^{(0)}) + \int_0^t \frac{\partial F\left(p^{(0)}\right)}{\partial \tau} m_{ij}(t-\tau) d\tau, \qquad (4.53)$$

where $\left\langle \sigma_{ij}^{T} \right\rangle$ is a total stress tensor for a multiphase medium, A_{ijkh} is an overall elasticity stiffness tensor, α_{ij} is the so-called Biot effective stress coefficient tensor and γ_{ij} is an overall sorption swelling coefficient tensor. These materials' constants are defined by the following formulae:

$$A_{ijkh} = (1-n\ )a_{ijkh} + a_{ijlm}\left\langle e^{y}_{lm}(\xi^{kh})\right\rangle$$

$$\alpha_{ij} = n\delta_{ij} - a_{ijkh}\left\langle e^{y}_{kh}(\eta)\right\rangle \tag{4.54}$$

$$\gamma_{ij} = \gamma_c(1-n)\delta_{ij} + \gamma_c a_{ijkh}\left\langle e^{y}_{kh}(\eta)\right\rangle$$

The memory function $m_{ij}(t)$ is defined as:

$$m_{ij}(t) = \gamma_c\left\langle G(t)\right\rangle\delta_{ij} + \gamma_c a_{ijkh}\left\langle e^{y}_{kh}(w(t))\right\rangle \tag{4.55}$$

Volume average of the VII local problem leads to the macroscopic mass conservation law for the free gas flowing through a network of macro-pores, i.e.:

$$n\frac{\partial\rho^{(0)}}{\partial t} + \frac{\partial}{\partial x_i}\left[\rho^{(0)}\left(\left\langle v_i^{(0)}\right\rangle - n\dot{u}_i^{(0)}\right)\ \right] + \frac{\partial\left\langle C^{(0)}\right\rangle}{\partial t} +$$

$$+\rho^{(0)}\left(\alpha_{ij}\dot{e}^{x}_{ij}(u) - \beta\dot{p} + \gamma_c\beta\dot{F}(p) + \int_0^t\frac{\partial F(p(\tau))}{\partial\tau}\beta_m(t-\tau)d\tau\right) = 0 \tag{4.56}$$

where $\beta_m(t)$ is an additional memory function defined as:

$$\beta_m(t) = -\gamma_c\left\langle e^{y}_{ii}(w)\right\rangle \tag{4.57}$$

4.3 Macroscopic Description

The macroscopic description of the processes considered is constituted by:

- the constitutive equations

$$\left\langle\sigma^{T}_{ij}\right\rangle = A_{ijkh}e_{khx}(u) - \alpha_{ij}p - \gamma_{ij}F(p) + \int_0^t\frac{\partial F(p(\tau))}{\partial\tau}m_{ij}(t-\tau)d\tau \tag{4.58}$$

- the balance equations

$$\frac{\partial\left\langle\sigma^{T}_{ij}\right\rangle}{\partial x_i} = 0 \tag{4.59}$$

- the dependence of a constrained gas concentration on applied pore pressure in macro-pores

$$\left\langle C\right\rangle = (1-n)F(p) - \int_0^t\frac{\partial F(p(\tau))}{\partial\tau}\left\langle G(t-\tau)\right\rangle d\tau \tag{4.60}$$

– the filtration law for a free gas:

$$\langle v_i \rangle - n\dot{u}_i = -K_{ij}\frac{\partial p}{\partial x_j} \tag{4.61}$$

– the mass conservation law for a free gas:

$$n\frac{\partial \rho}{\partial t} + \frac{\partial}{\partial x_i}\left[\rho\left(\langle v_i \rangle - n\dot{u}_i\right)\right] + \frac{\partial \langle C \rangle}{\partial t} +$$

$$+\rho\left(\alpha_{ij}\dot{e}_{ijx}(u) - \beta\dot{p} + \gamma_c\beta\dot{F}(p) + \int_0^t \frac{\partial F(p)}{\partial \tau}\beta_m(t-\tau)d\tau\right) = 0 \tag{4.62}$$

– the function $F(p)$ of a sorption isotherm.

The macroscopic description for the case 2° , distinguished previously, is composed of the same set of equations. It corresponds to the equations presented above with canceled terms involving the memory functions.

4.4 The Model Constants: Reciprocal Relations

The reciprocal relations between coefficients are obtained with the postulate that the porous skeleton is composed of a homogeneous and an isotropic material.

The relations between coefficients of the Biot's poroelasticity theory are presented in a few papers (see, for instance, Auriault (1991), Cheng (1997), Lydzba and Shao (2000)). They could be written as follows:

$$\beta = \frac{n - \frac{1}{3}\alpha_{ij}\delta_{ij}}{K_s} \tag{4.63}$$

$$\alpha_{ij} = \delta_{ij} - \frac{A_{ijkh}\delta_{kh}}{3K_s} \tag{4.64}$$

where: K_s denotes a value of the bulk modulus of the skeleton material.
Considering the definitions of γ_{ij} and of α_{ij} (see Eqs. 4.54) as well as the formula (4.64), one can immediately conclude that the following relations hold also true, i.e.:

$$\frac{\gamma_{ij}}{\gamma_c} + \alpha_{ij} = \delta_{ij} \quad \text{and} \quad \frac{\gamma_{ij}}{\gamma_c} = \frac{A_{ijkh}\delta_{kh}}{3K_s} \tag{4.65}$$

The last reciprocal relation involves the memory function. Multiply the both side of the definition (4.55) by δ_{ij} and use the identity $a_{ijkh}\delta_{ij} = 3K_s$. Compare the result obtained with the definition (4.57). The following reciprocal relation is then established, i.e.:

$$\beta_m(t) = \frac{3\gamma_c\langle G(t)\rangle - m_{ij}(t)\delta_{ij}}{3K_s} \tag{4.66}$$

5 Final Conclusions

The macroscopic descriptions of the gas transport process and of the mechanical behavior of porous medium considered have been obtained by a systematic use of the asymptotic expansion method. This approach enabled to formulate the macroscopic description through a so-called passage micro-macro. The analysis has been focused on the multiphase system of bituminous coal and carbon dioxide. This system characterizes a phenomenon of local mass exchange between medium components. Only a small part of the gas is in a free state. The remainder part of the gas occurs in the system as a constrained phase. Two different cases have been distinguished in the analysis. The first corresponds to a case when a diffusion process of a sorbed gas is much slower than the filtration one. Such difference in the characteristics times of the transport phenomena leads to a strong rheological charcter of the macroscopic description. For instance, the sorbed gas concentration and phenomenon of a sorption swelling depend on the history of the sorption isotherm first derivative. When the two characteristics times of the transport processes are comparable fast, then this strong rheological effect disappears. The mechanical behavior is governed by a mixed system of equations, i.e.: the equations of the diffuso-elasticity theory and of the Biot's poro-elasticity theory.

References

Auriault J. L. (1991). Dynamic behaviour of porous media, in *Transport Processes in Porous Media*, J. Bear and M. Y. Corapcioglu (eds.), Kluwer Academic Publishers, 471-519.

Bensoussan A.,Lions J.L., Papanicolau G. (1978). *Asymptotic analysis for periodic structures*, North-Holland Publishing Company, Amsterdam.

Cheng A.H.D.(1997). Material Coefficients of anisotropic poroelasticity, *Int.J.Rock Mech.Min.Sci.*, 34, 199-205.

Gałka A., Telega J.J., Wojnar R. (1994). Thermodiffusion in heterogeneous elastic solids and homogenization, *Arch. Mech*, 46, 3, 267-314.

Hornung U. (ed). (1997). *Homogenization and Porous Media*, Interdisciplinary Applied Mathematics, Vol.6, Springer-Verlag, Berlin.

Landau L.D., Lifszyc E.M. (1994). *Hydrodynamics*, PWN, Warszawa, seria Fizyka Techniczna, (in polish)

Lasoń M., (1988), Specific surface of porous materials, *Arch.Min.Sci.*, 33, 475-497.

Łydżba D., Auriault J.L. (1996). Gas filtration through porous coal medium. Effect of the gas constrained in micropores, *Arch. Mech*, 48 (3), 447-473

Łydżba D., Shao J.F. (2000). Study of poroelasticity material coefficients as response of microstructure, *Mechanics of Cohesive-Frictional Materials*, 5, 149-171.

Łydżba D. (2002). *Applications of asymptotic homogenization method in soil and rock mechanics*, Oficyna Wydawnicza PWr, Wroclaw, (in polish).

Nowacki W.. (1986). *Advances of Elasticity Theory,* PWN, Warszawa, (in polish).

Özişik M.N.. (1980). *Heat Conduction,* John Wiley&Sons, New York.

Sab K. (1992). On the homogenization and the simulation of random materials, *Eur.J.Mech.A/Solids*,11, 585-607.

Sab K. (1994). Homogenization of non-linear random media by duality method. Application to plasticity, *Asymptotic Analysis*, 9, 311-336.

Sanchez-Palencia E. (1980). *Non-homogeneous Media and Vibration Theory*, Lecture Notes in Physics, 127, Springer-Verlag, Berlin.

Diffusive transport in disordered media. Application to the determination of the tortuosity and the permeability of cracked materials.

Luc Dormieux * and Djimedo Kondo †

* Ecole Nationale des Ponts et Chaussées, Champs sur Marne, France.
† Laboratoire de Mécanique de Lille, Lille University of Technology, France.

Abstract This Chapter deals with molecular diffusion in saturated porous media. It provides a micromechanical basis to the phenomenological concept of tortuosity which is classically introduced in order to capture the barrier effect of the solid-fluid interface. Various Eshelby-based estimates of the effective diffusion tensor are derived in the isotropic case. The influence of cracks on the tortuosity is also investigated. Finally, it is shown that the permeability of a cracked porous medium can be estimated with the help of the same mathematical tools.

1 Introduction

Molecular diffusion of a solute in a liquid solvent is usually modelled by Fick's law (Atkins (1992)). Accordingly, the direction of the diffusive flux is that of the gradient of the solute concentration. Molecular diffusion also takes place in the pore space of porous media. In this case, the solid-fluid interface acts as a barrier which prevents the pathlines of solute molecules to be straightlines parallel to the concentration gradient. This leads to the phenomenological concept of tortuosity. The focus of this chapter is to present a micro-to-macro approach to the latter.
The periodic case has been considered in details by Auriault and Lewandowska (1996). We shall deal hereafter with molecular diffusion in disordered porous media. The pore space is saturated by a liquid composed of a solvent and a solute. For simplicity, it is assumed that the pore fluid is incompressible, and that the solid phase is rigid. The solute is characterized by the superscript γ.
The notations and concepts introduced in the chapter "Poroelasticity and strength of fully and partially saturated porous materials" are widely used in the sequel. In particular, $\underline{z}$ (resp. $\underline{x}$) denotes the position vector at the microscopic (resp. macroscopic) scale. The pore volume fraction is denoted by φ.

We consider steady state conditions, advection and adsorption phenomena being disregarded. ρ^γ, $\underline{v}^\gamma$ and $\underline{j}^\gamma = \rho^\gamma \underline{v}^\gamma$ denote the density, the velocity and the diffusive flux of the γ-component at the microscopic scale. Denoting by Ω^f and I^{fs} the fluid domain and the solid-fluid interface respectively, the diffusive flux $\underline{j}^\gamma$ and the solute concentration ρ^γ

must comply with the mass balance equation for the γ-component:

$$div_z \, \underline{j}^\gamma = 0 \quad (\Omega^f) \tag{1.1}$$

In the absence of adsorption, there is no mass exchange of the γ-component at the solid-fluid interface[1]:

$$\underline{j}^\gamma \cdot \underline{n} = 0 \quad (I^{fs}) \tag{1.2}$$

According to Fick's law of diffusion, the diffusive flux of the γ-component is proportional to the gradient of its density :

$$\underline{j}^\gamma = -D^\gamma \, \underline{grad}_z \, \rho^\gamma \quad (\Omega^f) \tag{1.3}$$

where D^γ is the diffusion coefficient that characterizes the diffusion of the solute in the solvent *in an infinite domain.* The diffusion coefficient is therefore independent of the morphology of the pore space, and depends only on the nature of the solute and of the solvent. The purpose of the micro-to-macro approach is the determination of the macroscopic counterpart of (1.3) in a way that incorporates the available informations concerning the microstructure of the porous medium.
From a physical point of view, ρ^γ and $\underline{j}^\gamma$ are defined on the fluid phase only. However, from a mathematical point of view, it will prove efficient to extend them as well as Fick's law to the solid phase. We therefore introduce a diffusion coefficient equal to 0 in the solid phase, which implies that the diffusive flux is extended to 0 in the solid phase :

$$(\Omega) \quad \underline{j}^\gamma(\underline{z}) \;=\; -D(\underline{z})\,\underline{\mathrm{grad}}_z \rho^\gamma \;\text{ with }\; D(\underline{z}) = \begin{cases} D^\gamma & \text{in the fluid} \\ 0 & \text{in the solid} \end{cases} \tag{1.4}$$

(1.4) describes Ω as an heterogeneous medium with two different values of the diffusion coefficient, depending on the location point $\underline{z}$. With this point of view, the mass balance equation (1.5) is now valid on the whole r.e.v. whereas the boundary condition (1.2) is replaced by a continuity of the normal component of the diffusive flux accross the solid-fluid interface :

$$(\Omega) \quad \mathrm{div}_z \, \underline{j}^\gamma = 0 \;\; ; \;\; [\underline{j}^\gamma].\underline{n} = 0 \text{ at } I^{fs} \tag{1.5}$$

where $[\underline{j}^\gamma]$ is the discontinuity of the diffusive flux at the solid-fluid interface. In turn, in the absence of advection and adsorption, the mass balance equation at the macroscopic scale reads :

$$div_x \, \underline{J}^\gamma = 0 \tag{1.6}$$

where $\underline{J}^\gamma$ is the macroscopic diffusive mass flux of the γ-component. It can be shown[2] that the consistency between (1.5) and (1.6) requires to relate the macroscopic diffusive flux $\underline{J}^\gamma$ to the microscopic field of diffusive flux $\underline{j}^\gamma$ through an average rule :

$$\underline{J}^\gamma = \overline{\underline{j}^\gamma} = \varphi \overline{\underline{j}^\gamma}^f \tag{1.7}$$

[1] $\underline{n}$ is the unit outward normal to the fluid phase at the solid-fluid interface I^{fs}.

[2] The reasoning is the same as in the chapter "Poroelasticity and strength of fully or partially saturated porous materials" (section 3.1). The microscopic and macroscopic diffusive fluxes respectively play the role of the microscopic and macroscopic stresses.

We now look for the macroscopic counterpart of the microscopic Fick's law (1.4). The homogenized diffusive properties will be derived from the solution to a boundary value problem defined on the r.e.v. in the framework of uniform boundary conditions.

2 The Local Boundary Value Problem and the Homogenized Diffusion Tensor

The uniform boundary conditions on the density ρ^γ consist in writing the latter as a linear function of the local position vector $\underline{z}$:

$$(\forall\, \underline{z} \in \partial\Omega) \quad \rho^\gamma(\underline{z}) = \underline{H} \cdot \underline{z} \tag{2.1}$$

where $\underline{H}$ represents the density gradient at the macroscopic scale. It is recalled that the density of the γ-component in the solid has no physical meaning. It is a continuous extension of the real density defined in the fluid. Note that the average rule :

$$\overline{\underline{\operatorname{grad}}\, \rho^\gamma} = \underline{H} \tag{2.2}$$

is a direct consequence of (2.1). The boundary value problem on the r.e.v. then reads:

$$\begin{array}{lll} \rho^\gamma(\underline{z}) = \underline{H} \cdot \underline{z} & (\partial\Omega) & (a) \\ div_z\, \underline{j}^\gamma = 0 & (\Omega) & (b) \\ \underline{j}^\gamma(\underline{z}) \;=\; -D(\underline{z})\, \underline{\operatorname{grad}}_z\, \rho^\gamma & (\Omega) & (c) \end{array} \tag{2.3}$$

The linear dependence of the solution to (2.3) with respect to $\underline{H}$ can be expressed through the concept of *concentration tensor*. This second order tensor, denoted here by $\boldsymbol{A}(\underline{z})$, relates the macroscopic density gradient to the microscopic one :

$$\underline{\operatorname{grad}}_z\, \rho^\gamma = \boldsymbol{A}(\underline{z}).\underline{H} \tag{2.4}$$

Combining (2.2) with (2.4) yields :

$$\overline{\boldsymbol{A}} = (1-\varphi)\overline{\boldsymbol{A}}^s + \varphi\overline{\boldsymbol{A}}^f = \mathbf{1} \tag{2.5}$$

Combining (2.3c) and (2.4) together with the average rule (1.7) allows us to relate the macroscopic mass flux $\underline{J}^\gamma = \overline{\underline{j}^\gamma}$ to the macroscopic concentration gradient $\underline{H}$:

$$\underline{J}^\gamma = -\boldsymbol{D}^{hom} \cdot \underline{H} \tag{2.6}$$

with

$$\boldsymbol{D}^{hom} = \overline{D(\underline{z})\boldsymbol{A}(\underline{z})} \tag{2.7}$$

$\boldsymbol{D}^{hom}$ can be interpreted as the homogenized or efffective diffusion tensor. Using the fact that $D(\underline{z})$ is equal to 0 in the solid phase also yields :

$$\boldsymbol{D}^{hom} = D^\gamma \varphi \overline{\boldsymbol{A}}^f = D^\gamma \left(\mathbf{1} - (1-\varphi)\overline{\boldsymbol{A}}^s\right) \tag{2.8}$$

(2.6) represents the macroscopic counterpart of (1.4) (Fick's law). It provides a macroscopic description of the diffusion process in a porous medium.
The macroscopic diffusion tensor is classically put in the form $\boldsymbol{D}^{hom} = \varphi D^{\gamma} \boldsymbol{T}$ (Bear and Bachmat (1990)). As compared to the diffusion coefficient in an infinite fluid domain, it contains a correcting factor $\varphi \boldsymbol{T}$ taking the presence of the solid domain into account. The porosity φ accounts for the fact that diffusion is prevented in a fraction $1-\varphi$ of the total volume. The tortuosity tensor $\boldsymbol{T}$ represents the deviation of the effective diffusive solute path with respect to the idealized one. It aims in particular at capturing the effect of the geometry of the microstructure on the overall diffusive properties of the porous medium. From (2.8) one obtains :

$$\boldsymbol{T} = \overline{\boldsymbol{A}}^f = \frac{1}{\varphi}\left(1 - (1-\varphi)\overline{\boldsymbol{A}}^s\right) \tag{2.9}$$

(2.9) provides a micromechanical definition of the tortuosity tensor which implicitly involves the geometry of the microstructure through $\boldsymbol{A}(\underline{z})$. From (2.8)-(2.9), it appears that the determination of estimates for $\boldsymbol{D}^{hom}$ and $\boldsymbol{T}$ requires estimates for $\overline{\boldsymbol{A}}^f$ or $\overline{\boldsymbol{A}}^s$. This is the purpose of the next section.

3 First estimates of the effective diffusion tensor

For clarity, we begin with the academic case of an infinitesimal volume fraction of spherical solid inclusions embedded in the fluid phase. This classically yields the so-called dilute estimate of the effective diffusion tensor (section 3.1). A non dilute estimate is then derived by means of an iterative algorithm (section 3.2). The case of ellipsoidal inclusions of solid is considered at section 5.

3.1 the dilute estimate

When the volume fraction of the solid is infinitesimal, it is reasonable to neglect the interaction between the solid grains. In this case, the microscopic density in the solid phase and the corresponding average $\overline{\boldsymbol{A}}^s$ can be approximated by their values in the auxiliary problem of a single spherical solid grain $\mathcal{S}$ of radius $r = R$ embedded in an infinite homogeneous fluid continuum, with the condition $\rho^{\gamma} = \underline{H} \cdot \underline{z}$ at infinity (see figure 1). As before, the diffusion coefficient in the pure fluid is denoted by D^{γ}. In spherical coordinates with the origin $r = 0$ at the center of $\mathcal{S}$, the equations of this auxiliary problem are:

$$\begin{array}{lll} r > R & \operatorname{div}_z \underline{j}^{\gamma} = 0 & (a) \\ r > R & \underline{j}^{\gamma} = -D^{\gamma} \underline{\operatorname{grad}}_z \rho^{\gamma} & (b) \\ r = R & \underline{j}^{\gamma} \cdot \underline{n} = 0 & (c) \\ r \to \infty & \rho^{\gamma} \to \underline{H} \cdot \underline{z} & (d) \end{array} \tag{3.1}$$

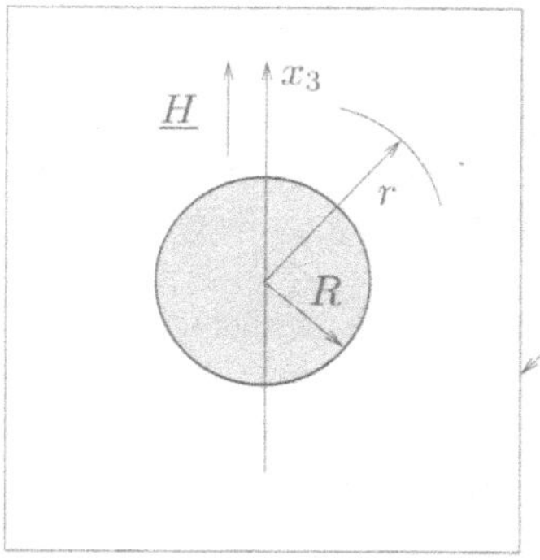

Figure 1. Spherical solid grain in an infinite fluid medium.

where R denotes the radius of the solid sphere. The average of the density gradient over the sphere is obtained from the value of ρ^γ on the boundary of the sphere:

$$\overline{\underline{\mathrm{grad}}\,\rho^\gamma}^s = \frac{1}{|\mathcal{S}|}\int_{\mathcal{S}} \underline{\mathrm{grad}}\,\rho^\gamma\, dV = \frac{1}{|\mathcal{S}|}\int_{\partial\mathcal{S}} \rho^\gamma \underline{n}\, dS \tag{3.2}$$

The value of the solution to (3.1) on $\partial\mathcal{S}$ thus yields an estimate of the isotropic tensor $\overline{\boldsymbol{A}}^s$:

$$\overline{\boldsymbol{A}}^s \cdot \underline{H} = \frac{1}{|\mathcal{S}|}\int_{\partial\mathcal{S}} \rho^\gamma \underline{n}\, dS \tag{3.3}$$

Consider for example that $\underline{H}$ is parallel to the unit vector $\underline{e}_3$: $\underline{H} = H\underline{e}_3$. The solution ρ^γ to (3.1) then reads :

$$\rho^\gamma = H z_3 \left(1 + \frac{R^3}{2r^3}\right) \tag{3.4}$$

and (3.2)-(3.3) show that:

$$\overline{\underline{\mathrm{grad}}\,\rho^\gamma}^s = \frac{3}{2}\underline{H} \Leftrightarrow \overline{\boldsymbol{A}}^s = \frac{3}{2}\mathbf{1} \tag{3.5}$$

Introducing this result into (2.8) yields the dilute estimate $\boldsymbol{D}_{dil}^{hom}$ of the macroscopic diffusion tensor in an isotropic form $D_{dil}^{hom}\mathbf{1}$, which is valid for $\varphi \approx 1$:

$$D_{dil}^{hom} = \left(1 - \frac{3}{2}(1-\varphi)\right) D^\gamma \tag{3.6}$$

The tortuosity tensor (see (2.9)) in the case of a dilute concentration of solid particles thus reads :

$$\boldsymbol{T}_{dil} = \frac{3\varphi - 1}{2\varphi}\mathbf{1} \tag{3.7}$$

3.2 The Differential Scheme

As underlined before, the dilute estimate (3.6) is valid only for an infinitesimal solid volume fraction, which of course does not correspond to usual porous media. It is therefore necessary to overcome this limitation. One way to do this is the differential scheme considered hereafter.
Starting from $\varphi = 1$ (pure fluid), the idea consists in introducing the solid phase by infinitesimal increments. After each increment, the solid-fluid mixture is homogenized in order to determine the estimate of $\boldsymbol{D}^{hom}$ corresponding to the current value of the porosity φ. During each step the volume fraction dx of homogenized material is removed and replaced by the same volume of solid. This process is associated with the variation $d\varphi = -\varphi dx$ of the porosity. The corresponding evolution of the effective diffusion coefficient can be determined in applying the dilute scheme to each infinitesimal step. More precisely, let D^j_{DS} denote the homogenized diffusion coefficient D^{hom} at the beginning of step $j+1$. At the end of this step, the homogenized diffusion coefficient is D^{j+1}_{DS}. The expression of the dilute estimate D^{hom}_{dil} given in (3.6) can be used for deriving the increment $dD_{DS} = D^{j+1}_{DS} - D^j_{DS}$ of the effective diffusion coefficient which is induced by the increment dx of the solid volume fraction. Indeed, we just have to replace D^γ and $1-\varphi$ in (3.6) by D^j_{DS} and dx respectively. This yields the following differential equation:

$$d\boldsymbol{D}_{DS} = -\frac{3dx}{2}\boldsymbol{D}_{DS} = \frac{3d\varphi}{2\varphi}\boldsymbol{D}_{DS} \tag{3.8}$$

The solution to (3.8) with $\boldsymbol{D}_{DS} = D^\gamma \mathbf{1}$ for $\varphi = 1$ is the estimate of $\boldsymbol{D}^{hom}$ given by the differential scheme and reads:

$$\boldsymbol{D}^{hom} = \varphi^{3/2} D^\gamma \mathbf{1} \tag{3.9}$$

4 Eshelby's approach for Linear Diffusion

4.1 Introduction

We now address the determination of the effective behavior in the light of a fundamental result of micromechanics known as Eshelby's problem : the latter goes back to J. Eshelby's classical paper (Eshelby (1957)). The interest of this new approach is that it is not restricted to spherical solid inclusions.
We again consider a porous material made up of a solid phase Ω^s and a fluid phase Ω^f (occupied by the fluid phase), through which a diffusive flux takes place. As previously, the concentration ρ^γ and the diffusive flux $\underline{j}^\gamma$ are extended to the solid phase, the diffusion coefficient D^s of the solid phase being equal to 0.
As before, the physics of the molecular diffusion problem is defined by the following set

of equations:

$$\begin{aligned}
&\operatorname{div} \underline{j}^{\gamma} = 0 && (a) \\
&\underline{j}^{\gamma} = -D(\underline{z})\underline{\operatorname{grad}}\,\rho^{\gamma} \quad \text{with} \begin{cases} D(\underline{z}) = D^{s} = 0 & \text{for} \quad \underline{z} \in \Omega^{s} \\ D(\underline{z}) = D^{\gamma} & \text{for} \quad \underline{z} \in \Omega^{f} \end{cases} && (b) \\
&\rho^{\gamma} = \underline{H} \cdot \underline{z} \text{ when } \underline{z} \in \partial\Omega && (c)
\end{aligned} \tag{4.1}$$

where $\underline{H}$ denotes the macroscopic concentration gradient, prescribed at the boundary $\partial\Omega$ of the r.e.v.. The set of equations (4.1) defines a problem in a bounded domain Ω. In view of determining estimates of $\overline{\underline{A}}^{s}$, it will prove useful to introduce an auxiliary problem [3] in which a bounded inhomogeneity I is embedded in an infinite homogeneous medium, the former representing a solid inclusion, the latter the fluid one. The boundary condition (4.1c) in the "real" problem is replaced by a condition written at infinity. The set of equations that define this so-called inhomogeneity problem thus reads:

$$\begin{aligned}
&\operatorname{div} \underline{j}^{\gamma} = 0 && (a) \\
&\underline{j}^{\gamma} = -D(\underline{z})\underline{\operatorname{grad}}\,\rho^{\gamma} \quad \text{with} \begin{cases} D(\underline{z}) = 0 & \text{for } \underline{z} \in I \\ D(\underline{z}) = D^{\gamma} & \text{for } \underline{z} \notin I \end{cases} && (b) \\
&\rho^{\gamma} \to \underline{H} \cdot \underline{z} \text{ when } |\, \underline{z} \,| \to \infty && (c)
\end{aligned} \tag{4.2}$$

We shall see at section 5 how to use the solution to (4.2) for the determination of the

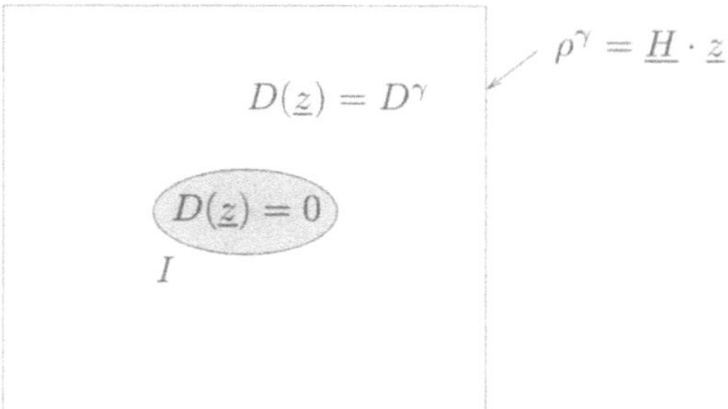

Figure 2. Inhomogeneity in an infinite fluid medium.

effective diffusion properties.
Introducing $\delta D = D^{s} - D^{\gamma} = -D^{\gamma}$, (4.2b) can then be rearranged in the form:

$$\begin{aligned}
&\operatorname{div} \underline{j}^{\gamma} = 0 && (a) \\
&\underline{j}^{\gamma} = -D^{\gamma}\underline{\operatorname{grad}}\,\rho^{\gamma} + \underline{j}^{p}(\underline{z}) \text{ with } \underline{j}^{p}(\underline{z}) = -\delta D\, \chi_{I}(\underline{z})\, \underline{\operatorname{grad}}\,\rho^{\gamma} && (b) \\
&\rho^{\gamma} \to \underline{H} \cdot \underline{z} \text{ when } |\, \underline{z} \,| \to \infty && (c)
\end{aligned} \tag{4.3}$$

[3] This kind of auxiliary problem and its solution are due to Eshelby

where χ_I denotes the characteristic function of domain I. By definition, $\underline{j}^p(\underline{z})$ is equal to 0 in the fluid phase. For solving (4.3), we shall see that it is convenient to introduce a second auxiliary problem in which the field $\underline{j}^p(\underline{z})$ is uniform over the solid domain I :

$$\begin{aligned} &\operatorname{div} \underline{j}^\gamma = 0 && (a) \\ &\underline{j}^\gamma = -D^\gamma \underline{\operatorname{grad}}\, \rho^\gamma + \underline{j}^I \chi_I(\underline{z}) && (b) \\ &\rho^\gamma \to \underline{H} \cdot \underline{z} \text{ when } \mid \underline{z} \mid \to \infty && (c) \end{aligned} \tag{4.4}$$

where $\underline{j}^I$ is a given constant vector. This set of equations defines the so-called inclusion problem. In the latter, the microscopic transport law is linear in the fluid and affine in the solid. As opposed to (4.2b), the diffusion coefficient is the same in both phases. As we shall see, it turns out that the solutions to the inhomogeneity and inclusion problems (4.2) and (4.4) are in fact identical when the domain I is an ellipsoid (section 4.4). This is what motivates the introduction of the inclusion problem.

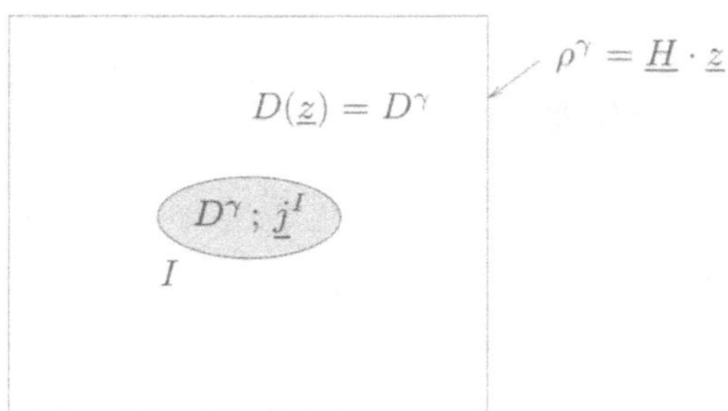

Figure 3. Eshelby's inclusion problem.

4.2 Eshelby's Inclusion Problem in the context of molecular diffusion

For simplicity, we first consider the case $\underline{H} = \underline{0}$. The combination of (4.4a) and (4.4b) yields:

$$-D^\gamma \Delta \rho^\gamma + \underline{j}^I \cdot \underline{\operatorname{grad}}\, \chi_I = 0 \tag{4.5}$$

$\underline{\operatorname{grad}}\, \chi_I$ deals with the differentiation of a discontinuous function, which calls for employing distribution theory. More precisely, according to the definition of the derivation of a distribution, one obtains :

$$< \underline{\operatorname{grad}}\, \chi_I, \psi > = - < \chi_I, \underline{\operatorname{grad}}\, \psi > = - \int_I \underline{\operatorname{grad}}\, \psi \, dV = - \int_{\partial I} \psi \underline{n} \, dS \tag{4.6}$$

where ψ is any function of $\mathcal{D}(\mathbb{R}^3)$ [4]. In (4.6), $\underline{n}$ denotes the outward unit normal to I. Let us introduce the Dirac distribution $\delta_{\partial I}$ associated with the boundary of I and

[4] $\mathcal{D}(\mathbb{R}^3)$ is the set of C^∞ functions that are equal to 0 out of a bounded domain

defined by :

$$< \delta_{\partial I}, \psi >= \int_{\partial I} \psi \, dS \tag{4.7}$$

From (4.6) and (4.7) it follows that:

$$\underline{\text{grad}}\, \chi_I = -\underline{n}\delta_{\partial I} \tag{4.8}$$

Introducing (4.8) in (4.5) yields:

$$-D^\gamma \Delta \rho^\gamma - \underline{j}^I \cdot \underline{n}\delta_{\partial I} = 0 \tag{4.9}$$

The solution to (4.9) goes back to the concept of Green function (Willis (1977), Mura (1987), Zaoui (2002)). The latter, denoted by $G(\underline{z})$, is defined as the elementary solution to the operator $D^\gamma \Delta$:

$$-D^\gamma \Delta_{\underline{z}} G + \delta_0 = 0 \tag{4.10}$$

Recalling that:

$$\Delta(\frac{1}{4\pi r}) + \delta_0 = 0 \quad \text{with} \quad r = \mid \underline{z} \mid \tag{4.11}$$

the solution of (4.10) reads:

$$G(\underline{z}) = -\frac{1}{4D^\gamma \pi r} \tag{4.12}$$

More generally, for any value of $\underline{z}'$, let $\delta_{\underline{z}'}$ denote the Dirac distribution at point $\underline{z}'$ defined by $< \delta_{\underline{z}'}, \psi >= \psi(\underline{z}')$ for any function $\psi \in \mathcal{D}(\mathbb{R}^3)$. The solution $\mathcal{G}(\underline{z}, \underline{z}')$ to

$$-D^\gamma \Delta_{\underline{z}} \mathcal{G} + \delta_{\underline{z}'} = 0 \tag{4.13}$$

reads:

$$\mathcal{G}(\underline{z}, \underline{z}') = G(\underline{z} - \underline{z}') \tag{4.14}$$

That of (4.9) is then obtained by superposition:

$$\rho^\gamma(\underline{z}) = -\int_{\partial I} G(\underline{z} - \underline{z}')\underline{j}^I \cdot \underline{n} \, dS_{z'} \tag{4.15}$$

where the integration in (4.15) is performed with respect to $\underline{z}'$. The above expression is first rearranged by means of an integration by parts

$$\rho^\gamma(\underline{z}) = -\int_I \frac{\partial}{\partial z'_j} (G(\underline{z} - \underline{z}'))\, j_j^I \, dV_{z'} \tag{4.16}$$

Taking advantage of the identity

$$\frac{\partial}{\partial z'_j} (G(\underline{z} - \underline{z}')) = -\frac{\partial}{\partial z_j} (G(\underline{z} - \underline{z}')) \tag{4.17}$$

one then obtains the density field in the form :

$$\rho^\gamma(\underline{z}) = \frac{\partial}{\partial z_j} \left(\int_I G(\underline{z} - \underline{z}') \, dV_{z'} \right) j_j^I \tag{4.18}$$

A differentiation with respect to $\underline{z}$ eventually yields the density gradient:

$$\frac{\partial \rho^\gamma}{\partial z_i}(\underline{z}) = \frac{\partial^2}{\partial z_i \partial z_j}\left(\int_I G(\underline{z}-\underline{z}')\,dV_{z'}\right) j_j^I \tag{4.19}$$

(4.19) can be put in the form:

$$\underline{\operatorname{grad}}\,\rho^\gamma = \boldsymbol{P}\cdot \underline{j}^I \tag{4.20}$$

where the components of the second order tensor $\boldsymbol{P}$ are given by

$$P_{ij}(\underline{z}) = \frac{\partial^2}{\partial z_i \partial z_j}\left(\int_I G(\underline{z}-\underline{z}')\,dV_{z'}\right) \tag{4.21}$$

(4.20) holds for $\underline{H} = \underline{0}$. In turn, the solution to (4.4) being linear with respect to $\underline{j}^I$ and $\underline{H}$, the density gradient in the general case ($\underline{H} \neq \underline{0}$) thus takes the form:

$$\underline{\operatorname{grad}}\,\rho^\gamma(\underline{z}) = \boldsymbol{P}(\underline{z})\cdot \underline{j}^I + \underline{H} \tag{4.22}$$

In summary, provided that $\underline{j}^I$ is constant, the determination of the solution to Eshelby's inclusion problem (4.4) reduces to that of the $\boldsymbol{P}$ tensor defined by (4.21).

4.3 Determination of the $\boldsymbol{P}$ Tensor

The form (4.12) of the Green function (respectively its more general expression (4.14)) motivates the introduction of the so-called potential $\Phi(\underline{z})$ of the domain I, which only depends on the geometry of I :

$$\Phi(\underline{z}) = \int_I \frac{1}{\mid \underline{z}-\underline{z}' \mid}\,dV_{z'} \tag{4.23}$$

Using the potential $\Phi(\underline{z})$, relations (4.12) and (4.21) immediately lead to:

$$P_{ij}(\underline{z}) = -\frac{1}{4\pi D^\gamma}\frac{\partial^2 \Phi}{\partial z_i \partial z_j}(\underline{z}) \tag{4.24}$$

In order to obtain the expression of the $\boldsymbol{P}$ tensor, the geometry of the inclusion I should now be specified. For illustrative purpose, we again consider the case where I is a sphere of radius R, centered at the origin. In this case, the restriction of $\Phi(\underline{z})$ to I reads:

$$\Phi(\underline{z}) = \frac{2\pi R^2}{3}\left(3 - \frac{r^2}{R^2}\right) \quad \text{with} \quad r = \mid \underline{z} \mid \leq R \tag{4.25}$$

Hence, in the case of a spherical particle, the potential $\Phi(\underline{z})$ turns out to be a quadratic function of the coordinate $r = |\underline{z}|$. Therefore, according to (4.24), the components P_{ij} of tensor $\boldsymbol{P}(\underline{z})$ for $\underline{z} \in I$ do not depend on $\underline{z}$:

$$(\forall \underline{z} \in I) \quad \boldsymbol{P} = \frac{1}{3D^\gamma}\mathbf{1} \tag{4.26}$$

More generally, this property of $\boldsymbol{P}$, observed for the spherical inclusion, actually holds for any inclusion I of ellipsoidal shape. Indeed, for $\underline{z} \in I$, the potential $\Phi(\underline{z})$ of any ellipsoid proves to be a quadratic polynomial function of the coordinates z_i. From (4.24), the $\boldsymbol{P}$ tensor is therefore uniform throughout the domain I. In turn, for $\underline{z} \notin I$, the $\boldsymbol{P}$ tensor depends on the position vector : $\boldsymbol{P} = \boldsymbol{P}(\underline{z})$.

4.4 Eshelby's Inhomogeneity Problem in the context of molecular diffusion

We now return to the inhomogeneity problem in the form (4.3) :

$$\begin{aligned} &\operatorname{div} \underline{j}^\gamma = 0 && (a) \\ &\underline{j}^\gamma = -D^\gamma \underline{\operatorname{grad}}\, \rho^\gamma - \delta D \chi_I(\underline{z})\, \underline{\operatorname{grad}}\, \rho^\gamma && (b) \\ &\rho^\gamma \rightarrow \underline{H} \cdot \underline{z} \text{ when } \mid \underline{z} \mid \rightarrow \infty && (c) \end{aligned} \tag{4.27}$$

Let us investigate the existence of a solution to (4.27) in which the density gradient is uniform in the domain I :

$$(\forall \underline{z} \in I) \quad \underline{\operatorname{grad}}\, \rho^\gamma(\underline{z}) = \underline{G}^I \tag{4.28}$$

With this assumption, the solution to (4.27) is given by that to the inclusion problem (4.4), provided that $\underline{j}^I$ is related to $\underline{G}^I$ by :

$$\underline{j}^I = -\delta D\, \underline{G}^I \tag{4.29}$$

In turn, $\underline{G}^I$ is related to $\underline{j}^I$ and $\underline{H}$ by (4.22) :

$$(\forall \underline{z} \in I) \quad \underline{G}^I = \boldsymbol{P}(\underline{z}) \cdot \underline{j}^I + \underline{H} \tag{4.30}$$

Eliminating $\underline{j}^I$ between (4.29) and (4.30) yields the local density gradient in I as a function of the macroscopic density gradient $\underline{H}$:

$$(\forall \underline{z} \in I) \quad \underline{G}^I = (\mathbf{1} + \delta D \boldsymbol{P}(\underline{z}))^{-1} \cdot \underline{H} \tag{4.31}$$

Clearly enough, (4.31) shows that the validity of the assumption (4.28) demands that the $\boldsymbol{P}(\underline{z})$ tensor be uniform over the domain I. As seen in section 4.3, this proves to be true when the domain I is an ellipsoid. We henceforth consider this particular geometry for which (4.31) becomes :

$$(\forall \underline{z} \in I) \quad \underline{\operatorname{grad}}\, \rho^\gamma(\underline{z}) = (\mathbf{1} + \delta D \boldsymbol{P}_I)^{-1} \cdot \underline{H} \tag{4.32}$$

where $\boldsymbol{P}_I$ denotes the uniform value of $\boldsymbol{P}(\underline{z})$ in the ellipsoidal domain I.
Recalling that $\delta D = D^s - D^\gamma$, it is convenient to introduce the so-called Eshelby tensor $\boldsymbol{S}_I$, defined as :

$$\boldsymbol{S}_I = D^\gamma \boldsymbol{P}_I \tag{4.33}$$

With this notation, (4.32) also reads :

$$(\forall \underline{z} \in I) \quad \underline{\operatorname{grad}}\, \rho^\gamma(\underline{z}) = (\mathbf{1} - \boldsymbol{S}_I)^{-1} \cdot \underline{H} \tag{4.34}$$

By way of application, consider again a spherical inhomogeneity I. Using expression (4.26) of $\boldsymbol{P}_I$ in (4.32) immediately yields the concentration gradient in the spherical inhomogeneity:

$$\underline{\operatorname{grad}}\, \rho_I^\gamma = \frac{3}{2} \underline{H} \tag{4.35}$$

We first observe that, given the constant value of $\underline{\operatorname{grad}\rho_I^\gamma}$ in I, we retrieve the result (3.5) obtained from the direct solution (3.4) of a single spherical solid grain embedded in an infinite homogeneous fluid continuum.

5 Eshelby-based estimates of the effective diffusion tensor

This further motivates the use of the Eshelby's solution for estimating the effective diffusive properties of porous media in the case of any ellipsoidal grains (for example prolate or oblate spheroid). Indeed, (2.8) suggests that estimates of the homogenized diffusion tensor $\boldsymbol{D}^{hom}$ can be obtained from estimates of $\overline{\boldsymbol{A}}^s$ which represents the average concentration tensor over the solid phase of the real r.e.v. . As already stated, the dilute scheme presented at section 3 is based on the assumption that the interactions between the solid grains in the r.e.v. are negligible. In this context, the solution of Eshelby's inhomogeneity problem is of great interest. The idea consists in approximating the average concentration gradient in the solid phase by the uniform concentration gradient that takes place in an ellipsoidal solid grain embedded in an infinite fluid phase, with the condition $\rho^\gamma = \underline{H} \cdot \underline{z}$ at infinity. In fact, combining (3.3) with (4.32) yields the dilute estimate of the concentration tensor in an ellipsoidal inclusion phase:

$$\overline{\boldsymbol{A}}^s_{dil} = \boldsymbol{A}^I = \left(\mathbf{1} - D^\gamma \boldsymbol{P}_I\right)^{-1} \tag{5.1}$$

where $\boldsymbol{P}_I$ is given by (4.24). Use of (5.1) in (2.8) provides a dilute estimate of the homogenized diffusion tensor:

$$\boldsymbol{D}^{hom}_{dil} = D^\gamma \left(\mathbf{1} - \varphi \left(\mathbf{1} - D^\gamma \boldsymbol{P}_I\right)^{-1}\right) \tag{5.2}$$

In order to improve the dilute estimate, the interaction between the solid particles must be taken into account. The so-called differential scheme presented in Section 3.2 is one way to address this issue. Depending on the morphology of the microstructure, the Mori-Tanaka and the self-consistent schemes provide two alternative approaches[5].

5.1 the Mori-Tanaka scheme

We begin with the Mori-Tanaka estimate. The latter is classically devoted to the matrix-inclusion morphology, in which one phase (the inclusions) is embedded in the other one (the matrix), the latter being connected. As far as transport properties are concerned, this description seems a reasonable way for expressing the connectedness of the fluid phase, the latter playing the role of the matrix and the solid, that of the inclusions.

As before, we aim at estimating the average density gradient in the solid phase from the solution to Eshelby's inhomogeneity problem. Still, according to the very definition, the single ellipsoidal particle I in this problem does not "see" the other particles of

[5]For a comprehensive presentation of the available homogenization schemes in relation to the microstructural morphology, see Zaoui (1997).

the solid phase. In order to capture the interaction between particles, the Mori-Tanaka scheme proposes to modify the boundary condition (4.27c) at infinity. The idea is to introduce a fictitious macroscopic density gradient $\underline{H}_o$ and to replace the condition $\rho^\gamma = \underline{H} \cdot \underline{z}$ by $\rho^\gamma = \underline{H}_o \cdot \underline{z}$. The value of $\underline{H}_o$ will have to meet the micro-macro compatibility condition, namely $\overline{\underline{grad}\,\rho^\gamma} = \underline{H}$:

$$\begin{array}{lr} \operatorname{div} \underline{j}^\gamma = 0 & (a) \\ \underline{j}^\gamma = -D(\underline{z})\underline{\operatorname{grad}}\,\rho^\gamma \quad \text{with} \left\{ \begin{array}{ll} D(\underline{z}) = 0 & \text{for } \underline{z} \in I \\ D(\underline{z}) = D^\gamma & \text{for } \underline{z} \notin I \end{array} \right. & (b) \\ \rho^\gamma \rightarrow \underline{H}_o \cdot \underline{z} \text{ when } \mid \underline{z} \mid \rightarrow \infty & (c) \end{array} \tag{5.3}$$

As in the dilute scheme, the uniform concentration gradient in the ellipsoid I is adopted as an estimate for the average density gradient in the solid phase of the real r.e.v. :

$$\overline{\underline{\operatorname{grad}}\,\rho^\gamma}^s = (\mathbf{1} + \delta D \boldsymbol{P}_I)^{-1} \cdot \underline{H}_o \tag{5.4}$$

Simultaneously, the method assumes that the average concentration gradient in the fluid can be estimated by the average concentration gradient in the (infinite) homogeneous fluid surrounding I in the inhomogeneity problem, which is equal to $\underline{H}_o$ since I is a bounded domain :

$$\overline{\underline{\operatorname{grad}}\,\rho^\gamma}^f = \underline{H}_o \tag{5.5}$$

This provides a physical interpretation to $\underline{H}_o$. Combining (5.4) and (5.5) with the average condition $\overline{\underline{\operatorname{grad}}\,\rho^\gamma} = \underline{H}$ yields :

$$\underline{H}_o = \left(\varphi \mathbf{1} + (1-\varphi)(\mathbf{1} + \delta D \boldsymbol{P}_I)^{-1}\right)^{-1} . \underline{H} \tag{5.6}$$

Introducing (5.6) into (5.4) leads to the following estimate of the average concentration tensor in the solid phase :

$$\overline{\boldsymbol{A}}^s = (\mathbf{1} + \delta D \boldsymbol{P}_I)^{-1} . \left(\varphi \mathbf{1} + (1-\varphi)\,(\mathbf{1} + \delta D \boldsymbol{P}_I)^{-1}\right)^{-1} \tag{5.7}$$

or, using $\delta D = -D^\gamma$:

$$\overline{\boldsymbol{A}}^s = (\mathbf{1} - \varphi \boldsymbol{S}_I)^{-1} \tag{5.8}$$

Finally, introducing (5.7) into (2.8) yields the following expression of the homogenized diffusion tensor:

$$\boldsymbol{D}^{hom} = D^\gamma \left(\mathbf{1} - (1-\varphi)\,(\mathbf{1} - \varphi \boldsymbol{S}_I)^{-1}\right) \tag{5.9}$$

The expression of the tortuosity tensor $\boldsymbol{T}$ that accounts for particle interactions is obtained from the definition $\boldsymbol{T} = \overline{\boldsymbol{A}}^f$ (see (2.9)):

$$\boldsymbol{T} = \frac{1}{\varphi}\left(\mathbf{1} - (1-\varphi)\,(\mathbf{1} - \varphi \boldsymbol{S}_I)^{-1}\right) \tag{5.10}$$

For illustrative purposes, consider an isotropic distribution of spherical solid particles embedded in the fluid phase. Use of (4.26) and (4.33) in (5.9) yields:

$$D^{hom} = \frac{2\varphi}{3-\varphi} D^{\gamma} \tag{5.11}$$

5.2 the self-consistent scheme

We now move to the self-consistent scheme. The latter deals with the so-called perfectly disordered microstructure. This means that the various phases[6] are dispersed in the r.e.v. in such a way that none plays any specific role from a morphological point of view. It is therefore appealing to assume that each phase reacts as if it were surrounded by the equivalent homogeneous medium which is looked for. In view of estimating the average density gradient in the solid (resp. fluid) phase, this suggests to introduce an inhomogeneity problem in which a bounded inclusion of solid (resp. fluid) is embedded in an infinite medium made up of the equivalent homogeneous material. We henceforth restrict to the isotropic situation. In particular, the self-consistent estimate of the effective diffusion tensor is an isotropic tensor : $\boldsymbol{D}_{sc} = D_{sc}\mathbf{1}$ ('sc' standing for self-consistent). The inhomogenity problem thus reads :

$$\begin{array}{ll} \operatorname{div} \underline{j}^{\gamma} = 0 & (a) \\ \underline{j}^{\gamma} = -D(\underline{z})\underline{\operatorname{grad}}\,\rho^{\gamma} \quad \text{with} \left\{ \begin{array}{ll} D(\underline{z}) = D^{\alpha} & \text{for } \underline{z} \in I \\ D(\underline{z}) = D_{sc} & \text{for } \underline{z} \notin I \end{array} \right. & (b) \\ \rho^{\gamma} \to \underline{H}_o \cdot \underline{z} \text{ when } |\, \underline{z} \,| \to \infty & (c) \end{array} \tag{5.12}$$

with $D^{\alpha} = D^{s}$ or $D^{\alpha} = D^{\gamma}$ according to the nature of the inclusion I. The essential difference with the inhomogeneity problem (5.3) introduced in the framework of the Mori-Tanaka scheme lies in the fact that the diffusion coefficient of the infinite medium surrounding the inclusion is not equal to D^{γ} and is actually the unknown of the homogenization process. Accordingly, the $\boldsymbol{P}$ tensor involved in the solution to (5.12) is still defined as in section 4.2, except for the fact that D^{γ} is replaced by D_{sc}. As such, it depends on D_{sc}. We still have to define the geometry of the solid (resp. fluid) inclusion I to be considered in (5.12). The assumed isotropy of the microstructure suggests to select the spherical shape for both the fluid and the solid. The $\boldsymbol{P}$ tensor inside such a spherical inclusion thus reads (see (4.26)) :

$$\boldsymbol{P}_{sc} = \frac{1}{3D_{sc}}\mathbf{1} \tag{5.13}$$

As regards the solid phase, (5.4) is still valid with $\delta D = D^{s} - D_{sc} = -D_{sc}$:

$$\overline{\underline{\operatorname{grad}}\,\rho^{\gamma}}^{s} = (\mathbf{1} + \delta D^{s}_{sc}\boldsymbol{P}_{sc})^{-1} \cdot \underline{H}_o \quad \text{with} \quad \delta D^{s}_{sc} = D^{s} - D_{sc} \tag{5.14}$$

Similarly, the average density gradient $\overline{\underline{\operatorname{grad}}\,\rho^{\gamma}}^{f}$ in the fluid is given by

$$\overline{\underline{\operatorname{grad}}\,\rho^{\gamma}}^{f} = \left(\mathbf{1} + \delta D^{f}_{sc}\boldsymbol{P}_{sc}\right)^{-1} \cdot \underline{H}_o \quad \text{with} \quad \delta D^{f}_{sc} = D^{\gamma} - D_{sc} \tag{5.15}$$

[6] namely the fluid and the solid in the present case

The average rule $\overline{\underline{\text{grad}}\,\rho^\gamma} = \underline{H}$ therefore yields[7] :

$$\underline{H}_o = \overline{(\mathbf{1} + \delta D^\alpha_{sc} \boldsymbol{P}_{sc})^{-1}}^{-1} \cdot \underline{H} \tag{5.16}$$

(with $\alpha = s$ or f). Introducing (5.16) into (5.14) and (5.15) provides an estimate of the average $\overline{\boldsymbol{A}}^\alpha$ ($\alpha = s$ or f) :

$$\overline{\boldsymbol{A}}^\alpha = (\mathbf{1} + \delta D^\alpha_{sc} \boldsymbol{P}_{sc})^{-1} \cdot \overline{(\mathbf{1} + \delta D^\alpha_{sc} \boldsymbol{P}_{sc})^{-1}}^{-1} \tag{5.17}$$

The last step consists in introducing (5.17) into the general expression of $\boldsymbol{D}^{hom}$ given in (2.7) :

$$D_{sc}\mathbf{1} = \overline{D^\alpha (\mathbf{1} + \delta D^\alpha_{sc} \boldsymbol{P}_{sc})^{-1}} \cdot \overline{(\mathbf{1} + \delta D^\alpha_{sc} \boldsymbol{P}_{sc})^{-1}}^{-1} \tag{5.18}$$

Taking advantage of the fact that the $\boldsymbol{P}$ tensor in the above equation does not depend on the phase (solid or fluid), it can be seen from (5.18) that $\overline{(\mathbf{1} + \delta D^\alpha_{sc} \boldsymbol{P}_{sc})^{-1}} = \mathbf{1}$, so that this equation reduces to :

$$D_{sc}\mathbf{1} = \overline{D^\alpha (\mathbf{1} + \delta D^\alpha_{sc} \boldsymbol{P}_{sc})^{-1}} \tag{5.19}$$

Introducing (5.13) into (5.19) yields the final expression of the self-consistent estimate

$$D_{sc} = D^\gamma \, \frac{3\varphi - 1}{2} \tag{5.20}$$

It should be observed that the above estimate is identical to the dilute one given in (3.6). This is in fact fortuitous, since the dilute estimate has no validity out of the domain $\varphi \approx 1$, whereas the self-consistent one aims at covering the whole interval $0 \leq \varphi \leq 1$. The negative values of D_{sc} obtained for $\varphi < 1/3$ are of course physically meaningless. Nevertheless, the value $\varphi = 1/3$ may be interpreted as a threshold above which the connectedness of the pore space is achieved, so that diffusion becomes possible at the macroscopic scale.

Figure 4 shows a comparison of the estimates derived from the Mori-Tanaka, differential and self-consistent schemes.

As expected, all these estimates coincide for infinitesimal solid volume fraction ($\varphi \approx 1$) with the dilute estimate (see (3.7)). The latter can be interpreted as an expansion to the first order with respect to the solid volume fraction $1 - \varphi$ in the vicinity of $\varphi = 1$ of the effective diffusion coefficient.

The methodology presented above of course in the isotropic case can be applied in anisotropic situations as well. In particular, some results obtained in the case of transverse anisotropy are presented in Dormieux and Lemarchand (2000).

We now would like to apply the general results concerning the determination of the effective diffusion properties of disordered porous media to the case of cracked porous materials. We assume that it is possible to define a representative elementary volume made up of the uncracked porous phase and of a set of opened cracks. Section 6 is devoted to the effective diffusive properties of such a material. In section 7, it will be shown

[7] In this 2-phase case, it is recalled that the notation $\overline{a^\alpha}$ means $\varphi a^f + (1 - \varphi) a^s$

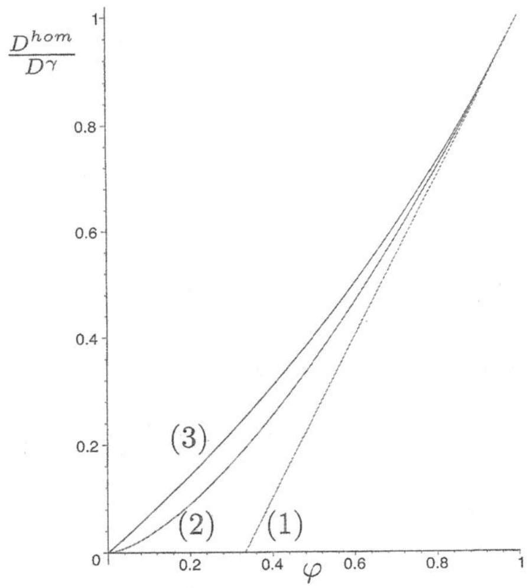

Figure 4. 3 different estimates of the effective diffusion coefficient. (1) : self-consistent scheme; (2) : differential scheme; (3) : Mori-Tanaka scheme.

that the determination of the effective permeability of the cracked porous material can be achieved with the same mathematical tools.

6 Diffusive properties of cracked porous media

For the cracks to influence the effective diffusive properties, it is necessary that their opening be large with respect to the pore size. This means that a scale separation exists between the pores of the uncracked material and the cracks. It is therefore relevant to replace the uncracked phase by an equivalent homogeneous medium and to model the cracks as fluid domains. In other words, the appropriate diffusion law in the cracks is the one which is used in an infinite fluid :

$$(\mathcal{C}_j) : \quad \underline{j}^\gamma = -D^\gamma \underline{\text{grad}}\,\rho^\gamma \tag{6.1}$$

where $\mathcal{C}_j$ refers to crack n°j. Furthermore, a homogenized diffusion tensor of the type derived in (2.6) may be used for describing the diffusive properties at the microscopic scale, within the uncracked domain Ω_u. It involves the pore volume fraction φ_u in the uncracked porous material and the tortuosity tensor $\boldsymbol{T}_u$:

$$\Omega_u : \quad \underline{j}^\gamma = -\boldsymbol{D}_u \cdot \underline{\text{grad}}\,\rho^\gamma \quad \text{with} \quad \boldsymbol{D}_u = D^\gamma \varphi_u \boldsymbol{T}_u \tag{6.2}$$

This means that the microscopic level used in the present model allows to separate the fluid in the cracks and the homogenized uncracked medium. But it does not consider the microstructure of the uncracked material.
In the sequel, $\boldsymbol{D}_u$ is assumed to be isotropic : $\boldsymbol{D}_u = D_u \mathbf{1}$. We note that the contrast

between the diffusive properties in the cracks and in the uncracked medium respectively can be measured by the ratio D_u/D^γ. Using for instance the estimate of D_u derived in (3.9), we have :

$$\frac{D_u}{D^\gamma} = {\varphi_u}^{3/2} \tag{6.3}$$

In short, the r.e.v. of the cracked porous material appears as a heterogeneous medium as regards the diffusive properties :

$$\underline{j}^\gamma = -D(\underline{z})\,\underline{\text{grad}}\,\rho^\gamma \quad \text{with} \quad D(\underline{z}) = \begin{cases} D^\gamma & (\underline{z} \in \mathcal{C}_j) \\ D_u & (\underline{z} \in \Omega_u) \end{cases} \tag{6.4}$$

6.1 micro-to-macro approach to the macroscopic diffusion tensor

The boundary value problem on the r.e.v. is strictly identical to (2.3), the diffusion coefficient $D(\underline{z})$ being defined according to (6.4).

$$\begin{array}{lll} div\,\underline{j}^\gamma = 0 & (\Omega) & (a) \\ \underline{j}^\gamma = -D(\underline{z})\,\underline{\text{grad}}_z\,\rho^\gamma & (\Omega) & (b) \\ \rho^\gamma = \underline{H}\cdot\underline{z} & (\partial\Omega) & (c) \end{array} \tag{6.5}$$

(2.7) then reads :

$$\boldsymbol{D}^{hom} = \overline{\boldsymbol{D}\,\boldsymbol{A}} = f_u D_u \overline{\boldsymbol{A}}^u + D^\gamma \sum_j f_j \overline{\boldsymbol{A}}^j \tag{6.6}$$

where f_j denotes the volume fraction of crack $\mathcal{C}_j$ and f_u that of the uncracked material $(f_u + \sum_j f_j = 1)$.
The different homogenization schemes aim at providing estimates of the averages $\overline{\boldsymbol{A}}^u$ and $\overline{\boldsymbol{A}}^j$ of the concentration tensor $\boldsymbol{A}$ over the uncracked porous domain and the cracks, respectively. As discussed in section 5, the choice of a scheme depends on the morphology that is considered. A first micromechanical model consists in describing the uncracked porous domain as a matrix in which the cracks are embedded. By definition, each crack is surrounded by the uncracked porous matrix, so that the connexion between two cracks is always indirect. With such a model, it can be shown that the presence of cracks enhances the effective diffusive properties, but does not modify their order of magnitude (Dormieux and Lemarchand (2001)). In fact, it is reasonable to expect that a stronger increase of the effective diffusive properties may be induced by the formation of a network of connected cracks. As opposed to the matrix-inclusion scheme, a direct connexion between cracks is possible in the case of a disordered morphology. It is therefore appealing to select the estimate of the diffusion tensor that is provided by the self-consistent scheme.
In order to apply the analytical estimates based on the solution to the Eshelby problem, the cracks are modelled as flat ellipsoids. Each inclusion is given the orientation and the characteristic dimensions of the corresponding crack (half opening c, radius a) with axisymmetry with respect to the normal to the crack plane. The aspect ratio $X = c/a$ for flat ellipsoids meets the condition $X \ll 1$. Assuming an isotropic repartition of the

crack orientations, we look for isotropic macroscopic properties. As in (5.18), the self-consistent estimate $\boldsymbol{D}_{sc} = D_{sc}\mathbf{1}$ of the macroscopic diffusion tensor is then defined by :

$$\boldsymbol{D}_{sc} = \overline{D(\mathbf{1} + (D - D_{sc})\boldsymbol{P}_{sc})^{-1}} \cdot \overline{\mathbf{1} + (D - D_{sc})\boldsymbol{P}_{sc})^{-1}}^{-1} \tag{6.7}$$

in which the value of $D(\underline{z})$ is that given in (6.4). Note that each crack orientation and aspect ratio define a specific phase as regards the homogenization process. Indeed, as opposed to the example of section 5.2, the tensor $\boldsymbol{P}_{sc}(\underline{z})$ in (6.7) takes different values according to the considered phase. It can be derived from (4.24) in which D^γ must be replaced by D_{sc}. Introducing the Eshelby tensor $\boldsymbol{S}_{sc} = D_{sc}\boldsymbol{P}_{sc}$, one obtains :

$$\begin{cases} z \in \mathcal{C}_j & \boldsymbol{S}_{sc} = \dfrac{1}{4}\pi X^j \mathbf{1} + (1 - \dfrac{3}{4}\pi X^j)\underline{n}^j \otimes \underline{n}^j \\ z \in \Omega_u & \boldsymbol{S}_{sc} = \dfrac{1}{3}\mathbf{1} \end{cases} \tag{6.8}$$

The above expression of the Eshelby tensor $\boldsymbol{S}_{sc}$ in $\mathcal{C}^j$ is a first order expansion with respect to X^j. Interestingly, the information concerning the orientation of the cracks is captured in (6.7) through the fact that $\boldsymbol{S}_{sc}$ in $\mathcal{C}_j$ depends on the normal $\underline{n}^j$. This is the way a possible anisotropy of the crack orientations could be taken into account in the homogenized diffusion tensor $\boldsymbol{D}_{sc}$.

6.2 Results of the self-consistent scheme

Assuming identical aspect ratios, crack radii and opening for all cracks, an analytical solution to (6.7), denoted by D^o_{sc}, is available in the particular case $D_u = 0$. This corresponds to the situation where the uncracked material has no connected porous space. Let $\mathcal{N}$ denote the crack density (number of cracks per unit volume). As expected from a dimensional analysis, the non dimensional ratio D^o_{sc}/D^γ is a function of two geometrical characteristics, namely the so-called crack density parameter $\epsilon = \mathcal{N}a^3$ and the aspect ratio $X = c/a$:

$$D^o_{sc} = \frac{\pi\,(16\,\epsilon + 27)\,(64\,\epsilon - 27)\,X}{27(64\,\epsilon + 108)} D^\gamma \tag{6.9}$$

The above estimate is actually the first order expansion with respect to the aspect ratio X.

According to (6.9), D^o_{sc} remains equal to 0 until the critical threshold $\epsilon^*_d = 27/64$. Above the latter ($\epsilon > \epsilon^*_d$), the existence of macroscopic diffusive properties appears as the counterpart of a connected crack network. The self-consistent scheme thus proves to be able to take into account the concept of percolation. D^o_{sc} can be interpreted as the effective diffusion coefficient of the crack network itself.

When $D_u \neq 0$, the order of magnitude of the variations of D_{sc} as a function of ϵ strongly depends on the ratio $D_u/D^\gamma = \varphi_u T_u$. As could be expected from the physical meaning of D^o_{sc}, the influence of the cracks is negligible if $D_u \gg D^o_{sc}$, that is, if $\varphi_u T_u \gg X$. Conversely, the order of magnitude of the effective diffusion coefficient can be significantly increased if $\varphi_u T_u \ll X$. Referring to (6.3), this amounts to the condition $\varphi_u \ll X^{2/3}$. In this case, for $\epsilon > \epsilon^*_d$, it can be shown numerically that the effective diffusion D_{sc}

coefficient can be approximated by D_{sc}^{o} (apart from the vicinity of ϵ_d^*). In other words, diffusion dominantly takes place in the crack network.
However, remembering that the aspect ratio itself meets the condition $X \ll 1$, it may be concluded that cracks affect the effective diffusive properties only for extremely small values of the porosity φ_u.

7 Permeability of cracked porous media

Experimental evidences as well as numerical simulations prove that the existence of microcracks in a saturated porous medium can modify the order of magnitude of its permeability (Pouya and Courtois (2002)). In this section, this question is addressed in the light of homogenization. We particularly focus on the critical damage threshold observed experimentally beyond which the permeability increases dramatically. The coupling between permeability and the mechanical loading is also considered.
Since the contributions of Ene and Sanchez-Palencia (1975) and Auriault (1987) in the framework of periodic homogenization, it is well known that the permeability concept is the macroscopic counterpart of the flow of a newtonian viscous fluid at the pore scale. From a mathematical point of view, the permeability refers to the second order tensor that linearly relates the macroscopic filtration vector $\underline{Q}$ to the macroscopic pressure gradient $\underline{\nabla P}$ (Darcy's law) :

$$\underline{Q} = -\frac{\boldsymbol{K}}{\mu^f} \cdot \underline{\nabla P} \tag{7.1}$$

where μ^f denotes the viscosity of the pore fluid. $\boldsymbol{K}$ only depends on the morphology of the microstructure and can be determined, at least theoretically, as the solution to an elliptic variational problem defined on the pore space. Still, the coupling between permeability and damage is not a straightforward application of this general framework (Lévy (1990)).

7.1 The concept of equivalent permeability for modelling the fluid flow in the cracks

As regards the effect of cracks on the macroscopic permeability of a porous medium, two difficulties arise. First, from a quantitative point of view, the effect of a disordered crack network on the overall permeability cannot be taken into account easily in a periodic framework. Second, there is a fundamental difference between the mathematical description of the flow in the uncracked porous domain Ω_u of the r.e.v. on the one hand, and in the cracks on the other hand. More precisely, the traditional Darcy law relating the microscopic filtration vector $\underline{q}(\underline{z})$ and the microscopic pressure gradient $\underline{\text{grad}}_z\, p\,(\underline{z})$ can be used in Ω_u, together with the mass balance equation :

$$\Omega_u \;:\; \left\{ \begin{array}{l} \underline{q} = -\boldsymbol{k}_u \cdot \underline{\text{grad}}_z\, p \\ \text{div}_z\, \underline{q} = 0 \end{array} \right. \tag{7.2}$$

In the sequel, it is assumed that the permeability tensor $\boldsymbol{k}_u$ in Ω_u is isotropic : $\boldsymbol{k}_u = k_u \mathbf{1}$. In contrast, each crack $\mathcal{C}_j$ appears as a fluid domain (see section 6) in which the velocity

$\underline{v}(\underline{z})$ and the pressure $p(\underline{z})$ are subjected to Stokes equations :

$$\mathcal{C}_j \; : \quad \left\{ \begin{array}{l} \underline{\text{grad}}_z \, p = \mu^f \, \Delta_z \underline{v} \\ \text{div}_z \, \underline{v} = 0 \end{array} \right. \tag{7.3}$$

The description of the flow in Ω_u through Darcy law may be regarded as a first homogenization step in which the Stokes flow in the pore space of the uncracked porous domain is replaced by (7.2). This technique is valid if the order of magnitude of the crack opening is large enough with respect to the pore size. This is also clearly a necessary condition for the cracks to affect the overall permeability.

Still, when described by (7.2) and (7.3), the flow in the r.e.v. refers to two distinct physical concepts, namely the filtration vector $\underline{q}(\underline{z})$ in Ω_u and the velocity $\underline{v}(\underline{z})$ in each crack $\mathcal{C}_j$, both involved in strongly different mathematical problems. This renders the determination of the overall permeability uneasy.

In order to overcome these difficulties, the idea explored herefater is inspired from the modelling of viscous flows in jointed rocks (Guéguen and Palciauskas (1994)). It consists in regarding a crack as a couple of two parallel planes. The real flow in the crack is then approximated by a Poiseuille flow between these planes. In the latter, the pressure gradient is uniform and the velocity profile solution to (7.3) is parabolic (figure 5) :

$$\mathcal{C}_j \; : \quad \underline{v}(z) = -\frac{1}{2\mu^f}(c^2 - z^2)\underline{\text{grad}}_z \, p \tag{7.4}$$

where c denotes the half distance between the planes (resp. between the crack faces) . The average velocity $\overline{v}^j$ in the pore space $|z| < c$ therefore reads

$$\overline{v}^j = \frac{1}{2c}\int_{-c}^{+c} \underline{v}(z)\, dz = -\frac{c^2}{3\mu^f}\, \underline{\text{grad}}_z \, p \tag{7.5}$$

It is therefore possible to define a fictitious porous medium with permeability $\boldsymbol{k}_j$, that is equivalent to the real crack $\mathcal{C}_j$ as regards the average flow induced by a given pressure gradient. $\underline{n}$ denoting the unit normal to the plane crack, $\boldsymbol{k}_j$ reads :

$$\boldsymbol{k}_j = \frac{c^2}{3}\mathbf{1} + (k_{nn} - \frac{c^2}{3})\underline{n} \otimes \underline{n} \tag{7.6}$$

The crack is expected to enhance the overall permeability in the direction of the crack plane if $c^2 \gg k_u$. From a physical point of view, the influence of the crack on the macroscopic permeability $\boldsymbol{K}$ is due to the fact that it provides an additional flow path in the crack plane. In contrast, its influence on the flow in the direction of $\underline{n}$ is expected to be negligible. In fact, for flat cracks, it can be shown that the value of the normal permeability k_{nn} has no influence on the overall permeability provided that $k_{nn} \neq 0$ and that $k_{nn} \ll c^2/3$. We therefore let $k_{nn} = k_u$.

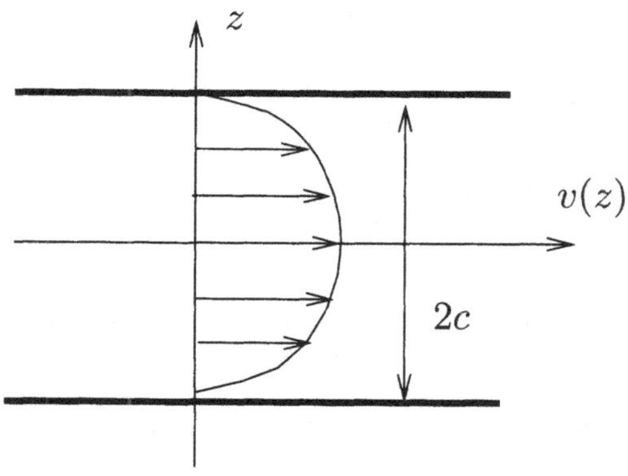

Figure 5. approximation of the flow in the crack by a parabolic velocity profile

7.2 micro-to-macro approach to the macroscopic permeability

From a mathematical point of view, substituting the fictitious porous medium with permeability $\boldsymbol{k}_j$ for the crack significantly simplifies the upscaling process. The homogenization of the Stokes flow within the cracks together with the Darcy flow in the uncracked porous medium is now replaced by the homogenization of a heterogeneous Darcy medium. In the latter, the permeability $\boldsymbol{k}(\underline{z})$ depends on the position vector $\underline{z}$ in the r.e.v. :

$$\begin{aligned} \underline{z} \in \Omega_u : \quad & \boldsymbol{k}(\underline{z}) = k_u \mathbf{1} \\ \underline{z} \in \mathcal{C}_j : \quad & \boldsymbol{k}(\underline{z}) = \boldsymbol{k}_j \end{aligned} \tag{7.7}$$

Note that $\boldsymbol{k}_j$ depends on both the crack opening through c and its orientation through $\underline{n}$. As opposed to the Stokes flow that is only defined in the fluid, the Darcy flow is defined on the whole r.e.v. by the microscopic field $\underline{q}(\underline{z})$ of filtration vectors. The same holds for the pressure field $p(\underline{z})$. By analogy to the uniform boundary conditions (2.1), we introduce the "uniform pressure gradient" boundary conditions defined in (7.8c). $\underline{q}(\underline{z})$ and $p(\underline{z})$ are solution to the following differential problem :

$$\begin{aligned} & div\, \underline{q} = 0 && (\Omega) \quad (a) \\ & \underline{q} = -\boldsymbol{k}(\underline{z}) \cdot \underline{\mathrm{grad}}_z\, p && (\Omega) \quad (b) \\ & p = \underline{\nabla P} \cdot \underline{z} && (\partial\Omega) \quad (c) \end{aligned} \tag{7.8}$$

Note that (7.8) is formally analogous to (6.5) : $\underline{j}^\gamma$ and ρ^γ are respectively replaced by $\underline{q}$ and p, whereas the solute mass balance condition and Fick's law are respectively replaced by the fluid mass balance condition (7.8a) and by the local Darcy's law (7.8b). The general ideas presented in section 5 can therefore be adapted to the present context. To begin with, the linearity of the solution $(p, \underline{q})$ to (7.8) with respect to $\underline{\nabla P}$ is expressed through the concept of concentration tensor : denoted by $\boldsymbol{A}(\underline{z})$, it relates the microscopic pressure gradient to the macroscopic one :

$$\underline{\text{grad}}_z\, p(\underline{z}) = \boldsymbol{A}(\underline{z}) \cdot \underline{\nabla P} \tag{7.9}$$

By analogy with the average rule $\underline{J}^\gamma = \overline{\underline{j}^\gamma}$, the macroscopic filtration vector $\underline{Q}$ is defined as the average of the field $\underline{q}(\underline{z})$:

$$\underline{Q} = \overline{\underline{q}} \tag{7.10}$$

Combining (7.8b) and (7.9) with (7.1), one obtains a micromechanical interpretation of the macroscopic permeability :

$$\boldsymbol{K} = \overline{\boldsymbol{k} \cdot \boldsymbol{A}} = f_u k_u \overline{\boldsymbol{A}}^u + \sum_j f_j \boldsymbol{k}_j \cdot \overline{\boldsymbol{A}}^j \tag{7.11}$$

which constitutes the hydraulic counterpart to (6.6).
As for diffusion, the micromechanical model in which the cracks are embedded in the uncracked porous domain cannot explain that the cracks may increase the permeability by several orders of magnitude (Dormieux and Lemarchand (2001)). In fact, it is reasonable to expect that the sudden increase of the permeability beyond a critical damage threshold is due to the formation of a network of connected cracks. We therefore again resort to the self-consistent scheme.
Similarly to the effective diffusion coefficient (see (6.7)), the self-consistent estimate $\boldsymbol{k}_{sc}$ of the macroscopic permeability is then given by :

$$\boldsymbol{k}_{sc} = \overline{\boldsymbol{k} \cdot (\mathbf{1} + \boldsymbol{P}_{sc} \cdot (\boldsymbol{k} - \boldsymbol{k}_{sc}))^{-1}} \cdot \overline{(\mathbf{1} + \boldsymbol{P}_{sc} \cdot (\boldsymbol{k} - \boldsymbol{k}_{sc}))^{-1}}^{\,-1} \tag{7.12}$$

Still, as opposed to the local diffusion tensor in the cracks (see (6.1)), it is worth stressing the fact that the local permeability tensor $\boldsymbol{k}(\underline{z})$ is not isotropic in the cracks (see (7.6) and (7.7)). The tensor $\boldsymbol{P}_{sc}(\underline{z})$ in (7.12) is related to the Eshelby tensor $\boldsymbol{S}_{sc}(\underline{z})$ of (6.8) by $\boldsymbol{P}_{sc} = \boldsymbol{S}_{sc} \cdot \boldsymbol{k}_{sc}{}^{-1}$.

7.3 Results of the self-consistent scheme

(7.12) is applied with the assumption that the distribution of the crack orientations is isotropic. The macroscopic permeability tensor $\boldsymbol{k}_{sc}$ is therefore isotropic as well ($\boldsymbol{k}_{sc} = k_{sc}\mathbf{1}$). For simplicity, we again consider uniform values of the crack radius (a) and opening (c) and we introduce the crack permeability $k_c = c^2/3$ (see (7.6)). As before, $\mathcal{N}$ denotes the crack density (number of cracks per unit volume). As expected from a dimensional analysis, the non dimensional ratio k_{sc}/k_c derived from (7.12) is a function of the ratio k_u/k_c, the crack density parameter $\epsilon = \mathcal{N}a^3$ and the aspect ratio $X = c/a$.
We now focus on the situation where the permeability of the uncracked porous media is very weak as compared to that of the cracks. For a given value of X, figure 6 presents the variations of k_{sc}/k_c as a function of ϵ, for 4 different values of k_u/k_c.

In all cases, the macroscopic permeability k_{sc} is an increasing function of ϵ. In the limit case $k_u = 0$, k_{sc} remains equal to 0 until the critical threshold $\epsilon_p^* = 9/16$. Above the

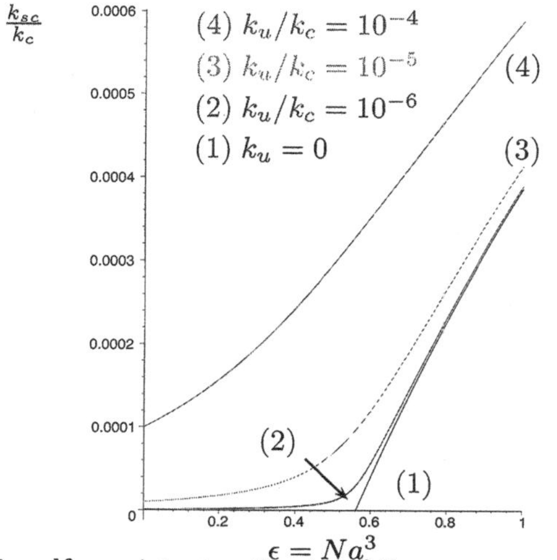

Figure 6. self-consistent estimate of the macroscopic permeability

latter ($\epsilon > \epsilon_p^*$), the existence of a macroscopic permeability (curve 1) can be interpreted as the hydraulic counterpart of a connected crack network. The self-consistent scheme thus proves to be able to take into account the concept of percolation threshold. An analytical expression of k_{sc} for $\epsilon > \epsilon_p^*$ is available :

$$k_u = 0 \Rightarrow k_{sc} = 3\pi k_c X \frac{16\epsilon - 9}{64\epsilon + 108} \tag{7.13}$$

When $k_u \neq 0$, the order of magnitude of the variations of k_{sc} as a function of ϵ strongly depends on the ratio k_u/k_c. For the higher values of the latter, k_{sc} is significantly affected by the value of ϵ but its order of magnitude is not modified (curve (4)). As expected, the coupling between damage and permeability is more pronounced for smaller values of k_u/k_c for which the macroscopic permeability gains one or several orders of magnitude when ϵ is increased beyond ϵ_p^* (curves (2) and (3)). For such values of k_u/k_c, the results of figure 6 sugests to neglect the variations of k_{sc} as a function of ϵ in the range $[0, \epsilon_p^*]$ and to approximate k_{sc} by the estimate corresponding to $k_u = 0$ for $\epsilon > \epsilon_p^*$:

$$\begin{aligned} \epsilon < \epsilon_p^* : &\quad k_{sc} = k_u \\ \epsilon > \epsilon_p^* : &\quad k_{sc} = k_u + 3\pi k_c X \frac{16\epsilon - 9}{64\epsilon + 108} \end{aligned} \tag{7.14}$$

Recalling that $k_c = c^2/3$, the above model yields $k_{sc} = \mathcal{O}(a^2X^3)$. This result reveals that the coupling between permeability and mechanical loading can be activated by two different phenomena, respectively related to the variations of a and X. On the one hand, a variation of the macroscopic stress state may lead to a reversible modification of the crack opening, that is, of the aspect ratio X. On the other hand, it may also induce crack propagation, that is, a variation of the crack radius a.

Bibliography

P.W. Atkins. *Physical chemistry.* Oxford University Press, 1992.

J.-L. Auriault. Comportement des milieux poreux saturés. In F. Darve, editor, *Manuel de rhéologie des géomatériaux.* Presses de l'Ecole Nationale des Ponts et Chaussées, 1987.

J.-L. Auriault and Y. Lewandowska. Diffusion/adsorption/advection macrotransport in soils. *Eur. J. Mech. A/Solids*, 15:681–704, 1996.

J. Bear and Y. Bachmat. *Introduction to the modelling of transport phenomena in porous media.* Kluwer Academic Publishers, 1990.

L. Dormieux and E. Lemarchand. Modélisation macroscopique du transport diffusif : apport des méthodes de changement d'échelle. *Oil Gas Sc. Tech.*, 55:15–34, 2000.

L. Dormieux and E. Lemarchand. An homogenization approach of advection and diffusion in cracked porous material. *J. Engrg. Mech.*, 127:1267–1274, 2001.

H. Ene and E. Sanchez-Palencia. Equations et phénomènes de surface pour l'écoulement dans un modèle de milieu poreux. *Journal de Mécanique*, 14:73–108, 1975.

J.D. Eshelby. The determination of the elastic field of an ellipsoidal inclusion, and related problems. *Proc. R. Soc. Lond. A*, 241:376–396, 1957.

Y. Guéguen and V. Palciauskas. *Introduction to Physics of Rocks.* Princeton University Press, 1994.

T. Lévy. Filtration in a porous fissured rock : influence of the fissures connexity. *Eur. J. Mech., B/Fluids*, 9:309–337, 1990.

T. Mura. *Micromechanics of defects in solids, 2nd edition.* Martinus Nijhoff Publ., 1987.

A. Pouya and A. Courtois. Définition de la perméabilité équivalente das massifs fracturés par des méthodes d'homogénéisation. *C. R. Acad. Sc. Paris Géoscience*, 334:975–979, 2002.

J.R. Willis. Bounds and self consistent estimates for the overall properties of anisotropic composites. *J. Mech. Phys. Solids*, 25:185–202, 1977.

A. Zaoui. In P. Suquet, editor, *Continuum micromechanics.* Springer, 1997.

A. Zaoui. Continuum micromechanics : survey. *J. Eng. Mech.*, 128:808–816, 2002.

Part II

Microporomechanics

Poroelasticity and strength of fully or partially saturated porous materials

Luc Dormieux

Ecole Nationale des Ponts et Chaussées, Champs-sur-Marne, France.

Abstract To begin with, a mathematical framework for upscaling and averaging operations is proposed. The classical strategies for the definition of the mechanical loading applied to the representative elementary volume in terms of boundary conditions are recalled. This framework is first applied to a saturated porous medium in which the solid phase is linear elastic. This yields Biot's poroelasticity theory together with estimates of the poroelastic coefficients and of the average strain level in the solid phase. The situation of a non linear elastic solid is then considered with the help of non linear homogenization techniques. A non linear theory of poroelasticity is derived as an application of the secant method which also allows to investigate the macroscopic strength. The existence of macroscopic effective stresses controlling the non linearity of the secant poroelastic coefficients and the strength criterion is discussed under various assumptions on the solid properties.To finish with, the capillary and surface tension effects are incorporated into a micromechanical approach to the poroelastic behavior and the strength of partially saturated porous media.

1 Representative Elementary Volume (r.e.v.)

Considering a three-dimensional macroscopic mechanical structure, let L and d respectively denote the characteristic length of the structure and that of the heterogeneities at the microstructural level. Roughly speaking, in the case of a porous medium, the characteristic length scale of the local heterogeneities is controlled typically by the pore size. Most of the time, L and d are separated by several orders of magnitude. This allows to introduce the concept of elementary volume on which Continuum Mechanics is based . By definition, the latter is an infinitesimal part of the three-dimensional material system under consideration, which characteristic length ℓ complies with the two following conditions :

$$\begin{aligned} &\ell \ll L \quad (a) \\ &\ell \gg d \quad (b) \end{aligned} \tag{1.1}$$

Condition $\ell \ll L$ guaranties that the tools of differential calculus offered by a continuous description can be used. In turn, $\ell \gg d$ aims at ensuring that the elementary volume is large enough for being representative of the constitutive material. This means that all the information concerning the geometrical and physical properties of the material are captured, in a statistical sense, in an elementary volume of size of the order of ℓ. We therefore refer to the concept of *representative elementary volume*, in short r.e.v. The conditions for an elementary volume to be representative have been discussed by many authors. The interested reader is referred to the works of Gilbert (1987), Bear and Bachmat (1990) and Zaoui (1996).

At the macroscopic scale, the r.e.v. is characterized by a position vector $\underline{x}$, the order of magnitude of the variation of $\underline{x}$ being equal to the size L of the macroscopic structure. Macroscopic poromechanics accounts for the multiphase nature of the porous material by considering the different phases present in the r.e.v. (solid, gas, liquids) as distinct macroscopic particles that are all located at the same macroscopic point $\underline{x}$. At this scale, the r.e.v. is regarded as the superposition of these particles in space.

By contrast, the micromechanical point of view consists in considering the r.e.v. as a heterogeneous structure in which the different physical phases are represented by separated domains. This requires a refinement of the length scale. At the microscopic scale, the position vector is denoted by $\underline{z}$. Depending on $\underline{z}$, the microscopic particle located at point $\underline{z}$ belongs to the solid phase or to a fluid phase. By definition, the characteristic order of magnitude of the variation of $\underline{z}$ is the size ℓ of the r.e.v.

The relevance of the r.e.v. concept is crucial for both micro and macro approaches. In a macroscopic approach, the concept of constitutive law relies upon the existence of an r.e.v. . In turn, in a micro-to-macro approach, the experiment that could be performed on a representative material sample in the laboratory is replaced by a thought experiment on the r.e.v. considered as a heterogeneous structure and subjected to appropriate boundary conditions.

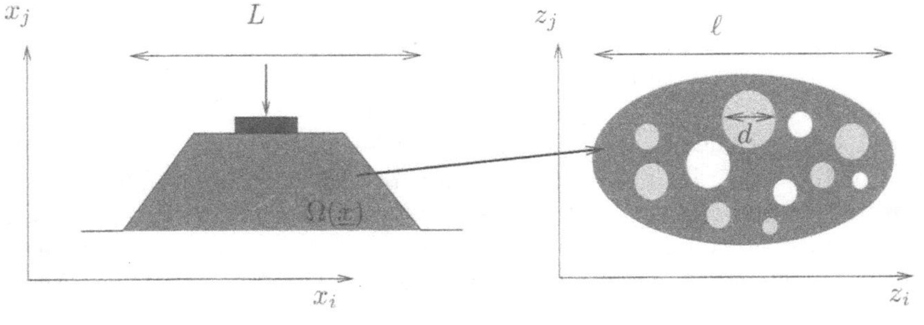

Figure 1. Elementary volume of porous medium in a macroscopic structure. Microscopic and macroscopic scales.

2 Averaging Operations

For extensive physical quantities such as mass, momentum or energy, it is natural to define the link between the microscopic and the macroscopic scales through averages defined on the r.e.v. . This is the focus of this section, which sets out a mathematical basis for forthcoming applications.

2.1 Apparent and Intrinsic Averages

Let $\Omega(\underline{0})$ be an r.e.v. centered at the origin O of the coordinate system. $\chi_o(\underline{z})$ denotes the characteristic function of $\Omega(\underline{0})$ at the microscopic scale :

$$\chi_o(\underline{z}) = \begin{cases} 0 & \text{if } \underline{z} \notin \Omega(\underline{0}) \\ 1 & \text{if } \underline{z} \in \Omega(\underline{0}) \end{cases} \tag{2.1}$$

We assume that the elementary volume $\Omega(\underline{x})$ obtained from $\Omega(\underline{0})$ by the translation $\underline{z} \to \underline{x} + \underline{z}$ is an r.e.v. for the material at the macroscopic point $\underline{x}$. The characteristic function of $\Omega(\underline{x})$ is $\chi_{\underline{x}}(\underline{z}) = \chi_o(\underline{z} - \underline{x})$. Note that $\Omega(\underline{0})$ and $\Omega(\underline{x})$ are purely geometrical domains which do not depend on time.

In the sequel, $\mathcal{P}^\alpha$ denotes the geometrical domain occupied by the α-phase (solid, liquid or gas, $\alpha = s,\ l,\ g$) and $\chi^\alpha(\underline{z}, t)$ its characteristic function.

$\Omega^\alpha(\underline{x}, t)$ (also simply denoted Ω^α) represents the domain occupied by the α-phase in the r.e.v. $\Omega(\underline{x})$, whereas Ω^p is the pore space contained in $\Omega(\underline{x})$. The characteristic function of $\Omega^\alpha(\underline{x}, t)$ reads $\chi^\alpha(\underline{z}, t)\chi_{\underline{x}}(\underline{z})$. As opposed to $\Omega(\underline{x})$, Ω^α a priori depends on time. So does its characteristic function. This time dependence may be due for instance to the deformation of the solid phase or to mass exchange at the fluid-solid interface.

The average $\overline{e}(\underline{x})$ at the macroscopic point $\underline{x}$ of a physical quantity represented in $\Omega(\underline{x})$ by the volume density $e(\underline{z})$ is defined by :

$$\overline{e}(\underline{x}, t) = \frac{1}{\mathcal{V}} \int_{\Omega(\underline{x})} e(\underline{z}, t)\, dV_z = \frac{1}{\mathcal{V}} \int \chi_o(\underline{z} - \underline{x}) e(\underline{z}, t)\, dV_z \tag{2.2}$$

where $\mathcal{V}$ is the volume of both $\Omega(\underline{0})$ and $\Omega(\underline{x})$:

$$\mathcal{V} = |\Omega(\underline{0})| = |\Omega(\underline{x})| \tag{2.3}$$

The porosity $\varphi(\underline{x}, t)$, usually defined as the volume fraction of the pore space in the r.e.v., appears as the average of the characteristic function $\chi^p(\underline{z}, t)$ of the pore space :

$$\varphi(\underline{x}, t) = \overline{\chi^p}(\underline{x}, t) = \frac{1}{\mathcal{V}} \int \chi^p(\underline{z}, t) \chi_o(\underline{z} - \underline{x})\, dV_z \tag{2.4}$$

More generally, the volume fraction φ^α of the α-phase in the r.e.v. is defined by

$$\varphi^\alpha(\underline{x}, t) = \overline{\chi^\alpha}(\underline{x}, t) = \mathcal{V}^\alpha(\underline{x}, t)/\mathcal{V} \tag{2.5}$$

where $\mathcal{V}^\alpha(\underline{x}, t)$ represents the volume of $\Omega^\alpha(\underline{x})$.

Let us now consider a physical quantity $e(\underline{z})$ attached to the α-phase and thus defined on the domain $\mathcal{P}^\alpha$ only. We shall encounter two different averages, namely the "apparent" one $< e >_\alpha (\underline{x}, t)$, and the "intrinsic" one $\overline{e}^\alpha(\underline{x}, t)$, respectively defined by :

$$< e >_\alpha (\underline{x}, t) = \overline{\chi^\alpha e}(\underline{x}, t) = \frac{1}{\mathcal{V}} \int \chi_o(\underline{z} - \underline{x}) \chi^\alpha(\underline{z}, t) e(\underline{z}, t)\, dV_z \tag{2.6}$$

and

$$\overline{e}^\alpha(\underline{x}, t) = \frac{1}{\varphi^\alpha} < e >_\alpha (\underline{x}, t) = \frac{1}{\mathcal{V}^\alpha} \int \chi_o(\underline{z} - \underline{x}) \chi^\alpha(\underline{z}, t) e(\underline{z}, t)\, dV_z \tag{2.7}$$

By way of illustration, let $\rho(\underline{z}, t)$ denote the mass density at the microscopic scale. The total mass dM^α of the α-phase contained in the r.e.v. reads $< \rho >_\alpha \mathcal{V}$. Alternatively, dM^α is also equal to $\varphi^\alpha(\underline{x}, t) \rho^\alpha_M(\underline{x}, t) \mathcal{V}$, where $\rho^\alpha_M(\underline{x}, t)$ is the intrinsic mass density at the macroscopic scale :

$$\rho^\alpha_M = \frac{1}{\varphi^\alpha} < \rho >_\alpha = \overline{\rho}^\alpha \tag{2.8}$$

Clearly enough, it is also possible to define apparent and intrinsic averages over the pore space. It amounts to replacing in (2.6) χ^α by the characteristic function χ^p of the pore space. The case of a porous medium saturated by a single fluid can be defined by the conditions $\chi^p(\underline{z}, t) = \chi^f(\underline{z}, t)$ or $\Omega^p = \Omega^f$ (f for fluid). In this situation, we note that the operators $< \cdot >_p$ and $< \cdot >_f$ are equivalent, as well as the operators $\overline{(\cdot)}^p$ and $\overline{(\cdot)}^f$.

2.2 Spatial Derivatives of $\overline{e}$

Let us first recall a mathematical result concerning the derivatives of a characteristic function. Since $\chi_o(\underline{z}, t)$ is discontinuous across the boundary of $\Omega(\underline{0})$, its spatial derivatives must be defined in the sense of the distribution theory. More precisely, let $\delta_{\partial\Omega(\underline{0})}$ be the Dirac distribution of support $\partial\Omega(\underline{0})$. It is defined by :

$$< \delta_{\partial\Omega(\underline{0})}, \psi > = \int \delta_{\partial\Omega(\underline{0})} \psi\, dV_z = \int_{\partial\Omega(\underline{0})} \psi\, dS_z \tag{2.9}$$

where ψ is any function of $\mathcal{D}(\mathbb{R}^3)$[1]. According to the definition of the derivation of a distribution, one obtains:

$$< \frac{\partial \chi_o}{\partial z_i}, \psi > = - < \chi_o, \frac{\partial \psi}{\partial z_i} > = - \int_{\Omega(\underline{0})} \frac{\partial \psi}{\partial z_i}\, dV_z = - \int_{\partial\Omega(\underline{0})} \psi n_i\, dS_z \tag{2.10}$$

where $\underline{n}$ is the unit normal of components n_i oriented out of $\Omega(\underline{0})$. The combination of (2.9) and (2.10) yields :

$$\frac{\partial \chi_o}{\partial z_i}(\underline{z}) = -n_i \delta_{\partial\Omega(\underline{0})} \tag{2.11}$$

Let us now consider the derivative of $\overline{e}$ with respect to x_i :

$$\frac{\partial}{\partial x_i}(\overline{e}) = \frac{1}{\mathcal{V}} \int \frac{\partial}{\partial x_i} (\chi_o(\underline{z} - \underline{x}))\, e(\underline{z}, t)\, dV_z \tag{2.12}$$

[1] $\mathcal{D}(\mathbb{R}^3)$ is the set of C^∞ functions which are equal to 0 out of a bounded domain.

We then observe that :

$$\frac{\partial}{\partial x_i}\left(\chi_o(\underline{z}-\underline{x})\right) = -\frac{\partial \chi_o}{\partial z_i}(\underline{z}-\underline{x}) = -\frac{\partial \chi_{\underline{x}}}{\partial z_i}(\underline{z}) \tag{2.13}$$

Besides, a combination of (2.11) and (2.13) yields :

$$\frac{\partial}{\partial x_i}\left(\chi_o(\underline{z}-\underline{x})\right) = n_i \delta_{\partial\Omega(\underline{x})} \tag{2.14}$$

Introducing this result into (2.12), one obtains :

$$\frac{\partial}{\partial x_i}(\overline{e}) = \frac{1}{\mathcal{V}}\int_{\partial\Omega(\underline{x})} n_i(\underline{z},t)e(\underline{z},t)\,dS_z = \frac{1}{\mathcal{V}}\int_{\Omega(\underline{x})} \frac{\partial e}{\partial z_i}(\underline{z},t)\,dV_z \tag{2.15}$$

that is :

$$\frac{\partial}{\partial x_i}(\overline{e}) = \overline{\frac{\partial e}{\partial z_i}} \tag{2.16}$$

2.3 Spatial Derivatives of $< e >_\alpha$

For the sake of clarity, we assume that the pore space is saturated by a single fluid ($\alpha = f$). The characteristic function χ^f of the fluid phase is discontinuous across the solid-fluid interface I^{sf}. Combining definition (2.6) and (2.16), the spatial derivative of $< e >_\alpha$ reads :

$$\frac{\partial}{\partial x_i}(< e >_f) = \overline{\frac{\partial}{\partial z_i}(e\chi^f)} \tag{2.17}$$

By analogy with (2.11), we note that :

$$\frac{\partial \chi^f}{\partial z_i}(\underline{z}) = -n_i \delta_{I^{sf}} \tag{2.18}$$

where the unit normal vector $\underline{n}$ to the solid-fluid interface is oriented towards the solid. Introducing this result in (2.17) and recalling (2.6), we then obtain :

$$\frac{\partial}{\partial x_i}(< e >_f) =< \frac{\partial e}{\partial z_i}(\underline{z},t) >_f - \frac{1}{\mathcal{V}}\int_{I^{sf}\cap\Omega(\underline{x})} e(\underline{z},t)\, n_i(\underline{z},t)\; dS_z \tag{2.19}$$

2.4 Time derivatives of $\overline{e}$ and $< e >_\alpha$

Observing that the characteristic function χ_o of $\Omega(\underline{0})$ does not depend on time, it follows immediatly from (2.2) that :

$$\frac{\partial}{\partial t}(\overline{e}) = \overline{\frac{\partial e}{\partial t}} \tag{2.20}$$

We then obtain from (2.6) that :

$$\frac{\partial}{\partial t}(< e >_f) = \overline{\frac{\partial (e\chi^f)}{\partial t}} =< \frac{\partial e}{\partial t} >_f + \overline{e\frac{\partial \chi^f}{\partial t}} \tag{2.21}$$

Let $\underline{u}$ denote the velocity of the solid-fluid interface. It is left as an exercise to show that $\partial\chi^f/\partial t = (\underline{u}\cdot\underline{n})\delta_{I^{sf}}$, where the unit normal vector $\underline{n}$ to the solid-fluid interface is oriented towards the solid. It then follows from (2.21) that :

$$\frac{\partial}{\partial t}(< e >_f) =< \frac{\partial e}{\partial t} >_f + \frac{1}{\mathcal{V}}\int_{I^{sf}\cap\Omega(\underline{x})} e\,\underline{u}\cdot\underline{n}\,dS_z \tag{2.22}$$

3 Application to Balance Laws

3.1 Momentum balance : the stress average rule

At the microscopic (resp. macroscopic) scale, the body forces acting on the r.e.v. $\Omega(\underline{x})$ are characterized by a density field $\underline{f}(\underline{z},t)$ (resp. $\underline{F}(\underline{x},t)$), such that the total body force acting on $\Omega(\underline{x})$ reads :

$$\overline{\rho}(\underline{x},t)\underline{F}(\underline{x},t)\mathcal{V} = \overline{\rho(\underline{z},t)\underline{f}(\underline{z},t)}\,\mathcal{V} \tag{3.1}$$

In turn, the concept of internal forces may be described at both scales. We look for the link between the macroscopic stress state defined at point $\underline{x}$ by the tensor $\mathbf{\Sigma}(\underline{x},t)$ and the microscopic stress field $\boldsymbol{\sigma}(\underline{z},t)$ defined on $\Omega(\underline{x})$.
The momentum balance equation may be written either at the microscopic scale in the form :

$$\frac{\partial\sigma_{ij}}{\partial z_j} + \rho f_i = 0 \tag{3.2}$$

or at the macroscopic scale :

$$\frac{\partial\Sigma_{ij}}{\partial x_j} + \overline{\rho}F_i = 0 \tag{3.3}$$

The link between $\mathbf{\Sigma}(\underline{x},t)$ and $\boldsymbol{\sigma}(\underline{z},t)$ is going to be derived from the compatibility of (3.2) and (3.3). Indeed, taking the average of (3.2) and using (2.16) yields :

$$\overline{\frac{\partial\sigma_{ij}}{\partial z_j}} + \overline{\rho f_i} = \frac{\partial}{\partial x_j}(\overline{\sigma_{ij}}) + \overline{\rho f_i} \tag{3.4}$$

Recalling (3.1), the comparison between (3.3) and (3.4) leads to the following average rule :

$$\mathbf{\Sigma} = \overline{\boldsymbol{\sigma}} \tag{3.5}$$

The two different origins of the variations of the microscopic stress field within the r.e.v. are the body forces appearing in (3.2) and the heterogeneity of the r.e.v. . Within the latter, the order of magnitude of the stress variations induced by the body forces is $|\rho\underline{f}|\ell$, while those induced by the heterogeneities are at least of the order of the average stress $\mathbf{\Sigma}$ (consider for instance an empty pore where $\boldsymbol{\sigma} = 0$). It thus appears that the contribution of the body forces can be neglected if

$$|\rho\underline{f}|\ell \ll |\mathbf{\Sigma}| \tag{3.6}$$

Note that this assumption is implicit in lab tests in which the contribution of the gravity forces is not taken into account. It is adopted from now on, so that the momentum balance equation at the microscopic scale reduces to :

$$\operatorname{div}_z \boldsymbol{\sigma} = 0 \tag{3.7}$$

3.2 Mass Balance

The transformation of the fluid phase can be described by the eulerian microscopic velocity field $\underline{v}(\underline{z})$ and by the macroscopic one $\underline{V}(\underline{x})$. Once again, we look for the appropriate link between the vector $\underline{V}(\underline{x})$ and the field $\underline{v}(\underline{z})$. The latter is going to be derived from the mass balance law.
At the microscopic scale, the mass balance equation reads :

$$(\forall \underline{z} \in \Omega^f) \quad \frac{\partial \rho}{\partial t} + \operatorname{div}_z (\rho \underline{v}) = 0 \tag{3.8}$$

In turn, the classical macroscopic formulation of the same principle is (see (2.8)) :

$$\frac{\partial}{\partial t}(\varphi^f \overline{\rho}^f) + \operatorname{div}_x (\varphi^f \overline{\rho}^f \underline{V}) = C(\underline{x}) \tag{3.9}$$

in which $C(\underline{x})$ represents the rate of mass exchange between the fluid and the solid phases. Of course, $C(\underline{x}) = 0$ when the fluid and the solid have no physico-chemical interactions. We now take the (apparent) average of (3.8) over the fluid phase, as defined in (2.6). From the differentiation rule (2.19), it is readily seen that :

$$< \operatorname{div}_z (\rho \underline{v}) >_f = \operatorname{div}_x (< \rho \underline{v} >_f) + \frac{1}{\mathcal{V}} \int_{I^{sf} \cap \Omega(\underline{x})} \rho \underline{v} \cdot \underline{n} \, dS_z \tag{3.10}$$

where $\underline{n}$ is the unit normal oriented towards the solid. In turn, from the differentiation rule (2.22), we obtain :

$$< \frac{\partial \rho}{\partial t} >_f = \frac{\partial}{\partial t}(\varphi^f \overline{\rho}^f) - \frac{1}{\mathcal{V}} \int_{I^{sf} \cap \Omega(\underline{x})} \rho \, \underline{u} \cdot \underline{n} \, dS_z \tag{3.11}$$

We then combine (3.10) and (3.11). Recalling (3.8), this yields :

$$\frac{\partial}{\partial t}(\varphi^f \overline{\rho}^f) + \operatorname{div}_x (< \rho \underline{v} >_f) + \frac{1}{\mathcal{V}} \int_{I^{sf} \cap \Omega(\underline{x})} \rho \, (\underline{v} - \underline{u}) \cdot \underline{n} \, dS_z = 0 \tag{3.12}$$

Clearly enough, the integral term in (3.12) represents the fluid mass flux through the (moving) solid-fluid interface and thus corresponds to $-C(\underline{x})$. The phenomenon of mass exchange between the solid and fluid phases takes place at the solid-fluid interface. At the microscopic scale, it is described by the surface density $\rho(\underline{u} - \underline{v}) \cdot \underline{n}$. In contrast, at the macroscopic scale, it is taken into account through the volume density $C(\underline{x})$:

$$C(\underline{x}) = \frac{1}{\mathcal{V}} \int_{I^{sf} \cap \Omega(\underline{x})} \rho \, (\underline{u} - \underline{v}) \cdot \underline{n} \, dS_z \tag{3.13}$$

For (3.9) and (3.12) to be compatible, we conclude that :

$$\varphi^f \overline{\rho}^f \underline{V} = < \rho \underline{v} >_f \tag{3.14}$$

or

$$\underline{V} = \frac{\overline{\rho \underline{v}}^f}{\overline{\rho}^f} \tag{3.15}$$

In particular, in the case of an incompressible fluid, the macroscopic velocity appears as the intrinsic average of the microscopic velocity :

$$\underline{V} = \overline{\underline{v}}^f \tag{3.16}$$

4 The macroscopic loading : uniform boundary conditions

In a lab experiment, a given macroscopic stress (resp. strain) state is applied to the sample and the macroscopic strain (resp. stress) thus induced is measured in order to derive the constitutive law. The micromechanical approach consists in replacing the real experiment by a theoretical one, in which the r.e.v. plays the role of the sample. To do so, we have to translate mathematically the fact that the r.e.v. is subjected to a macroscopic stress $\boldsymbol{\Sigma}$ or to a macroscopic strain $\boldsymbol{E}$ (Hashin (1983), Zaoui (2002)). What is the appropriate definition of such loadings ? This question is the focus of the present section. In section 4.1, we begin with some considerations concerning the stress state in the fluid phase (Ene and Sanchez-Palencia (1975)). We then present two possible definitions of a mechanical loading on the r.e.v. (sections 4.2 and 4.3).

4.1 The stress state in the fluid phase

We consider a porous medium saturated by a single fluid that is able to flow through the pore network. We assume that the fluid is a linear viscous incompressible material. The constitutive equation then reads :

$$\boldsymbol{\sigma} = -p\mathbf{1} + 2\mu^f \mathbf{d} \quad \text{with} \quad \operatorname{tr} \mathbf{d} = 0 \tag{4.1}$$

In (4.1), $p(\underline{z}, t)$ is the microscopic pressure defined in the pore space of the r.e.v., $\mathbf{d}$ is the strain rate (symmetric part of the velocity gradient), μ^f is the viscosity coefficient. We assume that inertia effects are negligible with respect to the viscous one. This corresponds to small values of the Reynolds number $\mathcal{R} = \rho V/\mu^f$, where V denotes the order of magnitude of the microscopic velocity. The momentum balance equation then takes the form (Stokes equation) :

$$\underline{\operatorname{grad}}_z\, p = \mu^f \Delta_z \underline{v} \tag{4.2}$$

Following (4.1), we note that the stress vector applied by the fluid on the solid boundary comprises two contributions, respectively related to the fluid pressure and to the viscous component $\underline{T}_v$ of $\boldsymbol{\sigma} \cdot \underline{n}$ (see figure 2).

The forthcoming developments are based on the physical intuition that the microscopic

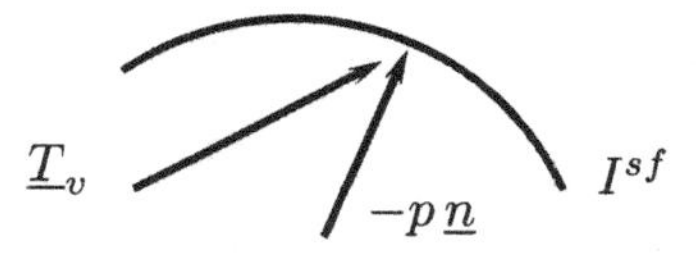

Figure 2. The two contributions to the stress vector acting on the solid-fluid interface

and macroscopic pressure gradients are of the same order of magnitude, say Π/L, where Π is the order of magnitude of the pressure variation at the scale of the macroscopic structure. This property can be guessed from the analysis of the Poiseuille flow in a cylindrical chanel, in which the pressure gradient has no component in the cross section of the chanel. It can be established rigorously using the technique of double scale asymptotic expansions (Auriault, in this book). It implies that the fluctuations of the microscopic pressure within the pore space of $\Omega(\underline{x})$ is of the order of $\Pi\ell/L \ll \Pi$ (see (1.1)). Accordingly, the contribution of the pressure to the surface forces applied by the fluid on the solid boundary may be represented by its intrinsic average $P = \overline{p}^f$.
The second contribution to the mechanical fluid-solid interaction is the viscous component $\underline{T}_v = 2\mu^f \mathbf{d} \cdot \underline{n}$. The condition $\underline{v} = 0$ on I^{sf} reveals that the characteristic length of the variations of $\underline{v}$ is d. Recalling that $\mathbf{d}$ is the symmetric part of the microscopic velocity field, the order of magnitude of the viscous force reads $T_v = \mu^f V/d$. Besides, the order of magnitude of the righthand side in (4.2) is $\mu^f V/d^2$ and is related to that of the pore pressure gradient according to :

$$\mu^f \frac{V}{d^2} = \frac{\Pi}{L} \tag{4.3}$$

It follows from (4.3) that :

$$\frac{T_v}{\Pi} = \frac{d}{L} \ll 1 \tag{4.4}$$

In other words, the viscous component of the surface force applied by the fluid to the solid boundary is negligible with respect to the pressure. As far as the solid-fluid mechanical interaction is concerned, the above results amount to approximating the microscopic stress state in the fluid by a uniform pressure state $\boldsymbol{\sigma} = -P\mathbf{1}$ with $P = \overline{p}^f$:

$$\begin{aligned} (\forall \underline{z} \in I^{sf}) \quad \boldsymbol{\sigma} \cdot \underline{n} &= -P\underline{n} \\ (\forall \underline{z} \in \Omega^p) \quad \boldsymbol{\sigma} &= -P\mathbf{1} \end{aligned} \tag{4.5}$$

It is worth observing that the average stress rule (3.5) then reads :

$$\boldsymbol{\Sigma} = (1-\varphi)\overline{\boldsymbol{\sigma}}^s - \varphi P\mathbf{1} \tag{4.6}$$

4.2 Uniform Stress Boundary Condition

The approach which is explored hereafter expresses that the r.e.v. is subjected to the macroscopic stress state represented by the tensor $\boldsymbol{\Sigma}$ through a stress boundary condition. This is the so-called uniform stress boundary condition :

$$(\forall \underline{z} \in \partial\Omega(\underline{x})) \quad \boldsymbol{\sigma}(\underline{z}) \cdot \underline{n}(\underline{z}) = \boldsymbol{\Sigma} \cdot \underline{n}(\underline{z}) \tag{4.7}$$

where $\underline{n}(\underline{z})$ is the outward unit normal to the boundary of the r.e.v. at point $\underline{z}$.
If the r.e.v. $\Omega(\underline{x})$ were homogeneous the microscopic stress field could be uniform and would meet the boundary conditions (4.7). In fact, the heterogeneities in $\Omega(\underline{x})$ are responsible for spatial fluctuations of $\boldsymbol{\sigma}(\underline{z}, t)$, so that the real surface forces that are applied on the boundary $\partial\Omega(\underline{x})$ are certainly not so simple. However, the underlying idea in (4.7) is that the particular form of the boundary conditions does not matter beyond a thin layer with a thickness of the order of d (see (1.1b)) provided that they are compatible with the average rule (3.5). We therefore need to check that this condition is satisfied.
To this end, we note the property :

$$\frac{\partial}{\partial z_k}\left(z_j \sigma_{ik}\right) = \sigma_{ij} \tag{4.8}$$

which is readily derived from the momentum balance condition (3.7). The integration of (4.8) over the whole r.e.v. yields :

$$\int_{\Omega(\underline{x})} \sigma_{ij}\, dV_z = \int_{\partial\Omega(\underline{x})} z_j \sigma_{ik} n_k\, dS_z = \int_{\partial\Omega(\underline{x})} z_j \Sigma_{ik} n_k\, dS_z \tag{4.9}$$

The above equation implies that the average $\overline{\sigma_{ij}}$ is identical for all stress fields that comply with (4.7). Observing that the uniform stress field $\boldsymbol{\sigma} = \boldsymbol{\Sigma}$ obviously satisfies this requirement, we conclude that $\overline{\sigma_{ij}} = \Sigma_{ij}$.
In the particular case of a porous medium, the boundary $\partial\Omega(\underline{x})$ a priori intersects some pores (Thompson and Willis (1991)). It may then be argued that the boundary conditions (4.7) are not compatible with a pressure stress state $\boldsymbol{\sigma} = -P\mathbf{1}$ in the fluid since it would imply $\boldsymbol{\Sigma}\cdot\underline{n} = -P\underline{n}$ at any point of the fluid part of $\partial\Omega(\underline{x})$. In order to avoid this difficulty, it is reasonable to assume that the boundary of the r.e.v. belongs to the solid phase. Indeed, this amounts to modifying the microstructure in a thin layer in the vicinity of $\partial\Omega(\underline{x})$. It will clearly not affect the overall response of the r.e.v. provided that the (average) pressure P in the fluid phase is regarded as a loading parameter independent of the macroscopic stress $\boldsymbol{\Sigma}$. The macroscopic loading applied to the r.e.v. is then completely characterized by the couple $(\boldsymbol{\Sigma}, P)$ and the boundary conditions (4.5)-(4.7) (figure 3).

4.3 Uniform Strain Boundary Condition

Instead of (4.7), we now explore the idea of defining a displacement boundary condition on $\partial\Omega(\underline{x})$. As in section 4.2, the latter is assumed to belong to the solid phase.
If the microscopic strain field $\boldsymbol{\varepsilon}$ were uniform within the r.e.v., it would be equal to the macroscopic strain $\mathbf{E}$ and the displacement $\underline{\xi}$ would be a linear function of the microscopic coordinates, that is $\underline{\xi}(\underline{z}) = \mathbf{E} \cdot \underline{z}$ (+ a rigid body motion). In the case of a heterogeneous r.e.v., the idea is to write the boundary conditions in the form :

$$(\forall \underline{z} \in \partial\Omega(\underline{x})) \quad \underline{\xi}(\underline{z}) = \mathbf{E} \cdot \underline{z} \tag{4.10}$$

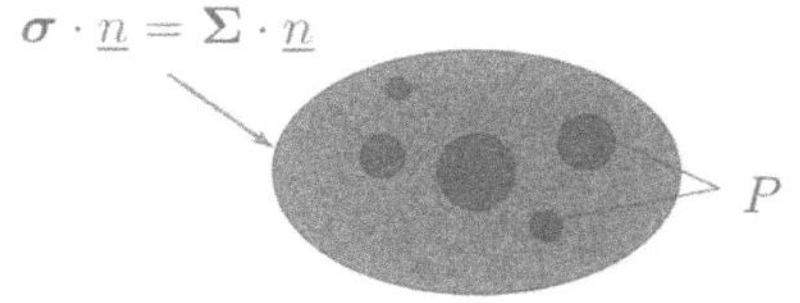

Figure 3. Uniform stress boundary conditions

As opposed to the homogeneous case, the material heterogeneity is responsible for microscopic fluctuations of the strain field ε within the r.e.v. However, such fluctuations are not seen in a macroscopic approach, which considers the r.e.v. from the outside. It is therefore natural to interpret $\mathbf{E}$ in (4.10) as the macroscopic strain tensor. From now on, it is assumed that $\mathbf{E}$ is symmetric.
From a physical point of view, the displacement field is defined on the solid phase only. Still, it may be useful to consider an extension of the latter to the pore space Ω^p. It is interesting to note that the average strain in Ω^p is entirely defined by the values of the displacement on the solid-fluid interface $\partial\Omega^p = I^{sf}$. Indeed, an integration by parts yields :

$$\int_{\Omega^p} \frac{\partial \xi_i}{\partial z_j} dV_z = \int_{I^{sf}} \xi_i n_j \, dS_z \tag{4.11}$$

from which one obtains the average strain in Ω^p in the form:

$$\overline{\varepsilon}^p = \frac{1}{\Omega^p} \int_{\Omega^p} \varepsilon \, dV_z = \frac{1}{2\Omega^p} \int_{I^{sf}} \left(\underline{\xi} \otimes \underline{n} + \underline{n} \otimes \underline{\xi}\right) dS_z \tag{4.12}$$

Hence, $\overline{\varepsilon}^p$ does not depend on the particular extension of $\underline{\xi}$ to Ω^p that is chosen, provided that the latter be continuous across I^{sf}.
Furthermore, considering now the average of ε over the whole r.e.v., an integration by parts together with (4.10) yields:

$$\int_{\Omega(\underline{x})} \frac{\partial \xi_i}{\partial z_j} dV_z = \int_{\partial\Omega(\underline{x})} \xi_i n_j \, dS_z = E_{ik} \int_{\partial\Omega(\underline{x})} z_k n_j \, dS_z = E_{ik} \delta_{kj} \mathcal{V} \tag{4.13}$$

or :

$$\overline{\frac{\partial \xi_i}{\partial z_j}} = E_{ij} \Leftrightarrow \overline{\mathbf{grad}\, \underline{\xi}} = \mathbf{E} \Rightarrow \overline{\varepsilon} = \mathbf{E} \tag{4.14}$$

where we have used that $\mathbf{E}$ is symmetric. Hence, the macroscopic strain $\mathbf{E}$ appears as the average of the heterogeneous microscopic strain field.
In the framework of uniform strain boundary conditions, the macroscopic loading applied to the r.e.v. is characterized by (4.5)-(4.10) (see figure 4).

4.4 The strain average rule and the Hill Lemma

Let us consider a microscopic stress field $\boldsymbol{\sigma}^*$ complying with the momentum balance condition $\text{div}_z\, \boldsymbol{\sigma}^* = 0$ and a microscopic strain field ε'. The latter is assumed to be the

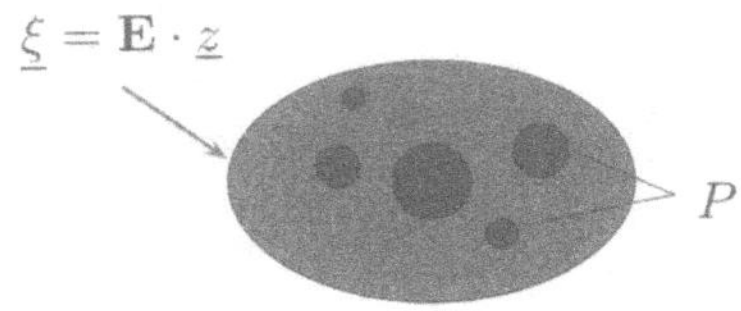

Figure 4. Uniform strain boundary conditions

symmetric part of the gradient of a vector field $\underline{\xi}'$:

$$\varepsilon'_{ij} = \frac{1}{2}\left(\frac{\partial \xi'_j}{\partial z_i} + \frac{\partial \xi'_i}{\partial z_j}\right) \tag{4.15}$$

In the sequel, $\boldsymbol{\sigma}^*$ and $\boldsymbol{\varepsilon}'$ are not a priori related by a constitutive behavior. We examine the virtual work $\mathcal{W}$ developed in the r.e.v. by the stress field $\boldsymbol{\sigma}^*$ in the strain field $\boldsymbol{\varepsilon}'$. To begin with, the microscopic derivation of $\mathcal{W}$ takes the form :

$$\mathcal{W} = \int_{\Omega(\underline{x})} \boldsymbol{\sigma}^* : \boldsymbol{\varepsilon}' \, dV_z = \overline{\boldsymbol{\sigma}^* : \boldsymbol{\varepsilon}'}\, \mathcal{V} \tag{4.16}$$

In turn, let $\boldsymbol{\Sigma}^*$ and $\mathbf{E}'$ respectively denote the macroscopic stress and strain tensors associated with the microscopic fields $\boldsymbol{\sigma}^*$ and $\boldsymbol{\varepsilon}'$. The macroscopic definition of $\mathcal{W}$ then reads :

$$\mathcal{W} = \boldsymbol{\Sigma}^* : \mathbf{E}' \, \mathcal{V} \tag{4.17}$$

The consistency of the micromechanical theory as regards an energy reasoning demands that the expressions of $\mathcal{W}$ derived respectively at the microscopic and macroscopic scales, namely (4.16) and (4.17), be consistent :

$$\boldsymbol{\Sigma}^* : \mathbf{E}' = \overline{\boldsymbol{\sigma}^* : \boldsymbol{\varepsilon}'} \tag{4.18}$$

We now look for an appropriate set of assumptions ensuring that it is satisfied. Two different situations are considered:

1. We assume that $\boldsymbol{\sigma}^*$ complies with a uniform stress boundary condition:

$$(\forall \underline{z} \in \partial\Omega(\underline{x})) \quad \boldsymbol{\sigma}^*(\underline{z}) \cdot \underline{n}(\underline{z}) = \boldsymbol{\Sigma}^* \cdot \underline{n}(\underline{z}) \tag{4.19}$$

The virtual work developped by $\boldsymbol{\sigma}^*$ in $\boldsymbol{\varepsilon}'$ reads:

$$\int_{\Omega(\underline{x})} \boldsymbol{\sigma}^* : \boldsymbol{\varepsilon}' \, dV_z = \int_{\partial\Omega(\underline{x})} \xi'_i \sigma^*_{ij} n_j \, dS_z = \Sigma^*_{ij} \int_{\partial\Omega(\underline{x})} \xi'_i n_j \, dS_z \tag{4.20}$$

An integration by parts shows that

$$\int_{\partial\Omega(\underline{x})} \xi'_i n_j \, dS_z = \int_{\Omega(\underline{x})} \frac{\partial \xi'_i}{\partial z_j} \, dV_z \tag{4.21}$$

(4.20) thus reads[2]:

$$\int_{\Omega(\underline{x})} \boldsymbol{\sigma}^* : \boldsymbol{\varepsilon}' \, dV_z = \Sigma^*_{ij} \overline{\frac{\partial \xi'_i}{\partial z_j}} \mathcal{V} = \boldsymbol{\Sigma}^* : \overline{\boldsymbol{\varepsilon}'} \mathcal{V} \tag{4.22}$$

Recalling that the stress average rule $\overline{\boldsymbol{\sigma}^*} = \boldsymbol{\Sigma}^*$ is a consequence of (4.19), (4.22) also takes the form :

$$\overline{\boldsymbol{\sigma}^* : \boldsymbol{\varepsilon}'} = \overline{\boldsymbol{\sigma}^*} : \overline{\boldsymbol{\varepsilon}'} \tag{4.23}$$

The consistency between (4.18) and (4.23) leads to define the macroscopic strain $\mathbf{E}'$ as the average of the microscopic strain field $\boldsymbol{\varepsilon}'$:

$$\mathbf{E}' = \overline{\boldsymbol{\varepsilon}'} \tag{4.24}$$

2. We assume that $\underline{\xi}'$ complies with a uniform strain boundary condition:

$$(\forall \underline{z} \in \partial\Omega(\underline{x})) \quad \underline{\xi}'(\underline{z}) = \mathbf{E}' \cdot \underline{z} \tag{4.25}$$

In this case, the average strain rule (4.24) follows from the boundary conditions (see section 4.3). No assumption is made concerning the stress vector $\boldsymbol{\sigma}^* \cdot \underline{n}$. The virtual work developped by $\boldsymbol{\sigma}^*$ in $\boldsymbol{\varepsilon}'$ reads:

$$\int_{\Omega(\underline{x})} \boldsymbol{\sigma}^* : \boldsymbol{\varepsilon}' \, dV_z = \int_{\partial\Omega(\underline{x})} \xi'_i \sigma^*_{ij} n_j \, dS_z = E'_{ij} \int_{\partial\Omega(\underline{x})} z_j \sigma^*_{ik} n_k \, dS_z \tag{4.26}$$

Use of (4.8) in (4.26) yields:

$$\int_{\Omega(\underline{x})} \boldsymbol{\sigma}^* : \boldsymbol{\varepsilon}' \, dV_z = E'_{ij} \overline{\sigma^*_{ij}} \mathcal{V} \Rightarrow \overline{\boldsymbol{\sigma}^* : \boldsymbol{\varepsilon}'} = \mathbf{E}' : \overline{\boldsymbol{\sigma}^*} \tag{4.27}$$

Finally, (4.23) is retrieved from (4.27) because of the strain average rule $\overline{\boldsymbol{\varepsilon}'} = \mathbf{E}'$.

In summary, in the framework of uniform boundary conditions (4.19) or (4.25), the volume average of the microscopic strain energy is equal to the macroscopic strain energy. The average strain rule is a consequence of the uniform strain boundary conditions. In the framework of uniform stress boundary conditions, the average strain rule is the definition of the link between the macroscopic strain and the microscopic strain field that ensures the energy consistency condition (4.18).

5 Linear poroelasticity

The theory of linear poroelasticity goes back to Biot's pionnering works, who addressed the poromechanical coupling within a thermodynamic framework formulated at the macroscopic scale (Biot (1941)). Later, Auriault and Sanchez-Palencia (1977) proposed a first micromechanical approach to linear poroelasticity in the framework of periodic homogenization. In the line of the previous sections, we hereafter deal with disordered

[2] $\boldsymbol{\Sigma}^*$ is a symmetric tensor.

porous media.
The r.e.v. $\Omega(\underline{x})$ located at the macroscopic point $\underline{x}$ is now simply denoted by Ω. We introduce a reference state of the porous medium. The latter is characterized by the conditions $\boldsymbol{\Sigma} = 0$ and $P = 0$. Ω_o^s and Ω_o^p denote the reference configurations of the solid and pore domains respectively. Accordingly, $\Omega_o = \Omega_o^s \cup \Omega_o^p$ represents the reference configuration of the r.e.v., with $\Omega_o = \Omega(\underline{x})$. $|\Omega_o| = \mathcal{V}$ is the corresponding volume.
Under a given loading, defined for instance by the current values at time t of the macroscopic stress $\boldsymbol{\Sigma}$ and pressure P, the initial solid and pore volumes Ω_o^s and Ω_o^p are transformed into Ω_t^s and Ω_t^p respectively. As before, I^{sf} is the time-dependent solid-fluid interface. Small strains are assumed, both at the macroscopic and microscopic levels.
For forthcoming purposes, it is useful to introduce the ratio $\phi = |\Omega_t^p|/|\Omega_o|$. Actually, ϕ aims at describing the pore volume change in a non dimensional way. In the reference configuration, $\phi = \phi_o$ is identical to the initial porosity φ_o. Strictly speaking, Ω_t^α is not identical to the domain $\Omega^\alpha(\underline{x}, t)$ introduced at section 2.1. Indeed, the latter is eulerian in nature in so far as it is made up of the phase α encountered in $\Omega(\underline{x})$ at time t. In contrast, Ω_t^α is obtained from the transformation of $\Omega_o^\alpha = \Omega^\alpha(\underline{x}, 0)$. This implies[3] that ϕ differs from the porosity φ introduced in (2.4). Nevertheless, in the framework of small strains, the geometrical changes are negligible : $\Omega_o^\alpha \approx \Omega_t^\alpha \approx \Omega^\alpha(\underline{x}, t)$ which amounts to saying that $\phi \approx \varphi \approx \varphi_o$. The index o or t will henceforth be omitted. However the variations in time of ϕ are not equal to those of the porosity φ, even for infinitesimal strain : $\phi - \phi_o \neq \varphi - \varphi_o$.[4]
In saturated conditions, we have seen that the mechanical fluid-solid interaction can be characterized by the average fluid pressure $P = \overline{p}^f$ (section 4.1). We seek the macroscopic counterpart of this interaction in the framework of a linear elastic behavior of the solid phase :

$$\boldsymbol{\sigma} = \mathbb{C}^s : \boldsymbol{\varepsilon} \; ; \; \boldsymbol{\varepsilon} = \mathbb{S}^s : \boldsymbol{\sigma} \tag{5.1}$$

It is first assumed that the elasticity tensors $\mathbb{C}^s$ and $\mathbb{S}^s = {\mathbb{C}^s}^{-1}$ are uniform throughout the solid phase. This assumption is removed at section 5.3.

5.1 Definition of a Mechanical Loading on the r.e.v.

In the line of the ideas presented at sections 4.2 and 4.3, there are two possible definitions of a mechanical problem on the solid part Ω^s of the r.e.v. . Adopting uniform stress boundary conditions, the macroscopic loading is characterized by the macroscopic

[3] According to (2.4), it is recalled that $\varphi = |\Omega^\alpha(\underline{x}, t)|/\mathcal{V}$, with $|\Omega_o| = \mathcal{V}$.
[4] Since it is reasonable to consider $\Omega_t = \Omega_t^s \cup \Omega_t^p$ as another possible representative elementary volume, the porosity φ also reads $|\Omega_t^p|/|\Omega_t|$. As opposed to $\phi - \phi_o$, the porosity change $\varphi - \varphi_o$ is therefore affected by that of $|\Omega_t|$.

stress $\mathbf{\Sigma}$ and the macroscopic pressure P :

$$\begin{aligned} &\mathrm{div}_z\, \boldsymbol{\sigma} = 0 && (\Omega^s) \\ &\boldsymbol{\sigma} = \mathbb{C}^s : \boldsymbol{\varepsilon} && (\Omega^s) \\ &\boldsymbol{\sigma} \cdot \underline{n} = \mathbf{\Sigma} \cdot \underline{n} && (\partial\Omega) \\ &\boldsymbol{\sigma} \cdot \underline{n} = -P\underline{n} && (I^{sf}) \end{aligned} \tag{5.2}$$

Alternatively, in the framework of uniform strain boundary conditions, the macroscopic loading parameters are the macroscopic strain $\mathbf{E}$ and the macroscopic pressure P :

$$\begin{aligned} &\mathrm{div}_z\, \boldsymbol{\sigma} = 0 && (\Omega^s) \\ &\boldsymbol{\sigma} = \mathbb{C}^s : \boldsymbol{\varepsilon} && (\Omega^s) \\ &\underline{\xi} = \mathbf{E} \cdot \underline{z} && (\partial\Omega) \\ &\boldsymbol{\sigma} \cdot \underline{n} = -P\underline{n} && (I^{sf}) \end{aligned} \tag{5.3}$$

In (5.2) as well as in (5.3), the material system is the solid domain. Still, it will prove efficient to extend the definition of the micromechanical problem to the whole elementary volume (solid and saturated pore space). The idea is to introduce a fictitious material in the pore space, which constitutive behavior is elastic with initial stress :

$$\boldsymbol{\sigma} = \mathbb{C}^p : \boldsymbol{\varepsilon} - P\mathbf{1} \quad \text{with} \quad |\mathbb{C}^p| \ll |\mathbb{C}^s| \tag{5.4}$$

The condition $|\mathbb{C}^p| \ll |\mathbb{C}^s|$ in fact amounts to $\boldsymbol{\sigma} \approx -P\mathbf{1}$ in Ω^p and thus captures the boundary condition on I^{sf} in (5.2) and (5.3). The interest of this point of view lies in the fact that it allows to formulate the local behavior in the whole r.e.v. in an unified way, of the form :

$$(\forall \underline{z} \in \Omega) \quad \boldsymbol{\sigma}(\underline{z}) = \mathbb{C}(\underline{z}) : \boldsymbol{\varepsilon}(\underline{z}) + \boldsymbol{\sigma}_o(\underline{z}) \tag{5.5}$$

with

$$\mathbb{C}(\underline{z}) = \left\{ \begin{array}{ll} \mathbb{C}^p & (\Omega^p) \\ \mathbb{C}^s & (\Omega^s) \end{array} \right. \qquad \boldsymbol{\sigma}_o(\underline{z}) = \left\{ \begin{array}{ll} -P\mathbf{1} & (\Omega^p) \\ 0 & (\Omega^s) \end{array} \right. \tag{5.6}$$

If uniform stress boundary conditions are considered, the mechanical problem (5.2) can be extended to the whole r.e.v. Ω in the form :

$$\begin{aligned} &\mathrm{div}_z\, \boldsymbol{\sigma} = 0 && (\Omega) \\ &\boldsymbol{\sigma} = \mathbb{C}(\underline{z}) : \boldsymbol{\varepsilon} + \boldsymbol{\sigma}_o(\underline{z}) && (\Omega) \\ &\boldsymbol{\sigma} \cdot \underline{n} = \mathbf{\Sigma} \cdot \underline{n} && (\partial\Omega) \end{aligned} \tag{5.7}$$

If uniform strain boundary conditions are prefered, (5.3) is now replaced by :

$$\begin{aligned} &\mathrm{div}_z\, \boldsymbol{\sigma} = 0 && (\Omega) \\ &\boldsymbol{\sigma} = \mathbb{C}(\underline{z}) : \boldsymbol{\varepsilon} + \boldsymbol{\sigma}_o(\underline{z}) && (\Omega) \\ &\underline{\xi} = \mathbf{E} \cdot \underline{z} && (\partial\Omega) \end{aligned} \tag{5.8}$$

5.2 Drained conditions

To begin with, we briefly address the loading cases defined by $(\mathbf{\Sigma} \neq 0, P = 0)$ and $(\mathbf{E} \neq 0, P = 0)$. This amounts to letting $\boldsymbol{\sigma}_o = 0$ in (5.7) and (5.8) :

$$\boldsymbol{\sigma} = \mathbb{C}(\underline{z}) : \boldsymbol{\varepsilon} \quad \Leftrightarrow \quad \boldsymbol{\varepsilon} = \mathbb{S}(\underline{z}) : \boldsymbol{\sigma} \quad \text{with } \mathbb{S}(\underline{z}) = \mathbb{C}(\underline{z})^{-1} \tag{5.9}$$

In both (5.7) and (5.8), the response linearly depends on the loading parameter, namely $\mathbf{\Sigma}$ and $\mathbf{E}$ respectively. This remark leads to the important concept of "concentration" tensor.
More precisely, let us first consider the uniform stress boundary conditions (problem (5.7)). The stress field $\boldsymbol{\sigma}(\underline{z})$ in the r.e.v. is proportional to the loading parameter $\mathbf{\Sigma}$. We therefore introduce the so-called stress concentration tensor $\mathbb{B}(\underline{z})$:

$$(\forall \underline{z} \in \Omega) \quad \boldsymbol{\sigma}(\underline{z}) = \mathbb{B}(\underline{z}) : \mathbf{\Sigma} \tag{5.10}$$

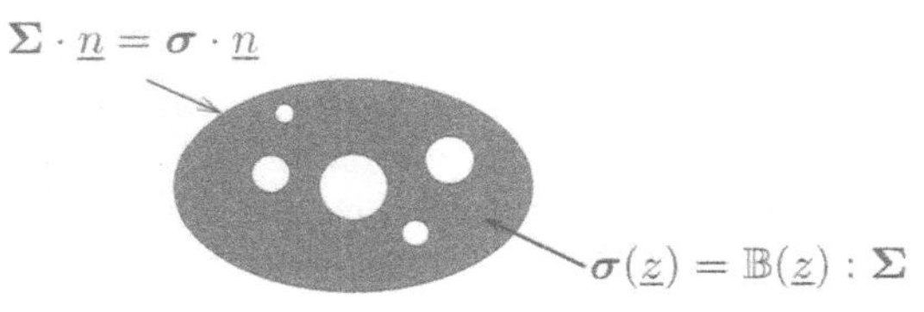

Figure 5. The stress concentration tensor $\mathbb{B}$

Note that the average rule (3.5) $\mathbf{\Sigma} = \overline{\boldsymbol{\sigma}}$ implies that $\overline{\mathbb{B}} = \mathbb{I}$, where $\mathbb{I}$ is the fourth order unit tensor :

$$I_{ijkl} = \frac{1}{2}\left(\delta_{il}\delta_{jk} + \delta_{ik}\delta_{jl}\right) \tag{5.11}$$

Owing to the condition $|\mathbb{C}^p| \ll |\mathbb{C}^s|$, the stress state in the pore space is $\sigma \approx 0$. This amounts to the condition $\mathbb{B} \approx 0$ in Ω^p and implies that :

$$\mathbb{I} = \overline{\mathbb{B}} = (1 - \varphi_o)\overline{\mathbb{B}}^s + \varphi_o \overline{\mathbb{B}}^p \approx (1 - \varphi_o)\overline{\mathbb{B}}^s \tag{5.12}$$

The macroscopic strain $\mathbf{E}$ being defined as the average $\overline{\boldsymbol{\varepsilon}}$ (see section 4.4), the combination of (5.9) with (5.10) yields :

$$\mathbf{E} = \mathbb{S}^{hom} : \mathbf{\Sigma} \tag{5.13}$$

with

$$\mathbb{S}^{hom} = \overline{\mathbb{S} : \mathbb{B}} \tag{5.14}$$

(5.13) shows that the macroscopic behavior is linear elastic and $\mathbb{S}^{hom}$ can be interpreted as the homogenized compliance tensor. The solid elastic compliance tensor $\mathbb{S}^s$ as well as the fictitious elastic compliance $\mathbb{S}^p$ of the pore space being uniform, $\mathbb{S}^{hom}$ reads :

$$\mathbb{S}^{hom} = (1 - \varphi_o)\mathbb{S}^s : \overline{\mathbb{B}}^s + \varphi_o \mathbb{S}^p : \overline{\mathbb{B}}^p \tag{5.15}$$

or, recalling (5.12) :

$$\mathbb{S}^{hom} = \mathbb{S}^s + \varphi_o \mathbb{S}^p : \overline{\mathbb{B}}^p \tag{5.16}$$

Interestingly, the homogenized compliance appears as the sum of that of the solid and of an additional contribution of the pores. Although $\overline{\mathbb{B}}^p \approx 0$, the latter does not vanish, since $\mathbb{S}^p \gg \mathbb{S}^s$. In fact, the order of magnitude of $\overline{\mathbb{B}}^p$ depends on the value of the fictitious stiffness $\mathbb{C}^p$ (resp. compliance $\mathbb{S}^p$) in such a way that the product $\mathbb{S}^p : \overline{\mathbb{B}}^p$ tends toward a finite limit when $\mathbb{C}^p \to 0$.
Similarly, in the framework of uniform strain boundary conditions (5.8), the microscopic strain field $\boldsymbol{\varepsilon}$ linearly depends on $\mathbf{E}$. This is taken into account through the strain concentration tensor $\mathbb{A}(\underline{z})$:

$$(\forall \underline{z} \in \Omega) \quad \boldsymbol{\varepsilon}(\underline{z}) = \mathbb{A}(\underline{z}) : \mathbf{E} \tag{5.17}$$

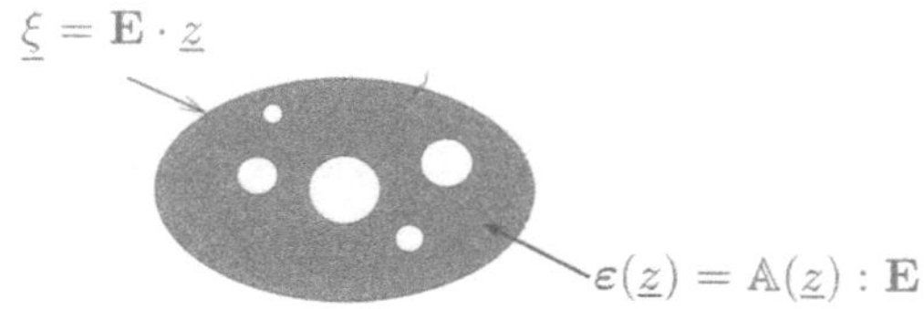

Figure 6. The strain concentration tensor $\mathbb{A}$

The strain average rule (4.14) $\mathbf{E} = \overline{\boldsymbol{\varepsilon}}$ implies that $\overline{\mathbb{A}} = \mathbb{I}$:

$$\mathbb{I} = (1 - \varphi_o)\overline{\mathbb{A}}^s + \varphi_o \overline{\mathbb{A}}^p \tag{5.18}$$

Owing to (5.9) and (5.17), the stress average rule $\boldsymbol{\Sigma} = \overline{\boldsymbol{\sigma}}$ reads :

$$\boldsymbol{\Sigma} = \mathbb{C}^{hom} : \mathbf{E} \quad \text{with } \mathbb{C}^{hom} = \overline{\mathbb{C} : \mathbb{A}} \tag{5.19}$$

Using (5.18) and recalling that $\mathbb{C}^p \to 0$, the macroscopic drained stiffness tensor also takes the form :

$$\mathbb{C}^{hom} = (1 - \varphi_o)\mathbb{C}^s : \overline{\mathbb{A}}^s = \mathbb{C}^s : (\mathbb{I} - \varphi_o \overline{\mathbb{A}}^p) \tag{5.20}$$

The $\varphi_o \overline{\mathbb{A}}^p$ term in the above expression of $\mathbb{C}^{hom}$ accounts for the reduction of stiffness due to the pore space.
The uniform stress boundary condition leads to the homogenized compliance tensor $\mathbb{S}^{hom}$. In turn, a macroscopic stiffness tensor $\mathbb{C}^{hom}$ has been derived from the uniform strain boundary condition. The consistency of these two different techniques can be justified from the fact that the corresponding boundary conditions both capture the same overal and local mechanical behavior, except in the vicinity of the boundary $\partial\Omega$. This amounts to saying that $\mathbb{S}^{hom} = {\mathbb{C}^{hom}}^{-1}$, provided that the scale separation rule (1.1b). More precisely, Hill (1967) and Mandel (1972) have proved that

$$\mathbb{S}^{hom} : \mathbb{C}^{hom} = \mathbb{I} + \mathcal{O}(\frac{d}{\ell})^3 \tag{5.21}$$

Let us assume that Ω^p is made up of a set of pores that can be described as inclusions of similar ellipsoidal shape. An estimate of $\overline{\mathbb{A}}^p$ can be derived from the solution to the Eshelby problem (Eshelby (1957))[5]. The idea consists in estimating the average strain in the pore space by the uniform strain that is induced in such an ellipsoidal pore ($\mathbb{C}^p \to 0$) embedded in an infinite matrix with stiffness $\mathbb{C}^s$, subjected to uniform strain boundary conditions at infinity : $\underline{\xi} \to \mathbf{E}_o \cdot \underline{z}$ when $\underline{z} \to \infty$. $\mathbf{E}_o$ represents the average strain $\overline{\varepsilon}^s$ in the solid matrix :

$$\begin{aligned} \overline{\varepsilon}^p &= (\mathbb{I} - \mathbb{S})^{-1} : \mathbf{E}_o \quad (a) \\ \overline{\varepsilon}^s &= \mathbf{E}_o \qquad\qquad\quad\;\; (b) \end{aligned} \tag{5.22}$$

In (5.22a), $\mathbb{S}$ is the Eshelby tensor that depends on $\mathbb{C}^s$ and on the (ellipsoidal) geometry of the considered inclusion. The average rule $\mathbf{E} = \overline{\varepsilon}$ then implies that :

$$\mathbf{E}_o = \left((1-\varphi_o)\mathbb{I} + \varphi_o(\mathbb{I} - \mathbb{S})^{-1}\right)^{-1} : \mathbf{E} \tag{5.23}$$

which can be introduced into (5.22a) and yields the following estimate of $\overline{\mathbb{A}}^p$:

$$\overline{\mathbb{A}}^p = (\mathbb{I} - \mathbb{S})^{-1} : \left((1-\varphi_o)\mathbb{I} + \varphi_o(\mathbb{I} - \mathbb{S})^{-1}\right)^{-1} = (\mathbb{I} - (1-\varphi_o)\mathbb{S})^{-1} \tag{5.24}$$

For illustrative purposes, we now consider spherical pores and we assume that the solid stiffness is isotropic. The isotropic tensor $\mathbb{C}^s$ may be splitted into its spherical and deviatoric components :

$$\mathbb{C}^s = 3k^s\mathbb{J} + 2\mu^s\mathbb{K} \tag{5.25}$$

where

$$J_{ijkl} = \frac{1}{3}\delta_{ij}\delta_{kl} \quad ; \quad \mathbb{K} = \mathbb{I} - \mathbb{J} \tag{5.26}$$

The Eshelby tensor then reads :

$$\mathbb{S} = \frac{3k^s}{3k^s + 4\mu^s}\mathbb{J} + \frac{6(k^s + 2\mu^s)}{5(3k^s + 4\mu^s)}\mathbb{K} \tag{5.27}$$

Introducing (5.27) into (5.24), the following estimates of the homogenized bulk and shear modulii are derived from (5.20) :

$$k^{hom} = \frac{4k^s\mu^s(1-\varphi_o)}{3k^s\varphi_o + 4\mu^s} \quad ; \quad \mu^{hom} = \mu^s \frac{(1-\varphi_o)(9k^s + 8\mu^s)}{9k^s(1 + \frac{2}{3}\varphi_o) + 8\mu^s(1 + \frac{3}{2}\varphi_o)} \tag{5.28}$$

5.3 The linear poroelastic response

first approach to the macroscopic state equations The mechanical problem (5.7) defined by $\mathbf{\Sigma}$ and P linearly depends on these two loading parameters. It is therefore convenient to split the loading $\mathcal{L} = (\mathbf{\Sigma}, P)$ into two components, namely $\mathcal{L}_1 = (\mathbf{\Sigma} + P\mathbf{1}, 0)$

[5] The technique described hereafter is similar to that presented in the context of molecular diffusion in the chapter "Diffusive transport in disordered media" of the present book (section 5.1).

and $\mathcal{L}_2 = (-P\mathbf{1}, P)$.
In $\mathcal{L}_1$, the macroscopic stress applied to the boundary $\partial\Omega$ is $\boldsymbol{\Sigma} + P\mathbf{1}$ and there is no pore pressure. The response to such a loading has been studied in section 5.2. The corresponding microscopic stress and strain fields read :

$$(\forall \underline{z} \in \Omega) \quad \boldsymbol{\sigma}_1 = \mathbb{B} : (\boldsymbol{\Sigma} + P\mathbf{1}) \quad ; \quad \boldsymbol{\varepsilon}_1 = \mathbb{S}(\underline{z}) : \mathbb{B} : (\boldsymbol{\Sigma} + P\mathbf{1}) \tag{5.29}$$

In $\mathcal{L}_2$, the same pressure P is applied on the external boundary $\partial\Omega$ as well as on the solid-fluid interface. Hence, the solid domain Ω^s is subjected to a uniform pressure on its boundary. Since the solid is homogeneous, the microscopic stress and strain fields in Ω^s are uniform and can be extended with the same values to the pore space :

$$(\forall \underline{z} \in \Omega) \quad \boldsymbol{\sigma}_2 = -P\mathbf{1} \quad ; \quad \boldsymbol{\varepsilon}_2 = -P\mathbb{S}^s : \mathbf{1} \tag{5.30}$$

Due to linearity, the response to the loading $\mathcal{L} = \mathcal{L}_1 + \mathcal{L}_2$ is :

$$(\forall \underline{z} \in \Omega) \quad \boldsymbol{\sigma} = \mathbb{B} : (\boldsymbol{\Sigma} + P\mathbf{1}) - P\mathbf{1} \quad ; \quad \boldsymbol{\varepsilon} = \mathbb{S} : \mathbb{B} : (\boldsymbol{\Sigma} + P\mathbf{1}) - P\mathbb{S}^s : \mathbf{1} \tag{5.31}$$

Owing to (5.14), the macroscopic strain $\mathbf{E} = \overline{\boldsymbol{\varepsilon}}$ then reads :

$$\mathbf{E} = \mathbb{S}^{hom} : (\boldsymbol{\Sigma} + P\mathbf{1}) - P\mathbb{S}^s : \mathbf{1} \tag{5.32}$$

or

$$\mathbf{E} = \mathbb{S}^{hom} : (\boldsymbol{\Sigma} + P\mathbf{B}) \Leftrightarrow \boldsymbol{\Sigma} = \mathbb{C}^{hom} : \mathbf{E} - \mathbf{B}P \tag{5.33}$$

with

$$\mathbf{B} = \mathbf{1} - \mathbb{C}^{hom} : \mathbb{S}^s : \mathbf{1} \tag{5.34}$$

(5.33) shows that $\boldsymbol{\Sigma} + P\mathbf{B}$ controls the macroscopic strain induced by the loading $\mathcal{L}$ and is therefore referred to as the macroscopic effective stress. It constitutes the first state equation of the poroelastic behavior. Note the isotropic form of (5.34) that involves the homogenized compression modulus k^{hom} and that of the solid, denoted by k^s :

$$\mathbf{B} = b\mathbf{1} \quad \text{with } b = 1 - \frac{k^{hom}}{k^s} \tag{5.35}$$

The second state equation is related to the variation of the pore volume which can be characterized by that of the ratio ϕ. The starting point is :

$$\phi - \phi_o = \varphi_o \mathbf{1} : \overline{\boldsymbol{\varepsilon}}^p \tag{5.36}$$

that is readily derived from (4.12). (5.31) then yields :

$$\varphi_o \mathbf{1} : \overline{\boldsymbol{\varepsilon}}^p = \varphi_o \mathbf{1} : \overline{\mathbb{S} : \mathbb{B}}^p : (\boldsymbol{\Sigma} + P\mathbf{1}) - \varphi_o P\mathbf{1} : \mathbb{S}^s : \mathbf{1} \tag{5.37}$$

$\varphi_o \mathbf{1} : \overline{\mathbb{S} : \mathbb{B}}^p$ can be eliminated by means of (5.16). Introducing (5.33) and (5.34) into (5.37), one eventually obtains :

$$\phi - \phi_o = \mathbf{B} : \mathbf{E} + \frac{P}{N} \tag{5.38}$$

where

$$\frac{1}{N} = (\mathbf{B} - \varphi_o \mathbf{1}) : \mathbb{S}^s : \mathbf{1} \tag{5.39}$$

Recalling (5.35), it is worth mentioning the isotropic form of (5.39) :

$$\frac{1}{N} = \frac{b - \varphi_o}{k^s} \tag{5.40}$$

(5.33) and (5.38) are the well-known equations of Biot's poroelasticity theory. Interestingly, (5.34) and (5.39) show that the poroelastic coefficients $\mathbf{B}$ and N can be determined from the solid elastic properties and from the homogenized drained stiffness tensor $\mathbb{C}^{hom}$. This stresses the fact that estimates of the poroelastic coefficients can be derived from estimates of $\mathbb{C}^{hom}$.

Combining (5.34) with (5.20), it is worth observing that $\mathbf{B}$ is related to the average volume strain in the pore space through :

$$\mathbf{B} = \varphi_o \mathbf{1} : \overline{\mathbb{A}}^p \tag{5.41}$$

With the same assumptions as for (5.24), (5.39) and (5.20) then read :

$$\mathbf{B} = \varphi_o \mathbf{1} : (\mathbb{I} - (1 - \varphi_o)\mathbb{S})^{-1} \tag{5.42}$$

and

$$\frac{1}{N} = \varphi_o \mathbf{1} : \left((\mathbb{I} - (1 - \varphi_o)\mathbb{S})^{-1} - \mathbb{I}\right) : \mathbb{S}^s : \mathbf{1} \tag{5.43}$$

generalization The previous reasoning has taken advantage of two simplifying assumptions. First, we have explicitely used the fact that the solid matrix is homogeneous for deriving the response to the loading component $\mathcal{L}_2$. Furthermore, the reasoning was based on the very simple structure of the "initial stress field" $\boldsymbol{\sigma}_o$, given in (5.6). It seems difficult to adapt it to a more general framework which will be necessary for instance to deal with the case of partially saturated porous media. We therefore propose an alternative approach that will be easier to generalize to such situations.

We consider the mechanical problem (5.8) defined on the r.e.v. and we take advantage of the linearity of the latter with respect to $\mathbf{E}$ and P. The loading $\mathcal{L} = (\mathbf{E}, P)$ is splitted into $\mathcal{L}_1 = (\mathbf{E}, 0)$ and $\mathcal{L}_2 = (0, P)$.

$\mathcal{L}_1 = (\mathbf{E}, 0)$ corresponds to a drained loading of the type that was studied at section 5.2. The microscopic strain field $\boldsymbol{\varepsilon}_1$ is :

$$(\forall \underline{z} \in \Omega) \quad \boldsymbol{\varepsilon}_1(\underline{z}) = \mathbb{A}(\underline{z}) : \mathbf{E} \tag{5.44}$$

and its average is equal to the loading parameter : $\overline{\boldsymbol{\varepsilon}_1} = \mathbf{E}$. The microscopic stress field induced by $\mathcal{L}_1$ is :

$$\boldsymbol{\sigma}_1 = \mathbb{C}(\underline{z}) : \boldsymbol{\varepsilon}_1 \tag{5.45}$$

and the associated macroscopic stress reads :

$$\overline{\boldsymbol{\sigma}_1} = \mathbb{C}^{hom} : \mathbf{E} \tag{5.46}$$

In $\mathcal{L}_2$, the macroscopic strain is prevented and fluid mass is injected into the r.e.v. until the fluid pressure is equal to P. Let $\boldsymbol{\sigma}_2$ (resp. $\boldsymbol{\varepsilon}_2$) denote the microscopic stress (resp. strain) field thus induced. The formal state equation (5.5) reads :

$$(\forall \underline{z} \in \Omega) \quad \boldsymbol{\sigma}_2(\underline{z}) = \mathbb{C}(\underline{z}) : \boldsymbol{\varepsilon}_2(\underline{z}) + \boldsymbol{\sigma}_o(\underline{z}) \tag{5.47}$$

with $\mathbb{C}(\underline{z})$ and $\boldsymbol{\sigma}_o(\underline{z})$ given in (5.6). We seek the average $\overline{\boldsymbol{\sigma}_2}$. To this end, the starting point consists in applying the Hill lemma[6] (section 4.4) to the stress field $\boldsymbol{\sigma}_2$ and the strain field $\boldsymbol{\varepsilon}_1$:

$$\overline{\boldsymbol{\sigma}_2 : \boldsymbol{\varepsilon}_1} = \overline{\boldsymbol{\sigma}_2} : \overline{\boldsymbol{\varepsilon}_1} = \overline{\boldsymbol{\sigma}_2} : \mathbf{E} \tag{5.48}$$

Introducing (5.47) and (5.45) into (5.48) yields :

$$\overline{\boldsymbol{\sigma}_2} : \mathbf{E} = \overline{\boldsymbol{\sigma}_o : \boldsymbol{\varepsilon}_1} + \overline{\boldsymbol{\varepsilon}_2 : \boldsymbol{\sigma}_1} \tag{5.49}$$

Recalling that $\overline{\boldsymbol{\varepsilon}_2} = 0$, a second application of the Hill lemma considering now the stress couple $(\boldsymbol{\sigma}_1, \boldsymbol{\varepsilon}_2)$ shows that $\overline{\boldsymbol{\varepsilon}_2 : \boldsymbol{\sigma}_1} = 0$. With (5.44), (5.49) now reads :

$$\overline{\boldsymbol{\sigma}_2} : \mathbf{E} = \overline{\boldsymbol{\sigma}_o : \mathbb{A}} : \mathbf{E} \Rightarrow \overline{\boldsymbol{\sigma}_2} = \overline{\boldsymbol{\sigma}_o : \mathbb{A}} \tag{5.50}$$

Combining (5.46) and (5.50) eventually yields the macroscopic stress associated with $\mathcal{L}$:

$$\boldsymbol{\Sigma} = \overline{\boldsymbol{\sigma}_1} + \overline{\boldsymbol{\sigma}_2} = \mathbb{C}^{hom} : \mathbf{E} + \overline{\boldsymbol{\sigma}_o : \mathbb{A}} \tag{5.51}$$

Note that this reasoning is general in the sense that it does not require to specify the form of the "initial" stress field $\boldsymbol{\sigma}_o$. With the value given in (5.6), we obtain the macroscopic stress induced by $\mathcal{L}_2$ in the form :

$$\overline{\boldsymbol{\sigma}_2} = -P\varphi_o \mathbf{1} : \overline{\mathbb{A}}^p = -P\mathbf{B} \tag{5.52}$$

where (5.41) has been used. We thus retrieve the state equation (5.33).

5.4 Energy Approach

The mechanical work $\dot{W}$ provided to the solid phase Ω^s during the loading rate $(\dot{\mathbf{E}}, \dot{P})$ comprises two contributions:

- The work of the pore pressure acting[7] on $I^{sf} = \partial\Omega^p \cap \partial\Omega^s$;
- The work of the surface forces $\boldsymbol{\sigma} \cdot \underline{n}$ acting on $\partial\Omega \cap \partial\Omega^s$ in the velocity $\dot{\mathbf{E}} \cdot \underline{z}$ prescribed on this boundary.

$$\dot{W} = \int_{\partial\Omega} (\underline{z} \cdot \dot{\mathbf{E}} \cdot (\boldsymbol{\sigma} \cdot \underline{n})\, dS_z + \int_{I^{sf}} -P\underline{n} \cdot \dot{\underline{\xi}}\, dS_z \tag{5.53}$$

[6]It is recalled that $\operatorname{div} \boldsymbol{\sigma}_2 = 0$ and that $\boldsymbol{\xi}_1$ is compatible with uniform strain boundary conditions.

[7]In saturated conditions, it is recalled that $\Omega^p = \Omega^f$.

where $\underline{n}$ is the outward unit normal vector to $\partial\Omega^s = \partial\Omega \cup I^{sf}$.
Using the momentum balance equation $\mathrm{div}_z\, \boldsymbol{\sigma} = 0$, the first integral in the righthand side can be rearranged in the form :

$$\dot{E}_{ij} \int_{\partial\Omega} z_j \sigma_{ik} n_k \, dS_z = \dot{E}_{ij} \int_{\Omega} \sigma_{ij} \, dV_z = \overline{\sigma_{ij}} \dot{E}_{ij} |\Omega_o| \tag{5.54}$$

Besides, the flux of the velocity $\dot{\underline{\xi}}$ through the solid-fluid interface in the second integral is equal to the pore volume rate:

$$\int_{I^{sf}} -\underline{n} \cdot \dot{\underline{\xi}} \, dS_z = |\Omega_o| \dot{\phi} \tag{5.55}$$

Combining (5.54) and (5.55) into (5.53), the rate of mechanical energy supplied to the solid domain Ω^s is :

$$\dot{W} = |\Omega_o| \left(\Sigma_{ij} \dot{E}_{ij} + P\, \dot{\phi} \right) = |\Omega_o| \left(\boldsymbol{\Sigma} : \dot{\mathbf{E}} + P\, \dot{\phi} \right) \tag{5.56}$$

Following Deudé et al. (2002), we now introduce the macroscopic elastic energy $\Psi(\mathbf{E}, P)$ and the potential energy $\Psi^*(\mathbf{E}, P)$ of the solid in the r.e.v., which are both functions of the loading parameters $\mathbf{E}$ and P :

$$\Psi(\mathbf{E}, P) = \frac{1}{2|\Omega_o|} \int_{\Omega^s} \boldsymbol{\varepsilon} : \mathbb{C}^s : \boldsymbol{\varepsilon} \, dV_z \quad ; \quad \Psi^* = \Psi - P(\phi - \phi_o) \tag{5.57}$$

In the absence of dissipative mechanisms, the increment of elastic energy is equal to the mechanical energy supply : $\dot{W} = |\Omega_o|\, \dot{\Psi}$. It follows that :

$$\dot{\Psi}^* = \boldsymbol{\Sigma} : \dot{\mathbf{E}} - (\phi - \phi_o)\dot{P} \quad \Rightarrow \left\{ \begin{array}{l} \boldsymbol{\Sigma} = \dfrac{\partial \Psi^*}{\partial \mathbf{E}}_{|P} \\[2ex] \phi - \phi_o = -\dfrac{\partial \Psi^*}{\partial P}_{|\mathbf{E}} \end{array} \right. \tag{5.58}$$

Relations (5.58) prove that the potential energy Ψ^* is the macroscopic thermodynamic potential for the macroscopic state variables $\mathbf{E}$ and P. Given the linearity of the macroscopic state equations (5.33) and (5.38), Ψ^* is a quadratic function of its arguments $\mathbf{E}$ and P:

$$\Psi^* = \frac{1}{2} \mathbf{E} : \mathbb{C}^{hom} : \mathbf{E} - \frac{P^2}{2N} - P\, \mathbf{B} : \mathbf{E} \tag{5.59}$$

Note that a combination of (5.57) with (5.59) also provides us with the expression of Ψ :

$$\Psi = \frac{1}{2} \mathbf{E} : \mathbb{C}^{hom} : \mathbf{E} + \frac{P^2}{2N} \tag{5.60}$$

5.5 Average Strain Level in the solid phase

For forthcoming applications (sections 6.1 and 7), it is interesting to estimate the strain level in the solid phase as a function of the macroscopic loading, defined by $\mathbf{E}$ and P, or by $\boldsymbol{\Sigma}$ and P.
Let us first consider the average strain $\overline{\boldsymbol{\varepsilon}}^s$, which is obtained from (5.1):

$$\overline{\boldsymbol{\varepsilon}}^s = \mathbb{S}^s : \overline{\boldsymbol{\sigma}}^s \tag{5.61}$$

Combining (5.61) with (4.6) gives the strain average as a function of the macroscopic stress and the pore pressure:

$$(1-\varphi_o)\overline{\boldsymbol{\varepsilon}}^s = \mathbb{S}^s : (\boldsymbol{\Sigma} + P\varphi_o\mathbf{1}) \tag{5.62}$$

Interestingly, we note that $\overline{\boldsymbol{\varepsilon}}^s$ is controlled by the effective stress $\boldsymbol{\Sigma} + P\varphi_o\mathbf{1}$.
Let us split the microscopic strain and its average in the solid into a spherical part and a deviatoric one :

$$\boldsymbol{\varepsilon} = \boldsymbol{\varepsilon}_d + \frac{1}{3}\varepsilon_v\mathbf{1} \quad ; \quad \overline{\boldsymbol{\varepsilon}}^s = \overline{\boldsymbol{\varepsilon}_d}^s + \frac{1}{3}\overline{\varepsilon_v}^s\mathbf{1} \tag{5.63}$$

where $\varepsilon_v = \operatorname{tr}\boldsymbol{\varepsilon}$. In the case of an isotropic behavior of the solid phase, (5.62) yields :

$$\begin{aligned} (1-\varphi_o)\overline{\varepsilon_v}^s &= \frac{1}{3k^s}\operatorname{tr}(\boldsymbol{\Sigma} + P\varphi_o\mathbf{1}) \\ (1-\varphi_o)\overline{\boldsymbol{\varepsilon}}^s_d &= \frac{1}{2\mu^s}\boldsymbol{\Sigma}_d \end{aligned} \tag{5.64}$$

where $\boldsymbol{\Sigma}_d = \mathbb{K} : \boldsymbol{\Sigma}$ is the deviatoric macroscopic stress.
(5.64) states that the average deviatoric strain $\overline{\boldsymbol{\varepsilon}_d}^s$ is only affected by the macroscopic deviatoric stress $\boldsymbol{\Sigma}_d$. This result can be illustrated on the example of a hollow sphere subjected to a uniform confining pressure on its external boundary. The shear strain induced in the solid phase is an essential component of the overall deformation mechanism of the sphere. Nevertheless, the average $\overline{\boldsymbol{\varepsilon}_d}^s$ is equal to 0 because of the spherical symmetry. This suggests that $\overline{\boldsymbol{\varepsilon}_d}^s$ may be inadequate for estimating the shear strain level in porous media.
In order to overcome this difficulty, let us then consider the equivalent shear strain ε_d :

$$\varepsilon_d = \sqrt{\frac{1}{2}\boldsymbol{\varepsilon}_d : \boldsymbol{\varepsilon}_d} \tag{5.65}$$

We restrict ourselves to the case of an isotropic solid phase (see (5.25)). In this case, the quadratic average $\overline{\varepsilon_d^2}^s$ is related to the shear component of the elastic energy of the solid. It thus provides an energetic measurement of the shear strain level in the solid. As such, it is expected to be free of the shortcomings of $\overline{\boldsymbol{\varepsilon}_d}^s$. Following Kreher (1990) and Dormieux et al. (2002), we start from the definition of the potential energy of the solid phase (see (5.57)):

$$|\Omega_o|\Psi^* = \frac{1}{2}\int_{\Omega^s} \boldsymbol{\varepsilon} : \mathbb{C}^s : \boldsymbol{\varepsilon}\, dV_z - P\int_{\Omega^p} \operatorname{tr}\boldsymbol{\varepsilon}\, dV_z \tag{5.66}$$

We now consider the derivative of the potential energy Ψ^* with respect to the shear modulus μ^s. Ψ^* depends on μ^s explicitely through the elastic stiffness tensor $\mathbb{C}^s$ and implicitely through the microscopic strain field $\boldsymbol{\varepsilon}$. Observing that $\partial\mathbb{C}^s/\partial\mu^s = 2\mathbb{K}$ (see (5.25)), it is then readily seen that,

$$|\Omega_o|\frac{\partial\Psi^*}{\partial\mu^s} = \int_{\Omega^s} \boldsymbol{\varepsilon}_d : \boldsymbol{\varepsilon}_d \, dV_z + \int_{\Omega^s} \frac{\partial\boldsymbol{\varepsilon}}{\partial\mu^s} : \mathbb{C}^s : \boldsymbol{\varepsilon} \, dV_z + \int_{\Omega^p} -P\mathbf{1} : \frac{\partial\boldsymbol{\varepsilon}}{\partial\mu^s} \, dV_z \qquad (5.67)$$

that can be re-arranged in the form:

$$\frac{\partial\Psi^*}{\partial\mu^s} = 2(1-\varphi_o)\overline{\varepsilon_d^2}^s + \overline{\boldsymbol{\sigma} : \frac{\partial\boldsymbol{\varepsilon}}{\partial\mu^s}} \qquad (5.68)$$

With the notations of section 4.4, the Hill Lemma can be applied to the couple $\boldsymbol{\sigma}^* = \boldsymbol{\sigma}$ and $\boldsymbol{\varepsilon}' = \partial\boldsymbol{\varepsilon}/\partial\mu^s$. Indeed, the displacement associated with $\partial\boldsymbol{\varepsilon}/\partial\mu^s$ is $\underline{\xi}' = \partial\boldsymbol{\xi}/\partial\mu^s$ and the latter meets the (uniform strain) boundary condition $\underline{\xi}' = 0$. This makes the second term on the right-hand-side vanish. Recalling (5.59), we then obtain:

$$2(1-\varphi_o)\overline{\varepsilon_d^2}^s = \frac{\partial}{\partial\mu^s}\left(\frac{1}{2}\mathbf{E} : \mathbb{C}^{hom} : \mathbf{E} - \frac{P^2}{2N} - P\,\mathbf{B} : \mathbf{E}\right) \qquad (5.69)$$

This expression can be further simplified in the case of macroscopic isotropy. Recalling (5.35) and (5.40), the derivatives of $\mathbf{B}$ and $1/N$ take the form:

$$\frac{\partial}{\partial\mu^s}(\mathbf{B}) = -\frac{1}{k^s}\frac{\partial k^{hom}}{\partial\mu^s}\mathbf{1} \quad ; \quad \frac{\partial}{\partial\mu^s}(\frac{1}{N}) = -\frac{1}{{k^s}^2}\frac{\partial k^{hom}}{\partial\mu^s} \qquad (5.70)$$

Introducing (5.70) into (5.69), we finally obtain:

$$2(1-\varphi_o)\overline{\varepsilon_d^2}^s = \frac{1}{2}\frac{\partial k^{hom}}{\partial\mu^s}(\mathrm{tr}\,\mathbf{E} + \frac{P}{k^s})^2 + \frac{\partial\mu^{hom}}{\partial\mu^s}\mathbf{E}_d : \mathbf{E}_d \qquad (5.71)$$

Alternatively, the state equation (5.33) can be used in order to derive an expression of $\overline{\varepsilon_d^2}^s$ as a function of the macroscopic stress and of the pore pressure. Taking the expression (5.35) of the Biot coefficient b into account, (5.33) is splitted into its spherical and deviatoric parts :

$$\Sigma_m + P = k^{hom}(\mathrm{tr}\,\mathbf{E} + \frac{P}{k^s}) \quad ; \quad \Sigma_d = 2\mu^{hom}E_d \qquad (5.72)$$

where the mean stress Σ_m and the equivalent deviatoric stress and strain are defined as :

$$\Sigma_m = \frac{1}{3}\mathrm{tr}\,(\boldsymbol{\Sigma}) \quad ; \quad \Sigma_d = \sqrt{\frac{1}{2}\boldsymbol{\Sigma}_d : \boldsymbol{\Sigma}_d} \quad ; \quad E_d = \sqrt{\frac{1}{2}\mathbf{E}_d : \mathbf{E}_d} \qquad (5.73)$$

Substituting (5.72) in (5.71) then yields :

$$4(1-\varphi_o)\overline{\varepsilon_d^2}^s = -\frac{\partial}{\partial\mu^s}(\frac{1}{k^{hom}})(\Sigma_m + P)^2 - \frac{\partial}{\partial\mu^s}(\frac{1}{\mu^{hom}})\Sigma_d^2 \qquad (5.74)$$

Whereas the average deviatoric strain tensor $\overline{\varepsilon_d}^s$ is independent of the pore pressure (see (5.64)), the quadratic (scalar) average $\overline{\varepsilon_d^2}^s$ appears to depend on the pore pressure through Terzaghi's effective stress $\mathbf{\Sigma} + P\mathbf{1}$.
The same methodology can be applied to the determination of the quadratic average $\overline{\varepsilon_v^2}^s$.
In particular, it can be readily seen that (5.69) is now replaced by

$$\frac{1}{2}(1-\varphi_o)\overline{\varepsilon_v^2}^s = \frac{\partial}{\partial k^s}\left(\frac{1}{2}\mathbf{E} : \mathbb{C}^{hom} : \mathbf{E} - \frac{P^2}{2N} - P\,\mathbf{B} : \mathbf{E}\right) \tag{5.75}$$

6 Non linear poroelasticity

6.1 The secant method

Section 5 has been devoted to the case of a linear elastic solid matrix. We now address the non linear elastic behavior of the latter in isothermal conditions. It can be characterized by a potential $\psi^s(\boldsymbol{\varepsilon})$ with which the state equation of the solid phase reads :

$$\boldsymbol{\sigma} = \frac{\partial \psi^s}{\partial \boldsymbol{\varepsilon}} \tag{6.1}$$

$\psi^s(\boldsymbol{\varepsilon})$ physically represents the free energy density of the solid.
We further assume that (6.1) can be put in the form :

$$\boldsymbol{\sigma} = \mathbb{C}^s(\boldsymbol{\varepsilon}) : \boldsymbol{\varepsilon} \tag{6.2}$$

$\mathbb{C}^s(\boldsymbol{\varepsilon})$ is classically referred to as secant stiffness tensor.
For instance, if $\psi^s(\boldsymbol{\varepsilon})$ depends on $\boldsymbol{\varepsilon}$ through the volume strain $\varepsilon_v = \operatorname{tr}\boldsymbol{\varepsilon}$ and the deviatoric invariant ε_d introduced in (5.65), that is $\psi^s = \psi^s(\varepsilon_v, \varepsilon_d)$, (6.1) reads :

$$\boldsymbol{\sigma} = \frac{\partial \psi^s}{\partial \varepsilon_v}\mathbf{1} + \frac{1}{2\varepsilon_d}\frac{\partial \psi^s}{\partial \varepsilon_d}\boldsymbol{\varepsilon}_d \tag{6.3}$$

(6.3) can be put in the form (6.2) with :

$$\mathbb{C}^s(\boldsymbol{\varepsilon}) = 3k^s(\varepsilon_v, \varepsilon_d)\mathbb{J} + 2\mu^s(\varepsilon_v, \varepsilon_d)\mathbb{K} \quad \text{with} \quad \begin{cases} k^s = \dfrac{1}{\varepsilon_v}\dfrac{\partial \psi^s}{\partial \varepsilon_v} \\[2ex] \mu^s = \dfrac{1}{4\varepsilon_d}\dfrac{\partial \psi^s}{\partial \varepsilon_d} \end{cases} \tag{6.4}$$

Conversely, it is worth observing that two functions $k^s(\varepsilon_v, \varepsilon_d)$ and $\mu^s(\varepsilon_v, \varepsilon_d)$ define a non linear elastic behavior only if :

$$\varepsilon_v \frac{\partial k^s}{\partial \varepsilon_d} = 4\varepsilon_d \frac{\partial \mu^s}{\partial \varepsilon_v} \tag{6.5}$$

The macroscopic loading applied to the r.e.v. is defined by the macroscopic strain tensor $\mathbf{E}$ and the pore pressure P. The microscopic stress, strain and displacement fields $\boldsymbol{\sigma}$, $\boldsymbol{\varepsilon}$,

$\underline{\xi}$ in the r.e.v. are therefore solutions to the following differential problem :

$$\begin{array}{lll} \operatorname{div} \boldsymbol{\sigma} = 0 & (\Omega) & (a) \\ \boldsymbol{\sigma}(\underline{z}) = \mathbb{C}(\underline{z}) : \boldsymbol{\varepsilon}(\underline{z}) + \boldsymbol{\sigma}_o(\underline{z}) & (\Omega) & (b) \\ \boldsymbol{\varepsilon} = \frac{1}{2}\left(\mathbf{grad}\,\underline{\xi} + {}^t\mathbf{grad}\,\underline{\xi}\right) & (\Omega) & (c) \\ \underline{\xi}(\underline{z}) = \mathbf{E} \cdot \underline{z} & (\partial\Omega) & (d) \end{array} \tag{6.6}$$

where

$$\begin{cases} \underline{z} \in \Omega^s : & \mathbb{C}(\underline{z}) = \mathbb{C}^s(\boldsymbol{\varepsilon}(\underline{z})) \\ \underline{z} \in \Omega^p : & \mathbb{C}(\underline{z}) = \mathbb{C}^p \end{cases} \qquad \boldsymbol{\sigma}_o(\underline{z}) = \begin{cases} 0 & (\Omega^s) \\ -P\mathbf{1} & (\Omega^p) \end{cases} \tag{6.7}$$

(6.6) is formally identical to (5.8), except for the fact that (5.6) is now replaced by (6.7), where it is recalled that $\mathbb{C}^p \to 0$.
There is however an essential difference between the linear problem (5.8) and the non linear one (6.6), that lies in the fact that the stiffness tensor $\mathbb{C}^s(\boldsymbol{\varepsilon}(\underline{z}))$ now depends on the location in the solid phase, instead of being uniform in Ω^s. Unfortunately, the classical homogenization techniques are not able to deal with such an heterogeneity. It is therefore necessary to simplify the problem.
The idea consists in assuming that it is possible to define a so-called effective strain which captures the non linearity of the matrix in an approximate way (Suquet (1997)). Denoted by ε^{ef}, this effective strain is expected to be an average of the microscopic strain field in the solid and, as such, to depend on the macroscopic loading $(\mathbf{E}, P)$ or $(\boldsymbol{\Sigma}, P)$. The simplification lies in the fact that the heterogeneous stiffness $\mathbb{C}^s(\boldsymbol{\varepsilon}(\underline{z}))$ is replaced by an homogeneous one :

$$(\forall \underline{z} \in \Omega^s) \quad \mathbb{C}^s(\boldsymbol{\varepsilon}(\underline{z})) \approx \mathbb{C}^s(\varepsilon^{ef}) \tag{6.8}$$

The non linear homogenization technique based on the effective strain concept then comprises three steps.

1. The first one is the resolution of the *l*inear homogenization problem, that is, the determination of the macroscopic response of the r.e.v. when the solid is linear elastic. This has already been presented at section 5. Let us recall the macroscopic state equations of the linear theory :

$$\boldsymbol{\Sigma} = \mathbb{C}^{hom} : \mathbf{E} - \mathbf{B}P \tag{6.9}$$

$$\phi - \phi_o = \mathbf{B} : \mathbf{E} + \frac{P}{N} \tag{6.10}$$

In particular, this step requires to relate the poroelastic coefficients $\mathbb{C}^{hom}$, $\mathbf{B}$ and N to the solid stiffness $\mathbb{C}^s$, by means of an appropriate linear homogenization scheme. The choice of the latter depends on the morphology at stake, formally represented hereafter by the symbolic parameter $\mathcal{M}$:

$$\mathbb{C}^{hom} = \mathbb{C}^{hom}(\mathbb{C}^s, \mathcal{M}) \; ; \; \mathbf{B} = \mathbf{B}(\mathbb{C}^s, \mathcal{M}) \; ; \; N = N(\mathbb{C}^s, \mathcal{M}) \tag{6.11}$$

2. The starting point of the second step is the choice of an effective strain ε^{ef}. In the framework of the linear homogenization scheme adopted in step 1, we then have to determine ε^{ef} as a function of the loading parameters $(\mathbf{E}, P)$ and of the solid stiffness $\mathbb{C}^s$, still regarded as a constant. Using (6.9) the couple $(\boldsymbol{\Sigma}, P)$ may also be used, instead of $(\mathbf{E}, P)$:
$$\varepsilon^{ef} = \mathcal{E}(\boldsymbol{\Sigma}, P, \mathbb{C}^s) \tag{6.12}$$
3. The last step deals with the non linearity of the problem. We now take into account the fact that the stiffness of the solid depends on the effective strain : $\mathbb{C}^s = \mathbb{C}^s(\varepsilon^{ef})$. (6.11) then implies that $\mathbb{C}^{hom}$, $\mathbf{B}$ and N also depend on ε^{ef} :
$$\mathbb{C}^{hom} = \mathbb{C}^{hom}(\varepsilon^{ef}) \ ; \ \mathbf{B} = \mathbf{B}(\varepsilon^{ef}) \ ; \ N = N(\varepsilon^{ef}) \tag{6.13}$$
The effective strain then appears as the solution of the following non linear problem :
$$\varepsilon^{ef} = \mathcal{E}(\boldsymbol{\Sigma}, P, \mathbb{C}^s(\varepsilon^{ef})) \tag{6.14}$$
The resolution of (6.14) yields ε^{ef} as a function of $\boldsymbol{\Sigma}$ and P : $\varepsilon^{ef} = \varepsilon^{ef}(\boldsymbol{\Sigma}, P)$. Introducing this determination of ε^{ef} into (6.13), the non linear counterpart of (6.9)-(6.10) reads :
$$\begin{aligned} \boldsymbol{\Sigma} &= \mathbb{C}^{hom}(\boldsymbol{\Sigma}, P) : \mathbf{E} - \mathbf{B}(\boldsymbol{\Sigma}, P)P \\ \phi - \phi_o &= \mathbf{B}(\boldsymbol{\Sigma}, P) : \mathbf{E} + \frac{P}{N(\boldsymbol{\Sigma}, P)} \end{aligned} \tag{6.15}$$
where the derivation of $\mathbf{B}(\boldsymbol{\Sigma}, P)$ as a function of $\mathbb{C}^{hom}(\boldsymbol{\Sigma}, P)$ and that of $N(\boldsymbol{\Sigma}, P)$ as a function of $\mathbf{B}(\boldsymbol{\Sigma}, P)$ take the same form as in the linear case :
$$\begin{aligned} \mathbf{B}(\boldsymbol{\Sigma}, P) &= \mathbf{1} : (\mathbb{I} - \mathbb{C}^s(\boldsymbol{\Sigma}, P)^{-1} : \mathbb{C}^{hom}(\boldsymbol{\Sigma}, P)) \\ \frac{1}{N(\boldsymbol{\Sigma}, P)} &= \mathbf{1} : \mathbb{C}^s(\boldsymbol{\Sigma}, P)^{-1} : (-\varphi_o \mathbf{1} + \mathbf{B}(\boldsymbol{\Sigma}, P)) \end{aligned} \tag{6.16}$$

A discussion of the effective strain concept in the light of the variational principles can be found in (Suquet (1997)) and (Ponte-Castaneda (1997)).

6.2 Application to the isotropic case

As in (6.4), we consider the isotropic case where ψ^s is a function of ε_v and ε_d. In the line of thought of the effective strain concept, the approximation (6.8) $\mathbb{C}^s(\boldsymbol{\varepsilon}) \approx \mathbb{C}^s(\boldsymbol{\varepsilon}^{ef})$ is adopted. We explore herein an isotropic modelling in which the effective strain $\boldsymbol{\varepsilon}^{ef}$ is characterized by two scalar invariants :
$$\varepsilon^{ef} = (\varepsilon_v^{ef}, \varepsilon_d^{ef}) \quad ; \quad \varepsilon_v^{ef} = tr\, \overline{\boldsymbol{\varepsilon}}^s \quad ; \quad \varepsilon_d^{ef} = \sqrt{\overline{\varepsilon_d^2}^s} \tag{6.17}$$
and the approximate solid stiffness reads :
$$\mathbb{C}^s(\varepsilon^{ef}) = 3k^s(\varepsilon_v^{ef}, \varepsilon_d^{ef})\mathbb{J} + 2\mu^s(\varepsilon_v^{ef}, \varepsilon_d^{ef})\mathbb{K} \tag{6.18}$$

ε_v^{ef} and ε_d^{ef} can be derived as functions of the macroscopic stress $\mathbf{\Sigma}$ and the pore pressure P from (5.64) and (5.74) respectively. More precisely, ε_v^{ef} depends on $\mathbf{\Sigma}+P\varphi_o\mathbf{1}$, whereas ε_d^{ef} is a function of $\mathbf{\Sigma}+P\mathbf{1}$. Recalling (6.13), the macroscopic stiffness $\mathbb{C}^{hom}$, as well as $\mathbf{B}$ and N, will depend on both $\mathbf{\Sigma}+P\varphi_o\mathbf{1}$ and $\mathbf{\Sigma}+P\mathbf{1}$. Hence, in the general case, the macroscopic non linearity is not controlled by an effective stress. Still, two exceptions deserve to be mentioned :

- k^s and μ^s do not depend on ε_d. In this case (6.5) shows that μ^s is necessarily a constant. ε_v^{ef} is determined from the solution to :

$$
\begin{aligned}
&(1-\varphi_o)\varepsilon_v^{ef} = \frac{1}{3k^s} tr\,(\mathbf{\Sigma}+P\varphi_o\mathbf{1}) && (a)\\
&k^s = k^s(\varepsilon_v^{ef}) && (b)
\end{aligned}
\tag{6.19}
$$

  ε_v^{ef} appears as a function of $tr\,\mathbf{\Sigma}+P\varphi_o\mathbf{1}$ that therefore controls the macroscopic non linearity (see (6.13)).

- k^s and μ^s do not depend on ε_v. In this case (6.5) shows that k^s is necessarily a constant. According to (5.74), ε_d^{ef} is the solution to :

$$
\begin{aligned}
&4(1-\varphi_o)\varepsilon_d^{ef\,2} = -\frac{\partial}{\partial\mu^s}\left(\frac{1}{k^{hom}}\right)(\Sigma_m+P)^2 - \frac{\partial}{\partial\mu^s}\left(\frac{1}{\mu^{hom}}\right)\Sigma_d^2 && (a)\\
&k^{hom} = K(k^s,\mu^s,\mathcal{M})\ ;\ \mu^{hom} = M(k^s,\mu^s,\mathcal{M}) && (b)\\
&\mu^s = \mu^s(\varepsilon_d^{ef}) && (c)
\end{aligned}
\tag{6.20}
$$

  This shows that ε_d^{ef} is a function of $tr\,(\mathbf{\Sigma}+P\mathbf{1})$ and $\mathbf{\Sigma}_d$. The effective stress $\mathbf{\Sigma}+P\mathbf{1}$ therefore controls the macroscopic non linearity.

7 Strength criterion of a porous material

We now assume that the microscopic strength of the solid phase is characterized by a strength criterion $f^s(\boldsymbol{\sigma})\leq 0$. The set G^s of strength-compatible microscopic stress states is thus defined by :

$$
\boldsymbol{\sigma}\in G^s \Leftrightarrow f^s(\boldsymbol{\sigma})\leq 0 \tag{7.1}
$$

7.1 The role of pore pressure on the macroscopic strength

In this section, we examine the question of the macroscopic strength. Following de Buhan and Dormieux (1996), a stress-pressure state $(\mathbf{\Sigma},P)$ is said to be strength-compatible if it is possible to find a microscopic stress field $\boldsymbol{\sigma}(\underline{z})$ that is compatible with the strength of the solid in the sense of (7.1), equal to $-P\mathbf{1}$ in Ω^p and that balances $\mathbf{\Sigma}$ in the sense of the average rule $\mathbf{\Sigma}=\overline{\boldsymbol{\sigma}}$. Let $G^{hom}(P)$ denote the set of strength-compatible stress states $\mathbf{\Sigma}$ when the pore fluid pressure is equal to P :

$$\mathbf{\Sigma} \in G^{hom}(P) \quad \Leftrightarrow \quad \begin{array}{l} \operatorname{div} \boldsymbol{\sigma} = 0 \\ \mathbf{\Sigma} = \overline{\boldsymbol{\sigma}} \\ (\forall \underline{z} \in \Omega^p)\ \boldsymbol{\sigma} = -P\mathbf{1} \\ (\forall \underline{z} \in \Omega^s)\ f^s(\boldsymbol{\sigma}) \leq 0 \end{array} \tag{7.2}$$

In particular, $G^{hom}(0)$ is the domain obtained in the non pressurized case ($P = 0$).
Let us consider a given macroscopic stress state $\mathbf{\Sigma} \in G^{hom}(P)$ and a microscopic stress field $\boldsymbol{\sigma}$ complying with (7.2). We then introduce $\tilde{\boldsymbol{\sigma}} = \boldsymbol{\sigma} + P\mathbf{1}$. It follows that :

$$\mathbf{\Sigma} \in G^{hom}(P) \Leftrightarrow \quad \exists\, \tilde{\boldsymbol{\sigma}} \left\{ \begin{array}{l} \operatorname{div} \tilde{\boldsymbol{\sigma}} = 0 \\ \mathbf{\Sigma} + P\mathbf{1} = \overline{\tilde{\boldsymbol{\sigma}}} \\ (\forall \underline{z} \in \Omega^p)\ \tilde{\boldsymbol{\sigma}} = 0 \\ (\forall \underline{z} \in \Omega^s)\ \tilde{\boldsymbol{\sigma}} \in G^s + P\mathbf{1} \end{array} \right. \tag{7.3}$$

where $G^s + P\mathbf{1}$ is the set obtained from G^s when the translation $\mathbf{a} \to \mathbf{a} + P\mathbf{1}$ is applied.

7.2 Von Mises or Tresca matrix

Let us assume that the solid strength is not influenced by an hydrostatic stress, that is, $G^s + P\mathbf{1} = G^s$. We split the microscopic stress into its spherical and deviatoric parts:

$$\boldsymbol{\sigma} = \boldsymbol{\sigma}_d + \sigma_m \mathbf{1} \quad ; \quad \sigma_m = \frac{1}{3} \operatorname{tr} \boldsymbol{\sigma} \quad ; \quad \operatorname{tr} \boldsymbol{\sigma}_d = 0 \tag{7.4}$$

σ_m is the mean stress and $\boldsymbol{\sigma}_d$ is the deviatoric stress tensor. In the sequel, we shall also refer to the equivalent deviatoric stress $\sigma_d = \sqrt{\boldsymbol{\sigma}_d : \boldsymbol{\sigma}_d / 2}$.
The assumption $G^s + P\mathbf{1} = G^s$ amounts to saying that the solid strength criterion only depends on the deviatoric stress : $f^s = f^s(\boldsymbol{\sigma}_d)$. According to (7.3), if $\mathbf{\Sigma} \in G^{hom}(P)$, the properties of $\tilde{\boldsymbol{\sigma}}$ ensure that $\mathbf{\Sigma} + P\mathbf{1} \in G^{hom}(0)$:

$$G^{hom}(0) = G^{hom}(P) + P\mathbf{1} \tag{7.5}$$

or, alternatively :

$$\mathbf{\Sigma} \in G^{hom}(P) \Leftrightarrow \mathbf{\Sigma} + P\mathbf{1} \in G^{hom}(0) \tag{7.6}$$

In other words, the macroscopic strength criterion of the pressurized porous material can be formulated as a function of Terzaghi's effective stress. (7.5) also shows that the determination of $G^{hom}(P)$ for any value of P amounts to that of $G^{hom}(0)$: $G^{hom}(P)$ is obtained from $G^{hom}(0)$ by a simple translation in the stress space (figure 8).
We are now going to present a method for deriving the macroscopic strength criterion in the case of a von Mises matrix for which the microscopic strength criterion reads :

$$f^s(\boldsymbol{\sigma}_d) = \frac{1}{2} \boldsymbol{\sigma}_d : \boldsymbol{\sigma}_d - k^2 = \sigma_d^2 - k^2 \tag{7.7}$$

According to the previous reasoning, it would be sufficient to consider the case of the empty pore space ($P=0$). Still, we hereafter propose a direct determination of $G^{hom}(P)$ which will be seen to be consistent with (7.5).
The starting point consists in looking for a non linear material, characterized by its secant stiffness tensor $\mathbb{C}^s(\boldsymbol{\varepsilon})$, such that the condition $f^s(\boldsymbol{\sigma})=0$ be asymptotically satisfied at the microscopic level, for large values of the deviatoric strain :

$$f^s(\lim_{\varepsilon_d \to \infty} \boldsymbol{\sigma}) = 0 \tag{7.8}$$

For a given value of P, we then consider radial macroscopic strain pathes parallel to various macroscopic strain states $\tilde{\mathbf{E}}$: $\lambda \to \lambda\tilde{\mathbf{E}}$. It is expected that large microscopic deviatoric strains are going to be induced by large macroscopic strains, i.e. large values[8] of λ. Accordingly, the corresponding macroscopic stress state $\boldsymbol{\Sigma} = \overline{\boldsymbol{\sigma}}$ should lie on the boundary of $G^{hom}(P)$:

$$\lim_{\lambda \to \infty} \boldsymbol{\Sigma} \in \partial G^{hom}(P) \tag{7.9}$$

In other words, the boundary $\partial G^{hom}(P)$ is made up of the extremities of the macroscopic stress pathes. In contrast, for small values of λ, the macroscopic stress path is expected to lie inside $G^{hom}(P)$.
Let us look for $\psi^s(\varepsilon_v, \varepsilon_d)$ in the form :

$$\psi^s(\varepsilon_v, \varepsilon_d) = \frac{1}{2} k^s {\varepsilon_v}^2 + \mathcal{F}(\varepsilon_d) \tag{7.10}$$

where k^s is a constant. (6.3) with (7.10) reads :

$$\boldsymbol{\sigma} = k^s \varepsilon_v \mathbf{1} + 2\mu^s(\varepsilon_d)\boldsymbol{\varepsilon}_d \quad with \quad 2\mu^s(\varepsilon_d) = \frac{1}{2\varepsilon_d}\mathcal{F}'(\varepsilon_d) \tag{7.11}$$

The condition (7.8) is then satisfied if $\mathcal{F}'(\varepsilon_d) \approx 2k$ for large deviatoric strains.This leads to choosing $\mathcal{F}(\varepsilon_d) \approx 2k\varepsilon_d$ in this strain range. In other words, the secant shear modulus $\mu^s(\varepsilon_d)$ is a decreasing function of the deviatoric strain that is equivalent to $k/(2\varepsilon_d)$ for large values of the deviatoric strain (figure 7). Asymptotically, we note that $\mu^s \to 0$ whereas k^s remains constant. This implies that $\mu^s/k^s \ll 1$ which means that the matrix then behaves as an incompressible material.
Let us now move to the response of an r.e.v. made up of such a matrix. We address the resolution of the non linear problem within the framework of the effective stress concept. Owing to the fact that k^s and μ^s do not depend on ε_v, this amounts to solving (6.20). Using the remark concerning the apparent incompressibility, the homogenized bulk and shear modulii (5.28) read :

$$k^{hom} = \frac{4(1-\varphi_o)}{3\varphi_o}\mu^s(\varepsilon_d^{ef}) \quad ; \quad \mu^{hom} = \frac{1-\varphi_o}{1+2\varphi_o/3}\mu^s(\varepsilon_d^{ef}) \tag{7.12}$$

[8] It is worth observing that a strain can be large with respect to a reference strain, while remaining in the domain of infinitesimal strains.

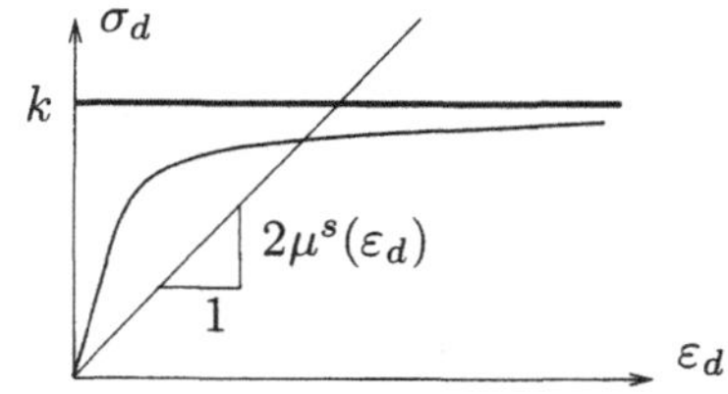

Figure 7. Typical non linear behavior meeting condition (7.8)

The effective deviatoric invariant ε_d^{ef} is derived from (6.20a) :

$$\left(2(1-\varphi_o)\varepsilon_d^{ef}\mu^s(\varepsilon_d^{ef})\right)^2 = \frac{3\varphi_o}{4}(\Sigma_m + P)^2 + (1+\frac{2}{3}\varphi_o){\Sigma_d}^2 \tag{7.13}$$

Recalling that $\mu^s(\varepsilon_d) \approx k/(2\varepsilon_d)$ in the domain of large strains, the macroscopic stress state asymptotically lies on an elliptic curve of the (Σ_m, Σ_d) plane :

$$\frac{3\varphi_o}{4(1-\varphi_o)^2}(\Sigma_m + P)^2 + \frac{1+2\varphi_o/3}{(1-\varphi_o)^2}\Sigma_d^2 = k^2 \tag{7.14}$$

We thus directly retrieve the fact that the macroscopic strength is controlled by Terzaghi's effective stress $\mathbf{\Sigma} + P\mathbf{1}$.

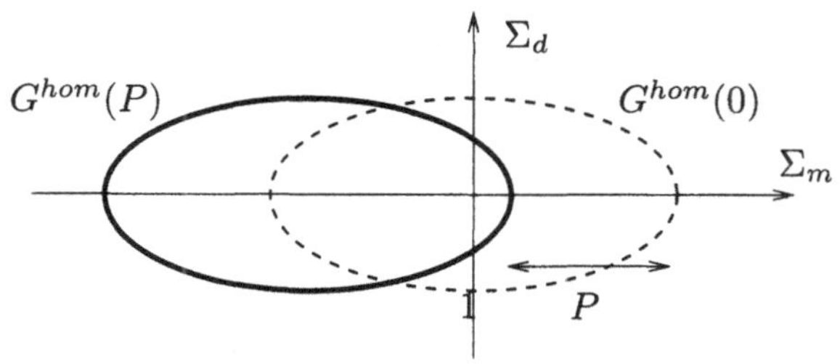

Figure 8. Admissible macroscopic stress states for $P = 0$ and $P \neq 0$

7.3 Drucker-Prager matrix

The above conclusion does not hold any more when the criterion of the matrix is sensitive to the mean stress. More precisely, let us now assume that the set of admissible stress states for the matrix is a cone. Its apex lies on the line $\sigma_1 = \sigma_2 = \sigma_3$ in the space of principal stresses and represents an isotropic tensile stress state $h\mathbf{1}$. This set is denoted by G_h^s. The scalar $h > 0$ can be referred to as the tensile strength. The corresponding set of macroscopic stress states in dry condition ($P = 0$) is denoted by $G_h^{hom}(0)$.

For example, the Drucker-Prager material (figure 9) which criterion reads :

$$f^s(\boldsymbol{\sigma}) = \alpha\left(\frac{1}{3}\mathrm{tr}\,\boldsymbol{\sigma} - h\right) + \sqrt{\frac{1}{2}\boldsymbol{\sigma}_d : \boldsymbol{\sigma}_d} = \alpha(\sigma_m - h) + \sigma_d \leq 0 \tag{7.15}$$

clearly belongs to this class of materials.
Interestingly, we observe that the sets G^s_h and $G^s_{h'}$ associated with two different values

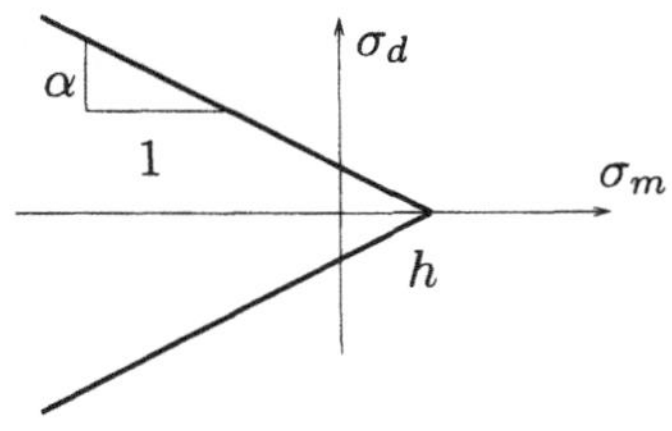

Figure 9. Drucker-Prager criterion : admissible microscopic stress states

h and h' of the tensile strength can be deduced from one another either by a translation or a homothety[9] :

$$(a)\quad G^s_{h'} = \frac{h'}{h} G^s_h \quad ; \quad (b)\quad G^s_{h'} = G^s_h + (h' - h)\mathbf{1} \tag{7.16}$$

According to (7.16a) and to the definition of G^{hom} given in (7.2) (with $P = 0$), we

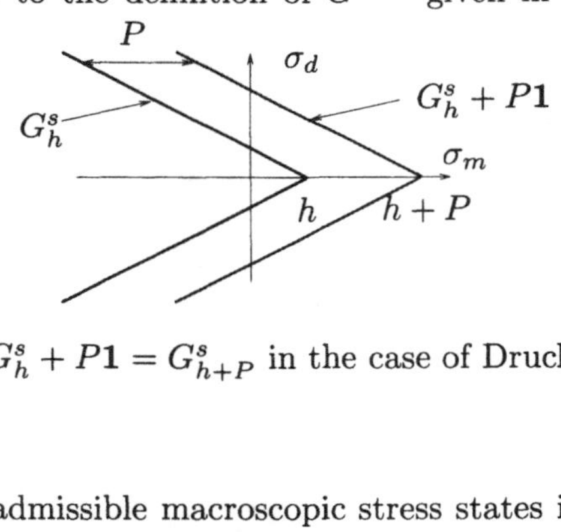

Figure 10. G^s_h and $G^s_h + P\mathbf{1} = G^s_{h+P}$ in the case of Drucker-Prager criterion

then note that the set of admissible macroscopic stress states in dry condition linearly depends on the microscopic tensile strength h :

$$G^{hom}_{h'}(0) = \frac{h'}{h} G^{hom}_h(0) \tag{7.17}$$

Besides, it is readily seen from (7.16b) that (figure 10) :

$$G^s_h + P\mathbf{1} = G^s_{h+P} \tag{7.18}$$

[9] λG^s is the image of G^s by the homothety which center is located at the origin, with a ratio equal to λ : $\underline{z} \rightarrow \lambda \underline{z}$

Introducing this result into (7.3) yields :

$$\boldsymbol{\Sigma} \in G_h^{hom}(P) \Leftrightarrow \boldsymbol{\Sigma} + P\mathbf{1} \in G_{h+P}^{hom}(0) \tag{7.19}$$

Combining (7.17) and (7.19) eventually reads :

$$G_h^{hom}(P) + P\mathbf{1} = (1 + \frac{P}{h})G_h^{hom}(0) \tag{7.20}$$

or, in other words :

$$\boldsymbol{\Sigma} \in G_h^{hom}(P) \Leftrightarrow \frac{\boldsymbol{\Sigma} + P\mathbf{1}}{1 + P/h} \in G_h^{hom}(0) \tag{7.21}$$

This means that the macroscopic strength is controlled by the effective stress :

$$\boldsymbol{\Sigma}'' = \frac{\boldsymbol{\Sigma} + P\mathbf{1}}{1 + P/h} \tag{7.22}$$

As opposed to the traditional concept of effective stress, we note that $\boldsymbol{\Sigma}''$ does not linearly depend on the pore pressure P. From a geometrical point of view, (7.20) states that $G_h^{hom}(P)$ can be determined from $G_h^{hom}(0)$ by application of two successive transformations. First, the homothety with ratio $1 + P/h$ is applied to $G_h^{hom}(0)$, followed by the translation $\boldsymbol{\Sigma} \to \boldsymbol{\Sigma} - P\mathbf{1}$ (figure 11).
We now consider the particular case of the Drucker-Prager solid matrix (7.15). Since

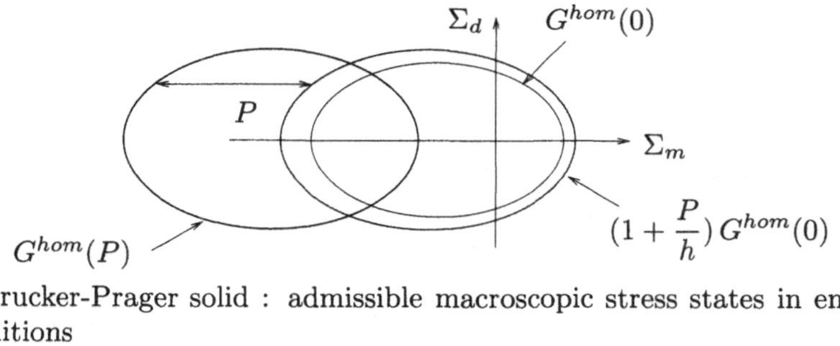

Figure 11. Drucker-Prager solid : admissible macroscopic stress states in empty and saturated conditions

the existence of an effective stress has been established, it is sufficient to determine the macroscopic strength in drained conditions ($P = 0$), that is characterized by the domain $G_h^{hom}(0)$. $G_h^{hom}(P)$ will then be derived from (7.21). To this end, we again use the technique implemented previously in the case of the von Mises matrix, which consists in introducing a non linear material having a state equation of the form (6.2) and meeting the condition (7.8). Looking for an isotropic solution to this problem, we obtain :

$$\mu^s(\varepsilon_v, \varepsilon_d) \approx \frac{\alpha}{2\varepsilon_d}(h - k^s\varepsilon_v) \tag{7.23}$$

in the range of large deviatoric strains. As opposed to that derived for the von Mises matrix ($\mu^s(\varepsilon_d) \approx k/(2\varepsilon_d)$), we note that μ^s in (7.23) depends on both ε_v and ε_d. This physically corresponds to the fact that the shear strength, namely $2\mu^s\varepsilon_d$, increases with the confining pressure represented by $-k^s\varepsilon_v$. As regards the bulk modulus, the simplest choice is a constant value k^s. It is worth observing that the corresponding model does not meet the condition (6.5) ensuring the existence of a potential, which however is not necessary for implementing the reasoning. The non linear problem (6.14) to be solved for the effective strain now comprises the two components ε_v^{ef} and ε_d^{ef}. Using (5.64) and (5.74) with $P = 0$, one obtains :

$$\begin{aligned}
&(1-\varphi_o)\varepsilon_v^{ef} = \frac{\Sigma_m}{k^s} && (a)\\
&4(1-\varphi_o)\varepsilon_d^{ef\,2} = -\frac{\partial}{\partial\mu^s}\left(\frac{1}{k^{hom}}\right)\Sigma_m^2 - \frac{\partial}{\partial\mu^s}\left(\frac{1}{\mu^{hom}}\right)\Sigma_d^2 && (b)\\
&k^{hom} = \frac{4(1-\varphi_o)}{3\varphi_o}\mu^s \quad ; \quad \mu^{hom} = \frac{1-\varphi_o}{1+2\varphi_o/3}\mu^s && (c)\\
&\mu^s = \frac{\alpha}{2\varepsilon_d^{ef}}(h - k^s\varepsilon_v^{ef}) && (d)
\end{aligned} \tag{7.24}$$

As for the von Mises case, we have used in (7.24c) that μ^s/k^s tends towards 0 for large deviatoric strains. The combination of (7.24b) and (7.24c) first yields :

$$\left(2(1-\varphi_o)\varepsilon_d^{ef}\mu^s(\varepsilon_d^{ef})\right)^2 = \frac{3\varphi_o}{4}\Sigma_m^2 + (1+\frac{2}{3}\varphi_o)\Sigma_d{}^2 \tag{7.25}$$

We then introduce (7.24a) and (7.24d) into the lefthand side of (7.25) :

$$\alpha^2h^2(1-\varphi_o)^2 = (\frac{3\varphi_o}{4} - \alpha^2)\Sigma_m^2 + (1+\frac{2}{3}\varphi_o)\Sigma_d{}^2 + 2\alpha^2h(1-\varphi_o)\Sigma_m \tag{7.26}$$

(7.26) characterizes the domain $G_h^{hom}(0)$. We thus retrieve the result obtained by Barthélémy and Dormieux (2003) by means of another non linear homogenization technique, in which the solid phase is given a fictitious non linear viscous constitutive behavior.

In turn, $G_h^{hom}(P)$ is obtained in replacing the macroscopic stress $\mathbf{\Sigma}$ by the effective stress (7.22). Introducing

$$\Sigma_m'' = \frac{\Sigma_m + P}{1+P/h} \quad ; \quad \Sigma_d'' = \frac{\Sigma_d}{1+P/h} \tag{7.27}$$

the elliptic boundary of $G_h^{hom}(P)$ in the (Σ_m, Σ_d) plane is defined by :

$$\alpha^2h^2(1-\varphi_o)^2 = (\frac{3\varphi_o}{4} - \alpha^2)\Sigma_m''^2 + (1+\frac{2}{3}\varphi_o)\Sigma_d''^2 + 2\alpha^2h(1-\varphi_o)\Sigma_m'' \tag{7.28}$$

(7.14) is retrieved as the limit of (7.28) when $h \to \infty$ and $\alpha \to 0$ with $\alpha h = k$.

$G_h^{hom}(0)$ has indeed an elliptic shape and the isotropic compressive strength is bounded if $3\varphi_o/4 > \alpha^2$. In contrast, for lower values of the porosity, namely if $3\varphi_o/4 < \alpha^2$, it should be observed that (7.26) predicts that the isotropic compressive strength is infinite. In other words, for such porosities, any macroscopic stress state of the form $\Sigma_m\mathbf{1}$ with $\Sigma_m < 0$ is admissible, so that $G_h^{hom}(0)$ is not a closed domain. Still, this conclusion is in contradiction with the fact that the pressure that can be applied to a hollow sphere is bounded, irrespective of its porosity (provided that $\alpha < \sqrt{3}/2$). Indeed, the exact value Σ^+ of the limit pressure can be derived from the classical techniques of limit analysis and reads (Barthélémy and Dormieux (2003)) :

$$\Sigma^+ = h\left(\varphi_o^{\frac{2\alpha/\sqrt{3}}{1-2\alpha/\sqrt{3}}} - 1\right) \tag{7.29}$$

This limitation of the result of the non linear homogenization process is due to a lack of accuracy of the effective strain. Clearly enough, a uniform estimate of the deviatoric strain level throughout the solid phase can become too crude for low values of the porosity. The difficulty can be overcome in dividing the solid phase into subdomains, each being associated with a specific value of the effective strain.

8 Poroelasticity in partially saturated conditions

We now assume that the pore space Ω^p is divided into a liquid domain Ω^ℓ and a gazeous one Ω^g. The pressures are uniform but different in each phase : p_ℓ in Ω^ℓ and p_g in Ω^g. For forthcoming purposes, we define the capillary pressure $p_c = p_g - p_\ell$, as well as the liquid saturation ratio $S_r = \varphi^\ell/\varphi$ that represents the liquid volume fraction in the pore space. We also introduce the average pressure $\overline{p}$ defined as :

$$\overline{p} = (1 - S_r)p_g + S_r p_\ell = p_g - S_r p_c \tag{8.1}$$

As opposed to section 5, we now consider the surface tension effects at the interfaces $I^{\alpha\beta}$ between the different physical phases (α, $\beta \in \{s, \ell, g\}$). This phenomenon can be taken into account by means of membrane stresses in these interfaces. From a physical point of view, the membrane stress aims at capturing the concentration of tensile internal forces in a layer of infinitesimal thickness. From a mathematical point of view, this concept goes back to the distribution theory : $\delta_{\alpha\beta}$ denoting the Dirac distribution associated with the interface $I^{\alpha\beta}$, and defined by :

$$(\forall\psi) \quad \int \delta_{\alpha\beta}(\underline{z})\psi(\underline{z})\, dV_z = \int_{I^{\alpha\beta}} \psi(\underline{z})\, dS_z \tag{8.2}$$

the membrane stress that takes place in $I^{\alpha\beta}$ reads $\gamma^{\alpha\beta}\mathbf{1}_T\delta_{\alpha\beta}$. $\mathbf{1}_T$ denotes the identity tensor of the plane T tangent to the interface $I^{\alpha\beta}$ at the considered point. $\gamma^{\alpha\beta}$ is the surface tension. The approach presented at section 5.1 can be generalized with a new definition of the field $\boldsymbol{\sigma}_o(\underline{z})$ appearing in (5.5) (Chateau and Dormieux (2002)) :

$$\boldsymbol{\sigma}_o(\underline{z}) = \begin{cases} 0 & (\Omega^s) \\ -p_\ell\mathbf{1} & (\Omega^\ell) \\ -p_g\mathbf{1} & (\Omega^g) \\ \gamma^{\alpha\beta}\mathbf{1}_T\delta_{\alpha\beta} & (I^{\alpha\beta}) \end{cases} \tag{8.3}$$

The macroscopic state equation (5.51) then takes the form :

$$\boldsymbol{\Sigma} = \mathbb{C}^{hom} : \mathbf{E} - p_\ell \mathbf{B}^\ell - p_g \mathbf{B}^g + \frac{\gamma^{\alpha\beta}}{|\Omega_o|} \int_{I^{\alpha\beta}} \mathbf{1}_T : \mathbb{A}\, dS_z \tag{8.4}$$

with

$$\mathbf{B}^\ell = \varphi^\ell \mathbf{1} : \overline{\mathbb{A}}^\ell \quad ; \quad \mathbf{B}^g = \varphi^g \mathbf{1} : \overline{\mathbb{A}}^g \tag{8.5}$$

In some cases, there is no difference between the morphologies of Ω^ℓ and Ω^g. This implies that $\overline{\mathbb{A}}^\ell = \overline{\mathbb{A}}^g$. From (5.41) and (8.5), we then obtain that $\mathbf{B}^\ell = S_r \mathbf{B}$ and $\mathbf{B}^g = (1-S_r)\mathbf{B}$. With this remark, (8.4) also reads :

$$\boldsymbol{\Sigma} + \overline{p}\mathbf{B} = \mathbb{C}^{hom} : \mathbf{E} + \frac{\gamma^{\alpha\beta}}{|\Omega_o|} \int_{I^{\alpha\beta}} \mathbf{1}_T : \mathbb{A}\, dS_z \tag{8.6}$$

The lefthand side is sometimes referred to as the Bishop effective stress. We note that it is actually an effective stress in the sense that it controls the macroscopic strain *only* if the contribution of the interfaces in the righthand side can be neglected.
(8.3) implicitely assumes that the initial state of the solid in the reference configuration ($\boldsymbol{\varepsilon} = 0$) is the natural state : $\boldsymbol{\sigma} = 0$. The macroscopic loading that is compatible with this assumption is characterized by $\boldsymbol{\Sigma} = p_c = p_g = 0$ as well as no surface tension effects. Strictly speaking, this implies that the pore space is empty. Still, for practical purposes, it may be more convenient to define the initial state (resp. the reference configuration) by the conditions $S_r = 1$ (liquid saturated) and $\boldsymbol{\Sigma} = p_l = 0$. In this state, the solid boundary is subjected to the solid-liquid surface tension $\gamma^{s\ell}$. It may be readily seen that this definition of the initial state amounts to replacing γ^{sg} by $\tilde{\gamma}^{sg} = \gamma^{sg} - \gamma^{s\ell}$ in the contribution of the solid-gas interface to (8.4) and (8.6) and to delete the contribution of the solid-liquid interface. For instance, (8.6) is replaced by :

$$\boldsymbol{\Sigma} + \overline{p}\mathbf{B} = \mathbb{C}^{hom} : \mathbf{E} + \frac{\gamma^{\ell g}}{|\Omega_o|} \int_{I^{\ell g}} \mathbf{1}_T : \mathbb{A}\, dS_z + \frac{\tilde{\gamma}^{sg}}{|\Omega_o|} \int_{I^{sg}} \mathbf{1}_T : \mathbb{A}\, dS_z \tag{8.7}$$

It is worth mentioning that the Young equation relates $\tilde{\gamma}^{sg}$, $\gamma^{\ell g}$ and the contact angle θ at the solid-liquid interface according to (figure 12) :

$$\tilde{\gamma}^{sg} = \gamma^{sg} - \gamma^{s\ell} = \gamma^{\ell g} \cos\theta \tag{8.8}$$

9 Partially saturated distribution of spherical pores

9.1 Linear poroelastic behavior

The implementation of (8.4) or (8.6) clearly requires informations concerning the morphology of the $I^{\alpha\beta}$ interfaces. In the sequel, we focus on the case where the pore space is made up of a set of isotropically distributed connected spherical pores. The statistical description of the pore radius distribution is characterized by a function $\alpha(r)$

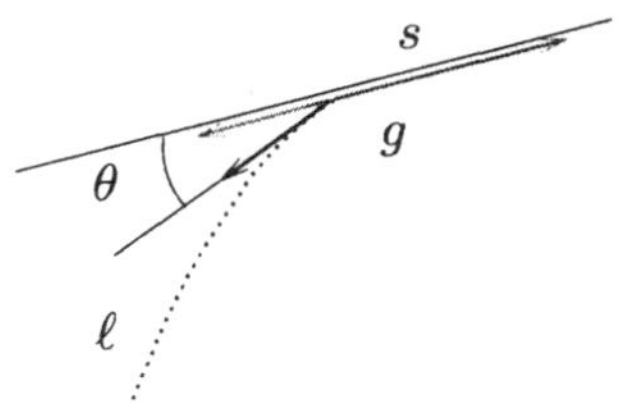

Figure 12. Mechanical equilibrium of surface tensions

such that $\alpha(r)\,dr$ is the volume fraction of the pores with radius $\rho \in [r, r+dr]$. In particular, we note that

$$\int_0^\infty \alpha(r)\,dr = \varphi_o \tag{9.1}$$

Each pore is assumed to be liquid- or gas-saturated (figure 13). For any value of the saturation ratio, it is assumed that it is possible to define a threshold $r^*(S_r)$ such that the pore is liquid-saturated if $r < r^*(S_r)$ and gas-saturated otherwise. During a drying process starting from the liquid-saturated state, this amounts to assuming that the larger pores are emptied first. In other words, $r^*(S_r)$ is expected to be a increasing function of S_r.

The existence of hydraulic connexion between the pores is essential for ensuring uniform pressures in the liquid and gazeous phases. Still, it is assumed that the contribution of the connexion channels to the macroscopic behavior can be disregarded with respect to that of the pores. Besides, the liquid-gas interface $I^{\ell g}$ is made up of meniscii located in the channels that relate the liquid-saturated pores to the gas-saturated ones. The contribution of the surface tension in $I^{\ell g}$ is also neglected with respect to those of $I^{s\ell}$ and I^{sg}. The initial state (resp. reference configuration) being defined by $S_r = 1$ and $\mathbf{\Sigma} = p_l = 0$, the macroscopic stress-strain state equation (8.7) reduces to :

$$\mathbf{\Sigma} + \overline{p}\mathbf{B} = \mathbb{C}^{hom} : \mathbf{E} + \frac{\tilde{\gamma}^{sg}}{|\Omega_o|} \int_{I^{sg}} \mathbf{1}_T : \mathbb{A}\, dS_z \tag{9.2}$$

Introducing the distribution function $\alpha(r)$ and the threshold $r^*(S_r)$, (9.2) also reads :

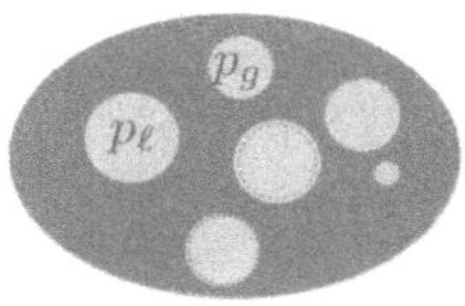

Figure 13. isotropic distribution of spherical liquid- or gas-saturated pores

$$\boldsymbol{\Sigma} + \overline{p}\mathbf{B} = \mathbb{C}^{hom} : \mathbf{E} + \tilde{\gamma}^{sg} \int_{r^*(S_r)}^{\infty} \frac{3\alpha(r)}{4\pi r^3} \left(\int_{\mathcal{S}(r)} \mathbf{1}_T : \mathbb{A} \, dS_z \right) dr \tag{9.3}$$

where $\mathcal{S}(r)$ denotes the spherical boundary of the pore with radius r. In the line of thought of the classical homogenization schemes, it is reasonable to estimate the strain concentration on the pore boundary in the framework of the Eshelby inclusion problem. In the latter, the pore is embedded in an infinite solid matrix, with appropriate uniform strain boundary conditions at infinity. In the case of spherical or ellipsoidal inclusions, it is found that the strain concentration tensor is uniform inside the pore. The corresponding value, denoted by $\mathbb{A}^{est}$, is adopted as an estimate for the real conditions :

$$(\forall \underline{z} \in \mathcal{S}(r)) \quad \mathbb{A}(\underline{z}) \approx \mathbb{A}^{est} \tag{9.4}$$

Recalling (5.41), the corresponding estimate of the Biot tensor $\mathbf{B}$ is :

$$\mathbf{B} \approx \varphi_o \mathbf{1} : \mathbb{A}^{est} \tag{9.5}$$

The integral on $\mathcal{S}(r)$ now reads :

$$\int_{\mathcal{S}(r)} \mathbf{1}_T : \mathbb{A} \, dS_z = \left(\int_{\mathcal{S}(r)} \mathbf{1}_T \, dS_z \right) : \mathbb{A}^{est} \tag{9.6}$$

Observing that $\operatorname{tr} \mathbf{1}_T = 2$, it is readily seen that

$$\int_{\mathcal{S}(r)} \mathbf{1}_T \, dS_z = \frac{8\pi r^2}{3} \mathbf{1} \tag{9.7}$$

Introducing (9.5) and (9.7) into (9.6) yields :

$$\int_{\mathcal{S}(r)} \mathbf{1}_T : \mathbb{A} \, dS_z = \frac{8\pi r^2}{3\varphi_o} \mathbf{B} \tag{9.8}$$

Returning to (9.3), one eventually obtains :

$$\boldsymbol{\Sigma} + \left(\overline{p} - \frac{1}{\varphi_o} \int_{r^*(S_r)}^{\infty} \frac{2\tilde{\gamma}^{sg}}{r} \alpha(r) dr \right) \mathbf{B} = \mathbb{C}^{hom} : \mathbf{E} \tag{9.9}$$

The contribution of the solid-gas interface to the macroscopic state equation appears as a function of the saturation ratio. However, for practical use, we would still have to characterize the threshold function $r^*(S_r)$.

9.2 average strain level in a linear elastic solid phase

We have seen that the specific mechanical effects to be considered in unsaturated conditions are two different pressures p_ℓ and p_g respectively in Ω^ℓ and Ω^g, as well as membrane tension effects in the interfaces (see (8.3)). For the sake of simplicity, we shall focus on the effects of the liquid and gas pressures on the macroscopic strength but disregard that of the membrane tensions. We look for an estimate of the deviatoric effective strain as defined in (6.17). As in section 5.5, the starting point is the potential energy Ψ^* of the solid phase. The latter comprises the term of elastic energy and the work of the fluid pressures. Generalizing (5.66), we obtain

$$|\Omega_o|\Psi^* = \frac{1}{2}\int_{\Omega^s} \boldsymbol{\varepsilon} : \mathbb{C}^s : \boldsymbol{\varepsilon}\, dV_z - p_\ell \int_{\Omega^\ell} \operatorname{tr}\boldsymbol{\varepsilon}\, dV_z - p_g \int_{\Omega^g} \operatorname{tr}\boldsymbol{\varepsilon}\, dV_z \tag{9.10}$$

(5.67) is now replaced by :

$$\begin{aligned} |\Omega_o|\frac{\partial \Psi^*}{\partial \mu^s} &= \int_{\Omega^s} \boldsymbol{\varepsilon}_d : \boldsymbol{\varepsilon}_d\, dV_z + \int_{\Omega^s} \frac{\partial \boldsymbol{\varepsilon}}{\partial \mu^s} : \mathbb{C}^s : \boldsymbol{\varepsilon}\, dV_z \\ &+ \int_{\Omega^g} -p_g \mathbf{1} : \frac{\partial \boldsymbol{\varepsilon}}{\partial \mu^s}\, dV_z + \int_{\Omega^\ell} -p_\ell \mathbf{1} : \frac{\partial \boldsymbol{\varepsilon}}{\partial \mu^s}\, dV_z \end{aligned} \tag{9.11}$$

Observing that the microscopic stress respectively reads $\mathbb{C}^s : \boldsymbol{\varepsilon}$ in Ω^s, $-p_g\mathbf{1}$ in Ω^g, and $-p_\ell\mathbf{1}$ in Ω^ℓ, we observe that (5.68) is still valid. Hill's lemma then yields :

$$\frac{\partial \Psi^*}{\partial \mu^s} = 2(1-\varphi_o)\overline{\varepsilon_d^2}^s \tag{9.12}$$

This remark also allows to write (9.10) as follows :

$$|\Omega_o|\Psi^* = \frac{1}{2}\left(\int_{\Omega} \boldsymbol{\sigma} : \boldsymbol{\varepsilon}\, dV_z - p_\ell \int_{\Omega^\ell} \operatorname{tr}\boldsymbol{\varepsilon}\, dV_z - p_g \int_{\Omega^g} \operatorname{tr}\boldsymbol{\varepsilon}\, dV_z\right) \tag{9.13}$$

and, applying Hill's lemma once more, we obtain :

$$\Psi^* = \frac{1}{2}\left(\boldsymbol{\Sigma} : \mathbf{E} - p_\ell\varphi^\ell \mathbf{1} : \overline{\boldsymbol{\varepsilon}}^\ell - p_g\varphi^g \mathbf{1} : \overline{\boldsymbol{\varepsilon}}^g\right) \tag{9.14}$$

In order to take advantage of (9.12), it appears that we need estimates for the average volume strains in both the liquid and gas domains. The latter can be derived from the solution of the Eshelby problem for prestressed inclusions (see for instance Dormieux et al. (2004)). Generalizing the reasoning presented for empty pores at the end of section 5.2 (see (5.22)), the average strain in Ω^α ($\alpha = \ell, g$) is assumed to be equal to the uniform strain that is induced in an ellipsoidal pore saturated by a fluid at the pressure p_α, embedded in an infinite solid matrix subjected to uniform strain boundary conditions at infinity : $\underline{\xi} \to \mathbf{E}_o \cdot \underline{z}$ when $\underline{z} \to \infty$ (see figure 14). $\mathbf{E}_o$ represents the average strain $\overline{\boldsymbol{\varepsilon}}^s$ in the solid matrix :

$$\begin{aligned} \overline{\boldsymbol{\varepsilon}}^\ell &= (\mathbb{I}-\mathbb{S})^{-1} : (\mathbf{E}_o + p_\ell \mathbb{P} : \mathbf{1}) \\ \overline{\boldsymbol{\varepsilon}}^g &= (\mathbb{I}-\mathbb{S})^{-1} : (\mathbf{E}_o + p_g \mathbb{P} : \mathbf{1}) \\ \overline{\boldsymbol{\varepsilon}}^s &= \mathbf{E}_o \end{aligned} \tag{9.15}$$

where $\mathbb{S}$ is the Eshelby tensor of a spherical inclusion in the solid matrix (see (5.27))

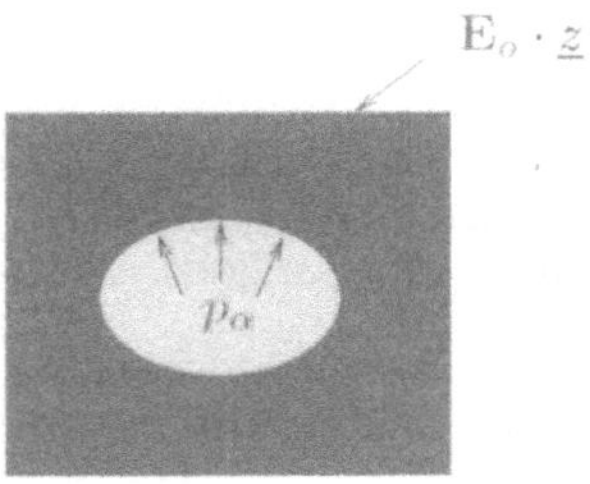

Figure 14. the generalized Eshelby problem

and $\mathbb{P} = \mathbb{S} : \mathbb{C}^{s-1}$. We now take advantage of the average rule $\mathbf{E} = \overline{\varepsilon}$ that provides $\mathbf{E}_o$ as a function of $\mathbf{E}$, p_ℓ and p_g :

$$\mathbf{E}_o = \left((1-\varphi_o)\mathbb{I} + \varphi_o(\mathbb{I}-\mathbb{S})^{-1}\right)^{-1} : \left(\mathbf{E} - \varphi_o \overline{p}(\mathbb{I}-\mathbb{S})^{-1} : \mathbb{P} : \mathbf{1}\right) \tag{9.16}$$

Introducing (9.16) into (9.15) and recalling the identity (5.24) then yields :

$$\overline{\varepsilon}^\alpha = p_\alpha(\mathbb{I}-\mathbb{S})^{-1} : \mathbb{P} : \mathbf{1} + (\mathbb{I}-(1-\varphi_o)\mathbb{S})^{-1} : \left(\mathbf{E} - \varphi_o\overline{p}(\mathbb{I}-\mathbb{S})^{-1} : \mathbb{P} : \mathbf{1}\right) \tag{9.17}$$

with $\alpha = \ell$ or g. Returning to (9.14), it appears that we have to estimate the total work of the fluid pressures. Using (9.17) and (5.42), we obtain[10] :

$$\begin{aligned} p_\alpha\varphi^\alpha \mathbf{1} : \overline{\varepsilon}^\alpha = \ & \overline{p}\mathbf{B} : \left(\mathbf{E} - \varphi_o\overline{p}(\mathbb{I}-\mathbb{S})^{-1} : \mathbb{P} : \mathbf{1}\right) \\ & + \varphi_o\overline{p^2}\mathbf{1} : (\mathbb{I}-\mathbb{S})^{-1} : \mathbb{P} : \mathbf{1} \end{aligned} \tag{9.18}$$

where $\overline{p^2} = (1-S_r)p_g^2 + S_r p_\ell^2$. After some algebra, the above quantity can also be rearranged into :

$$p_\alpha\varphi^\alpha\mathbf{1} : \overline{\varepsilon}^\alpha = \overline{p}\mathbf{B} : \mathbf{E} + \frac{\overline{p}^2}{N} + \varphi_o(\overline{p^2} - \overline{p}^2)\mathbf{1} : (\mathbb{I}-\mathbb{S})^{-1} : \mathbb{P} : \mathbf{1} \tag{9.19}$$

where the expression (5.43) of N has been used. The last step consists in introducing the state equation (8.6) (without the membrane tension term) and (9.19) into (9.14). This yields the following estimate of the potential energy in unsaturated conditions :

$$\Psi^* = \frac{1}{2}\mathbf{E} : \mathbb{C}^{hom} : \mathbf{E} - \overline{p}\mathbf{B} : \mathbf{E} - \frac{\overline{p}^2}{2N} - \frac{\varphi_o}{2}(\overline{p^2} - \overline{p}^2)\mathbf{1} : (\mathbb{I}-\mathbb{S})^{-1} : \mathbb{P} : \mathbf{1} \tag{9.20}$$

Interestingly, it is readily seen that $\overline{p^2} - \overline{p}^2 = S_r(1-S_r)p_c^2$. Considering spherical pores and recalling (5.27), the above expression therefore reduces to :

$$\Psi^* = \frac{1}{2}\mathbf{E} : \mathbb{C}^{hom} : \mathbf{E} - \overline{p}\mathbf{B} : \mathbf{E} - \frac{\overline{p}^2}{2N} - \frac{3\varphi_o}{8\mu^s}S_r(1-S_r)p_c^2 \tag{9.21}$$

[10] Summation over repeated subscript is assumed.

(9.21) appears as the extension of (5.59) to unsaturated conditions, the homogeneous pore pressure P being replaced by the average $\overline{p}$. Still, from an energy point of view, the last term in the above equation shows that the unsaturated conditions are not equivalent to saturated conditions with $P=\overline{p}$. We now return to (9.12) and we assume macroscopic isotropy. Using (5.70) and (9.21), we obtain :

$$2(1-\varphi_o)\overline{\varepsilon_d^2}^s = \frac{1}{2}\frac{\partial k^{hom}}{\partial \mu^s}(\operatorname{tr}\mathbf{E}+\frac{\overline{p}}{k^s})^2 + \frac{\partial \mu^{hom}}{\partial \mu^s}\mathbf{E}_d : \mathbf{E}_d + \frac{3\varphi_o}{8\mu^{s2}}S_r(1-S_r)p_c^2 \tag{9.22}$$

which extends (5.71) to unsaturated conditions. Alternatively, the state equation (9.2) can be used for deriving an expression of $\overline{\varepsilon_d^2}^s$ as a function of $\mathbf{\Sigma}$, $\overline{p}$ and p_c. If membrane tension effects are disregarded, this amounts to replacing P in (5.72) by $\overline{p}$:

$$4(1-\varphi_o)\overline{\varepsilon_d^2}^s = -\frac{\partial}{\partial \mu^s}(\frac{1}{k^{hom}})(\Sigma_m+\overline{p})^2 - \frac{\partial}{\partial \mu^s}(\frac{1}{\mu^{hom}})\Sigma_d^2 + \frac{3\varphi_o}{4\mu^{s2}}S_r(1-S_r)p_c^2 \tag{9.23}$$

9.3 strength in partially saturated conditions

In the line of thought of section 7, we now investigate the influence of partially saturated conditions on the strength of a porous medium. As in section 9.2, the membrane tension effects are disregarded.

von Mises solid matrix We first consider the case of a solid matrix of the von Mises type (see (7.7)). As in section 7.2, we shall look for the limit states of an r.e.v. which matrix is made up of the material defined by the potential introduced in (7.10) with $\mathcal{F}(\varepsilon_d)\approx 2k\varepsilon_d$.
In the framework of the effective strain concept, the macroscopic stress then reads :

$$\mathbf{\Sigma} + \overline{p}\mathbf{B}(\varepsilon_d^{ef}) = \mathbb{C}^{hom}(\varepsilon_d^{ef}) : \mathbf{E} \tag{9.24}$$

where the effective strain ε_d^{ef} defined in (6.17) is the solution to the following non linear problem :

$$4(1-\varphi_o){\varepsilon_d^{ef}}^2 = -\frac{\partial}{\partial \mu^s}(\frac{1}{k^{hom}})(\Sigma_m+\overline{p})^2 - \frac{\partial}{\partial \mu^s}(\frac{1}{\mu^{hom}})\Sigma_d^2 + \frac{3\varphi_o}{4\mu^{s2}}S_r(1-S_r)p_c^2 \tag{9.25}$$

$$k^{hom} = K(k^s,\mu^s,\mathcal{M})\ ;\ \mu^{hom} = M(k^s,\mu^s,\mathcal{M}) \tag{9.26}$$

$$\mu^s = \mu^s(\varepsilon_d^{ef}) \tag{9.27}$$

We look for the stress states reached asymptotically on radial strain paths $\lambda \to \lambda\tilde{\mathbf{E}}$ (see (7.9)). For large values of λ, the effective deviatoric strain is expected to be large as well and (9.26) can be replaced by the simplified expressions (7.12) which are valid if $\mu^s/k^s \ll 1$:

$$k^{hom} = \frac{4(1-\varphi_o)}{3\varphi_o}\mu^s(\varepsilon_d^{ef}) \quad ; \quad \mu^{hom} = \frac{1-\varphi_o}{1+2\varphi_o/3}\mu^s(\varepsilon_d^{ef}) \tag{9.28}$$

We then combine (9.25) and (9.28) :

$$\left(2\varepsilon_d^{ef}\mu^s(\varepsilon_d^{ef})\right)^2 = \frac{3\varphi_o}{4(1-\varphi_o)^2}(\Sigma_m+\overline{p})^2 + \frac{1+\frac{2}{3}\varphi_o}{(1-\varphi_o)^2}\Sigma_d^2 + \frac{3\varphi_o S_r(1-S_r)}{4(1-\varphi_o)}p_c^2 \quad (9.29)$$

Recalling that $\mu^s(\varepsilon_d) \approx k/(2\varepsilon_d)$ in the domain of large strains, the macroscopic stress state asymptotically lies on an elliptic curve of the (Σ_m, Σ_d) plane :

$$\frac{3\varphi_o}{4(1-\varphi_o)^2}(\Sigma_m+\overline{p})^2 + \frac{1+\frac{2}{3}\varphi_o}{(1-\varphi_o)^2}\Sigma_d^2 = k^2 - \frac{3\varphi_o S_r(1-S_r)}{4(1-\varphi_o)}p_c^2 \quad (9.30)$$

Strictly speaking, the strength predicted by the above criterion is not controlled by the effective stress $\mathbf{\Sigma}+\overline{p}\mathbf{1}$ since the capillary pressure p_c also appears in the righthand side of (9.30). Still, returning to (7.14), it is observed that the unsaturated conditions can be taken into account in replacing the pressure P in $\mathbf{\Sigma}+P\mathbf{1}$ by the average pressure $\overline{p}$, provided that the shear strength k of the solid be itself replaced by the corrected strength k' given by

$$k' = k\sqrt{1 - \frac{3\varphi_o S_r(1-S_r)}{4(1-\varphi_o)}\frac{p_c^2}{k^2}} \quad (9.31)$$

From a geometrical point of view, we start from the strength criterion in the dry case, that corresponds to the ellipse centered at the origin in figure 15. The strength criterion corresponding to the saturated state with fluid pressure $\overline{p}$ is obtained by a translation to the right, since we expect that $\overline{p} < 0$ (suction). The strength criterion for the unsaturated state is obtained from the previous ellipse by the homothety with ratio $k'/k < 1$.

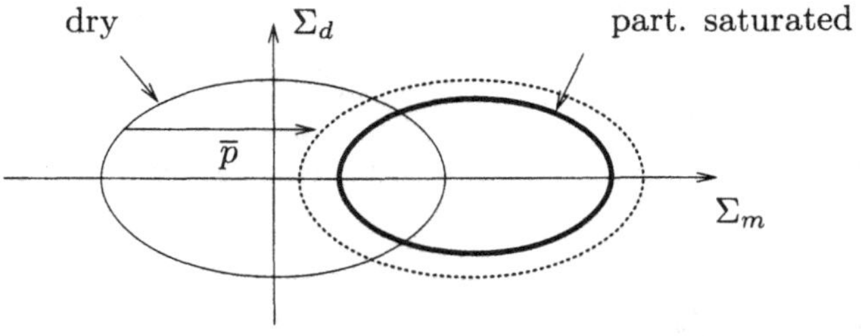

Figure 15. macroscopic strength criterion in the case of a von Mises solid matrix.

Drucker-Prager solid matrix In the case of a Drucker-Prager solid matrix (7.15), the same method can be used except for the fact that the shear modulus is now given by (7.23) in the range of large deviatoric strains. We then need an estimate for both ε_v^{ef} and ε_d^{ef}, still defined as in (6.17).
Disregarding membrane tension effects, the average stress rule reads $\mathbf{\Sigma} = (1-\varphi_o)\overline{\boldsymbol{\sigma}}^s - \varphi_o\overline{p}\mathbf{1}$. As regards $\overline{\varepsilon_v}^s$, we just have to replace P by $\overline{p}$ in (5.62). Hence, instead of the set

of equations (7.24) obtained in the saturated case, the non linear problem now reads :

$$\begin{array}{ll}
(1-\varphi_o)\varepsilon_v^{ef} = \dfrac{\Sigma_m+\varphi_o\overline{p}}{k^s} & (a)\\
4(1-\varphi_o){\varepsilon_d^{ef}}^2 = -\dfrac{\partial}{\partial\mu^s}(\dfrac{1}{k^{hom}})(\Sigma_m+\overline{p})^2-\dfrac{\partial}{\partial\mu^s}(\dfrac{1}{\mu^{hom}})\Sigma_d^2+\dfrac{3\varphi_o}{4\mu^{s2}}S_r(1-S_r)p_c^2 & (b)\\
k^{hom}=\dfrac{4(1-\varphi_o)}{3\varphi_o}\mu^s \quad ; \quad \mu^{hom}=\dfrac{1-\varphi_o}{1+2\varphi_o/3}\mu^s & (c)\\
\mu^s=\dfrac{\alpha}{2\varepsilon_d^{ef}}(h-k^s\varepsilon_v^{ef}) & (d)
\end{array} \qquad (9.32)$$

As in the von Mises case, (9.29) is retrieved from the combination of (9.32b) and (9.32c). Owing to (9.32a) and (9.32d), the following yield enveloppe is obtained :

$$\frac{3\varphi_o}{4(1-\varphi_o)^2}(\Sigma_m+\overline{p})^2+\frac{1+\frac{2}{3}\varphi_o}{(1-\varphi_o)^2}\Sigma_d^2+\frac{3\varphi_o S_r(1-S_r)}{4(1-\varphi_o)}p_c^2=\alpha^2(h-\frac{\Sigma_m+\varphi_o\overline{p}}{1-\varphi_o})^2 \quad (9.33)$$

Generalizing (7.27), we now introduce :

$$\boldsymbol{\Sigma}''=\frac{\boldsymbol{\Sigma}+\overline{p}\mathbf{1}}{1+\overline{p}/h} \quad ; \quad \Sigma''_m=\frac{\Sigma_m+\overline{p}}{1+\overline{p}/h} \quad ; \quad \Sigma''_d=\frac{\Sigma_d}{1+\overline{p}/h} \quad (9.34)$$

With these notations, (9.33) becomes :

$$\alpha^2h^2(1-\varphi_o)^2-\mathcal{Q}(p_c)=(\frac{3\varphi_o}{4}-\alpha^2){\Sigma''_m}^2+(1+\frac{2}{3}\varphi_o){\Sigma''_d}^2+2\alpha^2h(1-\varphi_o)\Sigma''_m \quad (9.35)$$

where

$$\mathcal{Q}(p_c)=\frac{3\varphi_o(1-\varphi_o)S_r(1-S_r)}{4(1+\overline{p}/h)^2}p_c^2 \quad (9.36)$$

P being replaced by $\overline{p}$, the above criterion is identical to (7.28), except for the correcting term $\mathcal{Q}(p_c)$.

10 acknowledgments

The author is grateful to Eric Lemarchand (Lille University of Technology, France) and Jean-François Barthélémy (Ecole Nationale des Ponts et Chaussées, France) for fruitful discussions.

Bibliography

J.-L. Auriault and E. Sanchez-Palencia. Etude du comportement macroscopique d'un milieu poreux saturé déformable. *Journal de Mécanique*, 16:575–603, 1977.

J.F. Barthélémy and L. Dormieux. Détermination du critère de rupture macroscopique d'un milieu poreux par homogénéisation non linéaire. *C.R. Mécanique*, 331:77–84, 2003.

J. Bear and Y. Bachmat. *Introduction to the modelling of transport phenomena in porous media.* Kluwer Academic Publishers, 1990.

M.A. Biot. A general theory of three-dimensional consolidation. *J. Appl. Phys.*, 12: 155–164, 1941.

X. Chateau and L. Dormieux. Micromechanics of saturated and unsaturated porous media. *Int. J. Num. Anal. Meth. Geomech.*, 26:831–844, 2002.

P. de Buhan and L. Dormieux. On the validity of the effective stress concept for assessing the strength of saturated porous materials : a homogenization approach. *J. Mech. Phys. Solids*, 44:1649–1667, 1996.

V. Deudé, L. Dormieux, D. Kondo, and S. Maghous. Micromechanical approach to nonlinear poroelasticity : application to cracked rocks. *J. Eng. Mech.*, 128:848–855, 2002.

L. Dormieux, E. Lemarchand, D. Kondo, and E. Fairbairn. Elements of poro-micromechanics applied to concrete. *Concrete Sc. Eng. Mater. Struct.*, 265:31–42, 2004.

L. Dormieux, A. Molinari, and D. Kondo. Micromechanical approach to the behavior of poroelastic materials. *J. Mech. Phys. Solids*, 50:2203–2231, 2002.

H. Ene and E. Sanchez-Palencia. Equations et phénomènes de surface pour l'écoulement dans un modèle de milieu poreux. *Journal de Mécanique*, 14:73–108, 1975.

J.D. Eshelby. The determination of the elastic field of an ellipsoidal inclusion, and related problems. *Proc. R. Soc. Lond. A*, 241:376–396, 1957.

F. Gilbert. *Description thermomécanique de milieux plusieurs constituants et application aux milieux poreux saturés.* PhD thesis, Paris VI, 1987.

Z. Hashin. Analysis of composite materials : a survey. *J. Appl. Mech.*, 50:481–505, 1983.

R. Hill. The essential structure of constitutive laws for metal composites and polycrystals. *J. Mech. Phys. Solids*, 15:79–95, 1967.

W. Kreher. Residual stresses and stored elastic energy of composites and polycristals. *J. Mech. Phys. Solids*, 38:115–128, 1990.

J. Mandel. *CISM Lecture notes No. 97*, chapter Plasticité classique et viscoplasticité. Springer, 1972.

P. Ponte-Castaneda. Non linear composites : effective constitutive behavior and microstructure evolution. In P. Suquet, editor, *Continuum micromechanics.* Springer, 1997.

P. Suquet. Effective behavior of non linear composites. In P. Suquet, editor, *Continuum micromechanics.* Springer, 1997.

M. Thompson and J.R. Willis. A reformulation of the equations of anisotropic poroelasticity. *J. Appl. Mech.*, 58:612–616, 1991.

A. Zaoui. *Matériaux hétérogènes et Composites.* Cours de l'Ecole Polytechnique, 1996.

A. Zaoui. Continuum micromechanics : survey. *J. Eng. Mech.*, 128:808–816, 2002.

Poroelasticity and damage theory for saturated cracked media

Luc Dormieux * and Djimedo Kondo †

* Ecole Nationale des Ponts et Chaussées, Champs sur Marne, France.
† Laboratoire de Mécanique de Lille, Lille University of Technology, France.

Abstract First, basic results of linear homogenization applied to porous media are presented and applied in order to derive a linear poroelasticity theory for saturated cracked materials. The emphasis is put on the geometrical modelling of the cracks as well as on the cracks-induced anisotropy. The effect of interactions between cracks and the role of the spatial distribution of cracks on the poroelastic properties are also analyzed. The non linear poroelastic behavior of cracked media due to the progressive cracks closure is also investigated. The last section is devoted to damage induced by crack propagation.

1 Introduction

We briefly recall some of the results established in the chapter "Poroelasticity and strength of fully or partially saturated porous materials" (section 5) which will be applied in the sequel for cracks-like pores. The notations already introduced in this chapter are adopted here. For two second order tensors $\boldsymbol{A}$ and $\boldsymbol{B}$, we also define the fourth order tensor $\boldsymbol{A}\underline{\overline{\otimes}}\boldsymbol{B}$ by :

$$(\boldsymbol{A}\underline{\overline{\otimes}}\boldsymbol{B})_{ijkl} = \frac{1}{2}(A_{ik}B_{jl} + A_{il}B_{jk}) \tag{1.1}$$

1.1 homogenized poroelastic behavior

The classical Biot theory can be derived from the response of a representative elementary volume (r.e.v.) to a mechanical loading defined by uniform strain boundary conditions (macroscopic strain $\mathbf{E}$) and by a fluid pressure P. The two macroscopic state equations which characterize the linear poroelastic behavior reads :

$$\boldsymbol{\Sigma} = \mathbb{C}^{hom} : \mathbf{E} - \mathbf{B}P \tag{1.2}$$

and

$$\phi - \phi_o = \frac{P}{N} + \mathbf{B} : \mathbf{E} \tag{1.3}$$

$\boldsymbol{\Sigma}$ is the macroscopic stress. $(\phi - \phi_o)$ is the normalized pore volume change. For many illustrative purposes, we shall consider the case of an homogeneous isotropic solid matrix.

$\mathbb{I}$ being the symmetric fourth order unit tensor and $\mathbf{1}$ the second order unit tensor, the solid stiffness tensor $\mathbb{C}^s$ then reads :

$$\mathbb{C}^s = 3k^s\mathbb{J} + 2\mu^s\mathbb{K} \quad \text{with } \mathbb{J} = \frac{1}{3}\mathbf{1}\otimes\mathbf{1}\ ;\ \mathbb{K} = \mathbb{I} - \mathbb{J} \tag{1.4}$$

The micromechanics analysis then provides a link between the Biot tensor $\mathbf{B}$ and the drained macroscopic stiffness tensor $\mathbb{C}^{hom}$ of the porous medium :

$$\mathbf{B} = \mathbf{1} - \mathbb{C}^{hom} : \mathbb{S}^s : \mathbf{1} = \mathbf{1} - \mathbf{1} : \mathbb{S}^s : \mathbb{C}^{hom} \tag{1.5}$$

in which $\mathbb{S}^s = (\mathbb{C}^s)^{-1}$ represents the compliance tensor of the solid matrix. Besides, the Biot modulus N is given by :

$$\frac{1}{N} = (\mathbf{B} - \varphi\mathbf{1}) : \mathbb{S}^s : \mathbf{1} \; = \; \left((1-\varphi)\mathbf{1} - \mathbb{C}^{hom} : \mathbb{S}^s : \mathbf{1}\right) : \mathbb{S}^s : \mathbf{1} \tag{1.6}$$

where φ denotes the porosity, that is, the pore volume fraction. These equations show that the determination of the poroelastic behavior of the cracked material only requires the calculation of the drained stiffness tensor $\mathbb{C}^{hom}$.
In the case of the empty (or drained) porous medium, the loading applied to the r.e.v. Ω is defined by the macroscopic strain tensor $\mathbf{E}$. By analogy with the diffusion problem[1], the microscopic displacement field $\underline{\xi}$ and stress field $\boldsymbol{\sigma}$ are extended into the pore space Ω^p which is regarded as an elastic material with stiffness $\mathbb{C}^p = 0$. The mechanical problem defined on the r.e.v. Ω therefore reads:

$$\begin{array}{lr} \operatorname{div}\boldsymbol{\sigma} = 0 & (a) \\ \boldsymbol{\sigma} = \mathbb{C}(\underline{z}) : \boldsymbol{\varepsilon} \quad \text{with} \left\{ \begin{array}{lll} \mathbb{C}(\underline{z}) = \mathbb{C}^p = 0 & \text{for} & \underline{z} \in \Omega^p \\ \mathbb{C}(\underline{z}) = \mathbb{C}^s & \text{for} & \underline{z} \in \Omega^s \end{array} \right. & (b) \\ \underline{\xi} = \mathbf{E}\cdot\underline{z} \text{ when } \underline{z} \in \partial\Omega & (c) \end{array} \tag{1.7}$$

The strain concentration tensor $\mathbb{A}$ relates the microscopic and macroscopic strain tensors $\boldsymbol{\varepsilon}$ and $\boldsymbol{E}$:

$$(\forall \underline{z} \in \Omega) \quad \boldsymbol{\varepsilon}(\underline{z}) = \mathbb{A}(\underline{z}) : \boldsymbol{E} \tag{1.8}$$

The macroscopic stiffness tensor is controlled by the average $\overline{\mathbb{A}}^p$ of $\mathbb{A}$ over Ω^p :

$$\mathbb{C}^{hom} = \mathbb{C}^s : (\mathbb{I} - \varphi\overline{\mathbb{A}}^p) \tag{1.9}$$

1.2 Eshelby's inhomogeneity problem

The solution to Eshelby's inhomogeneity problem has already been described in the chapter of the present book devoted to molecular diffusion. We now move back to the original problem treated by Eshelby (1957), namely a mechanical one. With respect to the problem of diffusive transport, the diffusive flux $\underline{j}^\gamma$ is replaced by the stress tensor $\boldsymbol{\sigma}$, the concentration ρ^γ by the displacement $\underline{\xi}$, Fick's law by the linear elastic constitutive

[1]See chapter "Diffusive transport in disordered porous media", section 1.

equation, and the mass balance law by the momentum balance law. Furthermore, the pore phase Ω^p ($\mathbb{C}^p = 0$) now plays the role of the solid phase Ω^s ($D^s = 0$).
Eshelby's elastic inhomogeneity problem consists in considering a bounded pore I embedded in an infinite elastic solid domain (stiffness tensor $\mathbb{C}^s$) with uniform strain boundary conditions at infinity. As already stated, the pore I is regarded as an elastic material with stiffness $\mathbb{C}^I = 0$. The governing equations of this 'ersatz'-problem then read:

$$\begin{aligned} &\operatorname{div}\boldsymbol{\sigma} = 0 && (a) \\ &\boldsymbol{\sigma} = \mathbb{C}(\underline{z}) : \boldsymbol{\varepsilon} \quad \text{with} \left\{ \begin{array}{ll} \mathbb{C}(\underline{z}) = \mathbb{C}^I = 0 & \text{for} \quad \underline{z} \in I \\ \mathbb{C}(\underline{z}) = \mathbb{C}^s & \text{for} \quad \underline{z} \notin I \end{array} \right. && (b) \\ &\underline{\xi} = \mathbf{E} \cdot \underline{z} \text{ when } \underline{z} \to \infty && (c) \end{aligned} \tag{1.10}$$

where $\mathbf{E}$ denotes the uniform strain tensor at infinity. The results presented hereafter being valid for any uniform value of $\mathbb{C}^I$, it is convenient to introduce $\delta\mathbb{C} = \mathbb{C}^I - \mathbb{C}^s$, with $\delta\mathbb{C} = -\mathbb{C}^s$ in the case of a pore. Using a similar reasoning as in the diffusion problem[2], it is found that the strain field in I solution to the inhomogeneity problem (1.10) is uniform when I is an ellipsoid. It is then related to the loading parameter $\boldsymbol{E}$ according to :

$$\boldsymbol{\varepsilon}^I = (\mathbb{I} + \mathbb{P} : \delta\mathbb{C})^{-1} : \mathbf{E} \tag{1.11}$$

The (fourth order) $\mathbb{P}$ tensor which appears in (1.11) is related to the elastic 2nd-order Green function $\mathbf{G}(\underline{z} - \underline{z}')$ (Willis (1977), Mura (1987), Zaoui (2002)):

$$P_{ijkl}(\underline{z}) = -\left(\frac{\partial^2}{\partial z_j \partial z_k}\left(\int_I G_{il}(\underline{z} - \underline{z}')\, dV_{z'}\right)\right)_{(ij),(kl)} \tag{1.12}$$

In the case of an ellipsoidal (empty) pore considered here, for which $\delta\mathbb{C} = -\mathbb{C}^s$, (1.11) reduces to :

$$\boldsymbol{\varepsilon}^I = (\mathbb{I} - \mathbb{S})^{-1} : \mathbf{E} \tag{1.13}$$

where $\mathbb{S}$ is the so-called Eshelby tensor,

$$\mathbb{S} = \mathbb{P} : \mathbb{C}^s \tag{1.14}$$

As indicated by (1.12), $\mathbb{P}$ and $\mathbb{S}$ depend on the shape and orientation of the ellipsoid I.

1.3 homogenization schemes for porous media

We briefly present the dilute scheme and the Mori-Tanaka scheme in the context of porous materials. For a comprehensive presentation of the available homogenization schemes in relation to the microstructural morphology, see Zaoui (1997).
The pore space is represented as a set of separated pores. The latter are approximated by ellipsoids of appropriate shape and orientation embedded in an elastic medium having

[2]See chapter "diffusive transport in disordered porous media"

the same stiffness as the solid phase. The r.e.v. is subjected to uniform strain boundary conditions.
In the framework of the dilute scheme, the pores do not interact with one another. For the empty porous medium, the estimate of the average strain in a pore is then provided by the solution of Eshelby's inhomogeneity problem (1.11)-(1.13). Accordingly, when the shape and orientation of the pores are uniform in the pore space, the average strain concentration tensor $\overline{\mathbb{A}}^p$ is estimated by:

$$\overline{\mathbb{A}}^p \approx (\mathbb{I} + \mathbb{P} : \delta\mathbb{C})^{-1} = (\mathbb{I} - \mathbb{S})^{-1} \tag{1.15}$$

The homogenized stiffness tensor being given by (1.9) and the Biot tensor by $\mathbf{B} = \varphi\mathbf{1} : \overline{\mathbb{A}}^p$, the dilute estimates of the poroelastic constants read:

$$\mathbb{C}^{hom} = \mathbb{C}^s : \left(\mathbb{I} - \varphi(\mathbb{I} - \mathbb{S})^{-1}\right) \tag{1.16}$$

$$\mathbf{B} = \varphi\mathbf{1} : (\mathbb{I} - \mathbb{S})^{-1} \tag{1.17}$$

When the shape and orientation are not uniform, the pore space should be divided into pore families associated with different values of the Eshelby tensor. The average strain concentration tensor now reads[3] :

$$\overline{\mathbb{A}}^p = \frac{\varphi_j}{\varphi}(\mathbb{I} - \mathbb{S}_j)^{-1} \tag{1.18}$$

where φ_j is the volume fraction of the pore family $\mathcal{F}_j$, $\mathbb{S}_j$ the corresponding Eshelby tensor and $\varphi = \sum_j \varphi_j$. The homogenized stiffness tensor and Biot tensor take the form :

$$\mathbb{C}^{hom} = \mathbb{C}^s : \left(\mathbb{I} - \varphi_j(\mathbb{I} - \mathbb{S}_j)^{-1}\right) \tag{1.19}$$

and

$$\mathbf{B} = \mathbf{1} : \varphi_j(\mathbb{I} - \mathbb{S}_j)^{-1} \tag{1.20}$$

Pore interaction can be captured for instance by means of the Mori-Tanaka scheme. By analogy with the technique presented in the context of molecular diffusion[4], the average $\overline{\varepsilon}^p$ in the interacting pores is obtained from a modified Eshelby inhomogeneity problem in which the real macroscopic strain $\boldsymbol{E}$ at infinity is replaced by a fictitious one $\boldsymbol{E}_o$. The latter can be interpreted as the average $\overline{\varepsilon}^s$ of the microscopic strain field in the solid phase. Considering first the case of a single pore family, one obtains :

$$\overline{\varepsilon}^p = (\mathbb{I} + \mathbb{P} : \delta\mathbb{C})^{-1} : \mathbf{E}_o = (\mathbb{I} - \mathbb{S})^{-1} : \mathbf{E}_o \tag{1.21}$$

The strain average rule[5] $\overline{\varepsilon} = (1 - \varphi)\overline{\varepsilon}^s + \varphi\overline{\varepsilon}^p = \boldsymbol{E}$ then implies

$$\mathbf{E}_o = \left((1 - \varphi)\mathbb{I} + \varphi(\mathbb{I} - \mathbb{S})^{-1}\right)^{-1} : \boldsymbol{E} \tag{1.22}$$

---

[3]Repeated subscript indicates summation.
[4]See chapter "Diffusive transport in disordered porous media", section 5.1.
[5]See chapter "Poroelasticity and strength of fully or partially saturated porous media"

Combining (1.22) with (1.21) yields

$$\overline{\mathbb{A}}^p = (\mathbb{I} - \mathbb{S})^{-1} : \left((1-\varphi)\mathbb{I} + \varphi(\mathbb{I} - \mathbb{S})^{-1}\right)^{-1} \tag{1.23}$$

The Mori-Tanaka estimates of $\mathbb{C}^{hom}$ and $\boldsymbol{B}$ are eventually derived from (1.9) and (1.5) :

$$\mathbb{C}^{hom} = (1-\varphi)\mathbb{C}^s : \left((1-\varphi)\mathbb{I} + \varphi(\mathbb{I} - \mathbb{S})^{-1}\right)^{-1} \tag{1.24}$$

$$\boldsymbol{B} = \mathbf{1} - (1-\varphi)\mathbf{1} : \left((1-\varphi)\mathbb{I} + \varphi(\mathbb{I} - \mathbb{S})^{-1}\right)^{-1} \tag{1.25}$$

The Mori-Tanaka scheme can be extended to the situation where different interacting pore families must be considered. The starting point is again the modified Eshelby inhomogeneity problem in which a pore of the family $\mathcal{F}_j$ is embedded in an infinite solid matrix. The fictitious macroscopic strain $\boldsymbol{E}_o$ is applied at infinity. The corresponding estimate of the average strain in $\mathcal{F}_j$ reads :

$$\overline{\varepsilon}^j = (\mathbb{I} - \mathbb{S})^{-1} : \boldsymbol{E}_o \tag{1.26}$$

As before, $\mathbf{E}_o$ is used as an estimate of the average strain $\overline{\varepsilon}^s$ in the solid phase and is determined from the strain average rule $\overline{\varepsilon} = \boldsymbol{E}$:

$$\mathbf{E}_o = \left((1-\varphi)\mathbb{I} + \varphi_j(\mathbb{I} - \mathbb{S}_j)^{-1}\right)^{-1} : \boldsymbol{E} \tag{1.27}$$

and the corresponding estimate of the average strain concentration tensor in $\mathcal{F}_j$ takes the form :

$$\overline{\mathbb{A}}^j = (\mathbb{I} - \mathbb{S}_j)^{-1} : \left((1-\varphi)\mathbb{I} + \varphi_i(\mathbb{I} - \mathbb{S}_i)^{-1}\right)^{-1} \tag{1.28}$$

Using (1.9) with $\overline{\mathbb{A}}^p = \varphi_j \overline{\mathbb{A}}^j$ eventually yields :

$$\mathbb{C}^{hom} = (1-\varphi)\mathbb{C}^s : \left((1-\varphi)\mathbb{I} + \varphi_i(\mathbb{I} - \mathbb{S}_i)^{-1}\right)^{-1} \tag{1.29}$$

2 Linear poroelasticity of cracked media : the dilute estimates

We now aim at deriving the estimates of the homogenized poroelastic properties of cracked media provided by the dilute scheme. Both opened and closed cracks will be considered. In each case, expressions of Biot tensor and Biot modulus will be given. In this section, the pore space is made up of cracks only. The case of cracked porous media will be examined in sections 3.4 and 4. From a geometrical point of view, the micro-cracks are modelled as oblate spheroids (Nemat-Nasser and Horii (1993)). Considering crack $\mathcal{C}_i$, let us introduce an orthonormal frame $(C_i, \underline{t}_1^i, \underline{t}_2^i, \underline{n}^i)$, in which $\underline{n}^i$ denotes the unit normal to the crack plane and C_i the center of the crack. This oblate spheroid occupies a domain defined by :

$$\frac{z_1^2 + z_2^2}{a^2} + \frac{z_3^2}{c^2} \leq 1. \tag{2.1}$$

with $z_1 = \underline{z} \cdot \underline{t}_1^i$, $z_2 = \underline{z} \cdot \underline{t}_2^i$ and $z_3 = \underline{z} \cdot \underline{n}^i$; a is the crack radius and c the half opening. The aspect ratio $X = c/a$ of such a penny-shaped microcrack is subjected to the condition

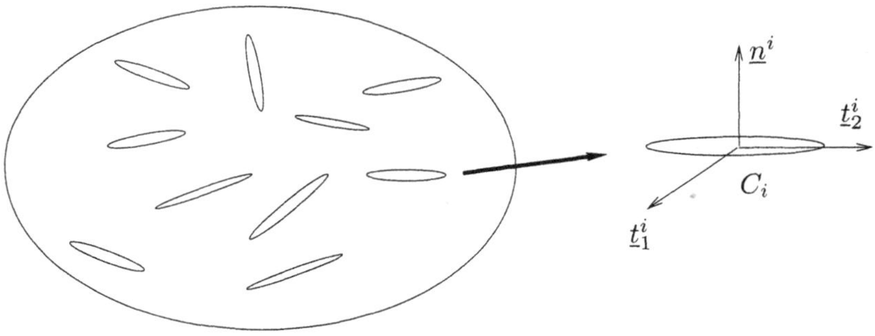

Figure 1. A representative elementary volume of cracked materials

$X \ll 1$.
From a mechanical point of view, a crack is represented as an inhomogeneity with an elastic stiffness tensor $\mathbb{C}^c$. As expected, the choice $\mathbb{C}^c = 0$ is made for opened cracks, in order to account for the cancellation of the stress vector on cracks faces. The particular case of closed cracks will be seen to correspond to $\mathbb{C}^c \neq 0$.

2.1 Materials weakened by parallel opened cracks

We begin with the situation of parallel cracks with identical radius and aspect ratios. The case of an elastic solid matrix weakened by a dilute concentration of opened microcracks is first considered. This amounts to assuming that there is no interaction between cracks and to using the dilute scheme for the determination of the homogenized properties of the material.

Dilute estimate of the homogenized drained stiffness tensor As stated in section 1, the determination of $\mathbb{C}^{hom}$ (and subsequently of the Biot tensor $\boldsymbol{B}$) requires an estimate of the average strain concentration tensor for cracks $\overline{\mathbb{A}}^c$:

$$\mathbb{C}^{hom} = \mathbb{C}^s : (\mathbb{I} - \varphi^c \overline{\mathbb{A}}^c) \tag{2.2}$$

in which φ^c represents the volume fraction of cracks. In the line of Eshelby-based estimates, application of (1.15) yields :

$$\overline{\mathbb{A}}^c = (\mathbb{I} - \mathbb{S}(X, \underline{n}))^{-1} \tag{2.3}$$

where $\mathbb{S}(X, \underline{n})$ is the Eshelby tensor associated with the considered crack family. The quantities X and $\underline{n}$, involved in the Eshelby tensor indicate the dependence of the latter on the aspect ratio of the oblate spheroid and on its orientation. It is recalled that it also depends on the elastic properties of the matrix. Note that $\mathbb{S}(X, \underline{n})$ is related to the tensor denoted $\mathbb{P}(X, \underline{n})$ by :

$$\mathbb{S}(X, \underline{n}) = \mathbb{P}(X, \underline{n}) : \mathbb{C}^s \tag{2.4}$$

For oblate spheroids, with radius a and aspect ratio X, the volume fraction reads :

$$\varphi^c = \frac{4}{3}\pi \mathcal{N} a^3 X \tag{2.5}$$

where $\mathcal{N}$ denotes the crack density (number of cracks per unit volume) of the considered set of parallel cracks.
Combining (2.2) and (2.3), and taking into account (2.5), the dilute estimate of the overall stiffness tensor takes the form :

$$\mathbb{C}^{hom} = \mathbb{C}^s : \left(\mathbb{I} - \frac{4}{3}\pi\epsilon X(\mathbb{I} - \mathbb{S}(X, \underline{n}))^{-1}\right) \tag{2.6}$$

where $\epsilon = \mathcal{N} a^3$ is the crack density parameter. This parameter was introduced by Budiansky and O'Connell (1976).

Comments and discussion The homogenized stiffness tensor $\mathbb{C}^{hom}$ given by (2.6) seems to depend on the current value of the aspect ratio X. Since X is expected to vary significantly as a function of the applied loading, such a dependence would be in contradiction with the assumption of linearity upon which relies the use of the strain concentration tensor in the homogenization procedure. The solution to this difficulty lies in the mathematical properties of the quantities involved in the expression (2.6) of $\mathbb{C}^{hom}$. In fact, for $X \ll 1$, which corresponds to the geometrical model of penny-shaped microcracks, $(\mathbb{I} - \mathbb{S}(X, \underline{n}))^{-1}$ is singular, of the order of $1/X$. In turn, this implies that $X(\mathbb{I} - \mathbb{S}(X, \underline{n}))^{-1}$ has a finite limit $\mathbb{T}$ when $X \to 0$. A constant value of $\mathbb{C}^{hom}$ is therefore obtained, as long as that the cracks remain opened ($X > 0$) :

$$\mathbb{C}^{hom} = \mathbb{C}^s : (\mathbb{I} - \frac{4}{3}\pi\epsilon\, \mathbb{T}(\underline{n})) \tag{2.7}$$

where

$$\mathbb{T}(\underline{n}) = \lim_{X \to 0} X(\mathbb{I} - \mathbb{S}(X, \underline{n}))^{-1} = \lim_{X \to 0} X(\mathbb{I} - \mathbb{P}(X, \underline{n}) : \mathbb{C}^s)^{-1} \tag{2.8}$$

However, this remark is not entirely satisfactory since the strain concentration tensor $\overline{\mathbb{A}}^c$ itself depends on the aspect ratio X (see (2.3)), and through X on the applied loading. We are actually faced with a problem of geometrical non linearity. In fact, a more rigorous reasoning based on a rate formulation is due.
In order to avoid the problem related to geometry changes, we consider a loading increment defined by $\dot{\boldsymbol{E}}$ on the current configuration. Due to the linearity of the relation between the microscopic strain rate $\dot{\boldsymbol{\varepsilon}}$ and $\dot{\boldsymbol{E}}$, we may refer to the concept of *strain rate* concentration tensor :

$$(\forall \underline{z} \in \Omega) \quad \dot{\boldsymbol{\varepsilon}}(\underline{z}) = \mathbb{A}(\underline{z}) : \dot{\boldsymbol{E}} \tag{2.9}$$

Adapting the reasoning presented in section 1, it is readily seen that the macroscopic stress and strain rates tensors are linearly related

$$\dot{\boldsymbol{\Sigma}} = \mathbb{C}_t^{hom} : \dot{\boldsymbol{E}} \quad \text{with} \quad \mathbb{C}_t^{hom} = \overline{\mathbb{C} : \mathbb{A}} = \mathbb{C}^s : (\mathbb{I} - \varphi^c \overline{\mathbb{A}}^c) \tag{2.10}$$

$\mathbb{C}_t^{hom}$ can be interpreted as the macroscopic *tangent* stiffness tensor, the subscript 't' standing for tangent. It a priori depends on the *current* configuration. In particular, φ^c refers to the *current* volume fraction of the cracks. The average strain rate concentration tensor $\overline{\mathbb{A}}_t^c$ over the crack may be estimated by :

$$\overline{\mathbb{A}}_t^c = (\mathbb{I} - \mathbb{S}(X, \underline{n}))^{-1} \tag{2.11}$$

which is formally identical to (2.3) but is free of the criticisms concerning the latter. The value of X appearing in (2.11) corresponds to the current configuration of the crack. Using (2.5) and taking advantage of (2.8), it is readily seen that $\mathbb{C}_t^{hom}$ is a constant independant of X, equal to the effective stiffness tensor $\mathbb{C}^{hom}$ derived in (2.7) :

$$\mathbb{C}_t^{hom} = \mathbb{C}^s : (\mathbb{I} - \frac{4}{3}\pi\epsilon\, \mathbb{T}(\underline{n})) \tag{2.12}$$

This result proves that the macroscopic behavior is indeed linear, despite of the variations of the crack aspect ratio and of the non linearity of the strain concentration rule. Note that the linearity of the macroscopic behavior requires that the cracks remain opened. Crack closure and a quantitative description of the evolution of the crack aspect ratio are considered at section 4.3.

From now on, the homogenized properties which are considered are related to the tangent state equations. Still, the subscript 't' is omitted for simplicity.

In the case of an isotropic solid, we note from (2.12) that a transversely isotropic macroscopic behavior is obtained. The components of the Eshelby tensor of an oblate spheroid can be found in Horii and Nemat-Nasser (1983) as well as in Mura (1987). $\underline{e}_3$ hereafter corresponds to the unit normal $\underline{n}$ of the crack plane. For $X \ll 1$ (penny-shaped cracks), it is sufficient to expand the components S_{ijkl} up to the first order in X. In the local orthonormal frame $(\underline{e}_1, \underline{e}_2, \underline{e}_3)$ with $\underline{n} = \underline{e}_3$, one obtains :

$$\begin{aligned}
&S_{1111} = S_{2222} = \frac{13 - 8\nu^s}{32(1-\nu^s)}\pi X; \qquad && S_{3333} = 1 - \frac{1 - 2\nu^s}{1 - \nu^s}\frac{\pi}{4}X \\
&S_{1122} = S_{2211} = \frac{8\nu^s - 1}{32(1-\nu^s)}\pi X; && S_{1133} = S_{2233} = \frac{2\nu^s - 1}{8(1-\nu^s)}\pi X \\
&S_{3311} = S_{3322} = \frac{\nu^s}{1-\nu^s}\left(1 - \frac{4\nu^s + 1}{8\nu^s}\pi X\right); && S_{1212} = \frac{7 - 8\nu^s}{32(1-\nu^s)}\pi X \\
&S_{1313} = S_{2323} = \frac{1}{2}\left(1 + \frac{\nu^s - 2}{1 - \nu^s}\frac{\pi}{4}X\right)
\end{aligned} \tag{2.13}$$

(with $S_{ijkl} = S_{jikl} = S_{ijlk}$). ν^s represents the Poisson ratio of the elastic solid matrix. Note that this Eshelby tensor is non symmetric. The non zero components of tensor $\mathbb{T}$ are :

$$T_{3311} = T_{3322} = \frac{4\nu^s(1-\nu^s)}{(1-2\nu^s)\pi} \; ; \; T_{3333} = \frac{4(1-\nu^s)^2}{(1-2\nu^s)\pi} \; ; \; T_{1313} = T_{2323} = \frac{2(1-\nu^s)}{(2-\nu^s)\pi} \tag{2.14}$$

with $T_{ijkl} = T_{jikl} = T_{ijlk}$. This yields the following intrinsic expression of $\mathbb{T}(\underline{n})$ (see section 6) :

$$\mathbb{T}(\underline{n}) = \frac{4(1-\nu^s)}{\pi} \left(\frac{\nu^s}{1-2\nu^s}\underline{n}\otimes\underline{n}\otimes\mathbf{1} + \frac{1}{2-\nu^s}(\mathbf{1}\overline{\underline{\otimes}}\underline{n}\otimes\underline{n} + \underline{n}\otimes\underline{n}\overline{\underline{\otimes}}\mathbf{1}) - \frac{\nu^s}{2-\nu^s}\underline{n}\otimes\underline{n}\otimes\underline{n}\otimes\underline{n} \right) \tag{2.15}$$

According to the definitions (2.3) and (2.8) of $\overline{\mathbb{A}}^c$ and $\mathbb{T}$, we note that the property $T_{ijkl} \neq 0$ implies that $\overline{A}^c_{ijkl} = O(1/X)$. Returning to (2.9), this means that the shear strain rates $\dot{\varepsilon}_{13}$ and $\dot{\varepsilon}_{23}$ as well as the normal strain rate $\dot{\varepsilon}_{33}$ are of the order of $|\dot{\boldsymbol{E}}|/X \gg |\dot{\boldsymbol{E}}|$. In other words, a crack is responsible for shear and normal strain amplification. Such amplification effects may be induced by macroscopic stretch rates $\dot{E}_{11}$, $\dot{E}_{22}$ and $\dot{E}_{33}$, or by macroscopic shear strain rates $\dot{E}_{13}$ and $\dot{E}_{23}$.
Finally, note that the dilute estimates of the Biot tensor and modulus are readily derived from (1.5) and (1.6) :

$$\boldsymbol{B} = \frac{4\pi\epsilon}{3}\mathbf{1} : \mathbb{T} = \frac{16(1-\nu^s)}{3(1-2\nu^s)}\epsilon\Big((1-2\nu^s)\underline{n}\otimes\underline{n} + \nu^s\mathbf{1}\Big) \quad ; \quad \frac{1}{N} = \frac{16(1-\nu^{s2})}{3E^s}\epsilon \tag{2.16}$$

where E^s is the Young modulus of the solid. As a remarkable consequence of this anisotropy of $\boldsymbol{B}$, the pore fluid pressure induces an anisotropic contribution to the macroscopic stress state. A micromechanical analysis of anisotropic poroelastic properties can also be found in Thompson and Willis (1991).

2.2 Randomly (isotropic) oriented opened cracks

We consider an isotropic distribution of the crack orientations. For simplicity, all cracks have the same radius. The crack network is divided into crack families corresponding to specific orientations. Starting from (1.19), and transforming the sum into an integral over the spherical angular coordinates $\theta \in [0,\pi]$ and $\psi \in [0, 2\pi]$ defining the orientation of the unit normal $\underline{n}$ yields :

$$\mathbb{C}^{hom} = \mathbb{C}^s : \left(\mathbb{I} - \int_0^\pi d\theta \int_0^{2\pi} \frac{4\pi a^3\mathcal{N}}{3} X\,(\mathbb{I} - \mathbb{S}(X,\theta,\psi))^{-1}\frac{\sin\theta d\psi}{4\pi}\right) \tag{2.17}$$

Considering the limit[6] of this quantity for $X \to 0$, it is then readily seen that the determination of $\mathbb{C}^{hom}$ amounts to that of the average $< \mathbb{T} >$ of $\mathbb{T}(\theta,\psi)$ over crack orientations :

$$\mathbb{C}^{hom} = \mathbb{C}^s : (\mathbb{I} - \frac{4}{3}\pi\epsilon < \mathbb{T} >) \tag{2.18}$$

where

$$< \mathbb{T} > = \int_0^\pi d\theta \int_0^{2\pi} \mathbb{T}(\theta,\psi)\frac{\sin\theta d\psi}{4\pi} \tag{2.19}$$

[6] In fact, it is not necessary that the aspect ratios X be identical for all cracks, provided that $X \ll 1$.

For deriving $<\mathbb{T}>$ from (2.15), the following identities prove to be useful:

$$<\underline{n}\otimes\underline{n}>=\int_{|\underline{n}|=1}\underline{n}\otimes\underline{n}\,\frac{dS}{4\pi}=\frac{1}{3}\mathbf{1} \tag{2.20}$$

and

$$<\underline{n}\otimes\underline{n}\otimes\underline{n}\otimes\underline{n}>=\int_{|\underline{n}|=1}\underline{n}\otimes\underline{n}\otimes\underline{n}\otimes\underline{n}\,\frac{dS}{4\pi}=\frac{1}{3}\mathbb{J}+\frac{2}{15}\mathbb{K} \tag{2.21}$$

in which dS is the surface element $\sin\theta\,d\theta\,d\psi$ on the unit sphere. It is convenient to introduce the (isotropic) fourth order tensor $\mathbb{Q}$ defined as :

$$\mathbb{Q}=\frac{4\pi}{3}<\mathbb{T}>=Q_1\mathbb{J}+Q_2\mathbb{K} \tag{2.22}$$

with :

$$Q_1=\frac{16}{9}\frac{1-\nu^{s2}}{1-2\nu^s}\quad ;\quad Q_2=\frac{32}{45}\frac{(1-\nu^s)(5-\nu^s)}{2-\nu^s} \tag{2.23}$$

(2.18) together with (2.22) and (2.23) yield :

$$\mathbb{C}^{hom}=\mathbb{C}^s:(\mathbb{I}-\epsilon\mathbb{Q})\quad ;\quad \boldsymbol{B}=\epsilon\mathbf{1}:\mathbb{Q}=b\mathbf{1}\ \text{ with } b=\epsilon Q_1=\frac{16}{9}\frac{1-\nu^{s2}}{1-2\nu^s}\epsilon \tag{2.24}$$

2.3 Anisotropic distribution of opened cracks

More generally, let $f(\theta,\psi)$ denote the density function describing the distribution of cracks orientation : $f(\theta,\psi)\sin\theta d\theta\,d\psi/4\pi$ then represents the number of cracks per unit volume for which the angular coordinates belong to $[\theta,\theta+d\theta]\times[\psi,\psi+d\psi]$. The density function $f(\theta,\psi)$ meets the condition :

$$\mathcal{N}=\int_{|\underline{n}|=1}f(\theta,\psi)\frac{dS}{4\pi} \tag{2.25}$$

(2.17) is now replaced by :

$$\mathbb{C}^{hom}=\mathbb{C}^s:\left(\mathbb{I}-\int_{|\underline{n}|=1}\frac{4\pi a^3 f(\theta,\psi)}{3}X\,(\mathbb{I}-\mathbb{S}(X,\theta,\psi))^{-1}\frac{dS}{4\pi}\right) \tag{2.26}$$

which limit when $X\to 0$ reads :

$$\mathbb{C}^{hom}=\mathbb{C}^s:\left(\mathbb{I}-\int_{|\underline{n}|=1}\frac{4\pi}{3}a^3 f(\theta,\psi)\mathbb{T}(\theta,\psi)\frac{dS}{4\pi}\right) \tag{2.27}$$

The above equation reveals that $\epsilon=f(\theta,\psi)a^3$ is the relevant micromechanical parameter controlling the influence of the defects. In fact, (2.27) implicitly assumes that the crack radius is either a constant or a function of θ and ψ. (2.18) is in turn replaced by :

$$\mathbb{C}^{hom}=\mathbb{C}^s:(\mathbb{I}-\frac{4\pi}{3}\int_{|\underline{n}|=1}\epsilon(\theta,\psi)\mathbb{T}(\theta,\psi)\frac{dS}{4\pi})=\mathbb{C}^s:(\mathbb{I}-\frac{4}{3}\pi<\epsilon\mathbb{T}>) \tag{2.28}$$

$\epsilon(\theta,\psi)=\epsilon(\underline{n})$ can be interpreted as the crack density parameter in the direction of $\underline{n}$. Note that a statistical description of the radius distribution is presented at section 4.4.

2.4 Effect of total crack closure on the overall stiffness

We now consider the case of a solid matrix containing a set of closed cracks[7] with unit normal $\underline{n}$. It is assumed that the contact between the two cracks lips is non frictional. As opposed to the case of opened cracks, we must now take into account the fact that a compressive normal stress can be transmitted through the contact of the crack lips. Still, the shear stress in the plane of the lips remains equal to zero. The idea therefore consists in replacing the closed crack by a flat ellipsoidal inhomogeneity filled with an isotropic fictitious material having a shear modulus $\mu^c = 0$ and a bulk modulus $k^c \neq 0$. The stiffness tensor of the latter thus reads :

$$\mathbb{C}^c = 3k^c \mathbb{J} \tag{2.29}$$

Considering for instance an isotropic compression in which the closed cracks clearly do not influence the macroscopic response of the r.e.v. (irrespective of their orientations), it seems natural to choose $k^c = k^s$. In fact, it can be shown that the value of k^c does not affect the homogenized properties of the composite made up of the solid and the fictitious inhomogeneities, provided that $k^c \neq 0$. This is due to the fact that the volume fraction φ_f^c and the corresponding aspect ratio X_f of the fictitious inhomogeneities tend asymptotically towards 0. In particular, it is emphasized that the latter have no real geometrical meaning and will not appear in the expression of the homogenized properties. A validation of the concept of fictitious material by means of a fracture mechanics-based approach can be found in Deudé et al. (2002b).

parallel closed cracks We seek the homogenized properties of a composite made up of the elastic solid matrix and of a second phase representing parallel closed cracks, the latter being characterized by the stiffness $\mathbb{C}^c$. The homogenized stiffness reads :

$$\mathbb{C}^{hom} = \overline{\mathbb{C} : \mathbb{A}} = (1 - \varphi_f^c)\mathbb{C}^s : \overline{\mathbb{A}}^s + \varphi_f^c \mathbb{C}^c : \overline{\mathbb{A}}^c = \mathbb{C}^s + \varphi_f^c (\mathbb{C}^c - \mathbb{C}^s) : \overline{\mathbb{A}}^c \tag{2.30}$$

Replacing now[8] $\mathbb{C}^c$ by $3k^s\mathbb{J}$, the above equation reduces to :

$$\mathbb{C}^{hom} = \mathbb{C}^s : (\mathbb{I} - \varphi_f^c \mathbb{K} : \overline{\mathbb{A}}^c) \tag{2.31}$$

In the framework of the dilute scheme, the strain concentration tensor $\overline{\mathbb{A}}^c$ is derived from (1.11) :

$$\overline{\mathbb{A}}^c = (\mathbb{I} + \mathbb{P} : (\mathbb{C}^c - \mathbb{C}^s))^{-1} = (\mathbb{I} - \mathbb{S}(X_f, \underline{n}) : \mathbb{K})^{-1} \tag{2.32}$$

where $\mathbb{S}(X_f, \underline{n})$ is the Eshelby tensor for the considered set of inhomogeneities (see (2.13)). The dilute estimate of $\mathbb{C}^{hom}$, obtained from (2.31) and (2.32), reads :

$$\mathbb{C}^{hom} = \mathbb{C}^s : \left(\mathbb{I} - \frac{4}{3}\pi\epsilon X_f \mathbb{K} : (\mathbb{I} - \mathbb{S}(X_f, \underline{n}) : \mathbb{K})^{-1}\right) \quad \text{with} \quad \mathbb{K} = \mathbb{I} - \mathbb{J} \tag{2.33}$$

[7]The modelling of crack closure in the framework of continuum damage mechanics still remains a difficult task (Chaboche (1992)).

[8]As already stated, the final result does not depend on the value of k^c, provided that $k^c \neq 0$

For flat ellipsoids ($X_f \ll 1$), $\mathbb{K} : (\mathbb{I} - \mathbb{S}(X_f, \underline{n}) : \mathbb{K})^{-1}$ proves to be $O(1/X_f)$. Accordingly, the quantity $X_f \mathbb{K} : (\mathbb{I} - \mathbb{S}(X_f, \underline{n}) : \mathbb{K})^{-1}$ can be replaced by its limit $\mathbb{T}'$ when $X_f \to 0$. The expression of $\mathbb{C}^{hom}$ for a set of parallel penny-shaped closed cracks then reads :

$$\mathbb{C}^{hom} = \mathbb{C}^s : (\mathbb{I} - \frac{4}{3}\pi\epsilon\mathbb{T}') \text{ with } \mathbb{T}' = \lim_{X_f \to 0} X_f \mathbb{K} : (\mathbb{I} - \mathbb{S}(X_f, \underline{n}) : \mathbb{K})^{-1} \quad (2.34)$$

Comparison between (2.34) and (2.7) shows that taking the total crack closure into account amounts to replacing $\mathbb{T}$ in (2.7) by $\mathbb{T}'$ defined in (2.34). For parallel cracks normal to $\underline{n} = \underline{e}_3$, one obtains the following components of the tensor $\mathbb{T}'$:

$$T'_{1313} = T'_{2323} = \frac{2(1-\nu^s)}{(2-\nu^s)\pi} \quad (2.35)$$

(with $T'_{ijkl} = T'_{jikl} = T'_{ijlk}$), the other components being equal to 0.
For parallel closed cracks with normal $\underline{n}$, $\mathbb{T}'$ reads:

$$\mathbb{T}'(\underline{n}) = \frac{4(1-\nu^s)}{\pi(2-\nu^s)}\Big(\mathbf{1}\overline{\underline{\otimes}}(\underline{n}\otimes\underline{n}) + (\underline{n}\otimes\underline{n})\overline{\underline{\otimes}}\mathbf{1} - 2(\underline{n}\otimes\underline{n})\otimes(\underline{n}\otimes\underline{n})\Big) \quad (2.36)$$

The corresponding expression of $\mathbb{C}^{hom}$ is:

$$\mathbb{C}^{hom} = \mathbb{C}^s - a_1\epsilon\Big(\mathbf{1}\overline{\underline{\otimes}}(\underline{n}\otimes\underline{n}) + (\underline{n}\otimes\underline{n})\overline{\underline{\otimes}}\mathbf{1} - 2(\underline{n}\otimes\underline{n})\otimes(\underline{n}\otimes\underline{n})\Big) \quad (2.37)$$

where the coefficient a_1 is given by (6.12).
As opposed to opened cracks, closed cracks normal to $\underline{n} = \underline{e}_3$ only affect the shear moduli C^{hom}_{1313} and C^{hom}_{2323} :

$$C^{hom}_{1313} = C^{hom}_{2323} = \mu^s\left(1 - \frac{16(1-\nu^s)}{3(2-\nu^s)}\epsilon\right) \quad (2.38)$$

However, since $T_{1313} = T'_{1313}$ and $T_{2323} = T'_{2323}$, these shear moduli are identical to those obtained for opened cracks.
For materials weakened by parallel closed cracks, $\boldsymbol{B}$ is obtained by replacing $\mathbb{T}$ in (2.16) by the expression (2.36) of tensor $\mathbb{T}'$. This leads to $\boldsymbol{B} = \frac{4\pi\epsilon}{3}\mathbf{1} : \boldsymbol{T}' = \mathbf{0}$. Consequently, it appears that the fluid pressure has no effect on the poroelastic behavior when cracks are closed. Actually, the question of the role of the fluid in a closed crack may be surprising. In fact, at a microscopic scale, the crack may be closed in the mechanical sense, while some fluid remains trapped between the two lips. Indeed, at the microscopic scale, the latter are not perfectly smooth. Therefore, a residual pore space may exist even if the lips are in contact.

randomly oriented closed cracks As previously, the case of randomly oriented closed cracks requires an averaging process on crack orientations. We just have to replace $< \mathbb{T} >$ by $< \mathbb{T}' >$ in (2.18). Starting from (2.36), and using (2.20) and (2.21), one obtains :

$$\mathbb{C}^{hom} = 3k^s\mathbb{J} + 2\mu^s(1 - \frac{32}{15}\frac{(1-\nu^s)}{(2-\nu^s)}\epsilon)\mathbb{K} \quad (2.39)$$

As expected, we note that only the homogenized shear modulus is affected. By comparison with (2.24), this effect is different from the one produced on shear modulus by randomly oriented opened cracks. Note also that a combination of (2.39) and (1.5) yields, as expected, a Biot tensor equal to zero.

3 The Mori Tanaka model applied to cracked media

We hereafter present the estimates of the overall elastic properties provided by the Mori-Tanaka scheme which was briefly introduced at section 1.

3.1 Parallel opened cracks

The starting point is (1.24). When the aspect ratio X of the cracks tends towards 0, the crack porosity $\varphi = 4\pi\epsilon X/3$ tends towards 0 as well, whereas $\varphi(\mathbb{I}-\mathbb{S})^{-1}$ tends towards $4\pi\mathbb{T}/3$ (see (2.8)). Accordingly , the Mori-Tanaka estimate of the effective stiffness takes the form :

$$\mathbb{C}^{hom} = \mathbb{C}^s : \left(\mathbb{I} + \frac{4}{3}\pi\epsilon\mathbb{T}\right)^{-1} \tag{3.1}$$

The corresponding homogenized compliance tensor is :

$$\mathbb{S}^{hom} = \left(\mathbb{I} + \frac{4}{3}\pi\epsilon\mathbb{T}\right) : \mathbb{S}^s \tag{3.2}$$

that is

$$\mathbb{S}^{hom} = \mathbb{S}^s + \frac{16(1-\nu^{s2})}{3(2-\nu^s)E^s}\epsilon\left\{\mathbf{1}\overline{\underline{\otimes}}(\underline{n}\otimes\underline{n}) + (\underline{n}\otimes\underline{n})\overline{\underline{\otimes}}\mathbf{1} - \nu^s\underline{n}\otimes\underline{n}\otimes\underline{n}\otimes\underline{n}\right\} \tag{3.3}$$

The latter appears to be a linear function of the crack density parameter ϵ. Interestingly, although the crack porosity $\varphi \to 0$ as $X \to 0$, the Mori-Tanaka estimate is different from the dilute one (see (2.12)). This underlines the fact that the appropriate measurement of the damage induced by the cracks is the crack density parameter ϵ and not the crack porosity. Indeed, the Mori-Tanaka and the dilute estimates are only equivalent when $\epsilon \ll 1$. In other words, the fact that the crack porosity is infinitesimal is not sufficient for the dilute scheme to be valid.

3.2 Parallel closed cracks

A good starting point is again (2.31). Using the kind of reasoning presented at section 1.3, the Mori-Tanaka estimate of $\mathbb{A}^c$ reads

$$\mathbb{A}^c = (\mathbb{I}-\mathbb{S}:\mathbb{K})^{-1} : \left((1-\varphi_f^c)\mathbb{I} + \varphi_f^c(\mathbb{I}-\mathbb{S}:\mathbb{K})^{-1}\right)^{-1} \tag{3.4}$$

Recalling that $\varphi_f^c = 4\pi\epsilon X_f/3$ and considering the limit when X_f tends towards 0, one obtains :

$$\varphi_f^c\mathbb{A}^c = \frac{4}{3}\pi\epsilon\mathbb{T}' : \left(\mathbb{I} + \frac{4}{3}\pi\epsilon\mathbb{T}'\right)^{-1} \tag{3.5}$$

where $\mathbb{T}'$ is the tensor introduced in (2.34). The effective stiffness and compliance tensors then take the form :

$$\mathbb{C}^{hom} = \mathbb{C}^s : \left(\mathbb{I} + \frac{4}{3}\pi\epsilon\mathbb{T}'\right)^{-1} \quad ; \quad \mathbb{S}^{hom} = \left(\mathbb{I} + \frac{4}{3}\pi\epsilon\mathbb{T}'\right) : \mathbb{S}^s \qquad (3.6)$$

In other words, with respect to the case of opened cracks, we just have to replace $\mathbb{T}$ by $\mathbb{T}'$ in (3.1) and (3.2).

3.3 Randomly oriented interacting cracks

In the case of randomly oriented opened cracks, the homogenized stiffness tensor $\mathbb{C}^{hom}$ is obtained from (1.29). When the aspect ratio of each crack tends towards 0, the limit of $\varphi_j(\mathbb{I} - \mathbb{S}_j)^{-1}$ is equal to[9] $\epsilon\mathbb{Q}$, while φ tends towards 0. This yields the following expression of $\mathbb{C}^{hom}$ and $\boldsymbol{B}$:

$$\mathbb{C}^{hom} = \mathbb{C}^s : (\mathbb{I} + \epsilon\mathbb{Q})^{-1} \quad ; \quad \boldsymbol{B} = \mathbf{1} : \left(\mathbb{I} - (\mathbb{I} + \epsilon\mathbb{Q})^{-1}\right) \qquad (3.7)$$

which also reads :

$$\mathbb{C}^{hom} = \frac{3k^s}{1+\epsilon Q_1}\mathbb{J} + \frac{2\mu^s}{1+\epsilon Q_2}\mathbb{K} \quad ; \quad \boldsymbol{B} = \epsilon\frac{Q_1}{1+\epsilon Q_1}\mathbf{1} \qquad (3.8)$$

A similar reasoning can be implemented for closed cracks as well and yields :

$$\mathbb{C}^{hom} = 3k^s\mathbb{J} + \frac{2\mu^s}{1 + \frac{32}{15}\frac{(1-\nu^s)}{(2-\nu^s)}\epsilon}\mathbb{K} \quad ; \quad \boldsymbol{B} = 0 \qquad (3.9)$$

3.4 Linear poroelasticity of a cracked porous medium

As opposed to sections 2, 3.1, 3.2 and 3.3, we now examine the situation in which the saturated pore space also comprises a set of saturated pores with the same fluid pressure P as in the cracks. We assume that there is no scale separation between pores and cracks[10] and we investigate their influence on the effective behavior in the framework of the Mori-Tanaka scheme.

We also assume that the cracks remain opened[11]. The total porosity in the *current* configuration is divided into the pore volume fraction φ^p and the crack volume fraction φ^c. In practice, we remember that $\varphi^c \ll 1$ so that $\varphi^p \approx \varphi$. Furthermore, as opposed to those of φ^c, the changes of the pore volume fraction φ^p are negligible.

For simplicity, we assume that the pore shape is spherical and we introduce the corresponding Eshelby tensor $\mathbb{S}_{sph}$:

$$\mathbb{S}_{sph} = \alpha^s\mathbb{J} + \beta^s\mathbb{K} \quad \text{with} \quad \alpha^s = \frac{3k^s}{3k^s + 4\mu^s} \; ; \; \beta^s = \frac{6}{5}\frac{k^s + 2\mu^s}{3k^s + 4\mu^s} \qquad (3.10)$$

[9] $\mathbb{Q}$ is the tensor introduced in (2.22)

[10] Otherwise, the effects of the cracks and those of the pores can be addressed in distinct homogenization steps

[11] See section 4 for a discussion on the effect of progressive crack closure.

The starting point is (1.29) :

$$\mathbb{C}^{hom} = (1-\varphi)\mathbb{C}^s : \left((1-\varphi)\mathbb{I} + \varphi(\mathbb{I}-\mathbb{S}_{sph})^{-1} + \sum_{i \in \mathcal{C}} \varphi_i^c(\mathbb{I}-\mathbb{S}_i)^{-1}\right)^{-1} \tag{3.11}$$

where $\mathcal{C}$ formally denotes the set of cracks. We then consider the limit of the above estimate when the aspect ratio tends towards 0 :

$$\mathbb{C}^{hom} = (1-\varphi)\mathbb{C}^s : \left((1-\varphi)\mathbb{I} + \varphi(\mathbb{I}-\mathbb{S}_{sph})^{-1} + \sum_{i \in \mathcal{C}} \frac{4\pi\epsilon^i}{3}\mathbb{T}^i\right)^{-1} \tag{3.12}$$

As in sections 2, 3.1, 3.2 and 3.3, the next step requires to specify the distribution of the crack orientations and to take advantage of (2.8). In the case of parallel cracks, we then obtain :

$$\mathbb{C}^{hom} = (1-\varphi)\mathbb{C}^s : \left((1-\varphi)\mathbb{I} + \varphi(\mathbb{I}-\mathbb{S}_{sph})^{-1} + \frac{4\pi\epsilon}{3}\mathbb{T}\right)^{-1} \tag{3.13}$$

Similarly, the Mori-Tanaka estimate for the case of spherical pores and randomly oriented cracks reads :

$$\mathbb{C}^{hom} = (1-\varphi)\mathbb{C}^s : \left((1-\varphi)\mathbb{I} + \varphi(\mathbb{I}-\mathbb{S}_{sph})^{-1} + \epsilon\mathbb{Q}\right)^{-1} \tag{3.14}$$

4 Non linear poroelastic behavior due to cracks closure

In the previous sections, all the cracks were simultaneously opened or closed, and a linear effective behavior was obtained. Still, a number of experimental evidences reveal that the elastic behavior of rocks under compressive loading may be non linear. This non linearity can be attributed to the progressive closure of microcracks. The micromechanical modelling of this phenomenon therefore requires that we consider intermediate situations in which some cracks are opened and others are closed. In this section, it is first shown that the non linear reversible poroelastic behavior is controlled by Terzaghi's effective stress. Then, we present a micromechanical approach to the effect of progressive crack closure on the effective tangent stiffness.

4.1 Origin of the reversible nonlinearity

The non linearity of the macroscopic elastic properties of rocks under compression has been widely investigated in rock mechanics and particularly in the case of sandstones (Bouteca et al. (1994), Zimmermann (1991)) and limestones (Lion et al. (2005)). During an isotropic compression test, the typical feature is a significant increase of the tangent bulk modulus and the existence of an asymptotic value of the latter, reached at high confining pressures.

Two different interpretations of the non-linear elastic behavior of rocks under compression have been proposed in the literature : (1) contact between deformable grains, and (2) progressive closure of microcracks leading to stiffer overall elastic properties. This second

type of explanation is widely adopted in the field of rock mechanics since it is consistent with the existence of an asymptotic tangent bulk modulus at high confinement (Dormieux et al. (2002)). The focus of this section is therefore to provide a micromechanical model of non linearity based on progressive crack closure.
As in section 3.4, the pore space is made up of a set of cracks and of a network of spherical pores. Accordingly, the total porosity φ is the sum of the pore volume fraction φ^p and of the crack volume fraction $\varphi^c \ll 1$: $\varphi = \varphi^p + \varphi^c \approx \varphi^p$. The solid volume fraction is $\varphi^s = 1 - \varphi$.
As can be expected from the comparison between the results obtained respectively with opened or closed cracks (see for instance sections 2.1 and 2.4), the evolution of the state of cracks (opened or closed) is an essential factor in the understanding of the macroscopic non linearity. For this reason, the variation of the crack volume fraction is a crucial parameter of the macroscopic non linearity. In contrast, as regards the spherical pores, the geometry changes are very small for the range of mechanical loadings considered here and can be neglected.
As regards the effect of the poromechanical coupling on non linearity, a brief discussion on the existence of a hydraulic connexion between cracks and pores is due. A simple experiment allows us to clarify whether such a connexion exists. It consists in applying an isotropic macroscopic stress $\mathbf{\Sigma} = -\Sigma \mathbf{1}$ to the sample, together with a pore pressure equal to Σ. Pore pressure and confining pressure are then simultaneously increased, generating a volume strain of the sample. An apparent modulus $\mathcal{M}$ can be defined by

$$\mathcal{M} = -\frac{dP}{d\Omega/\Omega} \tag{4.1}$$

If cracks and pores are connected and saturated by a fluid at pressure P, the solid phase appears as a homogeneous domain, at the boundary of which a uniform pressure P is applied. It follows that the microscopic stresses $\boldsymbol{\sigma}$ and strains $\boldsymbol{\varepsilon}$ are uniform in the solid, with $\boldsymbol{\sigma} = -P\mathbf{1}$ and $\boldsymbol{\varepsilon} = -P\mathbb{S}^s : \mathbf{1}$, and are equal to the macroscopic stress $\mathbf{\Sigma}$ and strain $\boldsymbol{E}$ respectively. In particular, this implies that the apparent modulus $\mathcal{M}$ is a constant equal to $(\mathbf{1} : \mathbb{S}^s : \mathbf{1})^{-1}$. In the case of an isotropic linear elastic solid with bulk modulus k^s, we obtain $\boldsymbol{\varepsilon} = -P/3k^s\mathbf{1}$ and $\mathcal{M} = k^s$. Experimental results on Fontainebleau sandstone are in good agreement with this conclusion (Bouteca et al. (1994)).
Conversely, if cracks and pores are not connected, the cracks being empty and the pores being saturated, the simultaneous increase of confining pressure and pore pressure will progressively close the cracks. An increase of $\mathcal{M}$ is therefore expected, with an asymptotic value equal to k^s, reached once all cracks are closed.
These two strongly different responses suggest that the existence of a hydraulic connexion between cracks and pores is a decisive parameter of the non linear poroelastic behavior. In what follows (section 4.2), the mathematical structure of the non linearity is investigated with the assumption that such a connexion exists.

4.2 differential formulation of the poroelastic response

In this section, cracks and pores are supposed to be fluid saturated, the fluid pressure P being identical in the pores and in the cracks. The loading can be characterized by two independent parameters, namely the macroscopic stress $\boldsymbol{\Sigma}$ (not necessarily isotropic) and the fluid pressure P. Dissipative phenomena, such as crack propagation[12] or frictional sliding on the cracks lips, are not considered. The response of the system being reversible, the mechanical state of the r.e.v. is independent of the loading path in the $(\boldsymbol{\Sigma}, P)$ plane. Therefore, following Dormieux et al. (2002), the following decomposition of the loading in two successive sequences can be considered.

In the first step, the macroscopic stress $\boldsymbol{\Sigma}_1 = \boldsymbol{\Sigma} + P\mathbf{1}$ is applied while keeping the fluid pressure at the value $P = 0$. We observe that $\boldsymbol{\Sigma}_1$ is equal to Terzaghi's effective stress $\boldsymbol{\Sigma}'$. Still, note that pores and cracks are not pressurized during this loading. The latter induces a change in the geometry of cracks including some crack closure. Therefore, the overall response of the microcracked medium is a priori non linear. In other words, the resulting macroscopic strain $\boldsymbol{E}_1$ is a nonlinear function of $\boldsymbol{\Sigma}_1 = \boldsymbol{\Sigma}'$.

In a second step, the macroscopic stress $\boldsymbol{\Sigma}_2 = -P\mathbf{1}$ is superimposed and the fluid pressure is increased simultaneously until the value P is reached. Following the conclusion[13] of section 4.1, the corresponding macroscopic strain is $\boldsymbol{E}_2 = -P\mathbb{S}^s : \mathbf{1}$. This result does not depend on the non-linearities generated during the first sequence of the loading path provided that cracks and pores remain connected. The loading state after these two successive sequences is characterized by the macroscopic stress $\boldsymbol{\Sigma}$ and by the fluid pressure P. The total macroscopic strain reads:

$$\boldsymbol{E} = \boldsymbol{E}_1(\boldsymbol{\Sigma}') - P\mathbb{S}^s : \mathbf{1} \tag{4.2}$$

It follows that the term $\boldsymbol{E} + P\mathbb{S}^s : \mathbf{1}$ is a function of the effective stress $\boldsymbol{\Sigma}'$. In differential form, this result can be expressed as:

$$\boldsymbol{\Sigma}' = \boldsymbol{\Sigma}'(\boldsymbol{E} + P\mathbb{S}^s : \mathbf{1}) \Rightarrow \dot{\boldsymbol{\Sigma}}' = \mathbb{C}_t^{hom}(\boldsymbol{\Sigma}') : (\dot{\boldsymbol{E}} + \dot{P}\mathbb{S}^s : \mathbf{1}) \tag{4.3}$$

where $\mathbb{C}_t^{hom} = \partial\boldsymbol{\Sigma}_1/\partial\boldsymbol{E}_1$ is the tensor of macroscopic tangent elastic moduli for the drained or empty porous material (no pore pressure). It turns out that its non linearity is controlled by Terzaghi's effective stress $\boldsymbol{\Sigma}'$. The relationship (4.3) can be also written as:

$$\dot{\boldsymbol{\Sigma}} = \mathbb{C}_t^{hom}(\boldsymbol{\Sigma}') : \dot{\boldsymbol{E}} - \dot{P}\boldsymbol{B}_t(\boldsymbol{\Sigma}') \tag{4.4}$$

where the following tangent Biot tensor has been introduced:

$$\boldsymbol{B}_t(\boldsymbol{\Sigma}') = (\mathbb{I} - \mathbb{C}_t^{hom}(\boldsymbol{\Sigma}') : \mathbb{S}^s) : \mathbf{1} \tag{4.5}$$

Using the same decomposition of the loading $(\boldsymbol{\Sigma}, P)$ into two sequences $(\boldsymbol{\Sigma}_1 = \boldsymbol{\Sigma}', 0)$ and $(\boldsymbol{\Sigma}_2 = -P\mathbf{1}, P)$, the second state equation of non linear poroelasticity giving the incremental pore volume change reads :

$$\dot{\phi} = \boldsymbol{B}_t(\boldsymbol{\Sigma}') : \dot{\boldsymbol{E}} + \frac{\dot{P}}{M_t(\boldsymbol{\Sigma}')} \tag{4.6}$$

[12]This phenomenon will be studied at section 5

[13]It is recalled that pores and cracks are connected.

with the Biot modulus N_t defined by:

$$\frac{1}{N_t(\boldsymbol{\Sigma}')} = \mathbf{1} : \mathbb{S}^s : (\boldsymbol{B}_t(\boldsymbol{\Sigma}') - \varphi_o \mathbf{1}) \tag{4.7}$$

(4.4)-(4.6) together with (4.5)-(4.7) represent the generalization of the classical relations (1.2)-(1.3) and (1.5)-(1.6) of Biot's linear poroelasticity to the non linear regime. With respect to Biot's theory, the constant poroelastic coefficients have been replaced in the incremental formulation by tangent coefficients. The latter are controlled by Terzaghi's effective stress $\boldsymbol{\Sigma}'$. This result provides a theoretical basis to a number of experimental findings. In particular, plotting the tangent bulk modulus as function of the effective confining pressure $\Sigma' = -\boldsymbol{tr}\boldsymbol{\Sigma}'/3$, Bouteca et al. (1994) have obtained a unique experimental curve for different values of the pore pressure.

4.3 Estimates of the tangent poroelastic properties

Once $\mathbb{C}_t^{hom}(\boldsymbol{\Sigma}')$ has been estimated, $\boldsymbol{B}_t(\boldsymbol{\Sigma}')$ and $N_t(\boldsymbol{\Sigma}')$ can be determined from the above equations (4.5)-(4.7). We therefore focus on $\mathbb{C}_t^{hom}(\boldsymbol{\Sigma}')$ which can be interpreted as the macroscopic tangent stiffness tensor in empty conditions under the macroscopic stress state $\boldsymbol{\Sigma} = \boldsymbol{\Sigma}'$. In other words, the dependence of $\mathbb{C}_t^{hom}$ with respect to the stress state, that is, the mathematical structure of the nonlinearity, can be analyzed in performing experiments without pore pressure ($P = 0$, $\boldsymbol{\Sigma} = \boldsymbol{\Sigma}'$). As long as all the cracks are opened, the effective response is linear and the Mori-Tanaka estimate derived in (3.11) is valid. We now address the non linear response due to progressive cracks closure (Deudé et al. (2002a)).

non linearity and crack closure criterion

As before, the change of geometry of the spherical pores is neglected. In contrast, the macroscopic confining pressure a priori closes some of the cracks and modifies the aspect ratio of the others. Perfect adhesion is assumed between crack lips when they are in contact. This implies that closed cracks do not contribute to damage.
The linear reasoning presented at section 3.4 is only valid up to the stress level at which the crack closure process begins. However, beyond this threshold, the linear reasoning can be adapted in the incremental form[14] in order to relate the macroscopic strain rate $\dot{\boldsymbol{E}}$ to the corresponding macroscopic stress rate $\dot{\boldsymbol{\Sigma}}$:

$$\dot{\boldsymbol{\Sigma}} = \mathbb{C}_t^{hom}(\boldsymbol{\Sigma}) : \dot{\boldsymbol{E}} \tag{4.8}$$

The expression of the tangent stiffness tensor $\mathbb{C}_t^{hom}(\boldsymbol{\Sigma})$ is obtained from (3.11), except for the fact that only opened cracks must be taken into account :

$$\mathbb{C}_t^{hom}(\boldsymbol{\Sigma}) = \varphi^s \mathbb{C}^s : \left(\varphi^s \mathbb{I} + \varphi^p (\mathbb{I} - \mathbb{S}_{sph})^{-1} + \sum_{j \in \mathcal{O}(\boldsymbol{\Sigma})} \varphi_j^c (\mathbb{I} - \mathbb{S}_j)^{-1} \right)^{-1} \tag{4.9}$$

[14] See "comments and discussion" in section 2.1.

where $\mathcal{O}(\mathbf{\Sigma})$ denotes the set of opened cracks. These cracks are to be considered in the current state, with the value of the aspect ratio reached for the stress state $\mathbf{\Sigma}$. It is worth recalling that this estimate corresponds to non sliding closed cracks, i.e. closed cracks with perfect adhesion. Therefore, once the crack is closed, it does not affect anymore the material properties.

For increasing compressive stresses, cracks are progressively closed, i.e. $\mathcal{O}(\mathbf{\Sigma})$ tends towards the empty set. (4.9) then shows that $\mathbb{C}_t^{hom}$ tends towards an asymptotic value $\mathbb{C}_\infty^{hom}$ which is equal to the stiffness tensor of a porous medium made up of an *uncracked* solid phase (stiffness $\mathbb{C}^s$) and of a porous domain (porosity φ^p) :

$$\mathbb{C}_\infty^{hom} = \varphi^s \mathbb{C}^s : \left(\varphi^s \,\mathbb{I} + \varphi^p \left(\mathbb{I} - \mathbb{S}_{sph}\right)^{-1}\right)^{-1} \tag{4.10}$$

The micromechanical modelling accounts for the non linearity of the macroscopic behavior through the way the set of opened cracks $\mathcal{O}(\mathbf{\Sigma})$ depends on the stress level. A quantitative modelling therefore requires a criterion for crack closure.

As long as a crack of family ω_i is opened, the strain rate $\dot{\boldsymbol{\varepsilon}}^i$ associated with the macroscopic strain rate $\dot{\boldsymbol{E}}$ can be derived from (2.9) together with (1.28) :

$$\dot{\boldsymbol{\varepsilon}}^i = \overline{\mathbb{A}}^{c,i} : \dot{\boldsymbol{E}} = \left(\mathbb{I} - \mathbb{S}_i\right)^{-1} : \left(\varphi^s \,\mathbb{I} + \varphi^p \left(\mathbb{I} - \mathbb{S}_{sph}\right)^{-1} + \sum_{j \in \mathcal{O}(\mathbf{\Sigma})} \varphi_j^c \left(\mathbb{I} - \mathbb{S}_j\right)^{-1}\right)^{-1} : \dot{\boldsymbol{E}} \tag{4.11}$$

Combining (4.11) with (4.9) yields an expression of $\dot{\boldsymbol{\varepsilon}}^i$ as a function of the macroscopic stress rate $\dot{\mathbf{\Sigma}} = \mathbb{C}_t^{hom} : \dot{\boldsymbol{E}}$

$$\dot{\boldsymbol{\varepsilon}}^i = \frac{1}{\varphi^s}\left(\mathbb{I} - \mathbb{S}_i(\mathbf{\Sigma})\right)^{-1} : \mathbb{S}^s : \dot{\mathbf{\Sigma}} \tag{4.12}$$

The above expression emphasizes the fact that the Eshelby tensor $\mathbb{S}_i$ depends on the stress level $\mathbf{\Sigma}$ through the value of the aspect ratio $X^i = c^i/a^i$ which progressively decreases as the crack is being closed. Let $\underline{n}^i$ be the unit normal to the crack midplane. Using (4.12), the crack opening/closure[15] rate is readily derived from (4.12) :

$$\dot{c}^i = c^i \underline{n}^i \cdot \dot{\boldsymbol{\varepsilon}}^i \cdot \underline{n}^i = \frac{a^i}{\varphi^s} \underline{n}^i \cdot \left(X^i \left(\mathbb{I} - \mathbb{S}_i\right)^{-1} : \mathbb{S}^s : \dot{\mathbf{\Sigma}}\right) \cdot \underline{n}^i \tag{4.13}$$

As already stated, when $X^i \to 0$, the tensor $X^i(\mathbb{I} - \mathbb{S}_s^i)^{-1}$ can be approximated by its a limit $\mathbb{T}^i$ (see (2.8)). We recall that $\mathbb{T}^i$ depends on the orientation of the crack and on the elastic properties of the solid. (4.13) can then be rewritten as :

$$\dot{c}^i = \frac{a^i}{\varphi^s} \underline{n}^i \cdot \left(\mathbb{T}^i : \mathbb{S}^s : \dot{\mathbf{\Sigma}}\right) \cdot \underline{n}^i \tag{4.14}$$

$\mathbb{T}^i$ as well as a^i and φ^s are constant, that is, are not affected by the crack closure process. The above expression of the crack closure rate can therefore be integrated with respect to time. Assuming that the initial macroscopic state is free of stress, we get :

$$\delta c^i = \frac{a^i}{\varphi^s} \underline{n}^i \cdot \left(\mathbb{T}^i : \mathbb{S}^s : \mathbf{\Sigma}\right) \cdot \underline{n}^i \tag{4.15}$$

[15] Closure is associated with $\dot{c}^i < 0$.

which states that crack closure is completed if $\delta c^i = -c^i_o$, where c^i_o stands for the value of c^i in the initial state ($\mathbf{\Sigma} = 0$). (4.15) therefore yields a stress-based criterion for crack closure in the form :

$$X^i_o \leq X_{c\ell}(\mathbf{\Sigma}, \underline{n}^i) = -\frac{1}{\varphi^s}\underline{n}^i \cdot \left(\mathbb{T}^i : \mathbb{S}^s : \mathbf{\Sigma}\right) \cdot \underline{n}^i \tag{4.16}$$

where $X^i_o = c^i_o/a^i$ is the value of the aspect ratio when the r.e.v. is free of stress. In other words, the crack n° i is closed by the macroscopic stress $\mathbf{\Sigma}$ if its initial aspect ratio X^i_o is below the threshold $X_{c\ell}(\mathbf{\Sigma}, \underline{n}^i)$. Accordingly, the set $\mathcal{O}(\mathbf{\Sigma})$ of opened cracks in the r.e.v. subjected to the macroscopic stress $\mathbf{\Sigma}$ can be determined with the above criterion :

$$\mathcal{O}(\mathbf{\Sigma}) = \{X^i_o \geq X_{c\ell}(\mathbf{\Sigma}, \underline{n}^i)\} \tag{4.17}$$

We now return to the estimate of the tangent stiffness derived in (4.9). Let $\mathcal{N}^i$ denote the crack density relative to family ω_i. The corresponding crack density parameter ϵ^i is defined as $\epsilon^i = (a^i)^3\mathcal{N}^i$. X^i denoting the aspect ratio in the current configuration, the volume fraction of ω_i is $\varphi^c_i = 4\pi\epsilon^i X^i/3$. Taking once again advantage of (2.8) yields a new expression of the tangent stiffness tensor :

$$\mathbb{C}^{hom}_t(\mathbf{\Sigma}) = \varphi^s\mathbb{C}^s : \left(\varphi^s\,\mathbb{I} + \varphi^p\,(\mathbb{I} - \mathbb{S}_{sph})^{-1} + \sum_{j\in\mathcal{O}(\mathbf{\Sigma})}\frac{4\pi}{3}\epsilon^i\,\mathbb{T}^i\right)^{-1} \tag{4.18}$$

The tensor $\mathbb{T}^i$ being constant, (4.18) emphasizes the fact that the tangent stiffness tensor depends on the stress level only through the evolution of the set $\mathcal{O}(\mathbf{\Sigma})$ induced by the crack closure phenomenon. Indeed, provided that a crack remains opened, the variations of its aspect ratio do not affect the value of $\mathbb{C}^{hom}_t(\mathbf{\Sigma})$.

4.4 Statistical description of the crack density parameter

Tangent elastic stiffness under isotropic compression The practical implementation of (4.18) requires a quantitative determination of the set $\mathcal{O}(\mathbf{\Sigma})$ of opened cracks. The principle of the analysis is presented in the case of an isotropic compression : $\mathbf{\Sigma} = -\Sigma\mathbf{1}$ ($\Sigma > 0$). Assuming that the solid phase is isotropic, the expression of the threshold $X_{c\ell}(\mathbf{\Sigma}, \underline{n})$ introduced in (4.16) significantly simplifies. It is readily seen that $X_{c\ell}$ does not depend on the crack orientation and reads :

$$X_{c\ell}(\Sigma) = \frac{T\Sigma}{3\,k^s\,\varphi^s} \quad \textit{with} \quad T = \underline{n}\,.\,(\mathbb{T} : \mathbf{1})\,.\,\underline{n} \tag{4.19}$$

In the above definition of T, it is recalled that the tensor $\mathbb{T}$ given in (2.8) depends on the orientation of the normal $\underline{n}$ of the crack, whereas T does not. Using (2.14), one obtains :

$$T = \frac{4\,(1-\nu^{s2})}{\pi\,(1-2\,\nu^s)} \qquad X_{cl}(\Sigma) = \frac{4\,(1-\nu^{s2})}{3\,k^s\,\varphi^s\,\pi\,(1-2\nu^s)}\Sigma \tag{4.20}$$

Furthermore, let us consider a crack which initial aspect ratio X_o is greater than the threshold $X_{c\ell}(\Sigma)$. Integration of (4.14) yields the evolution of the aspect ratio in the form of an affine function of the stress state :

$$X(\Sigma) = \frac{c}{a} = \frac{c_0 + \delta c}{a} = X_o - \frac{T\,\Sigma}{3\,k^s\,\varphi^s} \tag{4.21}$$

We now introduce a continuous description of the geometrical parameters which characterize the cracks : namely, the radius a, the initial aspect ratio X_o and two Euler angles $\theta \in [0, \pi]$ and $\varphi \in [0, 2\pi]$ which define the orientation of the unit normal $\underline{n}$ to the crack plane.

As far as the non linear behavior is concerned, we have to focus on the statistical distribution of the aspect ratios. For simplicity, the distribution of the orientations is therefore assumed to be isotropic. We characterize the distribution of a and X_o by means of a density $n(a, X_o)$: the number dN of cracks per unit volume for which the radius, the initial aspect ratio and the Euler angles respectively belong to $[a, a+da]$ $[X_o, X_o+dX_o]$, $[\theta, \theta+d\theta]$ and $[\varphi, \varphi+d\varphi]$ is :

$$dN = n(a, X_o)\frac{1}{4\,\pi}\,\sin\theta\;d\varphi\,d\theta\,da\,dX_o \tag{4.22}$$

The corresponding current volume fraction $d\varphi^c$ reads:

$$d\varphi^c = \frac{4}{3}\pi\,a^2 c\,dN = \frac{X}{3}a^3 n(a, X_o)\,\sin\theta da\,d\varphi\,d\theta\,dX_o \tag{4.23}$$

Note in (4.23) that X_o and X represent the aspect ratio in the initial and current configuration, respectively. Cracks being modelled as flat ellipsoids, only small values of the aspect ratio X_o are relevant. In other words, there is a threshold X_{max} such that $n(a, X_o) = 0$ if $X_o \geq X_{max}$.

In the present situation, the appropriate definition of the crack density parameter is statistical in nature. We first introduce the density function $\epsilon(X_o)$:

$$\epsilon(X_o) = \int_0^{\infty} n(a, X_o)\,a^3\,da \tag{4.24}$$

from which we define a generalized crack density parameter in the form

$$\mathcal{E}(\Sigma) = \int_{X_{c\ell}(\Sigma)}^{X_{max}} \epsilon(X_o)dX_o \tag{4.25}$$

The sum over $\mathcal{O}(\Sigma)$ in (4.18) now takes the form :

$$\sum_{j\in\mathcal{O}(\Sigma)} \frac{4}{3}\pi\epsilon^j\mathbb{T}^j = \mathcal{E}(\Sigma)\mathbb{Q} \tag{4.26}$$

where $\mathbb{Q}$ is the fourth order isotropic tensor defined in (2.22). Introducing (4.26) into (4.18) yields the following expression of the tensor of macroscopic elastic compliance $\mathbb{S}^{hom} = {\mathbb{C}^{hom}}^{-1}$:

$$\mathbb{S}^{hom}(\Sigma) = \frac{1}{\varphi^s}\,\left(\varphi^s\,\mathbb{I} + \varphi^p\,(\mathbb{I} - \mathbb{S}_{sph})^{-1} + \mathcal{E}(\Sigma)\mathbb{Q}\right) : \mathbb{S}^s \tag{4.27}$$

This equation appears as an extension of (3.14) taking into account the evolution of the set of opened cracks and the progressive crack closure phenomenon through the concept of stress-dependent crack density parameter.
The stress level $\Sigma_{c\ell}$ for which all cracks are closed is defined by the condition $X_{c\ell}(\Sigma_{c\ell}) = X_{max}$. Recalling (4.19), one obtains :

$$\Sigma_{c\ell} = \frac{3\,k^s\,\varphi^s}{T} X_{max} \tag{4.28}$$

According to (4.27), if $\Sigma \geq \Sigma_{c\ell}$ (without crack propagation) the elastic compliance tensor is that of a porous medium made up of an uncracked solid phase (stiffness $\mathbb{C}^s$) and a porous domain (porosity φ^p). The tangent stiffness is then given by (4.10).

In particular, if the radius and the aspect ratio are identical for all cracks, respectively equal to $\tilde{a}$ and $\tilde{X}_o$, we note that $n(a, X_o) = \mathcal{N}\delta(a - \tilde{a})\delta(X_o - \tilde{X}_o)$, where $\delta(x)$ is the classical Dirac distribution. In this case, it is readily seen that the density function $\epsilon(X_o)$ defined in (4.24) is equal to $\mathcal{N}\tilde{a}^3\delta(X_o - \tilde{X}_o)$. In this situation, the confining pressure that closes the cracks is derived from (4.19) : $\Sigma_{c\ell} = 3k^s\varphi^s\tilde{X}_o/T$. Accordingly, (4.25)-(4.27) predict two possible values of the macroscopic elastic properties. If $\Sigma < \Sigma_{c\ell}$, the cracks are opened and it is readily seen that $\mathcal{E}(\Sigma)$ is equal to the classical crack density parameter $\mathcal{N}\tilde{a}^3$ so that the stiffness (3.14) is retrieved. In contrast, if $\Sigma > \Sigma_{c\ell}$, the cracks are closed. (4.25) indicates that $\mathcal{E}(\Sigma) = 0$ and the stiffness, given by (4.10), is that of the uncracked material. This kind of behavior is symbolically depicted at figure 2.

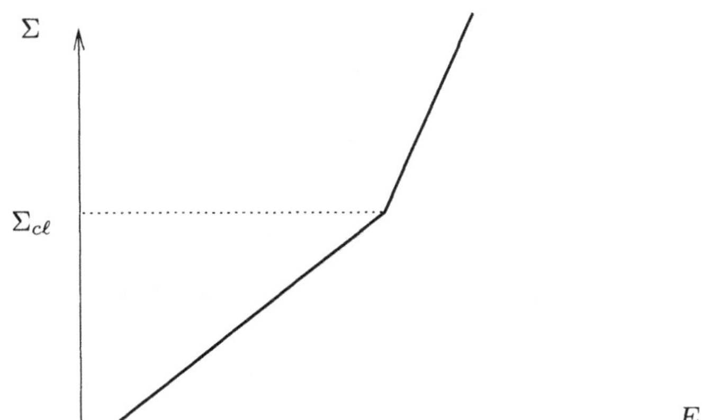

Figure 2. Bilinear elastic behavior obtained for a uniform value of the crack aspect ratio.

Experimental identification of the crack density parameter We now aim at determining the statistical description of the crack density parameter, i.e. the density function $\epsilon(X_o)$, from macroscopic experimental results. Differentiating (4.27) w.r.t. Σ yields :

$$\varphi^s \mathbb{Q}^{-1} : \frac{d\mathbb{S}_t^{hom}(\Sigma)}{d\Sigma} : \mathbb{C}^s = -\epsilon(X_{c\ell}(\Sigma))\frac{dX_{c\ell}(\Sigma)}{d\Sigma}\,\mathbb{I} \tag{4.29}$$

Recalling that $X_{c\ell}(\Sigma)$ is proportional to Σ (see equation (4.19)), (4.29) shows that the density function $\epsilon(X_o)$ can be identified from experimental data concerning the effect of the confining pressure on the macroscopic elastic compliance tensor $\mathbb{S}^{hom}(\Sigma)$. More precisely, $\mathbb{S}^{hom}(\Sigma)$ and $\mathbb{Q}$ being isotropic (see (2.22)), the tensorial equation (4.29) splits into two scalar relationships, which respectively relate $\epsilon(X_o)$ to the non linearity of the compression and shear moduli :

$$-\epsilon(X_{c\ell})\frac{T}{3\,k^s\,\varphi^s} = k^s\frac{\varphi^s}{Q_1}\frac{d}{d\Sigma}\left(\frac{1}{k^{hom}}\right)(\Sigma) = \mu^s\frac{\varphi_s}{Q_2}\frac{d}{d\Sigma}\left(\frac{1}{\mu^{hom}}\right)(\Sigma) \tag{4.30}$$

where $X_{c\ell} = T\Sigma/(3k^s\varphi^s)$. (4.30) means that the crack density parameter $\epsilon(X_o)$ can be determined from the measurement of the macroscopic tangent bulk or shear modulus. In practice, it seems that most of the experimental studies concerning non linear elastic behavior have been devoted to $k^{hom}(\Sigma)$.

According to the definition of $\epsilon(X_o)$ and $n(a, X_o)$, the initial value of the crack porosity φ_o^c reads:

$$\varphi_o^c = \int \frac{4\pi}{3}a^2 c_o\, n(a, X_o) da\, dX_o = \frac{4\pi}{3}\int_0^{X_{max}} X_o\epsilon(X_o)\, dX_o \tag{4.31}$$

Combining (4.30) and (4.31) yields:

$$\varphi_o^c = \frac{4\pi T}{9Q_1}\int_0^{\Sigma_{c\ell}} -\Sigma\frac{d}{d\Sigma}(\frac{1}{k^{hom}})\, d\Sigma \tag{4.32}$$

This equation shows that the initial crack porosity can be determined from the macroscopic measurement of the bulk modulus.

We now consider the particular case where the crack radius is uniform, equal to $\tilde{a}$. The density $n(a, X_o)$ thus reads :

$$n(a, X_o) = \nu(X_o)\delta(a - \tilde{a}) \tag{4.33}$$

In this case, $\epsilon(X_o)$ is given by

$$\epsilon(X_o) = \int_0^{\infty} a^3\mathcal{N}(a, X_o)\, da = \nu(X_o)\tilde{a}^3 \tag{4.34}$$

The crack density $\mathcal{N}$ is therefore related to $\epsilon(X_o)$:

$$\mathcal{N} = \int_0^{X_{max}} dX_o \int_0^{\infty} n(a, X_o)\, da = \frac{1}{\tilde{a}^3}\int_0^{X_{max}} \epsilon(X_o) dX_o \tag{4.35}$$

Combining (4.35) with (4.30) provides the link between the classical crack density parameter $\mathcal{N}\tilde{a}^3$ and the non linearity of macroscopic elastic properties, represented by the difference between the initial and asymptotic bulk moduli k_o and k_∞ :

$$\mathcal{N}\tilde{a}^3 = \frac{k^s\varphi^s}{Q_1}(\frac{1}{k_o} - \frac{1}{k_\infty}) \tag{4.36}$$

According to this result, if the crack radius can be estimated, the crack density $\mathcal{N}$ can be derived from a measurement of the difference between the initial and asymptotic bulk moduli. This result does not require any assumption about the shape of the function $k^{hom}(\Sigma)$.

5 Micromechanical analysis of damage propagation in saturated media

The inelastic deformation of a large class of civil engineering materials, e.g. concrete or rocks (sandstone, granite), is mainly due to microcracks growth and nucleation process. The most important consequence of microcracks growth is probably the deterioration of poroelastic properties and the damage-induced anisotropy. Continuum Damage Mechanics (CDM) provides a macroscopic strategy for investigating such deteriorating phenomena. However, several important questions cannot be addressed in the framework of a purely macroscopic approach, among which the choice of the relevant representation of damage, the expression of the thermodynamical potential for the derivation of the state equations and the damage evolution law.

In the sequel, we address the damage propagation induced by the growth of saturated cracks. The proposed approach combines some results of micromechanics and a reasoning of macroscopic thermodynamics (Andrieux et al. (1986), Pensée et al. (2002), Dormieux et al. (2004)). This yields a damage criterion based on the energy release rate. For simplicity, most of the analytical developments are performed in the case of parallel or randomly oriented cracks. A link with a macroscopic theory of poroelastic damage based on a continuous approximation of the crack density distribution by means of a second order tensor $\boldsymbol{D}$ is provided in section 7. To begin with, the r.e.v. is made up of solid and of connected opened saturated cracks and pores. As before, infinitesimal strains and isothermal evolutions are considered.

5.1 Potential energy for porous materials with saturated cracks

Let us first recall some results and notations of the energy analysis presented in the chapter "*Poroelasticity and strength of fully or partially saturated porous materials*". Let Ψ denote the macroscopic free energy of the solid phase in the r.e.v. per unit volume of porous medium. The potential energy of the solid phase is defined as $\Psi^* = \Psi - P(\phi - \phi_o)$. The condition of positivity of the macroscopic dissipation $\mathcal{D}$ reads :

$$\mathcal{D} = \left(\boldsymbol{\Sigma} : \dot{\boldsymbol{E}} + P\dot{\phi} - \dot{\Psi}\right) = \left(\boldsymbol{\Sigma} : \dot{\boldsymbol{E}} - (\phi - \phi_o)\dot{P} - \dot{\Psi}^*\right) \geq 0 \tag{5.1}$$

The macroscopic strain $\boldsymbol{E}$ and the pore pressure P being considered as the loading parameters, Ψ^* reads :

$$\Psi^*(\boldsymbol{E}, P) = \frac{1}{2}\boldsymbol{E} : \mathbb{C}^{hom} : \boldsymbol{E} - \frac{P^2}{2N} - P\boldsymbol{B} : \boldsymbol{E} \tag{5.2}$$

The potential energy $\Psi^*(\boldsymbol{E}, P)$ appears as a thermodynamic potential. The poroelastic state equations derived from the latter read :

$$\boldsymbol{\Sigma} = \frac{\partial \Psi^*}{\partial \boldsymbol{E}} = \mathbb{C}^{hom} : \mathbf{E} - \mathbf{B}P \tag{5.3}$$

and

$$\phi - \phi_o = -\frac{\partial \Psi^*}{\partial P} = \frac{P}{N} + \mathbf{B} : \mathbf{E} \tag{5.4}$$

5.2 General methodology for the description of damage propagation

The incorporation of damage in the standard Biot theory is based on the observation that the tensors $\mathbb{C}^{hom}$ and $\boldsymbol{B}$ as well as Biot's modulus N which intervene in (5.2) and in the state equations (5.3) and (5.4) can be regarded as functions of the microcracking level. Then, in view of a quantitative modelling of damage, the contribution of micromechanics is twofold. On one hand, it provides meaningful physical damage variables, typically the crack density parameter in the simple cases of parallel or randomly oriented cracks. On the other hand, estimates of the thermodynamic potential Ψ^* as function of these damage variables can be derived by means of the homogenization schemes presented in the previous sections.

For simplicity, we first present the methodology in the situation of parallel or randomly oriented opened cracks with uniform radius. In this case, the potential energy in (5.2) appears as a function of $\boldsymbol{E}$ and P but also of ϵ which represents the damage level in the poroelastic coefficients $\mathbb{C}^{hom}$, $\boldsymbol{B}$ and N : $\Psi^* = \Psi^*(\boldsymbol{E}, P, \epsilon)$. We look for a damage criterion and for the link between the damage rate $\dot{\epsilon}$ to the loading increments. The starting point is to consider (5.1) in an irreversible evolution :

$$\mathcal{D} = (\boldsymbol{\Sigma} - \frac{\partial \Psi^*}{\partial \boldsymbol{E}}) : \dot{\boldsymbol{E}} - (\phi - \phi_o + \frac{\partial \Psi^*}{\partial P})\dot{P} - \frac{\partial \Psi^*}{\partial \epsilon}\dot{\epsilon} \geq 0 \tag{5.5}$$

Introducing the state equations (5.3) and (5.4) into (5.5) yields :

$$\mathcal{D} = -\frac{\partial \Psi^*}{\partial \epsilon}\dot{\epsilon} \geq 0 \tag{5.6}$$

The scalar $G = -\partial\Psi^*/\partial\epsilon$ thus appears as the thermodynamic force associated with the damage variable ϵ, that is, the driving force of damage. G can be interpreted as the energy release rate associated with the damage growth. Similarly to Griffith criterion in Fracture Mechanics, the idea is to write the damage criterion in the form of a threshold on the driving force G of damage propagation :

$$\mathcal{F}(\boldsymbol{E}, P, \epsilon) = G - G_c = -\partial\Psi^*/\partial\epsilon - G_c \leq 0 \tag{5.7}$$

in which G_c represents the critical energy release rate characterizing the studied material. In turn, introducing (5.2) into (5.7), the above criterion can be expressed as a function of the mechanical loading parameters $\boldsymbol{E}$ and P :

$$G = -\frac{1}{2}\boldsymbol{E} : \frac{\partial \mathbb{C}^{hom}}{\partial \epsilon} : \boldsymbol{E} + \frac{P^2}{2}\frac{\partial}{\partial \epsilon}(\frac{1}{N}) + P\frac{\partial \boldsymbol{B}}{\partial \epsilon} : \boldsymbol{E} \leq G_c \tag{5.8}$$

Once initiated, the damage evolution law, that is, the link between $\dot{\epsilon}$ and the loading increment can be theoretically predicted from the condition $\dot{\mathcal{F}} = 0$. In fact, in a loading defined by the evolution $\boldsymbol{E}(t)$ and $P(t)$ of the macroscopic strain and the pore pressure, the damage process is stable if $\mathcal{F}(\boldsymbol{E}, P, \epsilon)$ is a decreasing function of ϵ :

$$\text{stability} \quad \Leftrightarrow \quad -\frac{\partial \mathcal{F}}{\partial \epsilon}(\boldsymbol{E}, P, \epsilon) = \frac{\partial^2 \Psi^*}{\partial \epsilon^2}(\boldsymbol{E}, P, \epsilon) > 0 \tag{5.9}$$

Apart from the situations in which the damage parameter has proved to be the scalar $\epsilon = \mathcal{N}a^3$, a statistical description of damage is due (see sections 2.3 and 4.4). We now consider the situation where the crack network is divided into crack families, each being associated with a crack density parameter[16] ϵ^i. An immediate extension of the methodology presented in section 5.2 consists in introducing a damage criterion for each family in the form

$$(\forall\, i) \qquad -\frac{\partial \Psi^*}{\partial \epsilon^i}(\boldsymbol{E}, P, \epsilon^1, \ldots, \epsilon^n) - G_c \leq 0 \tag{5.10}$$

Accordingly, (5.8) is replaced by :

$$-\frac{1}{2}\boldsymbol{E} : \frac{\partial \mathbb{C}^{hom}}{\partial \epsilon^i} : \boldsymbol{E} + \frac{P^2}{2}\frac{\partial}{\partial \epsilon^i}(\frac{1}{N}) + P\frac{\partial \boldsymbol{B}}{\partial \epsilon^i} : \boldsymbol{E} \leq G_c \tag{5.11}$$

We now take advantage of (1.5) :

$$\frac{\partial}{\partial \epsilon^i}(\boldsymbol{B}) = -\frac{\partial}{\partial \epsilon^i}(\mathbb{C}^{hom}) : \mathbb{S}^s : \mathbf{1} \tag{5.12}$$

and of (1.6) :

$$\frac{\partial}{\partial \epsilon^i}(\frac{1}{N}) = -\mathbf{1} : \mathbb{S}^s : \frac{\partial}{\partial \epsilon^i}(\mathbb{C}^{hom}) : \mathbb{S}^s : \mathbf{1} \tag{5.13}$$

Incorporating (5.12) and (5.13) into (5.11), the damage criterion relative to family n° i reads :

$$-\frac{1}{2}(\boldsymbol{E} + P\mathbb{S}^s : \mathbf{1}) : \frac{\partial}{\partial \epsilon^i}(\mathbb{C}^{hom}) : (\boldsymbol{E} + P\mathbb{S}^s : \mathbf{1}) \leq G_c \tag{5.14}$$

This constitutes the strain formulation of the damage criterion. $\boldsymbol{E} + P\mathbb{S}^s : \mathbf{1}$ appears as an "effective strain" controlling damage propagation. Alternatively, a stress formulation can be derived from the combination of (5.14) with the state equations. More precisely, introducing (1.5) into (1.2) reveals that :

$$\boldsymbol{\Sigma} + P\mathbf{1} = \mathbb{C}^{hom} : (\boldsymbol{E} + P\mathbb{S}^s : \mathbf{1}) \tag{5.15}$$

Eliminating $\boldsymbol{E} + P\mathbb{S}^s : \mathbf{1}$ between (5.14) and (5.15) yields a stress formulation of the damage criterion :

$$-\frac{1}{2}(\boldsymbol{\Sigma} + P\mathbf{1}) : \mathbb{S}^{hom} : \frac{\partial}{\partial \epsilon^i}(\mathbb{C}^{hom}) : \mathbb{S}^{hom} : (\boldsymbol{\Sigma} + P\mathbf{1}) \leq G_c \tag{5.16}$$

[16] The radius is assumed to be uniform in each family, but can be different from one family to the other.

(5.16) shows that the damage criterion is controlled by Terzaghi's effective stress $\boldsymbol{\Sigma}+P\mathbf{1}$. For illustrative purposes, let us apply this general methodology with the Mori-Tanaka estimate (3.12). The latter includes the effect of spherical pores (volume fraction $\varphi^p \approx \varphi$) and those of various crack families. From (3.12), we first note that :

$$\frac{\partial}{\partial \epsilon^i}(\mathbb{S}^{hom}) = \frac{4\pi}{3(1-\varphi)}\mathbb{T}^i : \mathbb{S}^s \tag{5.17}$$

Using the fact that $\mathbb{S}^{hom} : \mathbb{C}^{hom} = \mathbb{I}$, it is readily seen that

$$\frac{\partial}{\partial \epsilon^i}(\mathbb{C}^{hom}) = -\mathbb{C}^{hom} : \mathbb{H}^i : \mathbb{C}^{hom} \tag{5.18}$$

with $\mathbb{H}^i = 4\pi/(3(1-\varphi))\mathbb{T}^i : \mathbb{S}^s$. Introducing the above result into (5.16) yields

$$(\forall i) \quad \frac{1}{2}(\boldsymbol{\Sigma}+P\mathbf{1}) : \mathbb{H}^i : (\boldsymbol{\Sigma}+P\mathbf{1}) \leq G_c \tag{5.19}$$

We observe that the propagation of the crack family n° i is described by an elliptic stress criterion. Clearly enough, the orientation of the crack family is taken into account through the anisotropic tensor $\mathbb{H}^i$.

5.3 The role of the homogenization scheme on the damage criterion

We now investigate the influence of the homogenization scheme on the structure of the damage criterion. For simplicity, we restrict this discussion to the isotropic case : the distribution of cracks is isotropic and the macroscopic strain is of the form $\boldsymbol{E} = E\mathbf{1}$. The loading defined by E and P is expected to induce an isotropic stress state $\boldsymbol{\Sigma} = 1/3 \operatorname{tr} \boldsymbol{\Sigma}\mathbf{1}$. The damage growth is therefore controlled by the crack density parameter $\epsilon = \mathcal{N}a^3$. We also assume that the pore space is made up of cracks only.

dilute scheme The dilute estimates of the poroelastic coefficients are given by (2.24) with $1/N = b/k^s$. Accordingly, the onset of damage is characterized by the following criterion :

$$\frac{k^s Q_1}{2}\left(\operatorname{tr}\boldsymbol{E} + \frac{P}{k^s}\right)^2 - G_c \leq 0 \tag{5.20}$$

where Q_1 has been given in (2.23). Interestingly, from the second poroelastic state equation, it follows that the quantity $\epsilon Q_1(\operatorname{tr}\boldsymbol{E} + P/k^s)$ is equal to the normalized pore volume change $\phi - \phi_o$ and (5.20) now reads :

$$\frac{k^s}{2Q_1}\left(\frac{\phi-\phi_o}{\epsilon}\right)^2 - G_c = \frac{3\,E^s}{32(1-\nu^{s2})}\left(\frac{\phi-\phi_o}{\epsilon}\right)^2 - G_c \leq 0 \tag{5.21}$$

At first sight, the above equation means that a loading in which the evolution $\phi(t)$ of the pore volume is prescribed induces a stable damage process. In fact, we note that the value of $\mathcal{F}(\boldsymbol{E}, P, \epsilon)$ predicted by the dilute scheme actually does not depend on ϵ, so that the stability of the damage process in an evolution defined by $\boldsymbol{E}(t)$ and $P(t)$ is not ensured (see (5.9)). Furthermore, (5.20) states that these loading parameters are no more independant during the damage process. These two observations reveal serious shortcomings of the dilute scheme, as regards its use for damage prediction.

The Mori-Tanaka scheme Estimates given by the Mori - Tanaka scheme (see (3.8)) leads to:

$$k^{hom}(\epsilon) = \frac{k^s}{1+\epsilon Q_1} \quad ; \quad \boldsymbol{B}(\epsilon) = b\mathbf{1} = \frac{\epsilon Q_1}{1+\epsilon Q_1}\mathbf{1} \quad ; \quad \frac{1}{N} = \frac{b}{k^s} \tag{5.22}$$

The damage criterion (5.14) then reads :

$$\mathcal{F}(\boldsymbol{E}, P, \epsilon) = \frac{k^s Q_1}{2(1+Q_1\epsilon)^2}\left(\text{tr}\,\boldsymbol{E} + \frac{P}{k^s}\right)^2 - G_c \leq 0 \tag{5.23}$$

As opposed to the conclusion drawn with the dilute scheme, it is readily seen that the stability condition (5.9) is now fulfilled. Furthermore, once the criterion is reached, we can derive the value of the crack density parameter as a function of the macroscopic strain and the pore pressure.
The damage criterion (5.23) can be also formulated as a function of the macroscopic stress tensor $\boldsymbol{\Sigma} = tr\,\boldsymbol{\Sigma}/3\mathbf{1}$:

$$\frac{Q_1}{2k^s}(\frac{1}{3}\text{tr}\,\boldsymbol{\Sigma} + P)^2 - G_c \leq 0 \tag{5.24}$$

From a purely phenomenological point of view, this stress-based formulation of the damage criterion indicates a response which presents some similitude with a perfect elasto-plastic behavior, as long as no unloading is considered (figure 3).

The Ponte-Castaneda and Willis estimate The approach proposed by Ponte-Castaneda and Willis (1995) aims at incorporating the spatial distribution of the cracks in the micromechanical analysis. It consists in separating the effects of inclusion shape and spatial distribution. In the case of an isotropic distribution of crack orientation and of an isotropic spatial ditribution of cracks, this line of reasoning leads to the following expression of the effective bulk modulus :

$$\frac{k^{PCW}}{k^s} = 1 - \frac{\dfrac{16}{9}\dfrac{(1-\nu^{s2})}{(1-2\nu^s)}\epsilon}{1+\dfrac{16}{27}\dfrac{(1+\nu^s)^2}{(1-2\nu^s)}\epsilon} \tag{5.25}$$

from which estimates of the Biot coefficient b and modulus N can be derived from (1.5) and (1.6). As done above for the dilute and Mori-Tanaka schemes, the isotropic damage criterion can be formulated in term of strain :

$$\mathcal{F}(\boldsymbol{E}, P, \epsilon) = 648k^s\frac{(1-2\nu^s)(1-\nu^{s2})}{(27(1-2\nu^s)+16(1+\nu^s)^2\epsilon)^2}\left(tr\,\boldsymbol{E} + \frac{P}{k^s}\right)^2 - G_c \leq 0 \tag{5.26}$$

and of stress :

$$648k^s \frac{(1-\nu^{s2})}{(1-2\nu^s)(27-32(1+\nu^s)\epsilon)^2}\left(\frac{1}{3}tr\,\mathbf{\Sigma}+P\right)^2 - G_c \leq 0 \qquad (5.27)$$

As for the Mori-Tanaka scheme, it is readily seen from (5.26) that the stability condition (5.9) is satisfied. In contrast, as opposed to the Mori-Tanaka scheme, the effective stress $\frac{1}{3}tr\,\mathbf{\Sigma}+P$ predicted by the PCW scheme decreases as cracks propagate. The predictions of the dilute, MT and PCW schemes are summarized on figure 3.

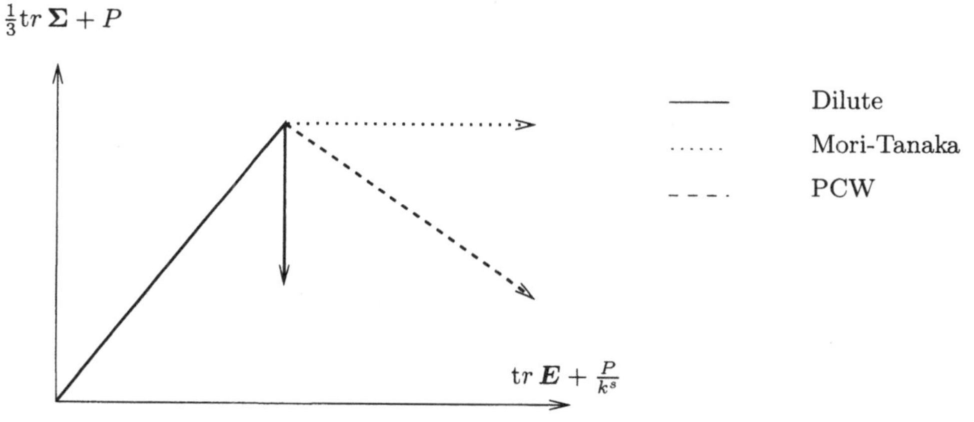

Figure 3. Effective stress response of a damaging cracked porous material.

Bibliography

S. Andrieux, Y. Bamberger, and J.-J. Marigo. Un modèle de matériau microfissuré pour les roches et les bétons. *J. Mech. Théor. Appl.*, 5(3):471–513, 1986.

M. Bouteca, D. Bary, J.-M. Piau, N. Kessler, M. Boisson, and D. Fourmaintraux. Contribution of poroelasticity to reservoir engineering : lab experiments, application to core decompression and implication in hp-ht reservoirs depletion. In *Proceedings of Eurock'94*, Delft, 1994. Balkema, Rotterdam.

B. Budiansky and R. O'Connell. Elastic moduli of a cracked solid. *Int. J. Solids Struct.*, 12:81–97, 1976.

J.-L. Chaboche. Damage induced anisotropy : on the difficulties associated with the active/passive unilateral conditions. *Int. J. Damage Mech.*, 2:311–329, 1992.

V. Deudé, L. Dormieux, D. Kondo, and S. Maghous. Micromechanical approach to nonlinear poroelasticity : application to cracked rocks. *J. Eng. Mech.*, 128:848–855, 2002a.

V. Deudé, L. Dormieux, D. Kondo, and V. Pensée. Non linear elastic properties of a mesocracked medium. *C. R. Acad. Sci. Paris, série IIb*, t. 330:587–592, 2002b.

L. Dormieux, E. Lemarchand, D. Kondo, and E. Fairbairn. Elements of poro-micromechanics applied to concrete. *Concrete Sc. Eng. Mater. Struct.*, 265:31–42, 2004.

L. Dormieux, A. Molinari, and D. Kondo. Micromechanical approach to the behavior of poroelastic materials. *J. Mech. Phys. Solids*, 50:2203–2231, 2002.

J.D. Eshelby. The determination of the elastic field of an ellipsoidal inclusion, and related problems. *Proc. R. Soc. Lond. A*, 241:376–396, 1957.

H. Horii and S. Nemat-Nasser. Overall moduli of solids with microcracks : load-induced anisotropy. *J. Mech. Phys. Solids*, 31(2):155–171, 1983.

D. Krajcinowic. *Damage mechanics.* North Holland, 1996.

M. Lion, F. Skoczylas, and B. Ledésert. Determination of the main hydraulic and poro-elastic properties of a limestone from bourgogne, france. *Int. J. Rock Mech. Mining Sci.*, to appear, 2005.

V.A. Lubarda and D. Krajcinovic. Damage tensors and the crack density distribution. *Int. J. Solids Structures*, 30(20):2859–2877, 1993.

T. Mura. *Micromechanics of defects in solids, 2nd edition.* Martinus Nijhoff Publ., 1987.

S. Murakami and K. Kamiya. Constitutive and damage evolution equations in elastic-brittle materials based on irreversible thermodynamics. *Int. J. Mech. Sci.*, 39(4), 1997.

S. Nemat-Nasser and M. Horii. *Micromechanics : overall properties of heterogeneous materials.* North Holland, 1993.

V. Pensée, D. Kondo, and L. Dormieux. Micromechanical analysis of anisotropic damage in brittle materials. *J. Engng. Mech., ASCE*, 128(8):889–897, 2002.

P. Ponte-Castaneda and J.R. Willis. The effect of spatial distribution on the effective behavior of composite materials and cracked media. *J. Mech. Phys. Solids*, 43(12): 1919–1951, 1995.

M. Thompson and J.R. Willis. A reformulation of the equations of anisotropic poroelasticity. *J. Appl. Mech.*, 58:612–616, 1991.

L.J. Walpole. Elastic behavior of composite materials : theoretical foundations. *Advances in Applied Mechanics*, 21:169–242, 1981.

J.R. Willis. Bounds and self consistent estimates for the overall properties of anisotropic composites. *J. Mech. Phys. Solids*, 25:185–202, 1977.

A. Zaoui. In P. Suquet, editor, *Continuum micromechanics.* Springer, 1997.

A. Zaoui. Continuum micromechanics : survey. *J. Eng. Mech.*, 128:808–816, 2002.

R.W. Zimmermann. *Compressibility of sandstones.* Elsevier, 1991.

6 Appendix 1 : algebra for transverse isotropy and applications

The homogenized properties of the medium made up of an isotropic solid matrix with parallel cracks exhibit transversely isotropic symmetry. For this reason, it is interesting to introduce standard notations and the corresponding simplified algebra (Walpole (1981)) for fourth order transversely isotropic tensors. This algebra is particularly useful to carry out easily inner product of transversely isotropic tensors or tensor inversion. Denoting by $\boldsymbol{i}_N = \underline{n} \otimes \underline{n}$ and $\boldsymbol{i}_T = \mathbf{1} - \underline{n} \otimes \underline{n}$, let us define:

$$\mathbb{E}_1 = \frac{1}{2}\boldsymbol{i}_T \otimes \boldsymbol{i}_T \; ; \; \mathbb{E}_2 = \boldsymbol{i}_N \otimes \boldsymbol{i}_N \; ; \; \mathbb{E}_3 = \boldsymbol{i}_T \overline{\underline{\otimes}} \boldsymbol{i}_T - \mathbb{E}_1 \; ; \; \mathbb{E}_4 = \boldsymbol{i}_N \overline{\underline{\otimes}} \boldsymbol{i}_T + \boldsymbol{i}_T \overline{\underline{\otimes}} \boldsymbol{i}_N \tag{6.1}$$

We recall that :

$$\mathbb{E}_1 + \mathbb{E}_2 + \mathbb{E}_3 + \mathbb{E}_4 = \mathbb{I} \tag{6.2}$$

and

$$\mathbb{E}_p : \mathbb{E}_q = \mathbb{E}_p \quad \text{if p=q and;} \; \mathbb{E}_p : \mathbb{E}_q = 0 \quad \text{if } p \neq q; \tag{6.3}$$

Two other elementary tensors are also defined :

$$\mathbb{E}_5 = \boldsymbol{i}_N \otimes \boldsymbol{i}_T \; ; \; \mathbb{E}_6 = \boldsymbol{i}_T \otimes \boldsymbol{i}_N \; ; \tag{6.4}$$

From these identities, it is shown that any transversely isotropic (not necessarily symmetric) fourth order tensor can be decomposed as :

$$\mathbb{L} = c\mathbb{E}_1 + d\mathbb{E}_2 + e\mathbb{E}_3 + f\mathbb{E}_4 + g\mathbb{E}_5 + h\mathbb{E}_6 \tag{6.5}$$

which may be expressed in the form :

$$\mathbb{L} = [c, d, e, f, g, h] \tag{6.6}$$

If $\mathbb{L}$ is symmetric ($\mathbb{L} =^t \mathbb{L}$, as for stiffness or compliance tensor), then $g = h$.
With this notation, considering $\mathbb{L}' = [c', d', e', f', g', h']$, the tensor product $\mathbb{L} : \mathbb{L}'$ reads :

$$\mathbb{L} : \mathbb{L}' = [cc' + 2hg', dd' + 2gh', ee', ff', gc' + dg', hd' + ch'] \tag{6.7}$$

and the inverse of $\mathbb{L}$ is

$$\mathbb{L}^{-1} = [\frac{d}{l}, \frac{c}{l}, \frac{1}{e}, \frac{1}{f}, -\frac{g}{l}, -\frac{h}{l}] \tag{6.8}$$

with $l = cd - 2gh$.
We note that the non symmetric tensor $(\mathbb{I} - \mathbb{S}(X, \underline{n}))$ can be put in the form $[1 - c, 1 - d, 1 - e, 1 - f, -g, -h]$ with :

$$c = S_{1111} + S_{1122}; \qquad d = S_{3333}; \qquad e = S_{1111} - S_{1122};$$
$$f = 2S_{3131}; \qquad g = S_{3311}; \qquad h = S_{1133} \tag{6.9}$$

the above components of S_{ijkl} being defined in (2.13) for $\underline{n} = \underline{e}_3$. Then, taking the inverse of $(\mathbb{I} - \mathbb{S}(X, \underline{n}))$ and performing the limit calculation in (2.8), one obtains the following expression of tensor $\mathbb{T}(\underline{n})$ associated to the set of parallel cracks normal to $\underline{n}$:

$$\mathbb{T}(\underline{n}) = \frac{4(1 - \nu^s)}{\pi} \left(\frac{1 - \nu^s}{1 - 2\nu^s}\mathbb{E}_2 + \frac{1}{2 - \nu^s}\mathbb{E}_4 + \frac{\nu^s}{1 - 2\nu^s}\mathbb{E}_5 \right) \tag{6.10}$$

or equivalently in the explicit form given in (2.15) Next, the combination of (2.7) and (2.15) yields the following expression for the homogenized stiffness tensor $\mathbb{C}^{hom}$ of the material weakened by a set of opened parallel cracks :

$$\begin{aligned}\mathbb{C}^{hom} &= \mathbb{C}^s - a_1\epsilon\Big(2a_2\mathbf{1}\otimes\mathbf{1} + a_3\Big(\mathbf{1}\otimes(\underline{n}\otimes\underline{n}) + (\underline{n}\otimes\underline{n})\otimes\mathbf{1}\Big) + \mathbf{1}\overline{\underline{\otimes}}(\underline{n}\otimes\underline{n}) \\ &+(\underline{n}\otimes\underline{n})\overline{\underline{\otimes}}\mathbf{1} - \nu^s(\underline{n}\otimes\underline{n})\otimes(\underline{n}\otimes\underline{n})\Big)\end{aligned} \quad (6.11)$$

for which it is recalled that $\epsilon = \mathcal{N}a^3$ is the crack density parameter. Coefficients a_i depend only on Poisson's ratio of the solid matrix, except a_1 which depends also on Young modulus E^s :

$$a_1 = \frac{16E^s(1-\nu^s)}{3(2-\nu^s)(1+\nu^s)}\ ;\ a_2 = \frac{(2-\nu^s)\nu^{s2}}{2(1-2\nu^s)^2}\ ;\ a_3 = \frac{(2-\nu^s)\nu^s}{(1-2\nu^s)} \quad (6.12)$$

7 Appendix 2 : macroscopic formulation of an anisotropic poroelastic damage model

7.1 Damage representation by a second order tensor $\boldsymbol{D}$

The classical approach of continuum damage mechanics first postulates the tensorial form of the damage variable, and then, the mathematical expression of the thermodynamics potential. The purpose of this section is to provide a link between the micromechanical approach and such a macroscopic theory of anisotropic damage. To do so, we now consider a continuous representation of the crack density distribution by means of a second order tensor $\boldsymbol{D}$. The description of directional distribution of defects by using tensors of different orders can be found in Lubarda and Krajcinovic (1993) and Krajcinowic (1996), from which the present analysis is inspired. For simplicity, as indicated before, we restrict ourselves to the use of second order tensor. Since any symmetric second order tensor admits three principal directions and at most three distinct principal real values, the representation of cracking by a second order tensor reduces any distribution of cracks to a set of three orthogonal families of cracks and cannot allow to describe materials symmetry more than orthotropy.
The idea of the continuous description considered here is first to approximate the crack density parameter[17] $\epsilon(\underline{n})$ associated to a normal $\underline{n}$ by a quadratic function of the components of $\underline{n}$. It is then assumed that there exists a second order tensor $\boldsymbol{d}$ such that:

$$\epsilon(\underline{n}) = \boldsymbol{d} : (\underline{n}\otimes\underline{n}) \quad (7.1)$$

In the line of a macroscopic reasoning, the next step is to express the second order tensor $\boldsymbol{d}$ by means of an overall tensor $\boldsymbol{D}$, depending on the crack density distribution :

$$\boldsymbol{D} = \frac{1}{4\pi}\int_{|\underline{n}|=1} \epsilon(\underline{n})(\underline{n}\otimes\underline{n})dS \quad (7.2)$$

[17] see section 2.3.

$\boldsymbol{d}$ being considered independent of $\underline{n}$, one has :

$$\boldsymbol{D} = \frac{1}{4\pi}\int_{|\underline{n}|=1} \boldsymbol{d} : (\underline{n}\otimes\underline{n}\otimes\underline{n}\otimes\underline{n})dS = \frac{1}{15}[(tr\boldsymbol{d})\mathbf{1} + 2\boldsymbol{d}] \quad (7.3)$$

Moreover, the definition (7.2) is such that :

$$tr\boldsymbol{D} = \frac{1}{4\pi}\int_{|\underline{n}|=1} \boldsymbol{d} : (\underline{n}\otimes\underline{n})dS = \frac{1}{3}tr\boldsymbol{d} \quad (7.4)$$

Reporting this result in (7.3), it follows that :

$$\boldsymbol{d} = \frac{3}{2}[5\boldsymbol{D} - (tr\boldsymbol{D})\mathbf{1}] \quad (7.5)$$

and therefore (7.1) reads :

$$\epsilon(\underline{n}) = \frac{15}{2}\boldsymbol{D} : (\underline{n}\otimes\underline{n}) - \frac{3}{2}tr\boldsymbol{D} = \frac{3}{2}[5\boldsymbol{D} : (\underline{n}\otimes\underline{n}) - tr\boldsymbol{D}] \quad (7.6)$$

7.2 Macroscopic formulation

Only opened cracks are considered. The determination of the homogenized stiffness tensor is done by integration of (2.28) with the expression of $\mathbb{T}$ given in (2.15). For this purpose, $\epsilon(\underline{n})$ is replaced by the expression given by (7.6). This calculation requires the determination of the integral $\frac{1}{4\pi}\int_{|\underline{n}|=1} \epsilon(\underline{n})(\underline{n}\otimes\underline{n}\otimes\underline{n}\otimes\underline{n})dS$ for which the following identities are useful (see Lubarda and Krajcinovic (1993)) :

$$\frac{1}{4\pi}\int_{S^2} n_i n_j n_k n_l n_\alpha n_\beta dS = \frac{1}{7}\big(\delta_{ij}L_{kl\alpha\beta} + \delta_{ik}L_{jl\alpha\beta} + \delta_{il}L_{jk\alpha\beta} + \delta_{i\alpha}L_{jkl\beta} + \delta_{i\beta}L_{jkl\alpha}\big) \quad (7.7)$$

where $\mathbb{L} = \frac{1}{4\pi}\int_{|\underline{n}|=1} \underline{n}\otimes\underline{n}\otimes\underline{n}\otimes\underline{n}\, dS$ is given by (2.21).
It appears that the homogenized stiffness can be put in the form :

$$\mathbb{C}^{hom}(\boldsymbol{D}) = \mathbb{C}^0 + \alpha(\mathbf{1}\otimes\boldsymbol{D}+\boldsymbol{D}\otimes\mathbf{1}) + \beta(\boldsymbol{D}\underline{\overline{\otimes}}\mathbf{1}+\mathbf{1}\underline{\overline{\otimes}}\boldsymbol{D}) + \eta tr\boldsymbol{D}(\mathbf{1}\otimes\mathbf{1}) + \gamma tr\boldsymbol{D}(\mathbf{1}\underline{\overline{\otimes}}\mathbf{1}) \quad (7.8)$$

where α, β, η and γ only depend on the elastic properties of the solid matrix.
(7.8) defines an approximation of the macroscopic stiffness tensor in the framework of a dilute scheme; it is based on the representation of $\epsilon(\underline{n})$ by means of the second order tensor $\boldsymbol{D}$ (see 7.6). The mathematical form of $\mathbb{C}^{hom}$ obtained from the present analysis is exactly the one provided by a macroscopic reasoning based on the representation theory (Murakami and Kamiya (1997)). However, in the macroscopic approach, the parameters α, β, η and γ have to be determined experimentally. Next, Biot tensor $\boldsymbol{B}(\boldsymbol{D})$ and Biot modulus $\frac{1}{N}(\boldsymbol{D})$ can be deduced respectively with the help of (1.5) and (1.6). Consequently, the potential energy (see 5.2) can be expressed as $\Psi^*(\boldsymbol{E}, P, \boldsymbol{D})$ from which is derived the complete anisotropic poroelastic state equations. The model has to be completed by the choice of a damage criterion which, using the thermodynamics framework,

can be based on the damage conjugated force (the second order tensor $-\frac{\partial \Psi^*}{\partial \boldsymbol{D}}$ in the present case).
Finally, it is worth observing that a more general description of various crack density distributions can be achieved by means of higher (4th) order tensor $\mathbb{D}$.

Modeling of Plastic Deformation of Saturated Porous Materials: Effective Stress Concept

Dariusz Łydżba[1] and Jian-Fu Shao[2]

[1] Institute of Geotechnics and Hydrotechnics, Wrocław Uniwersity of Technology, Wrocław, Poland
[2] Laboratoire de Mécanique de Lille, Ecole Polytechnique de Lille, Villeneuve d'Ascq, France

Abstract. The chapter presents some aspects of a mathematical modeling of the plastic behavior of saturated porous materials by a micromechanics approach. The general format of poro-plasticity is concluded by splitting the local elastic-plastic problem, stated for a representative volume element, into two ones corresponding to the elastic- and residual state of the solid, respectively. Making use of an appropriately defined comparison analysis, the so-called effective stress concept is verified.

1 Introduction

A micro-macro passage, in case one considers plastic deformations, could only lead to some approximate models, in general. This is caused by a fact that an actual macroscopic behavior of the medium being considered depends on an entire field of plastic strains within a microstructure. For instance, it results in a dependence of an overall loading function on a distribution and actual values of plastic micro-strains, i.e. on infinite number of variables (for details, see Suquet (1987)). Therefore, in order to propose a complete macroscopic description of such media some additional postulates, usually of a phenomenological character, have to be made.

Mathematical modeling of saturated porous media is usually simplified using the effective stress concept. This concept enables, by postulating the so-called stress- and strain equivalence principles to be valid, to extend the constitutive equations and complementary plastic laws of dry material to saturated one. Validity of this concept in the inelastic behavior range is, however, still an open problem, particularly for cohesive materials like rocks and concrete.

The chapter presents some aspects of mathematical modeling of the plastic behavior of saturated porous material by a micro-mechanics approach. The analysis starts with a general consideration of the problem studied. By splitting the local elastic-plastic problem into the two different ones, corresponding to the elastic- and the residual state of the solid within the representative volume element, the equations of poro-elasticity and the definitions of the plastic macro-strain and the plastic change of porosity are derived. Making use of the stress localization law, the closure of the elastic domain expressed by the appropriate macro-stress and the pore pressure is presented. The formula obtained involved, however, the residual micro-stress field, so the infinite number of variables, and is of no use, from practical point of view. Then by considering some particular microstructures as these composed of: rigid-ideally plastic constituents or a uniform material, some approximate models are proposed. Final comments end the chapter.

The analysis is based on the micromechanics approach for periodic structures. It has to be marked, however, that the other boundary conditions stated for representative volume element, as the uniform strain or the uniform stress, will not affect the results presented in the paper.

Throughout the chapter an index notation with a summation convention over repeated indices is used. The sign convention of the continuum mechanics (tension as positive) is applied. The hypothesis of small strains and small displacements is postulated.

2 Plasticity Format for Saturated Porous Materials

The analysis performed in this section follows the main strategy presented, for the elastic-plastic composite solid media, by Suquet (1987).

Consider a porous medium, composed of elastic-plastic constituents, saturated with fluid. The local boundary-value problem, governing the behavior of a solid phase within the unit cell, consists of the following equations, i.e.:

$$\begin{cases} \dfrac{\partial \sigma_{ij}^{s}}{\partial y_i} = 0 & \text{in } V_s \\ \sigma_{ij}^{s} N_i = -p\ \delta_{ij} N_i & \text{on } \Gamma \\ \sigma_{ij}^{s}(y) = a_{ijkh} e_{kh}(u\ ) & \text{in } V_s \\ u_i\,(y) = E_{ij} y_j + u_i^{*}(y) & \text{in } V_s \\ f\left(y,\sigma_{ij}^{s}\right) \le 0;\ \dot{\varepsilon}_{ij}^{P} = \lambda \dfrac{\partial g\left(y,\sigma_{ij}^{s}\right)}{\partial \sigma_{ij}^{s}} & \text{in } V_s \\ u_i^{*}(y) - Y \text{ periodic};\ \sigma_{ij}^{s} N_i - Y \text{ antiperiodic} & \end{cases} \tag{2.1}$$

where: $\sigma_{ij}^{s}(y)$ are components of the solid micro-stress tensor, $u_i(y)$ - the displacement field, u_i^{*} - the fluctuating part of the displacement field, $\varepsilon_{ij}^{P}(y)$ - the plastic micro-strain tensor, E_{ij} – the macro-strain tensor, p – the pore pressure, a_{ijkl} - the elastic stiffness tensor of the skeleton material, The functions $f(y,\sigma_{ij})$ and $g(y,\sigma_{ij})$ represent a yield function and a plastic potential, respectively. V_s is the volume of the solid material within the unit cell, Γ - the phase separation surface, N_i – component of a unit normal to the phase separation surface, λ – Lagrange multiplier. The symbol $e_{kh}(u)$ stands for the strain operator, i.e.:

$$e_{kh}(u) = \frac{1}{2}\left(\frac{\partial u_k}{\partial y_h} + \frac{\partial u_h}{\partial y_k}\right) \tag{2.2}$$

The system (2.1), due to a presence of a plastic deformation, is nonlinear. Its solution depends on entire history of the macro-strain and of the pore pressure. It could be characterized by the following relation:

$$\left[\tau \in (-\infty, t] \mapsto \left\{E_{ij}(\tau), p(\tau)\right\}\right] \mapsto \left\{e_{ij}\left(u(y,t)\right), \varepsilon_{ij}^{P}(y,t)\right\} \tag{2.3}$$

The left side of the expression involves only the macroscopic variables and represents the history of a macroscopic loading whereas the right side characterizes the actual micro-strain fields induced by this history. The stress field, in this formulation, is dependent function of both micro-strain fields and is determined through the constitutive equations of elasticity. The terms: a micro-

strain and a micro-stress are used in order to distinguish the strain and stress fields within a representative volume element from that being their averaged values and representing the macroscopic variables, i.e. the macro-strain and the macro-stress, respectively.
Let a set:

$$\left\{e_{ij}\left(u(y,t)\right),\varepsilon_{ij}^{p}(y,t),\sigma_{ij}^{s}(y,t)\middle|\left\{E_{ij}(t),p(t)\right\}\right\} \tag{2.4}$$

characterizes: an actual microscopic state, i.e. the state of strains and stress in the unit cell, and corresponding to it, at the instant time t, a macroscopic state characterized by the macro-strain and the pore pressure. This set could also be considered as a superposition of the two following ones, i.e.:

$$\begin{aligned}&\left\{e_{ij}\left(u(y,t)\right),\varepsilon_{ij}^{p}(y,t),\sigma_{ij}^{s}(y,t)\middle|\left\{E_{ij}(t),p(t)\right\}\right\}=\\&\quad=\left\{e_{ij}\left(u^{e}(y,t)\right),0,\sigma_{ij}^{e}(y,t)\middle|\left\{E_{ij}(t)-E_{ij}^{p}(t),p(t)\right\}\right\}+\\&\qquad+\left\{e_{ij}\left(u^{res}(y,t)\right),\varepsilon_{ij}^{p}(y,t),\sigma_{ij}^{res}(y,t)\middle|\left\{E_{ij}^{p},0\right\}\right\}\end{aligned} \tag{2.5}$$

The first set on the right side of the equation represents micro-fields as could be induced only by an elastic loading (or equivalently by freezing the plastic micro-strain and then loading) whereas the second one represents the residual states corresponding to null values: of the macro-stress and of the pore pressure. So, the total macro-strain for this state is simply the plastic one, i.e.: E_{ij}^{p}, whereas the micro-stress - the residual one. These sets correspond to the following local problems stated for the unit cell, i.e.:

$$\begin{cases}\dfrac{\partial\sigma_{ij}^{e}}{\partial y_i}=0 & \text{in } V_s\\ \sigma_{ij}^{e}N_i=-p\ \delta_{ij}N_i & \text{on } \Gamma\\ \sigma_{ij}^{e}(y)=a_{ijkh}e_{kh}(u^{e}) & \text{in } V_s\\ u_i^{e}(y)=\left(E_{ij}-E_{ij}^{p}\right)y_j+u_i^{*e}(y) & \text{in } V_s\\ u_i^{*e}(y)-Y\text{ periodic; } \sigma_{ij}^{e}N_i-Y\text{ antiperiodic}\end{cases} \tag{2.6}$$

$$\begin{cases}\dfrac{\partial\sigma_{ij}^{res}}{\partial y_i}=0 & \text{in } V_s\\ \sigma_{ij}^{res}N_i=0 & \text{on } \Gamma\\ \sigma_{ij}^{res}(y)=a_{ijkh}\left[e_{kh}(u^{res})-\varepsilon_{kh}^{p}\right] & \text{in } V_s\\ u_i^{res}(y)=E_{ij}^{p}y_j+u_i^{*res}(y) & \text{in } V_s\\ \dfrac{1}{\|V\|}\displaystyle\int_{V_s}\sigma_{ij}^{res}(y)dy=0 & \\ u_i^{*res}(y)-Y\text{ periodic; } \sigma_{ij}^{res}N_i-Y\text{ antiperiodic}\end{cases} \tag{2.7}$$

The superscripts e and res are used to mark the variables corresponding to the elastic state and to the residual one, respectively.

The micro-stress field has to verify, in addition, the plastic criterion, i.e.:,

$$f\left(\sigma_{ij}^{e}(y)+\sigma_{ij}^{res}(y)\right)-\kappa \leq 0 \qquad \forall y \in V_s \tag{2.8}$$

The local problem (2.6) enables to define constitutive equations of poro-elasticity whereas the problem (2.7) enables to define the plastic macro-strain and the plastic change of material porosity.

2.1 Constitutive Equations of Poro-Elasticity

Derivation of the constitutive equations of poro-elasticity is, in this section, only sketched. For details, the reader is referred to the lecture given by Dormieux (chapter in this book).

The system (2.6) implies the following relations, i.e.:

$$\begin{cases} e_{ij}\left(u^{e}(y)\right)=\left[I_{ijkh}+e_{ij}\left(\xi^{kh}(y)\right)\right]\left(E_{kh}-E_{kh}^{p}\right)+e_{ij}(\eta)\,p & \text{in } V_s \\ \sigma_{ij}^{e}(y)=a_{ijkh}\left[I_{khlm}+e_{kh}\left(\xi^{lm}(y)\right)\right]\left(E_{lm}-E_{lm}^{p}\right)+a_{ijkh}e_{kh}(\eta)\,p & \text{in } V_s \end{cases} \tag{2.9}$$

where: I_{ijkh} denotes a fourth order unit tensor.

The field $\xi_i^{kh}(y)$ and $\eta_i(y)$ stand for the solutions of the local problem (2.6) corresponding to the macroscopic state characterized by $\left\{\left(E_{ij}-E_{ij}^{p}\right)=\delta_{ik}\delta_{jh}, p=0\right\}$ and $\left\{\left(E_{ij}-E_{ij}^{p}\right)=0, p=1\right\}$, respectively.

Defining, for the porous medium saturated with fluid, a total stress tensor as:

$$\sigma_{ij}{}^{T}=h(y)\sigma_{ij}^{s}-(1-h(y))p\delta_{ij} \quad \text{where} \quad h(y)=\begin{cases}1 \text{ iff} & y\in V_s \\ 0 \text{ iff} & y\in V_l\end{cases} \qquad \text{in } V_s+V_l \tag{2.10}$$

and taking into account (2.5), one gets:

$$\sigma_{ij}^{T}=h(y)\left(\sigma_{ij}^{e}+\sigma_{ij}^{res}\right)-\left(1-h(y)\right)p\delta_{ij} \qquad \text{in } V_s+V_l \tag{2.11}$$

where: V_l denotes a domain, within the unit cell, occupied by the pores.

Volume average of the above equation results in a constitutive relation of the poro-elasticity theory, i.e.:

$$\Sigma_{ij}^{T}=A_{ijkh}\left(E_{kh}-E_{kh}^{p}\right)-\alpha_{ij}\,p \tag{2.12}$$

The symbol $\Sigma_{ij}^{T}\left(=\left\langle\sigma_{ij}^{T}\right\rangle\right)$ stands for a total macro-stress tensor. The overall material constants: A_{ijkh} and α_{ij} are determined by formulae:

$$A_{ijkh}=\left\langle h(y)\,a_{ijlm}(I_{lmkh}+e_{lm}\left(\xi^{kh}\right)\right\rangle; \quad \alpha_{ij}=n\delta_{ij}-\left\langle a_{ijkh}e_{kh}(\eta)\right\rangle \tag{2.13}$$

where n denotes a medium porosity.

2.2 Plastic Macro-Strain and Plastic Change of Porosity

The macro-variables corresponding to a plastic behavior of the porous medium result from the local problem (2.7). Two kinematic variables are involved in the mathematical description of saturated porous media plastic behavior, i.e. the plastic macro-strain and the plastic change of the porosity. The so-called localization operators play a crucial role in a derivation of the final definitions of these kinematic variables. Let us, therefore, state a localization formula, i.e. the relation which links the actual values of the micro-stress field with the actual values of: the macro-stress and the pore pressure.

Stress localization formula. The relation is stated, at first, for the elastic response of the microstructure. It is concluded, after some simple algebraic transformation, from the equation (2.9_2) and (2.12), i.e.:

$$\sigma^e_{ij}(y) = L_{ijkh}(y)\left(\Sigma^T_{kh} + \alpha_{kh}p\right) + M_{ij}(y)\,p \tag{2.14}$$

The above operators are defined as:

$$L_{ijkh}(y) = a_{ijrs}\left[I_{rslm} + e_{rs}\left(\xi^{lm}(y)\right)\right]S_{lmkh}; \quad M_{ij}(y) = a_{ijkh}e_{kh}(\eta) \tag{2.15}$$

where the fourth order tensor S_{lmkh} is the overall elastic compliance of porous medium considered. The localization operator $L_{ijkh}(y)$ is a classical one used in the analysis of composite solid materials. The operator M_{ij} reflects the effect of pore pressure on the micro-stress field.

The field $M_{ij}(y)$ can be exchanged, in the relation (2.15), by any others field being a linear combination of it and the operator L_{ijkh}, i.e.: $P_{ij}= c_1L_{ijkh}\delta_{kh}+c_2M_{ij}+\delta_{ij}$ (c_1 and c_2 – arbitrary constants). In a following such linear combination is applied, i.e.:

$$P_{ij}(y) = L_{ijkh}(y)\left(\alpha_{kh} - \delta_{kh}\right) + M_{ij}(y) + \delta_{ij} \tag{2.16}$$

Using the above operator, the localization relation (2.15) can be expressed as:

$$\sigma^e_{ij}(y) + p\delta_{ij} = L_{ijkh}(y)\Sigma^{eT}_{kh} + P_{ij}(y)\,p \quad \text{where} \quad \Sigma^{eT}_{ij} = \Sigma^T_{ij} + p\delta_{ij} \tag{2.17}$$

Volume average of the above formula leads to:

$$\Sigma^{eT}_{ij} = \left\langle L_{ijkh}(y)\right\rangle \Sigma^{eT}_{kh} + \left\langle P_{ij}(y)\right\rangle p \tag{2.18}$$

The above equation has to be valid for any elastic loading. Considering the case when pore pressure is null and then the case when medium is completely immersed in a fluid with pore pressure p and no other external loading is applied to the medium, one can conclude that the following relations have to be fulfilled, i.e.:

$$\left\langle h(y)L_{ijkh}\right\rangle = I_{ijkh} \quad \text{and} \quad \left\langle P_{ij}\right\rangle = 0 \tag{2.19}$$

The relation (2.17) shows also that the operator P_{ij} could be interpreted as a pressure polarization one since reflects, in any arbitrary point y of the solid, a deviation of the micro-stress from the hydrostatic state corresponding to the pore pressure p=1 applied to the medium. It can be also concluded, by inspecting the equation (2.16) as well as the definitions (2.15), that P_{ij} exhibits the same general properties as the residual micro-stress field, defined by the system (2.7), i.e. it is a self-equilibrated micro-stress field with its averaged value equal to null and fulfils, on phase separation surface Γ, a classical boundary condition as that for the porous medium without any fluid:

$$P_{ij}(y) \in \left\{ \tau_{ij}(y) \left| \frac{\partial \tau_{ij}}{\partial y_i} = 0 \text{ in } V_s,\ \tau_{ij} N_i = 0 \text{ on } \Gamma, \langle \tau_{ij} \rangle = 0 \right. \right\} \tag{2.20}$$

Finally, the actual micro-stress field, including its residual part, is related with macro-variables by the following formula, i.e.:

$$\sigma_{ij}(y) - \sigma_{ij}^{res}(y) + p\delta_{ij} = L_{ijkh}(y)\Sigma_{kh}^{eT} + P_{ij}(y)p \tag{2.21}$$

Using the plastic micro-strain field, the residual field can be also determined as (Suquet (1987)):

$$\sigma_{ij}^{res}(y) = -\int_{V_s} R_{ijkh}(y, y`)\varepsilon_{kh}^{p}(y`)dy` \tag{2.22}$$

where $R_{ijkh}(y, y')$ depends only on the geometry and elastic properties of microstructure considered.

The plastic macro-strain formula. Derivation of the plastic macro-strain formula is a quite easy task. The way as it could be done has been presented by Suquet (1987), nearly twenty years ago.

The crucial role in a derivation of this definition is played by a so-called macro-homogeneity condition. Let the set:

$$KA = \{u_i(y) \mid u_i(y) = E_{ij} y_j + w_i(y),\ w_i(y)\ \text{- Y periodic}\} \tag{2.23}$$

denotes a kinematically admissible displacement field and the set:

$$\overline{SA}(p) = \left\{ \tau_{ij}(y) \left| \frac{\partial \tau_{ij}}{\partial y_i} = 0 \text{ in } V_s, \tau_{ij} N_i = -p\delta_{ij} N_i \text{ on } \Gamma, \tau_{ij}(y) N_i - \text{Y antiperiodic} \right. \right\} \tag{2.24}$$

a statically admissible micro-stress field, at a prescribed value of pore pressure p. It can be proved, by simple algebraic transformation, that the following lemma holds true, i.e.:

$$\forall \sigma_{ij}(y) \in \overline{SA}(p) \quad \wedge \quad \forall u_i(y) \in KA \qquad \langle h(y)(\sigma_{ij}^{s} + p\delta_{ij}) e_{ij}^{y}(u) \rangle = \Sigma_{ij}^{eT} E_{ij} \tag{2.25}$$

The above relation, valid for the saturated porous medium, is a counterpart of the macro-homogeneity condition for composite solid media. Using the above macro-homogeneity condition, it is immediately seen that:

$$\langle h(y) L_{ijkh}(y) e_{ij}(u) \rangle = I_{ijkh} E_{ij} = E_{kh} \tag{2.26}$$

According to the system (2.7), the displacement field within the unit cell is described by: $u_i^{res}(y) = E_{ij}^{p} y_j + u_i^{*res}$, therefore:

$$\langle h(y) L_{ijkh}(y) e_{ij}(u^{res}) \rangle = E_{kh}^{p} \tag{2.27}$$

The relation (2.27) can be further transformed by using the elasticity relation for the constituents, i.e.:

$$E^p_{kh} = \langle h(y) L_{ijkh}(y) e_{ij}(u^{res}) \rangle = \langle h(y) L_{ijkh}(y) \left(s_{ijlm} \sigma^{res}_{lm} + \varepsilon^p_{ij} \right) \rangle \tag{2.28}$$

Bearing in mind the symmetry property of the compliance tensor (s_{ijlm}) as well as fact that both $s_{lmij}(y) L_{ijkh}$ and σ^{res}_{ij} are the kinematically- and statically admissible fields, respectively, one can write:

$$\langle h(y) L_{ijkh}(y) s_{ijlm}(y) \sigma^{res}_{lm} \rangle = \langle h(y) \sigma^{res}_{lm} s_{lmij}(y) L_{ijkh}(y) \rangle = \langle \sigma^{res}_{lm} \rangle \langle L_{rslm} s_{rsij}(y) L_{ijkh} \rangle = 0 \tag{2.29}$$

The final result of the equation (2.29) is a consequence of a null value of volume average of the residual micro-stress field. Comparing the equations (2.28) and (2.29), the formula for the plastic macro-strain is obtained, i.e.:

$$E^p_{kh} = \langle h(y) L_{ijkh}(y) \varepsilon^p_{ij} \rangle \tag{2.30}$$

The porosity plastic change formula. When one considers a porous material then the classical definition of strain macro-variable, used for composite solid media, does not hold true, in general, i.e.:

$$\langle h(y) L_{khij} e_{kh}(u) \rangle - \langle h(y) e_{ij}(u) \rangle \neq 0 \tag{2.31}$$

It appears that the left side of the above relation, when multiplied by δ_{ij}, represents a change of the medium porosity. To make it clear let us extend a displacement field within the solid phase on the entire space of the unit cell, i.e.:

$$u_i(y) = h(y)\left[E_{ij}\, y_j + u^*_i(y) \right] + (1 - h(y))\left[E_{ij}\, y_j + w_i(y) \right] \qquad \text{in } V_s + V_l$$
$$w_i(y) \text{ - Y-periodic}, \qquad u^*_i(y) = w_i(y) \quad \text{on } \Gamma \tag{2.32}$$

Then:

$$\langle e_{ii}(u) \rangle = E_{ii} \tag{2.33}$$

Taking into account the periodicity condition, the above volume average can be expressed as:

$$\langle h(y) e_{ii}(u) \rangle + \frac{1}{\|V\|} \int_{\partial V_l} u_i(y) N^p_i dS = E_{ii} \tag{2.34}$$

where ∂V_l denotes a boundary of pores domain and N^p a unit normal vector outward to this boundary.

It is easy to conclude that the integral represents the change of pores' volume in the medium with respect to its initial configuration. Therefore, denoting by δn the porosity change, one can writes:

$$\delta n = \langle h(y) L_{ijkh} e_{ij}(u) \rangle \delta_{kh} - \langle h(y) e_{ii}(u) \rangle \tag{2.35}$$

The plastic porosity change corresponds to the residual micro-strain field, therefore:

$$\delta n^p = \langle h(y) L_{ijkh} e_{ij}(u^{res}) \rangle \delta_{kh} - \langle h(y) e_{ii}(u^{res}) \rangle \tag{2.36}$$

In order to express the above formula by the plastic micro-strain field only, the localization operator P_{ij} has to be used, in addition to the operator L_{ijkh}. It appears that the equivalent definition of the plastic change of porosity can be also formulated:

$$\delta n^p = \left\langle h(y)\left(P_{ij} + L_{ijkh}\delta_{kh} - \delta_{ij}\right)e_{ij}\left(u^{res}\right)\right\rangle \tag{2.37}$$

The above relation is a direct consequence of the properties (2.20) of the pressure polarization operator, i.e.: it is the statically admissible stress field with volume averaged value equal null. Therefore:

$$\left\langle h(y)P_{ij}(y)e_{ij}\left(u^{res}\right)\right\rangle = 0 \tag{2.38}$$

and the equation (2.37) reduces to:

$$\delta n^p = \left\langle h(y)\left(L_{ijkh}\delta_{kh}e_{ij}\left(u^{res}\right)\right)\right\rangle - \left\langle \delta_{ij}e_{ij}\left(u^{res}\right)\right\rangle \tag{2.39}$$

which is exactly the definition (2.36) of the plastic change of porosity. Furthermore:

$$\left\langle h(y)\left(P_{ij} + L_{ijkh}\delta_{kh} - \delta_{ij}\right)e_{ij}\left(u^{res}\right)\right\rangle = \left\langle h(y)\left(P_{ij} - \delta_{ij}\right)e_{ij}\left(u^{res}\right)\right\rangle + \left\langle h(y)L_{ijkh}\delta_{kh}e_{ij}\left(u^{res}\right)\right\rangle \tag{2.40}$$

Since the field $s_{khij}(P_{ij}-\delta_{ij})$ is the kinematically admissible one, $P_{ij}-\delta_{ij}$ is the micro-stress field corresponding to the unit value of pore pressure and null value of the Therzaghi's effective stress (Eq. 2.17), therefore one can write:

$$\left\langle h(y)\left(P_{ij} - \delta_{ij}\right)e_{ij}\left(u^{res}\right)\right\rangle = \left\langle h(y)\left(P_{ij} - \delta_{ij}\right)\left(s_{ijkh}\sigma_{kh}^{res} + \varepsilon_{ij}^{p}\right)\right\rangle = \tag{2.41}$$

$$= \left\langle h(y)\sigma_{kh}^{res}s_{khij}\left(P_{ij} - \delta_{ij}\right)\right\rangle + \left\langle h(y)\left(P_{ij} - \delta_{ij}\right)\varepsilon_{ij}^{p}\right\rangle = \left\langle h(y)P_{ij}(y)\varepsilon_{ij}^{p}\right\rangle - \left\langle h(y)\varepsilon_{ij}^{p}\delta_{ij}\right\rangle$$

Comparing the equations (2.37) and (2.41), the following relation can be established, i.e.:

$$\delta n^p = \left\langle h(y)P_{ij}(y)\varepsilon_{ij}^{p}\right\rangle - \left\langle h(y)\varepsilon_{ij}^{p}\delta_{ij}\right\rangle + \left\langle h(y)L_{ijkh}(y)e_{ij}\left(u^{res}\right)\delta_{kh}\right\rangle \tag{2.42}$$

Bearing in mind the definition (2.30) of the plastic macro-strain, the final formula of the plastic change of porosity is obtained, i.e.:

$$\delta n^p = \left\langle h(y)P_{ij}(y)\varepsilon_{ij}^{p}\right\rangle - \left\langle h(y)\varepsilon_{ij}^{p}\delta_{ij}\right\rangle + \left\langle h(y)L_{ijkh}(y)\varepsilon_{ij}^{p}\delta_{ij}\right\rangle \tag{2.43}$$

Let us mark also that:

$$\delta n^e = \delta n - \delta n^p \tag{2.44}$$

Using the solutions of the elastic problems, i.e. Eq. (2.9), by similar arguments as applied above, one arrives at the following relation, i.e.:

$$\delta n - \delta n^p = \alpha_{ij}\left(E_{ij} - E_{ij}^{p}\right) + \beta p \tag{2.45}$$

or equivalently:

$$\dot{n} - \dot{n}^p = \alpha_{ij}\left(\dot{E}_{ij} - \dot{E}_{ij}^{p}\right) + \beta \dot{p} \tag{2.46}$$

where $\beta = -\left\langle h(y)e_{ii}(\eta)\right\rangle$ is the material constant of the Biot's poro-elasticity theory.

2.3 The closure of the elastic domain

Following Suquet (1987), the closure of the elastic domain in the macro-stress space, at prescribed distribution of residual stress field, can be expressed as:

$$\mathbb{E}_{\mathbf{S}}\left(p,\{\sigma_{ij}^{res}\}\right)=\left\{\ \Sigma_{ij}^{eT}\ \left|f\left(L_{ijkh}(y)\Sigma_{kh}^{eT}-p\delta_{ij}+P_{ij}(y)p+\sigma_{ij}^{res}(y)\right)\leq 0\quad\forall y\in V_s\ \right.\right\}\tag{2.47}$$

The set $\mathbb{E}_{\mathbf{S}}\left(p;\{\bar{\sigma}_{ij}^{res}\}\right)$ denotes, at prescribed residual stress field and value of p, the closure of the elastic domain in the Terzaghi's effective stress space. The above definition clearly shows that determination of the elastic domain requires the knowledge of the whole field of the residual stress $\{\sigma_{ij}^{res}(y)\}$.

In practice, some approximate models are used. For instance, for the porous material without any fluid, the closure of the elastic domain is approximated, assuming the strain-hardening rule to be valid, as:

$$\mathbb{E}_{\mathbf{D}}\left(E_{ij}^{p}\right)=\left\{\ \Sigma_{ij}^{s}\ \left|F_{D}\left(\Sigma_{ij}^{s};\chi\left(E_{ij}^{p}\right)\right)\leq 0\right.\right\}\tag{2.48}$$

By analogy, for the porous medium saturated with a fluid, the approximate model of the closure of the elastic domain can be proposed as:

$$\mathbb{E}_{\mathbf{S}}\left(E_{ij}^{p},\delta n^{p}\right)=\left\{\ \{\Sigma_{ij}^{eT},p\}\ \left|F_{S}\left(\Sigma_{ij}^{eT},p;\chi\left(E_{ij}^{p},\delta n^{p}\right)\right)\leq 0\right.\right\}\tag{2.49}$$

where $F_D\left(\Sigma_{ij}^{s};\chi\left(E_{ij}^{p}\right)\right),F_S\left(\Sigma_{ij}^{eT},p;\chi\left(E_{ij}^{p},\delta n^{p}\right)\right)\leq 0$ represent the macroscopic loading function for the dry and saturated state, respectively. $\mathbb{E}_{\mathbf{D}}\left(E_{ij}^{p}\right)$ stands for the closure of the elastic range in the dry state of the porous material.

2.4 Plastic Flow Rule: Rigid-Ideally Plastic Skeleton Case

As for the closure of the elastic domain, in order to propose a plastic flow rule useful for engineering practice, some approximate model has to be used. The one possibility to derive such model is by considering the particular microstructure. In a following, it is assumed that the porous skeleton is composed of rigid-ideally plastic constituents. Hence, the plastic behavior of the constituents is governed by the principle of maximum plastic dissipation. This principle states that for *given* plastic strain ε^p, among all possible stresses τ satisfying the yield criterion, the plastic dissipation:

$$\mathrm{d^p}\left[\tau,\varepsilon^{p}\right]=\tau_{ij}\varepsilon_{ij}^{p}\tag{2.50}$$

attains its maximum for the actual stress tensor σ.

Let a pair $\{\varepsilon_{ij}^{p}(y),\sigma_{ij}(y)\}$ represents the actual micro-strain and micro-stress field occurring in the solid skeleton, at the pore pressure p. Therefore, according to the principle of maximum plastic dissipation, the following inequality holds true:

$$\forall\tau_{ij}(y)\in\overline{SA}(\tilde{p})\qquad\dot{\varepsilon}_{ij}^{p}(y)\left[\tau_{ij}(y)-\sigma_{ij}(y)\right]\leq 0\quad\forall y\in V_s\tag{2.51}$$

or equivalently:

$$\forall \tau_{ij}(y) \in \overline{SA}(\tilde{p}) \quad \dot{\varepsilon}_{ij}^{p}(y)\left[\left(\tau_{ij}(y)+\tilde{p}\delta_{ij}\right)-\left(\sigma_{ij}(y)+p\delta_{ij}\right)\right]-$$
$$-\dot{\varepsilon}_{ij}^{p}(y)\delta_{ij}\left[\tilde{p}-p\right]\leq 0 \qquad \forall y \in V_s \tag{2.52}$$

Volume average of the above inequality leads to:

$$\forall\left\{\tilde{\Sigma}_{ij}^{eT},\tilde{p}\right\}\in\tilde{\mathbb{E}}_{\mathbf{S}} \quad \dot{E}_{ij}^{p}\left[\tilde{\Sigma}_{ij}^{eT}-\Sigma_{ij}^{eT}\right]-\left\langle h(y)\dot{\varepsilon}_{ij}^{p}\delta_{ij}\right\rangle\left[\tilde{p}-p\right]\leq 0 \tag{2.53}$$

Using the definition (2.34) of the porosity change, the above inequality can be further transformed to:

$$\forall\left\{\tilde{\Sigma}_{ij}^{T},\tilde{p}\right\}\in\mathbb{E}_{\mathbf{S}} \quad \dot{E}_{ij}^{p}\left[\tilde{\Sigma}_{ij}^{T}-\Sigma_{ij}^{T}\right]+\dot{n}^{p}\left[\tilde{p}-p\right]\leq 0 \tag{2.54}$$

The set $\tilde{\mathbb{E}}_{\mathbf{S}}$ stands for the plastically admissible pairs of: the Terzaghi's effective stress and the pore pressure whereas the set $\mathbb{E}_{\mathbf{S}}$ - for the pairs of: the total stress and the pore pressure.

The inequality (2.54) expresses the macroscopic principle of the maximum plastic dissipation. This principle implies also the associated plastic flow rule, therefore one gets finally:

$$\dot{E}_{ij}^{p}=\lambda\frac{\partial F_S\left(\Sigma_{kh}^{T},p\right)}{\partial\Sigma_{ij}^{T}} \quad \wedge \quad \dot{n}^{p}=\lambda\frac{\partial F_S\left(\Sigma_{kh}^{T},p\right)}{\partial p} \tag{2.55}$$

For the porous medium composed of the elastic plastic constituents, similarly as for the closure of the elastic domain, an approximate model of the plastic flow can be postulated in the following form, i.e.:

$$\dot{E}_{ij}^{p}=\lambda\frac{\partial G\left(\Sigma_{kh}^{T},p\right)}{\partial\Sigma_{ij}^{T}} \quad \wedge \quad \dot{n}^{p}=\lambda\frac{\partial G\left(\Sigma_{kh}^{T},p\right)}{\partial p} \tag{2.56}$$

where $G\left(\Sigma_{kh}^{T},p\right)$ represents the plastic potential.

2.5 Governing set of equations: approximate description

The governing set of equations of the macroscopic description is constituted of the following equations, i.e.:

- the balance equations:

$$\frac{\partial\Sigma_{ij}^{T}}{\partial x_i}=0 \tag{2.57}$$

- the constitutive equations of elasticity:

$$\Sigma_{ij}^{T}=A_{ijkh}^{eff}\left(E_{kh}-E_{kh}^{p}\right)-\alpha_{ij}p \tag{2.58}$$

- the constitutive law of the porosity change:

$$\dot{n}-\dot{n}^{p}=\alpha_{ij}\left(\dot{E}_{ij}-\dot{E}_{ij}^{p}\right)+\beta\dot{p} \tag{2.59}$$

- the (approximate) loading function and the (approximate) plastic flow rule:

$$F\left(\Sigma_{ij}^{T}, p^{(0)}; E_{ij}^{p}, n^{p}\right) \leq 0 ; \qquad \dot{E}_{ij}^{p} = \lambda \frac{\partial G\left(\Sigma_{kh}^{T}, p\right)}{\partial \Sigma_{ij}^{T}} \quad \wedge \quad \dot{n}^{p} = \lambda \frac{\partial G\left(\Sigma_{kh}^{T}, p\right)}{\partial p} \tag{2.60}$$

The above system is completed by the Darcy law and the incompressibility condition for the liquid in the pores (derivation of both is not presented in this chapter), i.e.:

- the filtration law

$$V_i - n\dot{u}_i(x,t) = -K_{ij} \frac{\partial p(x,t)}{\partial x_j} \tag{2.61}$$

- the incompressibility condition:

$$\dot{n} + \frac{\partial\left(V_i - n\dot{u}_i^{(0)}\right)}{\partial x_i} = 0 \tag{2.62}$$

2.6 Effective Stress Concept

Mathematical modeling of saturated porous media, as it is announced in the introduction, is usually simplified using the effective stress concept. This concept enables to extend the constitutive equations and complementary plastic laws of dry material to saturated one. It postulates that the following two equivalence principles hold true (see , for instance, R. de Boer and P.V. Lade (1997), Łydżba and Shao (2000)), i.e.:

- *Strain Equivalence:* the overall stress-strain relation for the saturated porous medium is obtained from the corresponding relation of the dry porous medium by substituting the effective stress tensor for the applied nominal stress;
- *Stress Equivalence:* the overall yield function for the saturated porous medium is obtained from the overall yield function of the dry porous medium by substituting the effective stress tensor for the applied nominal stress.

Quite recently, some results concerning validity of this concept have been obtained based on a micro-mechanics (de Buhan and Dormieux (1996), de Buhan and Dormieux (1999), Łydżba and Shao (2002)). For instance, de Buhan and Dormieux (1996) have shown that, according to the failure criterion, this concept holds true for some class of materials and for the other – it does not. Łydżba and Shao (2002) extended the analysis for the full plastic-hardening domain. With postulate that the porous skeleton is composed of a uniform material they arrived to the similar conclusions as those by de Buhan and Dormieux (1996).

This section recalls the results obtained by the authors and published in (Łydżba and Shao (2002)). During the analysis, it is assumed that the macroscopic behavior of the dry material is entirely determined. This lets us to proceed with the analysis based on a comparison of appropriately stated local problems: for dry and saturated material, respectively. Any prerequisite assumptions concerning pores geometry and arrangement are not needed.

The following simplifying assumption on the microstructure of porous medium is, however, postulated, i.e.: *the porous skeleton is composed of the uniform material.*

Dry material. For the porous medium without any liquid in the pores, a local boundary value problem, with a prescribed distribution of plastic strain ε_{ij}^{p} within the unit cell, can be presented as:

$$\begin{cases} \sigma_{ij,i}^{s} = 0 & \text{in } V_s \\ \sigma_{ij}^{s} n_i = 0 & \text{on } \Gamma \\ \sigma_{ij}^{s}(y) = a_{ijkh}\left[e_{kh}(u) - \varepsilon_{kh}^{p}\right] & \text{in } V_s \\ u_i(y) = E_{ij} y_j + u_i^{*}(y) & \text{in } V_s \\ f(\sigma_{ij}^{s}) \leq 0,\ u_i^{*}(y) - Y \text{ periodic}, \sigma_{ij}^{s} n_i - \text{antiperiodic} & \end{cases} \tag{2.63}$$

The above system implies the following localization law, i.e.:

$$\sigma_{ij}^{s}(y) - \sigma_{ij}^{res}(y) = L_{ijkh}(y)\Sigma_{kh}^{s} \tag{2.64}$$

The closure of the elastic domain is defined as:

$$\mathbb{E}_{\mathbb{D}}\left(\left\{\sigma_{ij}^{res}\right\}\right) = \left\{ \Sigma_{ij}^{s} \ \middle|\ f\left(L_{ijkh}(y)\Sigma_{kh}^{s} + \sigma_{ij}^{res}(y)\right) \leq 0 \quad \forall y \in V_s \right\} \tag{2.65}$$

where Σ_{ij}^{s} stands for the macro-stress tensor for the dry material. $\sigma_{ij}^{res}(y)$ represents the residual micro-stress field corresponding to the dry state of the material and is determined by the functional relation (2.22).

Fully saturated material. For the porous material saturated with a liquid pressure p and with a prescribed distribution of plastic strain $\bar{\varepsilon}_{ij}^{p}(y)$, a local boundary value problem reads:

$$\begin{cases} \sigma_{ij,i}^{s} = 0 & \text{in } V_s \\ \sigma_{ij}^{s} n_i = -p\delta_{ij} n_i & \text{on } \Gamma \\ \sigma_{ij}^{s}(y) = a_{ijkh}\left[e_{kh}(\bar{u}) - \bar{\varepsilon}_{kh}^{p}\right] & \text{in } V_s \\ \bar{u}_i(y) = \bar{E}_{ij} y_j + \bar{u}_i^{*}(y) & \text{in } V_s \\ f(\sigma_{ij}^{s}) \leq 0,\ \bar{u}_i^{*}(y) - Y \text{ periodic}, \sigma_{ij}^{s} n_i - \text{antiperiodic} & \end{cases} \tag{2.66}$$

The stress localization law, as it has been derived earlier, is expressed by the following formula, i.e.:

$$\sigma_{ij}^{s}(y) - \bar{\sigma}_{ij}^{res}(y) + p\delta_{ij} = L_{ijkh}(y)\Sigma_{kh}^{eT} + P_{ij}(y)\,p \tag{2.67}$$

where: $\bar{\sigma}_{ij}^{res}(y)$ represents the residual micro-stress field and corresponds to the saturated state of the material. It is also determined by the functional relation (2.22).

Making use of the postulate that the porous skeleton is composed of the uniform material, it can be proved that:

$$P_{ij}(y) = 0 \qquad \forall y \in V_s \tag{2.68}$$

The above statement is a consequence of the following identity, i.e.:

$$\left\langle P_{ij}(y)s_{ijkh}\left(P_{kh}(y)-\delta_{kh}\right)\right\rangle=0 \tag{2.69}$$

which, in turn, is a consequence of the macro-homogeneity condition.
Now, since for the uniform material $s_{ijkh}\delta_{kh}$=const. , therefore:

$$\left\langle P_{ij}(y)s_{ijkh}\delta_{kh}\right\rangle=s_{ijkh}\delta_{kh}\left\langle P_{ij}(y)\right\rangle=0 \tag{2.70}$$

and consequently:

$$\left\langle P_{ij}(y)s_{ijkh}P_{kh}(y)\right\rangle=0 \tag{2.71}$$

The above form is a positive defined. Therefore, it implies the identity (2.68). The relation (2.68) simplifies the localization law and the definition of the closure of the elastic domain for the saturated material, i.e.:

$$\mathbb{E}_{\mathbf{S}}\left(p,\{\bar{\sigma}_{ij}^{res}\}\right)=\left\{\ \Sigma_{ij}^{eT}\ \middle|f\left(L_{ijkh}(y)\Sigma_{kh}^{eT}-p\delta_{ij}+\bar{\sigma}_{ij}^{res}(y)\right)\leq 0\quad\forall y\in V_s\ \right\} \tag{2.72}$$

Form of the effective stress. In the following, the concept of effective stress is examined by considering two qualitatively different plastic yield criteria for the skeleton material, namely: a pressure insensitive and a pressure sensitive one.
The Von-Mises criterion: This criterion is the pressure insensitive. Defining for the saturated material an equivalent stress as:

$$\sigma_{ij}^{eq}(y)=\sigma_{ij}^{s}(y)+p\delta_{ij} \tag{2.73}$$

one gets the following identities:

$$(\forall p)\begin{cases}\sigma_{ij}^{eq}(y)=a_{ijkh}\left(e_{kh}^{eq}(\tilde{u})-\bar{\varepsilon}_{kh}^{p}\right) & \text{with}\quad e_{kh}^{eq}(\tilde{u})=e_{kh}(\bar{u})+\dfrac{p\delta_{kh}}{3K_s}\\ f\left(\sigma_{ij}^{s}\right)=f\left(\sigma_{ij}^{eq}\right)\ , & \dfrac{\partial f\left(\sigma_{ij}^{s}\right)}{\partial\sigma_{kh}^{s}}=\dfrac{\partial f\left(\sigma_{ij}^{eq}\right)}{\partial\sigma_{kh}^{eq}}\end{cases} \tag{2.74}$$

The property (2.74$_2$) together with the definition of the elastic domain (2.72) implies:

$$\mathbb{E}_{\mathbf{S}}\left(p,\{\bar{\sigma}_{ij}^{res}\}\right)=\left\{\ \Sigma_{ij}^{eT}\ \middle|f\left(L_{ijkh}(y)\Sigma_{kh}^{eT}+\bar{\sigma}_{ij}^{res}(y)\right)\leq 0\quad\forall y\in V_s\ \right\} \tag{2.75}$$

The identities (2.74) enable also to transform the local problem (Eqs. (2.66)) to the equivalent one:

$$\begin{cases}\sigma_{ij,i}^{eq}=0 & \text{in } V_s\\ \sigma_{ij}^{eq}n_i=0 & \text{on } \Gamma\\ \sigma_{ij}^{eq}(y)=a_{ijkh}\left[e_{kh}^{eq}(\tilde{u})-\bar{\varepsilon}_{kh}^{p}\right] & \text{in } V_s\\ \tilde{u}_i(y)=E_{ij}^{eq}y_j+u_i^{*}(y)\quad\text{with}\ E_{ij}^{eq}=\bar{E}_{ij}+\dfrac{p\delta_{ij}}{3K_s} & \text{in } V_s\\ f(\sigma_{ij}^{eq})\leq 0,\ u_i^{*}(y)-Y\,\text{periodic},\sigma_{ij}^{eq}n_i-\text{antiperiodic} & \end{cases} \tag{2.76}$$

where K_s denotes a bulk modulus of the skeleton material.
Assuming that the plastic deformation of the skeleton material is governed by an associative plastic flow rule, the properties (2.74_2) lead to:

$$\dot{\bar{\varepsilon}}_{ij}^{p}(y)=\begin{cases}\dfrac{\dfrac{\partial f}{\partial\sigma_{kl}^{eq}}a_{klmn}\dot{e}_{mn}^{eq}(y)}{\dfrac{\partial f}{\partial\sigma_{kl}^{eq}}a_{klmn}\dfrac{\partial f}{\partial\sigma_{mn}^{eq}}}\dfrac{\partial f}{\partial\sigma_{ij}^{eq}} & \text{for } f\left(\sigma_{ij}^{eq}(y)\right)=0\ \wedge\dot{f}\left(\sigma_{ij}^{eq}(y)\right)=0\\ 0 & \text{otherwise}\end{cases}\tag{2.77}$$

The system of equations (2.76)-(2.77) represents the local elasto-plastic problem for the saturated material expressed by the equivalent fields introduced. It is of the same form as the corresponding local problem for the dry material, except it contains the equivalent strain and stress, instead of the solid's strain and stress as for the dry state.
Consider now the saturated porous material, initially free of the plastic micro-strain field, subject to a loading history $\left[\tau\in(-\infty,t]\mapsto\left\{\bar{E}_{ij}(\tau),p(\tau)\right\}\right]$. The pair:

$$\left[\tau\in(-\infty,t]\mapsto\left\{\bar{E}_{ij}(\tau),p(\tau)\right\}\right]\mapsto\left\{e_{ij}\left(\bar{u}(y,t)\right),\bar{\varepsilon}_{ij}^{p}(y,t)\right\}\tag{2.78}$$

characterizes a solution of the local elasto-plastic problem for the saturated material, i.e. the loading history and corresponding micro-strain fields induced in the unit cell. Furthermore, as a direct consequence of the system (2.76)-(2.77), the following equivalences can be written:

$$\left[\tau\in(-\infty,t]\mapsto\left\{E_{ij}^{eq}(\tau)\right\}\right]\mapsto\left\{e_{ij}^{eq}\left(\tilde{u}(y,t)\right),\bar{\varepsilon}_{ij}^{p}(y,t)\right\}\tag{2.79}$$

or:

$$\left[\tau\in(-\infty,t]\mapsto\left\{E_{ij}(\tau)=E_{ij}^{eq}(\tau)\right\}\right]\mapsto\left\{e_{ij}\left(u(y,t)\right)=e_{ij}^{eq}\left(\tilde{u}(y,t)\right),\varepsilon_{ij}^{p}(y,t)=\bar{\varepsilon}_{ij}^{p}(y,t)\right\}\tag{2.80}$$

The latter relation represents a solution of the local problem for the dry material. It indicates that the solution for the saturated material can be recovered from the solution of the local problem for the dry material by imposing the following loading history:

$$\left[\tau\in(-\infty,t]\mapsto\left\{E_{ij}(\tau)\right\}=\left\{\bar{E}_{ij}(\tau)+\frac{p(\tau)}{3K_s}\delta_{ij}\right\}\right]\tag{2.81}$$

Furthermore, as a consequence of the functional relations (2.22)-(2.30) as well as the constitutive equations, one gets also:

$$\begin{cases}\sigma_{ij}^{s}(y,t)=\sigma_{ij}^{eq}(y,t);\ \ \sigma_{ij}^{res}(y,t)=\bar{\sigma}_{ij}^{res}(y,t) & \forall y\in V_s\\ E_{ij}^{p}(t)=\bar{E}_{ij}^{p}(t);\ \ \Sigma_{ij}^{s}(t)=\Sigma_{ij}^{eT}(t)\end{cases}\quad\text{for}\quad E_{ij}(t)=\bar{E}_{ij}(t)+\frac{p(t)}{3K_s}\delta_{ij}\tag{2.82}$$

where the variables on the left side of the equations (2.82) correspond to the dry material whereas on the right side to the saturated material.
The final step consists in comparison of the corresponding definitions of the elastic domains (2.65) and (2.72), for dry and saturated states, respectively. Taking into account the relation (2.82) one gets:

$$\mathbb{E}_{\mathbf{S}}\left(p,\left\{\bar{\sigma}_{ij}^{res}\right\}\right)=\mathbb{E}_{\mathbf{D}}\left(\left\{\bar{\sigma}_{ij}^{res}\right\}\right) \tag{2.83}$$

which clearly indicates that the Terzaghi's effective stress fulfills the stress equivalence principle for the material considered. This statement can be extended also for porous materials composed of a uniform material obeying any arbitrary pressure insensitive yield criterion with an associative or even a non-associative plastic flow rule. For the case of non-associative plasticity, the plastic potential has to be, however, of the form:

$$\tilde{g}\left(\sigma_{ij}\right)=aI_1+g\left(J_2,J_3\right) \tag{2.84}$$

where I_1 is the first invariant of stress tensor, J_3 is the third invariant of deviatoric stress tensor, $g\left(J_2,J_3\right)$ is an arbitrary function of J_2 and J_3 ,'a' is a parameter (could be a=0). Such the form of the plastic potential allows expressing the rate of the plastic micro-strain by the equivalent fields introduced above and the formula is of the same form as for the dry material. Therefore, the relations (2.82) as well as the identity (2.83) are still hold true.

According to the approximate models, if a description for the dry material uses, in a plastic range, the loading function (2.48) and a plastic flow rule as:

$$\dot{E}_{ij}^{p}=\lambda\frac{\partial G\left(\Sigma_{kh}^{s}\right)}{\partial\Sigma_{ij}^{s}} \tag{2.85}$$

therefore, it follows from (2.82) that:

$$\mathbb{E}_{\mathbf{S}}\left(p,\bar{E}_{ij}^{p}\right)=\left\{\ \Sigma_{ij}^{eT}\ \middle|F_D\left(\Sigma_{ij}^{eT};\chi\left(\bar{E}_{ij}^{p}\right)\right)\leq 0\right\} \tag{2.86}$$

$$\dot{\bar{E}}_{ij}^{p}=\lambda\frac{\partial G\left(\Sigma_{kh}^{eT}\right)}{\partial\Sigma_{ij}^{eT}} \tag{2.87}$$

The equations (2.86) and (2.87) are as good approximation of plastic behavior of the saturated material as the approximation (2.48) and (2.85) of the dry material. The results (2.86) and (2.87) prove also that the Terzaghi's effective stress fulfills, in the plastic range, the stress as well as the strain equivalence principles for saturated media. It has to be marked, however, that the result has been obtained with the assumption that the yield function was the pressure insensitive.

The Coulomb-Mohr criterion: The criterion is the pressure-sensitive. This time a particular loading history is only considered, namely the case corresponding to a so-called drained condition, i.e. $\left\{\bar{E}_{ij}(t),p(t)=\text{const.}\right\}$. The equivalent stress is now defined as:

$$\sigma_{ij}^{eq}=\frac{\sigma_{ij}^{s}+p\delta_{ij}}{1+p\,\mathrm{tg}\varphi/c} \tag{2.88}$$

The stress fulfills the identity:

$$f\left(\sigma_{ij}^{s}\right)=\left(1+p\,\mathrm{tg}\varphi/c\right)f\left(\sigma_{ij}^{eq}\right) \tag{2.89}$$

where: φ represents a friction angle $(\varphi\geq 0)$, c is the internal cohesion of the skeleton material $(c>0$ is assumed).

Since $p\geq 0$, the equation (2.89) implies:

$$f\left(\sigma_{ij}^{s}\right)\leq 0 \text{ if } f\left(\sigma_{ij}^{eq}\right)\leq 0\,;\ \frac{\partial f\left(\sigma_{ij}^{s}\right)}{\partial\sigma_{kh}}=\frac{\partial f\left(\sigma_{ij}^{eq}\right)}{\partial\sigma_{kh}^{eq}} \tag{2.90}$$

The relations (2.90) enable to transform the local problem (Eqs.(2.66)) to the equivalent one described by the system (2.76). The equations (2.76_3) and (2.77_4) are, however, expressed by the new equivalent variables, i.e.:

$$\begin{cases} \sigma_{ij}^{eq}(y)=a_{ijkh}\left(e_{kh}^{eq}(\tilde{u})-\dfrac{\varepsilon_{kh}^{p}}{1+p\mathrm{tg}\varphi/c}\right) & \text{with } e_{kh}^{eq}(\tilde{u})=\dfrac{e_{kh}(\overline{u})+\dfrac{p\delta_{kh}}{3K_s}}{1+p\mathrm{tg}\varphi/c} \\ \tilde{u}_i(y)=E_{ij}^{eq}y_j+u_i^{*}(y) & \text{with } E_{ij}^{eq}=\dfrac{\overline{E}_{ij}+\dfrac{p\delta_{ij}}{3K_s}}{1+p\mathrm{tg}\varphi/c} \end{cases} \tag{2.91}$$

Furthermore, a rate of the plastic micro-strain can be expressed as (the associative plasticity is assumed):

$$\frac{\dot{\overline{\varepsilon}}_{ij}^{p}(y)}{1+p\mathrm{tg}\varphi/c}=\begin{cases} \dfrac{\dfrac{\partial f}{\partial\sigma_{kl}^{eq}}a_{klmn}\dot{e}_{mn}^{eq}(y)}{\dfrac{\partial f}{\partial\sigma_{kl}^{eq}}a_{klmn}\dfrac{\partial f}{\partial\sigma_{mn}^{eq}}}\dfrac{\partial f}{\partial\sigma_{ij}^{eq}} & \text{for } f\left(\sigma_{ij}^{eq}(y)\right)=0\ \wedge\ \dot{f}\left(\sigma_{ij}^{eq}(y)\right)=0 \\ 0 & \text{otherwise} \end{cases} \tag{2.92}$$

Once more, the local elasto-plastic problem for the saturated material, expressed by the equivalent fields, is of the same form as for the dry material. Therefore, considering the pair:

$$\left[\tau\in(-\infty,t]\mapsto\left\{\overline{E}_{ij}(\tau),p(\tau)\right\}\right]\mapsto\left\{e_{ij}\left(\overline{u}(y,t)\right),\overline{\varepsilon}_{ij}^{p}(y,t)\right\} \tag{2.93}$$

one gets also:

$$\left[\tau\in(-\infty,t]\mapsto\left\{E_{ij}(\tau)=E_{ij}^{eq}(\tau)\right\}\right]\mapsto\left\{e_{ij}\left(u(y,t)\right)=e_{ij}^{eq}\left(\tilde{u}(y,t)\right),\varepsilon_{ij}^{p}(y,t)=\frac{\overline{\varepsilon}_{ij}^{p}(y,t)}{1+p\mathrm{tg}\varphi/c}\right\} \tag{2.94}$$

The relation (2.94) implies:

$$\begin{cases} \sigma_{ij}^{s}(y,t)=\sigma_{ij}^{eq}(y,t);\ \ \sigma_{ij}^{res}(y,t)=\dfrac{\overline{\sigma}_{ij}^{res}(y,t)}{1+p\mathrm{tg}\varphi/c} & \forall y\in V_s \\ E_{ij}^{p}(t)=\dfrac{\overline{E}_{ij}^{p}(t)}{1+p\mathrm{tg}\varphi/c};\ \ \Sigma_{ij}^{s}(t)=\Sigma_{ij}^{eq}(t) & \end{cases} \tag{2.95}$$

where the variables on the left side correspond to the dry material whereas on the right side to the saturated one.

The macroscopic equivalent stress tensor is the volume averaged equivalent micro-stress tensor (2.88):

$$\Sigma_{ij}^{eq} = \frac{\Sigma_{ij}^{eT}}{1+p\mathrm{tg}\varphi/c} = \frac{1}{\|V\|}\int_{V_s} \sigma_{ij}^{eq}(y)\,dy \tag{2.96}$$

Using the above tensor, the elastic domain for the saturated material can be defined as:

$$\mathbb{E}^{eq}\left(p,\{\bar{\sigma}_{ij}^{res}\}\right) = \left\{ \Sigma_{ij}^{eq} \;\middle|\; f\left(L_{ijkh}(y)\Sigma_{kh}^{eq} + \frac{\bar{\sigma}_{ij}^{res}(y)}{1+p\mathrm{tg}\varphi/c}\right) \le 0 \quad \forall y \in V_s \right\} \tag{2.97}$$

which together with the definition of the elastic domain (2.65) and the relations (2.95) result in:

$$\mathbb{E}^{eq}\left(p,\{\bar{\sigma}_{ij}^{res}\}\right) = \mathbb{E}_{\mathbb{D}}\left(\left\{\frac{\bar{\sigma}_{ij}^{res}}{1+p\mathrm{tg}\varphi/c}\right\}\right) \tag{2.98}$$

The equivalent macro-stress defined by (2.96) fulfills the stress equivalence principle, at the drained condition. It is clear that this statement is also valid for a non-associative plastic flow rule described by the potential (2.84) with a function $g(J_2, J_3)$ being a homogeneous function of deviatoric stress of the degree one.

According to the approximate model, it follows from (2.95) that:

$$\mathbb{E}^{eq}\left(p,\bar{E}_{ij}^{P}\right) = \left\{ \Sigma_{ij}^{eq} \;\middle|\; F_D\left(\Sigma_{ij}^{eq}; \chi\left(\frac{\bar{E}_{ij}^{P}}{1+p\mathrm{tg}\varphi/c}\right)\right) \le 0 \right\} \tag{2.99}$$

$$\frac{\dot{\bar{E}}_{ij}^{P}}{1+p\mathrm{tg}\varphi/c} = \lambda \frac{\partial G\left(\Sigma_{kh}^{eq}\right)}{\partial \Sigma_{ij}^{eq}} \tag{2.100}$$

3 Conclusions

The consideration shows that in a case of porous medium composed of the elastic-plastic constituents, a complete macroscopic description could not be derived based on the micromechanics approach. This method enables, however, identification of the variables involved in the macroscopic description. By postulating some particular microstructures, an approximate description could be formulated.

The effective stress concept has been validated for saturated porous media composed of micro-homogeneous and micro-isotropic skeleton material. Particularly, it has been shown that the appropriate form of the effective stress tensor depends on the form of a yield function used at the micro-level. Two different form of the effective stress have been determined. It has to be marked, however, that the assumption of the skeleton micro-homogeneity is the crucial one. Without this assumption one can only prove that the Terzaghi's effective stress gives only a potentially safe approximation of failure criterion for saturated porous medium. This case is carefully studied in Łydżba (2002).

References

de Boer R. , Lade P.V. (1997). The concept of effective stress for soil, concrete and rock, *Geotechnique*, 47, 61-78.

de Buhan P., Dormieux L. (1996). On the validity of the effective stress concept for assessing the strength of saturated porous materials: a homogenization approach, *J. Mech. Phys. Solids*, 44, 1649-1667.

de Buhan P., Dormieux L. (1999). A micromechanics-based approach to the failure of saturated porous media, *Transport in Porous Media*, 34, 47-62.

Łydżba D., Shao J.F. (2000). Study of poroelasticity material coefficients as response of microstructure, *Mechanics of Cohesive-Frictional Materials*, 5, 149-171.

Łydżba D. (2002). *Applications of asymptotic homogenization method in soil and rock mechanics*, Oficyna Wydawnicza PWr, Wrocław, (in polish).

Łydżba D., Shao J.F. (2002). Stress equivalence principle for saturated porous media, *C.R. Mecanique*, 330, 297-303.

Suquet P. (1987). Elements of homogenization for inelastic solid mechanics, in *Homogenization Techniques for Composite Media*, Lecture Notes in Physics, 272, Springer-Verlag, Berlin.

Part III

Materials Applications

Experimental Microporomechanics

Franz-Josef Ulm,* Antoine Delafargue, Georgios Constantinides
Massachusetts Institute of Technology, Cambridge MA, U.S.A.

Abstract Is it possible to break down natural porous material systems down to a scale where materials no longer change from one material to another, and upscale ('nanoengineer') the behavior from the nanoscale to the macroscale of engineering applications? – This is the challenging question we address in these lecture notes through a review of tools and methods of experimental microporomechanics. The combination of advanced experimental indentation techniques and microporomechanics theory provides a unique opportunity to understand and assess nanoproperties and microstructure, as a new basis for the engineering prediction of macroscopic poromechanical properties of natural composites. This is illustrated for cement-based materials and shales.

1 Introduction

Inarguably, natural composite materials (concrete, soils, rocks, bones, etc.) are multiphase and multiscale material systems. The multiphase composition of such materials is permanently evolving over various scales of time and length, creating in the course of this process the most heterogeneous classes of materials in existence, with heterogeneities that manifest themselves from the nanoscale to the macroscale. The most prominent heterogeneity of such natural composite materials is the porosity which is the space left in between the different solid phases at various scales, ranging from interlayer spaces in between minerals filled by a few water molecules, to the macropore space in between microstructural units of the material in the micrometer to millimeter range; and which are all detrimental to the mechanical behavior of natural composite materials.

Recent advances in both experimental and theoretical microporomechanics provides a rational means to quantitatively address multiphase and multiscale material systems. The underlying idea of continuum micro(poro)mechanics is that it is possible to separate a heterogeneous material into phases with on-average constant material properties (Suquet, 1997; Zaoui, 2002). A phase, in the sense of continuum micromechanics, is not necessarily a material phase such as a specific mineral as used in physical chemistry, but

*In Collaboration with Prof. Y. Abousleiman (Oklahoma University at Norman); Dr. Paul Acker and Jean-Francois Batoz (Lafarge Corp.), Drs. R. Ewy, L. Duranti, D. K. McCarty (Chevron Texaco); and the support of Nanolab@MIT team, A. Schwartzman and Prof. K. Van Vliet. Financial support by the Lafarge Corporation and Chevron Texaco is gratefully acknowledged.

a material domain that can be identified, at a given scale, with on-average constant material properties, so that a continuum mechanics analysis can be performed, with some confidence and accuracy, at the considered material scale. Such phases are referred to as microhomogeneous phases. Provided the existence of such microhomogeneous phases, a combination of novel nanomechanical testing of materials and upscaling methods is a powerful tool to 'break the code' of natural composites: identify material invariant properties and explain, by upscaling, the macroscopic diversity of natural composite materials systems.

The first part of these lecture notes provides an introduction to theory of instrumented indentation, and the application to heterogeneous material systems. By way of application, two case studies are presented in a second and third part, dealing with two natural material systems: cement-based materials and shales.

2 Nanoindentation

Indentation tests go a long way back, as early as 1722 (Réaumur, 1722), and started out with hardness measurements (for a review see Borodich and Keer (2004) and references cited herein). More recently, thanks to progress in hardware and software control, depth sensing techniques were introduced that allow a continuous monitoring of the displacement of the indenter into the specimen surface for both loading and unloading. The idea of depth sensing techniques and its implementation down to the nanoscale appears to have developed first in the former Soviet Union from the mid 1950ies on throughout the 1970ies, and received considerable attention world-wide, ever since Doerner and Nix (1986) and Oliver and Pharr (1992) in the late 1980ies and early 1990ies, also identified this technique for analysis and estimation of mechanical properties of materials. While the chronology of events of discovery may still be in debate[1], there is little doubt, at least as far as the elastic behavior is concerned, that it is the Hertz type contact problem that forms much of the theoretical background of modern indentation analysis. An indentation test provides a continuous record of the variation of the indentation load, P, as a function of the depth of indentation, h, into the indented specimen, and the extraction of material properties, from the $P-h$ curve, is achieved by inverse analysis.

2.1 Hardness

To motivate the forthcoming developments, we consider a three-dimensional, rigid, conical indenter of a given half-angle α, indenting normally into a homogeneous elastic perfectly plastic cohesive-frictional material half-space, $x_i \geq 0$. The origin (O) of the Cartesian system is put at the point of the initial contact between the conical indenter and the half-space (Fig. 1). The indentation test consists of (at least) two phases, a loading phase and an unloading phase, during which either the force, P, or the rigid displacement of the indenter, h, is prescribed. Conical indentation leads to geometrically self-similar indentation states. That is, for a given half-angle, the average pressure below

[1] The chronology of events of discovery of depth-sensing indentation and indentation anlaysis has only recently been revealed by Borodich in several remarkable publications (Borodich and Keer, 2004).

the indenter is independent of the indentation load and the true contact area. Using the projected current contact area, which is proportional to the true contact area for geometrically self-similar indenters (*ie.* $A_M = A/\sin\alpha$ for conical indenters), yields the classical definition of hardness H, which can be determined at any point along the $P-h$ curve for which the contact area is known:

$$H \stackrel{def}{=} \frac{P_1}{A_1} = \frac{P_2}{A_2} = \ldots = \frac{P}{A} \tag{2.1}$$

where $A = \pi a^2$ is the projected contact area, and $a = h_c \tan\alpha$ is the contact radius.

The main problem in the analysis is that the contact surface A (respectively the contact depth h_c) is not known *a priori*, but is a solution of a boundary value problem. In fact, the rigid displacement h of the indenter is generally not the contact depth, h_c (Fig. 1), corresponding to the maximum projected contact surface of the indenter with the deformed half-space surface: $h_c/h < 1$ corresponds to what is referred to, in the indentation literature, as sinking-in; and $h_c/h > 1$ as piling-up (Fig. 1). Hence, there are *a priori* two independent measurements to be carried out: the force P and the projected contact surface A. It is instructive to perform a dimensional analysis (Chen and Chen, 2004): The two dependent quantities of interest that define the hardness, force P and contact area A (respectively the contact depth h_c), depend on the material properties (stiffness C_{ijkl}, cohesion c, friction angle φ), the indenter geometry (which in the case of conical indentation reduces to the half-apex angle α), and the indentation depth h:

$$P = f(C_{ijkl}, c, \varphi, \alpha, h) \tag{2.2a}$$

$$A = g(C_{ijkl}, c, \varphi, \alpha, h) \tag{2.2b}$$

From a straightforward application of dimensional analysis (or more precisely the Pi-Theorem) to relations (2.2), it is readily found that the two dimensionless relations,

$$\frac{P}{ch^2} = \Pi_\alpha\left(\frac{C_{ijkl}}{c}, \varphi, \alpha\right) \tag{2.3a}$$

$$\frac{A}{h^2} = \Pi_\beta\left(\frac{C_{ijkl}}{c}, \varphi, \alpha\right) \tag{2.3b}$$

define a unique third dimensionless relation, the hardness-to-cohesion ratio as a unique function of the stiffness-to-cohesion ratio, the friction angle and the half-apex angle:

$$\frac{H}{c} = \frac{\Pi_\alpha}{\Pi_\beta} = \mathcal{H}\left(\frac{C_{ijkl}}{c}, \varphi, \alpha\right) \tag{2.4}$$

The dimensionless relations have two limits:

For conical indentation into a perfectly elastic material ($C_{ijkl}/c \to 0, \varphi = 0$), it is readily recognized that the dimensionless relation (2.3b) is independent of the elasticity constants, and that the contact depth to indentation depth ratio is a constant (Swadener and Pharr, 2001):

$$\lim_{C_{ijkl}/c \to 0} \frac{h_c}{h} = \frac{2}{\pi} \tag{2.5}$$

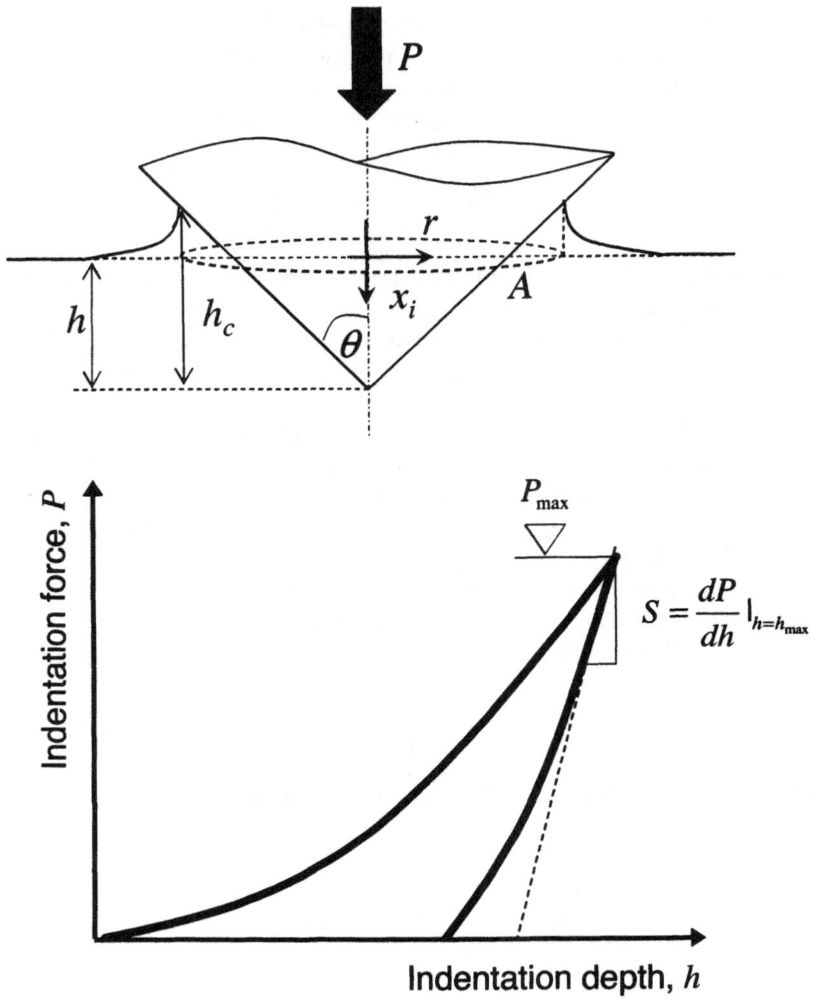

Figure 1. Principle of Indentation Test.

Relation (2.5), which is a general property of the Hertz contact problem into a perfectly elastic material (whether isotropic or anisotropic), states that perfect elasticity always produces sink-in, $h_c/h < 1$.

For conical indentation into a rigid plastic material ($c/C_{ijkl} \to 0$), yield design approaches are suitably employed to determine the H/c ratio for specific strength criteria. Yield design is based on the assumption, that plastic failure occurs when the material system has exhausted its capacity to store the externally supplied work rate, *ie.* $P\dot{h}$ into recoverable elastic energy, so that the supplied work rate is entirely dissipated within the material bulk and along surfaces of discontinuity through kinematically admissible velocity fields $\underline{U}$ (see e.g. Salençon (1983), Ulm and Coussy (2003) Chapter 9); *ie.* for the indentation test:

$$P\overset{\circ}{h} = \int_{\Omega} \pi(\mathbf{d})\, d\Omega + \int_{\Gamma} \pi([[\underline{U}]])\, d\Gamma \tag{2.6}$$

where $\pi(\mathbf{d}) = \sup \boldsymbol{\sigma} : \mathbf{d}$ and $\pi([[\underline{U}]]) = \sup \underline{T} \cdot [[\underline{U}]]$ is the maximum dissipation capacity the material can develop in the material bulk Ω and along surfaces of discontinuity Γ; $\boldsymbol{\sigma}$ is the statically and plastically admissible stress field, and $\underline{T} = \boldsymbol{\sigma} \cdot \underline{n}$ is the stress vector, satisfying:

$$\boldsymbol{\sigma} = {}^t\boldsymbol{\sigma};\ \operatorname{div} \boldsymbol{\sigma} = 0;\ [[\underline{T}]] = [[\boldsymbol{\sigma} \cdot \underline{n}]] = 0 \tag{2.7a}$$

$$f(\boldsymbol{\sigma}) \leq 0;\ f(\underline{T}) \leq 0 \tag{2.7b}$$

where superscript t stands for transpose; and $f(\boldsymbol{\sigma})$ and $f(\underline{T})$ are the yield function defining the strength domain of the material system respectively in continuous material sub-domains and on surfaces of discontinuity; while $\mathbf{d}$ is the solution strain rate field in continuous material sub-domains, and $[[\underline{U}]]$ is the velocity jump over surfaces of discontinuity Γ, which are kinematically compatible with the velocity field $\underline{U}$, and compatible with the plastic flow rule of the material:

$$\mathbf{d} = \frac{1}{2}\left(\operatorname{grad} \underline{U} + {}^t \operatorname{grad} \underline{U}\right) = \overset{\circ}{\lambda} \frac{\partial f}{\partial \boldsymbol{\sigma}} \tag{2.8a}$$

$$[[\underline{U}]] = \underline{U}^+ - \underline{U}^- = \overset{\circ}{\lambda} \frac{\partial f}{\partial \underline{T}} \tag{2.8b}$$

For a cohesive-frictional material of the Mohr-Coulomb type, the problem to be solved reads (Ganneau et al., 2004):

$$P\overset{\circ}{h} = \frac{c}{\tan\varphi} \int_{\partial\Omega} \underline{U} \cdot \underline{n} da \tag{2.9}$$

where $\underline{U} \cdot \underline{n}$ is the normal component of the velocity field at the surface $\partial\Omega$ of the half-space, which includes the cone mantel $A_M = A/\sin\alpha$, and the stress-free surface outside the contact radius $r \geq a = \sqrt{A/\pi}$. The previous relation can be recast in the form of (2.4):

$$\lim_{c/C_{ijkl} \to 0} \frac{H}{c} = \mathcal{H}(\varphi, \theta) = \frac{1}{A \tan\varphi} \int_{\partial\Omega} \overline{\underline{U}}' \cdot \underline{n} da \tag{2.10}$$

with $\overline{\underline{U}}' = \underline{U}'/\overset{\circ}{h}$ the normalized surface velocity field. Exact axi-symmetric yield design solutions for indentation in Mohr-Coulomb materials are scarce: Hopkins et al. (1961)

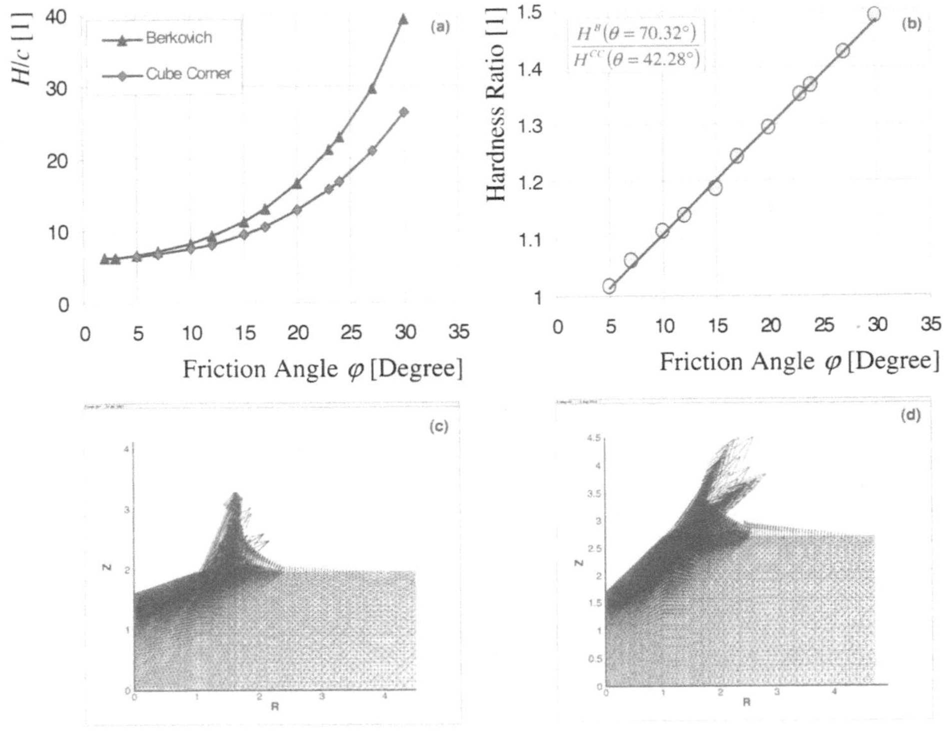

Figure 2. Upper-Bound Yield Design Approach (compiled from Ganneau et al., 2004): Hardness-Cohesion ratio as a function of friction angle for different semi-apex angles: $\alpha = 70.32°$ is representative of a Berkovich indenter; and $\alpha = 42.28°$ is representative of a Cube Corner indenter, when assimilating the three-sided pyramidal indenters to cones of same projected contact area.

provided a yield design solution for the smooth punch and Matar and Salençon (1982) provided a heuristic solution for the perfectly rough punch. Not surprisingly, as one reduces the cone angle, the amount of energy that is dissipated at plastic failure decreases (Figure 2). Making use of this property, it is possible to determine from two indentation tests with different apex angles the two strength properties of rigid-plastic cohesive frictionless materials. Figure 2 displays the hardness-to-cohesion ratio as a function of the friction angle for two apex angles ($\alpha = 70.32°$ and $\alpha = 42.28°$) obtained from a computational implementation of the upper bound limit theorem (Ganneau and Ulm, 2004; Ganneau et al., 2004). The figure also displays the hardness ratio vs. the friction angle, displaying an almost linear relation.

Finally, in between these two limit cases (purely elastic – rigid plastic), the con-

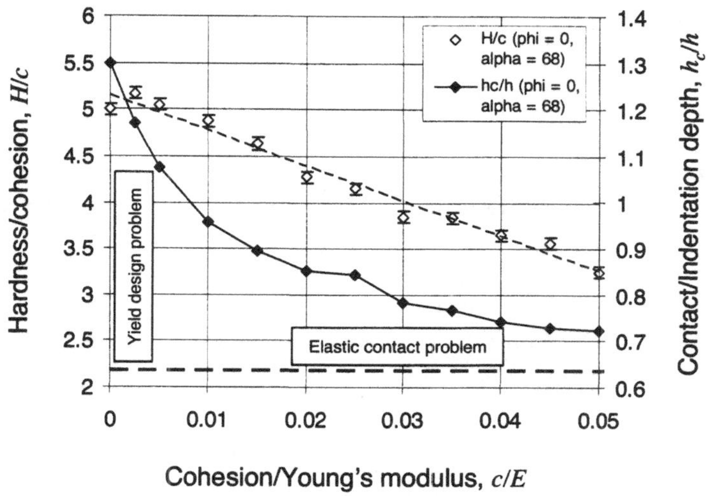

Figure 3. Hardness-to-cohesion ratio vs. strength-to-stiffness ratio for pure cohesive materials, $\varphi = 0$ (generated from Cheng and Cheng (1999)).

tact depth – to indentation depth ratio needs to be solved for any particular material behavior. Closed form solutions are generally not available, so that use of numerical methods to solve the nonlinear elastoplastic contact problem becomes necessary. For elastoplastic non-frictional materials ($\varphi = 0$) with and without strain hardening, Chen and Chen (1999,2004) showed, by means of finite element simulations, that h_c/h is insensitive to the Poisson's ratio, but dependent on the strength-to-stiffness ratio c/E and on the power-strain coefficient. For this class of materials, the dependence of h_c/h on the material properties translates into the dependence of the hardness-to-strength ratio on the strength-to-stiffness ratio (Figure 3). A similar dependence is expected to characterize cohesive-frictional materials, but to our knowledge, the dependence of h_c/h on the strength properties of frictional materials has not been studied.

For geomaterials, for which typically $c/E \sim 10^{-3} - 10^{-6}$ it is most likely that yield design solutions of the form $H/c = \mathcal{H}(\varphi, \alpha)$ are highly relevant. In return, it requires an independent measurement of the projected contact surface, in order to determine –with high accuracy– the exact value of the hardness, and hence the strength parameters, c and φ.

2.2 Indentation Stiffness Measurements

The quantity that is measured upon unloading is the slope S that is fitted to the $P-h$ curve at the maximum indentation depth $h_{\max}$ (Fig. 1), and the mechanical property that is extracted is the indentation modulus M, from the so-called Bulychev–Alekhin–

Shoroshorov equation, in short BASh-equation (see e.g. Borodich and Keer, 2004):

$$S = \frac{dP}{dh}|_{h_{\max}} \stackrel{def}{=} \frac{2}{\sqrt{\pi}} M \sqrt{A} \tag{2.11}$$

Again, it is instructive to perform a dimensional analysis. Provided a pure elastic behavior upon unloading, slope S depends on the elastic constants, as well as on the force $P_{\max}$, the maximum indentation depth $h_{\max}$, and the half-cone angle α. Dimensional analysis yields:

$$\frac{S h_{\max}}{P_{\max}} = \Pi_{\gamma}\left(\frac{C_{ijkl} h_{\max}^2}{P_{\max}}, \alpha\right) \tag{2.12}$$

or equivalently, using definition (2.11) and the invariants (2.4) that characterize the loading indentation response right upon unloading:

$$\frac{S}{H\sqrt{A}} = \frac{2}{\sqrt{\pi}} \frac{M}{H} = \sqrt{\Pi_{\beta}} \times \Pi_{\gamma}\left(\frac{C_{ijkl}}{H} \Pi_{\beta}^{-1}, \alpha\right) \tag{2.13}$$

where H is the hardness during loading. The r.h.s. of (2.13) is readily found to be independent of indentation depth $h_{\max}$, at which unloading takes place; as is the M/H ratio, which depends on the stiffness-to-cohesion ratio, the friction angle and the half-apex angle. For instance, in the case of a perfect elastic behavior, the elastic Hertz contact solution yields:

$$\lim_{C_{ijkl}/c \to 0} \left(\frac{M}{H}\right) = 2 \tan \alpha \tag{2.14}$$

In this case, $\lim_{C_{ijkl}/c \to 0} \Pi_{\beta} = \frac{4}{\pi} \tan^2 \alpha$ and $\lim_{C_{ijkl}/c \to 0} \Pi_{\gamma} = 2$. In general, however, M/H is a dimensionless number that characterizes, for a given cone angle, the stiffness-to-strength ratio of the intended (homogeneous) material.

Similar to the extraction of strength properties from hardness measurements, the key to a successful determination of the elastic properties from indentation relies on the determination of the correct projected contact area. This is not an easy task. Traditionally, the contact area has been determined by direct optical measurement of the size of the residual hardness impression after a complete unloading. For practical reasons, however, some means other than direct observation of the hardness impressions is needed to measure contact areas, since imaging very small indentations is both time-consuming and difficult (Fig. 4). Making use of the BASh relation (2.11), the contact surface can be scaled, for a given material (same M), from two unloading slopes:

$$\frac{A}{A_1} = \left(\frac{S}{S_1}\right)^2 \tag{2.15}$$

For instance, if the contact area for a specific indentation depth or indentation geometry (half-apex angle α) is measured, it is possible to either determine the contact area or contact depth for another indentation depth for the same indenter (same α), or scale the contact area from one indenter geometry to another. In the latter case, use of (2.15) allows one to determine e.g. the hardness ratio from:

$$\frac{H}{H_1} = \frac{P}{P_1}\left(\frac{S}{S_1}\right)^{-2} \tag{2.16}$$

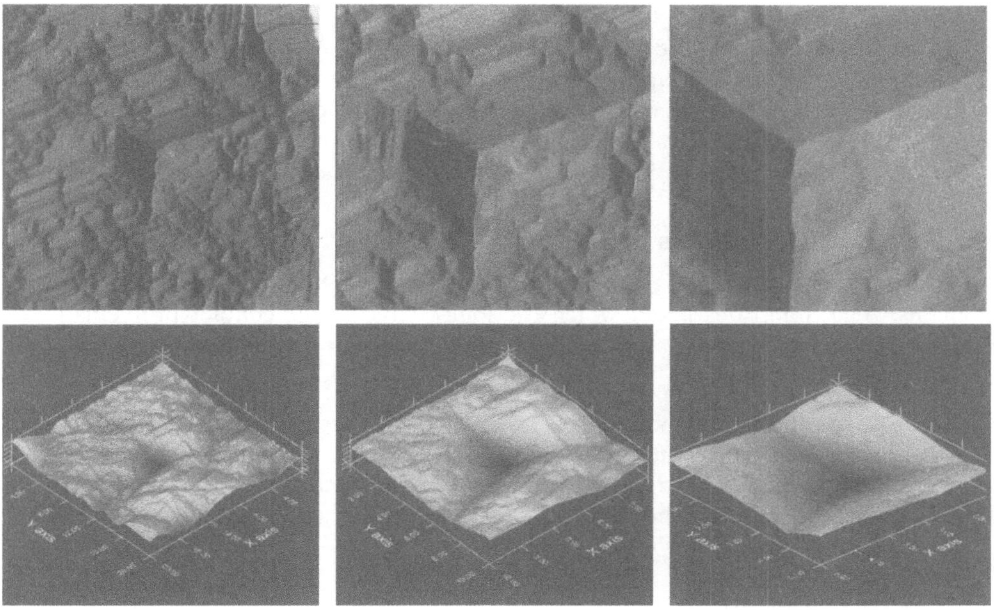

Figure 4. Gradient Images (Top) and AFM Images (Bottom) of residual impressions after microindentation in a shale material.

An original method for conical indentation that circumvents the necessity to measure the contact area, was suggested by Oliver and Pharr (1992), by applying relation (2.5) to the pure elastic recovery during unloading:

$$h_c - h_f = \frac{2}{\pi}\left(h_{\max} - h_f\right) \tag{2.17}$$

where h_f is the residual indentation depth. Since $h_{\max}$ is measured, the key to the analysis becomes the determination of the residual displacement h_f. This is achieved by considering that the elastic unloading is scaled, for conical indentation, by a power-two function:

$$P = c\left(h - h_f\right)^2;\ h - h_f = 2\frac{P}{S} \tag{2.18}$$

Finally, a combination of (2.17) and (2.18) yields:

$$\frac{h_c}{h_{\max}} = 1 - \epsilon\frac{P}{Sh_{\max}} \tag{2.19}$$

where $\epsilon = 2\left(1 - 2/\pi\right) = 0.73$. Relation (2.19) is also valid for other indenter shapes: $\epsilon = 1$ for a flat punch ($P \propto h$), and $\epsilon = 0.75$ for a parabola of revolution ($P \propto h^{3/2}$; Sneddon, 1977).

The main problem of the Oliver and Pharr method is that is assumes the shape of the deformed solid outside the area of contact to be elastic. This is not true when plastic deformations occur around the indenter to form pile-ups, $h_c/h > 1$, which cannot be predicted by (2.19). Several modifications have been proposed to the original method (Pharr, 2002; Chen and Chen, 2004, Oliver and Pharr, 2004), and the topic continues to be focus of intensive research in the indentation mechanics community.

2.3 Indenter Geometries

An essential feature of indentation analysis is the self-similarity of Hertz-type contact problems. The conditions under which frictionless Hertz type contact problems possess classical self-similarity were stated by Borodich (1988 according to Borodich et al., 2003); and include:

1. The shape of the indenter is described by a homogeneous function whose degree is greater or equal to unity. Using a Cartesian coordinate system $Ox_1x_2x_3$ whose origin O is at the indenter tip and x_3 is the orientation of the indentation, the shape of the indenter (height) is defined by:

$$f(\lambda x_1, \lambda x_2) = \lambda^d f(x_1, x_2) \tag{2.20}$$

for arbitrary positive λ. Here d is the degree of the homogenous function f; in particular $d = 1$ for a cone and $d = 2$ for the elliptic paraboloid considered by Hertz. Such axi-symmetric indenters can be described by monomial functions of the form (first introduced by Galin, 1946 according to Borodich and Keer, 2004):

$$f = f(x_1 = r\cos\theta, x_2 = r\sin\theta) = Br^d \tag{2.21}$$

where B is the shape function of the indenter at unit radius, and d is the degree of the homogeneous function. For a conical indenter ($d = 1$), having a semi-vertical angle α; $B = \cot\alpha$. For a spherical indenter of radius R, $d = 2$ and $B = 1/(2R)$. The previous expression was recently extended to indenters of non-axi-symmetric shape, such as pyramidal indenters that are frequently employed in depth-sensing indentation tests (Borodich et al., 2003):

$$f = B(\theta)\, r^d \tag{2.22}$$

where $B(\theta)$ describes the height of the indenter at a point $(\theta, r = 1)$. For a three-sided pyramid, $d = 1$, and making use of the triple symmetry:

$$B(\theta) = \cot\alpha_0 \sin(\pi/6 + \theta) \tag{2.23}$$

where α_0 is the angle in vertical cross-sections. For a Berkovich indenter, having a face angle of 115.13°, $\alpha_0 = 65.3°$; and for a cube corner indenter of 90° face angle, $\alpha_0 = 35.26°$.

2. The operators of constitutive relations F for the indented material is a homogeneous functions of degree κ with respect to the components of the strain tensor:

$$F(\lambda\varepsilon) = \lambda^\kappa F(\varepsilon) \tag{2.24}$$

Evidently, a linear elastic law satisfies this relation since $\kappa = 1$; as does any nonlinear secant elastic formulation of the form $\boldsymbol{\sigma} = \mathbb{C}(\boldsymbol{\varepsilon}) : \boldsymbol{\varepsilon}$ for which the secant elastic stiffness tensor satisfies:

$$\mathbb{C}(\lambda\boldsymbol{\varepsilon}) = \lambda^{\kappa-1}\mathbb{C}(\boldsymbol{\varepsilon}) \tag{2.25}$$

A similar reasoning applies to the dissipation function $\pi(\lambda\mathbf{d}) = \lambda^{\kappa}\pi(\mathbf{d})$ for yield design solutions based on expression (2.6).

Then provided the homogeneity of material properties and that the stress-strain relation remains the same for any depth of indentation, the whole load-displacement curve in a depth-sensing test can be scaled by (Borodich et al., 2003):

$$\frac{P}{P_1} = \left(\frac{h}{h_1}\right)^{\frac{2+\kappa(d-1)}{d}}; \frac{h}{h_1} = \left(\frac{A}{A_1}\right)^{d/2} \tag{2.26}$$

where A is the projected contact area, which appears to be not affected by the constitutive relation. In return, the hardness is scaled with the indentation depth by:

$$\frac{H}{H_1} = \left(\frac{h}{h_1}\right)^{\frac{\kappa(d-1)}{d}} \tag{2.27}$$

We verify that the load-displacement relation is scaled by $P \propto h^2$ for conical and ideal (sharp) pyramidal indentation, for which $d = 1$, irrespective of the constitutive relation (2.24). As a consequence, the hardness is a constant over the loading process, as defined by (2.1). On the other hand, the constitutive relation power κ significantly affects the load-displacement relation for spherical or elliptic paraboloid indenter geometries, for which $d = 2$, and must be known in advance in order to analyze the load-displacement curve. For instance, for an elastic behavior, $P \propto h$ (which corresponds to $d \to \infty$) is indicative of flat indentation, and is described by Boussinesq's solution; $P \propto h^{1.5}$ is indicative of linear elastic spherical or paraboloids of revolution.

Finally, given the same $d = 1$ degree of the homogeneous shape function of three-sided pyramidal (Berkovich, corner cube) and conical indentation, it is common practice to consider, instead of the original three-dimensional pyramidal shape, an equivalent cone of revolution in sharp indentation analysis, such that the projected contact area with respect to indentation depth of the cone is the same as that for the real indenter, *ie.* from (2.26):

$$A(h) = C_0 h^2 = \pi (h \tan \alpha_{eq})^2 \Longrightarrow \tan \alpha_{eq} = \sqrt{\frac{C_0}{\pi}} \tag{2.28}$$

where C_0 is a constant characterizing the specific pyramidal indenter, and α_{eq} is the equivalent half-apex cone angle. Using (2.28), the flat Berkovich indenter ($\alpha_0 = 65.3°$), for which $C_0 = 24.56$, can be assimilated to an equivalent cone of semi-apex angle $\alpha_{eq}^B = 70.32°$; and a Corner Cube Corner indenter ($\alpha_0 = 35.26°; C_0 = 2.598$) to one with $\alpha_{eq}^{CC} = 42.28°$. The hardness-to-cohesion ratio vs. friction angle relation displayed in Figure 2, are the equivalent conical indentation results for a Berkovich and a Corner Cube indenter.

2.4 Link between Indentation Modulus and Elastic Properties of Transverse Isotropic Materials

Provided the contact area is known (as a function of the indentation depth), the hardness H and the indentation modulus M can be determined. As we have seen above, except for very specific behaviors, these two quantities are not material properties, but provide rather a snapshot of some strength behavior (H) and elasticity behavior (M), that needs to be linked to actual material properties: cohesion c, friction angle φ, elastic stiffness tensor $\mathbb{C}$, etc.

We restrict ourselves to linear elastic indentation analysis, for which the measured initial unloading slope S and the measured (or estimated) contact surface give access to the indentation modulus through the BASh equation (2.11). The analytical link between the indentation modulus and the elastic constants is provided through elastic solutions of the Hertz-type contact problem for specific indenter geometries. For rigid indenter, a convenient way of solving this problem consists of two steps: (1) determination of the Green's surface function, that is the displacement field corresponding to a concentrated unit load; and (2) integration of the Green's function to find the displacement field resulting from an assumed pressure distribution under the indenter, and verifying that the result matches with the boundary conditions of a rigid indenter. These boundary conditions are defined by the shape of the rigid indenter.

To illustrate the procedure, we consider first the conical indentation into a linear elastic isotropic half-space. The Green's surface function reads:

$$\eta\left(r\right)=\frac{\mathcal{H}}{r} \tag{2.29}$$

The contact area is a circle of radius a, and the pressure distribution is of the form:

$$p(r)=p_0\cosh^{-1}\left(\frac{a}{r}\right) \tag{2.30}$$

where $p_0 = P/\left(\pi a^2\right)$ is the average pressure w.r.t. the projected contact area. The displacement at any point $Q(r_1,\theta)$ situated on the projected contact surface is:

$$u_1(r_1,\theta)=\int\int_S p(r)\ \eta(|r_1-r|)\ rdrd\theta \tag{2.31}$$

The indentation depth h is equal to the displacement u_1 at the cone tip. Evaluating (2.31) for $r=0$ gives:

$$h=u_1(r_1=0)=2\pi\mathcal{H}p_0\int_0^a\cosh^{-1}\left(a/r\right)dr=\pi\mathcal{H}\frac{P}{a} \tag{2.32}$$

or equivalently, after rearrangement:

$$P=\frac{2}{\pi}\frac{h^2\tan\alpha}{\pi\mathcal{H}};\ S=\frac{2}{\sqrt{\pi}}\sqrt{A}\frac{1}{\pi\mathcal{H}} \tag{2.33}$$

A comparison of the expression (2.33) with the BASh equation (2.11) allows us to identify:

$$M=\frac{1}{\pi\mathcal{H}} \tag{2.34}$$

The simple example shows that the key to the establishment of the link between the indentation modulus M and the elastic constants of the indented half-space is the Green's surface function expression (2.29). In the general case, the Green's surface function reads (Vlassak and Nix, 1993, 1994):

$$\eta(\underline{y}) = \frac{1}{8\,\pi^2\left|\underline{y}\right|}\left[\alpha_k\ B^{-1}_{km}\left(\frac{\underline{y}}{\left|\underline{y}\right|}\right)\ \alpha_m\right] = \eta(r,\theta) = \frac{\mathcal{H}(\theta)}{r} \tag{2.35}$$

where $\underline{y}$ is the position vector of Q relative to the load point P; (r,θ) are polar coordinates of Q on the surface; $\alpha_1, \alpha_2, \alpha_3$ are the cosines of the direction normal to the indented surface; $\mathbf{B}$ is a second order tensor defined by (Barnett and Lothe, 1975):

$$B_{js}(\underline{t}) = B_{sj}(\underline{t}) = \frac{1}{8\pi^2}\int_0^{2\pi}((\mathbf{mm})_{js} - (\mathbf{mn})_{jk}\ (\mathbf{nn})^{-1}_{kr}\ (\mathbf{nm})_{rs})d\phi \tag{2.36}$$

where $(\mathbf{ab})_{jk} = a_i C_{ijkl} b_l$ (with the summation of repeated indices); $\underline{t} = \underline{y}/\left|\underline{y}\right|$ is the normalized form of $\underline{y}$; $(\underline{m}, \underline{n}, \underline{t})$ forms a right hand cartesian system, and ϕ is the angle between vector $\underline{m}$ and the unit normal outward to the surface (Figure 5a). And finally, $\eta(r,\theta)$ is the surface Green's function, homogenous in r^{-1}. For isotropic materials, it is convenient to employ the following form of the elastic stiffness tensor:

$$\mathbb{C} = (C_{11} + 2C_{12})\,\mathbb{J} + (C_{11} - C_{12})\,\mathbb{K} \tag{2.37}$$

where $\mathbb{J} = \frac{1}{3}\boldsymbol{\delta}\otimes\boldsymbol{\delta}$ and $\mathbb{K} = \mathbb{I} - \mathbb{J}$ are the spherical and the deviator tensor projection of the 4th order unit tensor, and $C_{11} = C_{1111}$ and $C_{12} = C_{1122}$ are the elasticity constants, related to the bulk and shear modulus by:

$$3K = C_{11} + 2C_{12};\ 2G = C_{11} - C_{12} \tag{2.38}$$

Use of (2.37) in (2.36) and (2.35) yields after some transformations,

$$\mathcal{H} = \frac{1}{\pi}\frac{C_{11}}{C_{11}^2 - C_{12}^2} \Rightarrow M = \frac{C_{11}^2 - C_{12}^2}{C_{11}} = \frac{E}{1-\nu^2} \tag{2.39}$$

Things are more complicated when it comes to anisotropic materials. Much of the recent contributions to the analysis of the Hertzian contact for anisotropic solids can be traced back to the 1966 work of Willis, who reduced the problem to the evaluation of contour integrals for parabolic indenters. Vlassak and Nix (1993, 1994) simplified the solution using the surface Green's function expression (2.35), and provided implicit solution schemes for other indenter shapes. To our knowledge, the most refined (recent) solution schemes for general anisotropic materials are those proposed by Swadener and Pharr (2001) for conical and parabolic indenters and by Vlassak *et al.* (2003) for conical and spherical indenters. Both solution schemes involve computational demanding operations even in their approximated versions. For orthotropic materials indented in the axes of symmetry, some easy implementable explicit expressions can be given, based on some approximations of the Green's function (Delafargue and Ulm, 2004). The principle is briefly evoked below for transversely isotropic materials.

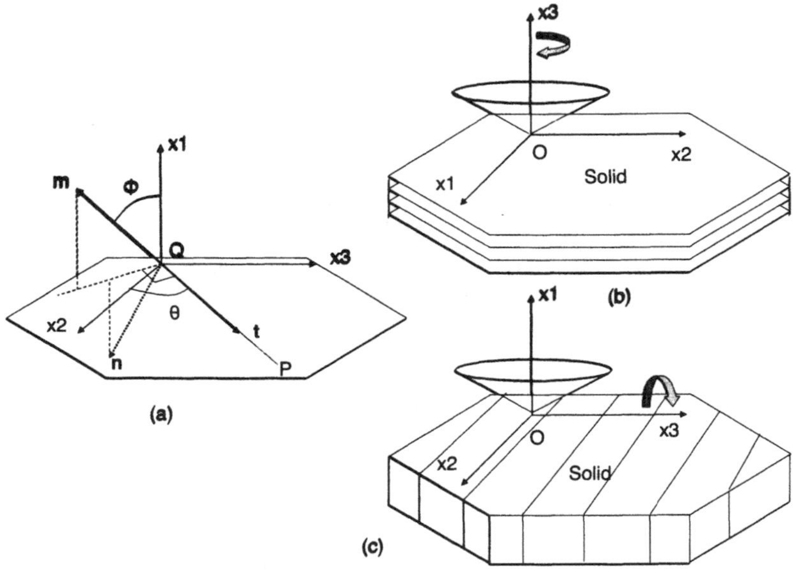

Figure 5. (a) Coordinate system; Indentation into (b) the axis of symmetry, and (c) into the plane of symmetry of a transversely isotropic material.

Let direction $x3$ be normal to the planes of isotropy of a transversely isotropic material; and directions $x1$ and $x2$ be parallel to the planes of isotropy so that the resulting coordinate system is a right-hand cartesian one, with the first indented point as origin O (Figure 5b). In this coordinate system, the elastic stiffness tensor is written in the form:

$$C_{ijkl} = \begin{bmatrix} C_{11} & C_{12} & C_{13} & 0 & 0 & 0 \\ & C_{11} & C_{13} & 0 & 0 & 0 \\ & & C_{33} & 0 & 0 & 0 \\ & & & 2C_{44} & 0 & 0 \\ & & & & 2C_{44} & 0 \\ sym & & & & & C_{11}-C_{12} \end{bmatrix} \tag{2.40}$$

where we use the reduced notations:[2]

$$\begin{aligned} & C_{11} = C_{1111} \\ & C_{12} = C_{1122} = C_{2211} \\ & C_{13} = C_{1133} = C_{3311} \\ & C_{11} - C_{12} = 2C_{1212} \\ & C_{44} = C_{2323} = C_{1313} \\ & C_{31} = \sqrt{C_{11}\, C_{33}} > C_{13};\ \text{with } C_{33} = C_{3333} \end{aligned} \tag{2.41}$$

[2]The last condition in (2.41), $C_{31} = \sqrt{C_{11}\, C_{33}} > C_{13}$, ensures that the stiffness tensor is symmetric positive definite.

We are first interested in an indentation in direction $x3$, *ie.* normal to the planes of isotropy. For this case, an analytical solution of the Hertz contact problem for conical indentation is available. The Elliot-Hanson solution reads (Elliot, 1949; Hanson, 1992):

$$P = \frac{2}{\pi^2 \mathcal{H}}\, h^2\, \tan(\alpha) \tag{2.42}$$

where $\mathcal{H}$ is the constant:

$$\mathcal{H} = \frac{1}{2\,\pi}\sqrt{\frac{C_{11}}{C_{31}^2 - C_{13}^2}\left(\frac{1}{C_{44}} + \frac{2}{C_{31} + C_{13}}\right)} \tag{2.43}$$

A derivation of (2.42) w.r.t. to h yields the BASh equation (2.11), together with the indentation modulus in the axis of symmetry $x3$:

$$M_3 = \frac{1}{\pi \mathcal{H}} = 2\sqrt{\frac{C_{31}^2 - C_{13}^2}{C_{11}}\left(\frac{1}{C_{44}} + \frac{2}{C_{31} + C_{13}}\right)^{-1}} \tag{2.44}$$

It is interesting to note that the solution does only depend on four of the five elastic constants of a transversely isotropic material $(C_{11}, C_{33}, C_{44}, C_{13})$. The fifth independent constant $C_{12} = C_{1122} = C_{1111} - 2\,C_{1212}$ does not appear in the expression of $\mathcal{H}$.

When the half-space surface is orthogonal to the material's planes of isotropy, the problem is no longer axi-symmetric and the projected contact area is no more circular. We consider an indentation into axis $x1$ of the half-space. The indentation axis ($x1$) belongs to two planes of symmetry (Fig. 5c): $(x1, x3)$ is orthogonal to the planes of isotropy, and $(x1, x2)$ is parallel to them. In this case, an evaluation of (2.35) for $(\alpha_1, \alpha_2, \alpha_3) = (1, 0, 0)$ yields the exact values of the the Green's function η in the $x2$–direction ($\theta = 0$), and in the $x3$–direction ($\theta = \pi/2$):

$$\mathcal{H}_2 = \mathcal{H}(\theta = 0) = \frac{1}{2\,\pi}\sqrt{\frac{C_{33}}{C_{31}^2 - C_{13}^2}\left(\frac{1}{C_{44}} + \frac{2}{C_{31} + C_{13}}\right)} \tag{2.45}$$

$$\mathcal{H}_3 = \mathcal{H}(\theta = \frac{\pi}{2}) = \frac{1}{\pi}\,\frac{C_{11}}{C_{11}^2 - C_{12}^2} \tag{2.46}$$

It is interesting to note, from a comparison of (2.43) and (2.45), that $\mathcal{H}_2 = \sqrt{C_{33}/C_{11}}\,\mathcal{H}$; while $\mathcal{H}_3$ turns out to be the Green's function constant for an isotropic material with stiffness constants $C_{11} = C_{1111}$ and $C_{12} = C_{1122}$, as defined by (2.39). Furthermore, since the Green's function is π-periodic by definition and even by symmetry, $\mathcal{H}_2$ and $\mathcal{H}_3$ are extreme values of $\mathcal{H}(\theta)$. Restricted to indentation in principal material axes, it seems appropriate to consider a first order approximation $\widetilde{\mathcal{H}}(\theta)$ that interpolates the Green's function $\eta(r, \theta) \simeq \widetilde{\mathcal{H}}(\theta)/r$ so that $\widetilde{\mathcal{H}}(\theta = 0) = \mathcal{H}_2$ and $\widetilde{\mathcal{H}}(\theta = \frac{\pi}{2}) = \mathcal{H}_3$:

$$\widetilde{\mathcal{H}}(\theta) = \frac{\mathcal{H}_2 + \mathcal{H}_3}{2} + \frac{\mathcal{H}_2 - \mathcal{H}_3}{2}\cos(2\theta) \tag{2.47}$$

Finally, to complete the determination, we need to specify the shape of the contact area and the pressure distribution. It is generally assumed that the projected contact area is elliptical for conical indentation of general anisotropic materials. By symmetry, the axes of the elliptical contact area must coincide with $x2$ and $x3$. If a_2 and a_3 are the ellipse dimensions in the respective directions $x2$ and $x3$, then the ellipse eccentricity is $e = \sqrt{1 - \left(\frac{a_2}{a_3}\right)^2}$ if $a_2 < a_3$, and $e = \sqrt{1 - \left(\frac{a_3}{a_2}\right)^2}$ otherwise. We assume that the pressure field $p(y_2, y_3)$ at point P of coordinates (y_2, y_3) has the form (Swadener and Pharr, 2001):

$$p(y_2, y_3) = p_0 \cosh^{-1} \left(\frac{y_2^2}{a_2^2} + \frac{y_3^2}{a_3^2} \right)^{-1/2} \tag{2.48}$$

The displacement in any point $Q(z_2, z_3)$ situated on the projected contact surface is:

$$u_1(z_2, z_3) = \int \int_S p(y_2, y_3)\ \eta(z_2 - y_2, z_3 - y_3)\ dy_2 dy_3 \tag{2.49}$$

Substituting in η the Green's function approximation (2.47), and proceeding as in the isotropic case, we evaluate the indentation depth h from the displacement u_1 at the cone tip, and express it as a function of the load $P = \pi a_2 a_3 p_0$. Deriving this expression yields together with the BASh equation (2.11) the following expression of the indentation modulus M_1:

$$M_1 = \Psi(e) \frac{1}{\pi \sqrt{\mathcal{H}_2\, \mathcal{H}_3}};\ \Psi(e) = \pi \frac{(1-e^2)^{1/4}}{2\, \mathcal{E}(e)} \tag{2.50}$$

where $\mathcal{E}(e)$ is the complete elliptic integral of the second kind, and e the eccentricity, which for the Green's function approximation (2.47) reads:

$$e = \left\{ \sqrt{1 - \frac{\mathcal{H}_2}{\mathcal{H}_3}},\ \text{if } \mathcal{H}_2 < \mathcal{H}_3\ ,\ \sqrt{1 - \frac{\mathcal{H}_3}{\mathcal{H}_2}},\ \text{else} \right\} \tag{2.51}$$

We note that $0.99 \leq \Psi(e) \leq 1$ if $e \leq 0.6$, which corresponds to an ellipse axis ratio smaller that 1.25. Hence, using $\Psi(e) \approx 1$ in (2.50) simplifies the expression of the indentation modulus normal to the axis of symmetry:

$$M_1 \simeq \frac{1}{\pi \sqrt{\mathcal{H}_2 \mathcal{H}_3}} = \sqrt{\sqrt{\frac{C_{11}}{C_{33}}} \frac{C_{11}^2 - C_{12}^2}{C_{11}} M_3} \tag{2.52}$$

where M_3 is the indentation modulus in the axis of symmetry.

In summary, relations (2.44) and (2.52) highlight that indentation moduli provide an intriguing snapshot of the elasticity activated in an indentation test. They also highlight the difficulty to directly determine for non-isotropic materials the elastic constants from indentation tests, as it would require five tests in different directions in order to determine five 'snapshots' of the transversely isotropic elasticity. The feasibility of such an endeavour can be checked using the two indentation moduli, M_1 and M_3, in the principal material axes, as they are expected to provide the highest contrast in elastic behavior of the material.

2.5 Indentation on Natural Composites

Is it possible to apply continuum indentation analysis to heterogeneous materials, and if – how?

Continuum indentation analysis is generally based on the assumption of homogeneity of material properties and that the stress-strain relation remains the same for any depth of indentation. Like all continuum analysis, continuum indentation analysis is based on the concept of a representative elementary volume (r.e.v.) of characteristic size $\mathcal{L}$ that needs to obey to the scale separability condition:

$$d \ll \mathcal{L} \ll \max(h, a) \tag{2.53}$$

where (h, a) is the indentation depth and the indentation radius that define the order of magnitude of the variation of the position vector $\underline{x}$; and d is the characteristic size of the (largest) heterogeneity contained in the r.e.v. Provided that (2.53) is satisfied, an indentation test operated to a penetration depth h gives access to the material properties that are characteristic of a materials system at a length scale of $\mathcal{L}$. On the other hand, since most indentation solutions are based on the similarity approach, and derived from the infinite half-space model (which by definition has no length scale), and from the assumption of uniform material properties, the properties extracted –by means of an inverse analysis– from indentation tests are averaged quantities characteristic of a material length scale defined by the indentation depth or the indentation radius. A good estimate is that the characteristic size of the material domain sensed by an indentation is on the order of $\max(h, a)$; roughly $3h$ for Berkovich indentation and h for the corner cube. Given the self-similarity of the indentation test, choosing the indentation depth, therefore, comes to choose the length scale of material investigation.

The heterogeneity of most natural composite materials down to very fine scales calls for a statistical analysis of indentation results. This requires that a large number of indentation tests be performed on a surface. A convenient way is to employ a grid indentation technique, defined by a grid size ℓ that should be larger than the characteristic size of the indentation impression, so to avoid interference in between individual indentation tests (Fig. 6).

Thought Experiment *Consider a material to be composed of two individual phases of characteristic size D_1 and D_2, and different material properties. If the indentation depth is much smaller than the characteristic size of the phases, $h \ll (D_1, D_2)$, then a single indentation test gives access to the material properties of either phase 1 or phase 2. If, in addition, a large number of tests is carried out on a grid size that is much larger than the characteristic size of the two phases, so that the locus of indentation has no statistical bias w.r.t. the spatial distribution of the two phases, the probability of encountering one or the other phase is equal to the surface fraction occupied by the two phases on the indentation surface. Provided that a similar distribution is found equally on other surfaces, the surface fraction can be assimilated with the volume fraction of the two phases present in the material.*

Consider next an indentation test performed to a maximum indentation depth that is much larger than the characteristic size of the individual phases, $h \gg (D_1, D_2)$. It

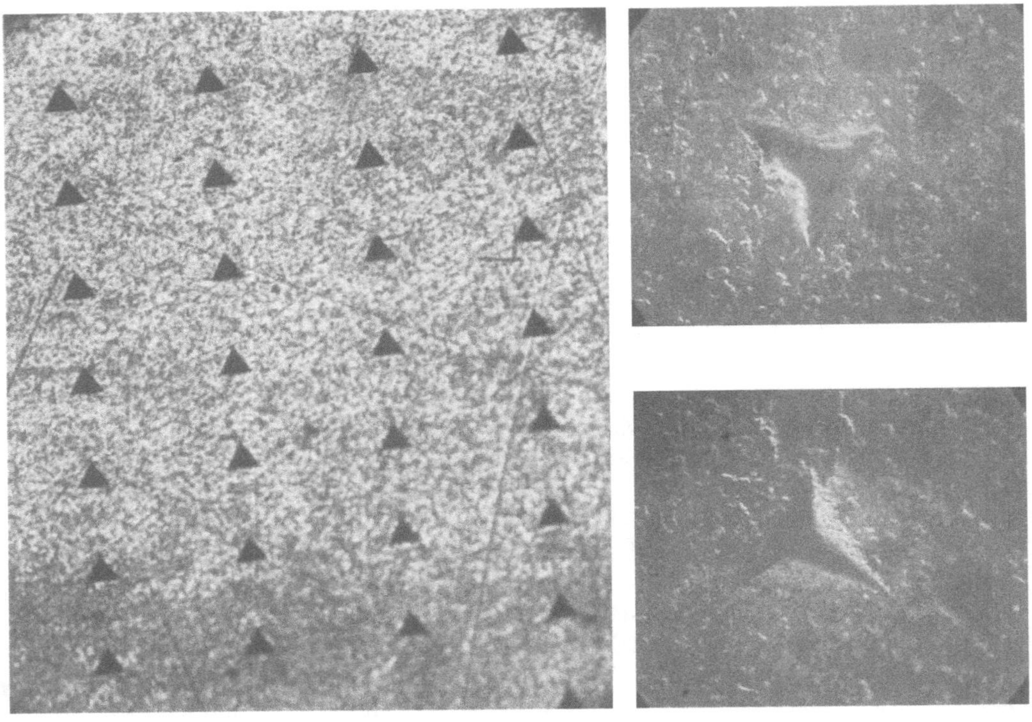

Figure 6. Optical microscope images of indentation grid: (left) example of microindentation grid on a cement paste; (right) residual impression after unloading (magnification x100).

is readily understood, from the scale separability condition (2.53), that the properties extracted from such an indentation test are representative in a statistical sense of the average properties of the composite material.

The simple thought-experiment has all the ingredients of statistical indentation analysis that need to be performed when it comes to natural composite materials. The key results of such analysis are distributions and their derivatives (*ie.* histograms or frequency plots) of mechanical properties determined by a large number of indentation tests at a specific scale of material observation defined by the indentation depth. Generally speaking, small indentation depths give access to mechanical phase properties, and potentially to volume fractions, from (Constantinides et al., 2003):

$$f_i = \frac{N_i}{N}; \quad \sum_{i=1}^{n} N_i = N \tag{2.54}$$

where N_i is the number of indentations on material phase i, that can be identified by

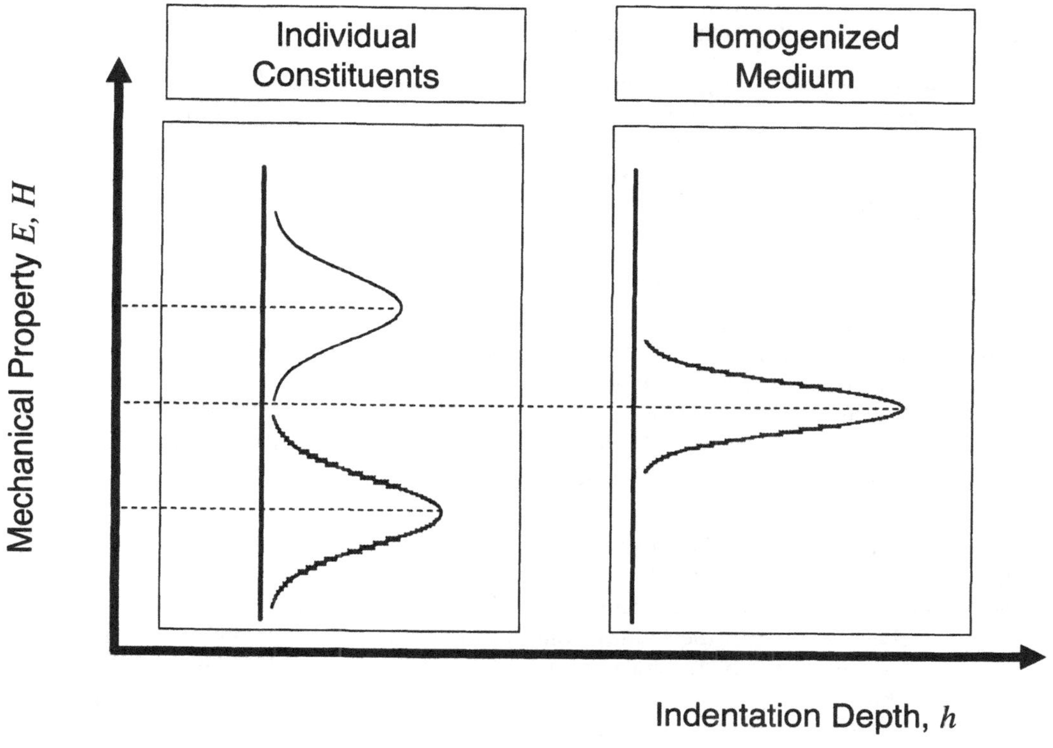

Figure 7. Principle of statistical indentation analysis of natural composites: small indentation depths give access to properties of individual phases; large indentation depths yield a homogenized response of the material.

the difference in material properties; that is f_i is the volume fraction of a 'mechanically' identifiable material phase. In turn, greater indentation depths give access to homogenized material properties of the composite (Fig. 7). Finally, a mapping of mechanical properties allows one to identify (if any) characteristic morphologies within the resolution defined by the grid size. These are all inputs required for microporomechanics analysis.

3 Experimental Microporomechanics of Cement-Based Materials

A nanoindentation investigation starts with an understanding of the microstructure of the investigated material, and *vice versa*, nanoindentation provides a means to identify the microstructure and the link between microstructure and material properties; and ultimately material invariant material properties. This is illustrated here for cement-based materials.[3]

3.1 Multiscale Microstructural Think-Model of Cement-Based Materials

Cement-based materials, like many geomaterials, are highly heterogeneous materials, with heterogeneities that manifest themselves from the nanoscale to the macroscale. For purpose of a poromechanics analysis, it is convenient to break this highly heterogeneous material system down into different scales defined by characteristic sizes of heterogeneities which are often the porosity. Figure 8 displays a four-level microstructural think-model of cement-based materials, which is detailed below.

Level '0': Nanoscale The lowest level of a mechanical representation of the complex microstructure of cement-based materials is the largest scale at which the material properties do not change from one cement-based material to another. It is the scale, where physical chemistry meets mechanics; that is the material properties are solely defined by the physical chemistry of the formation process of the material. This scale is typically situated above the atomic scale. In the case of cement-based materials, it is the scale of the C-S-H solid that forms at early ages by the hydration of C_3S and C_2S.[4] We refer to this scale as level '0', since this scale is currently not accessible to mechanical testing. This scale has been the focus of many theoretical and experimental cement chemistry investigations, starting with the groundbreaking works of Powers and his colleagues at the Portland Cement Association (Powers and Brownyard, 1948) recognizing the colloidal and gel-like properties of the C-S-H component. Powers attributed a porosity of roughly 28% to this solid phase. Ever since, many models have been devised to characterize the microstructure with an emphasis on a layered structure of C-S-H. To cite a few, the well known Feldman-Sereda model postulated the existence of 'interlayer spaces' in the C-S-H gel, containing strongly adsorbed water (Feldman and Sereda, 1970). Wittmann's Munich model is based on some concepts of colloidal science, relying particularly on the idea of a disjoining pressure that develops in the interlayer space as a consequence of hindered adsorption (Wittmann, 1974, 1988). Similar conclusions were arrived at by other authors as well (see e.g. Bažant (1972)): given the characteristic size of the interlayer space of

[3]The investigation of the fundamental properties of cement-based materials was carried out in a collaborative research project with the Lafarge Corp., with Dr. Paul Acker as Scientific Advisor and Jean-Francois Batoz as manager. The financial support of this study by the Lafarge Corporation is gratefully acknowledged. The investigated heat-cured specimen were provided by the Northwestern University team of Profs. Hamlin Jennings and Jeffrey Thomas (Civil Engineering and Materials Science), whose collaboration, comments and suggestions is gratefully acknowledged.

[4]The cement's chemistry abbreviation will be used in this paper ($C_3S = 3{\cdot}CaO{\cdot}SiO_2$, $C_2S = 2{\cdot}CaO{\cdot}SiO_2$, $C_3A = 3{\cdot}CaO{\cdot}Al_2O_3$, $C_4AF = 4{\cdot}CaO{\cdot}Al_2O_3{\cdot}Fe_2O_3$).

LEVEL	Density $[kg/m^3]$		Porosity [%]	Source
0: C-S-H solid	ρ_{sat}	ρ_{dry}	φ_0, ϕ_0	
Basic Building Block	2,800		–	Jennings, 2004[(a)]
Globules	2,480	2,300	18[(b)]	Jennings, 2004[(a)]
I: C-S-H matrix				
C-S-H$_{LD}$	1,930	1,440	37.3 ± 0.1[(c)]	Jennings, 2004[(a)]
C-S-H$_{HD}$	2,130	1,750	23.7 ± 0.1[(c)]	Jennings, 2004[(a)]
II: Cement paste				
C3S-Clinker	$\rho_c \frac{m_{C_3S}}{\sum_X m_X}$ (d)		–	
C2S-Clinker	$\rho_c \frac{m_{C_2S}}{\sum_X m_X}$ (d)		–	
C3A-Clinker	$\rho_c \frac{m_{C_3A}}{\sum_X m_X}$ (d)		–	
C4AF-Clinker	$\rho_c \frac{m_{C_4AF}}{\sum_X m_X}$ (d)		–	
CH	2,240		–	
III: Mortar				
Sand	2,650		–	Heukamp and Ulm, 2002

Table 1. Intrinsic properties of cement paste and mortar constituents:

[a] Density values of C-S-H as predicted from the quantitative colloidal model of C-S-H by Jennings (2000, 2004).

[b] Nanoporosity (intra-globules porosity) filled by structural water;

[c] Gelporosity (inter-globules porosity) of low density and high density C-S-H (excludes nanoporosity);

[d] p = total porosity = change in mass content due to drying at 105° C; includes structural water in nanoporosity and bulk water in gelporosity;

[e] Clinker density determined from cement density $\rho_c = 3,150$ kg/m^3, and the mass proportions m_X of the four clinker phases (X = C_3S, C_2S, C_3A, C_4AF) in the cement which are provided by the cement producer.

less than ten water molecules in size (elementary dimension of water is distance between O-atoms 0.284×10^{-9} m), it was quickly recognized that the water in this space cannot be considered to be a bulk water phase, to which e.g. Stokes equations apply. Instead, the mechanical response at this scale is dominated by the surface properties of the C-S-H gel, and the water present at this scale is structural water. Very recently, Jennings in a number of papers (Jennings, 2000; Tennis and Jennings, 2000; Jennings, 2004), provided qualitative and quantitative evidence of an amorphous colloidal structure of the C-S-H, organized in 'globules' (see Fig. 9), composed of basic building blocks and an intra-globules porosity. The way by which the porosity can be assessed, at this scale, is from mass density measurements (see Table 1). This porosity which manifests itself at a scale smaller than the characteristic solid dimension of 2.2×10^{-9} m, is $\varphi_0 = 18\%$ irrespective of the type of C-S-H. Instead, it is intrinsic to the C-S-H solid phase, and can be associated with the nanoporosity filled by structural water (and not bulk wa-

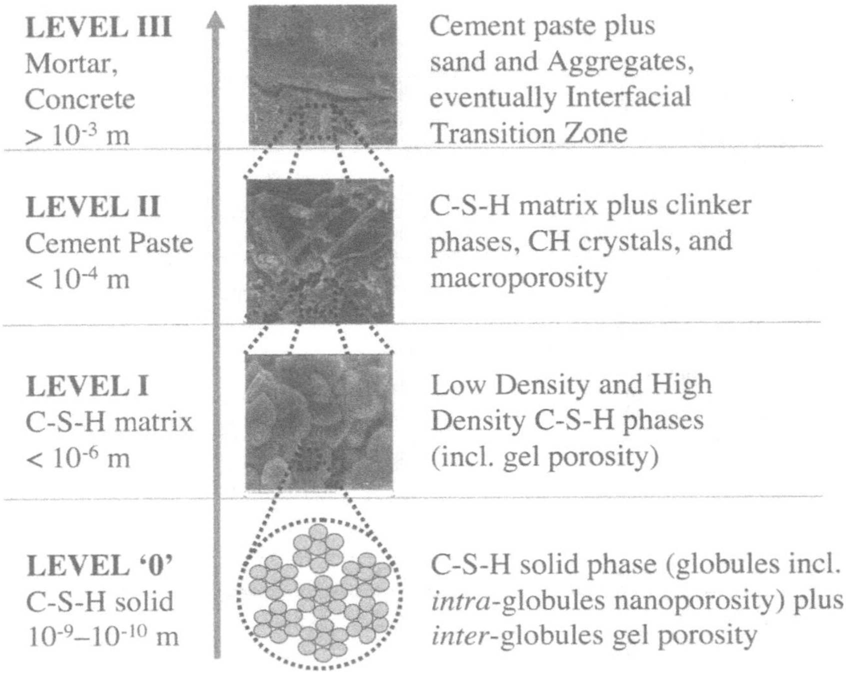

Figure 8. Four-level microstructural think-model of cement-based materials.

ter). Above this scale, there is a second type of porosity, the gel porosity, but which was found to differ from one type of C-S-H to another, as detailed below. Hence, from a poromechanics point of view, it is appropriate to consider this solid phase ('globules' in Jennings terminology) which includes a 18% intra-solid porosity filled by structural water, as the elementary solid phase of a poromechanics representation of cement-based materials. This solid phase has been found to be of a characteristic size of 5.6×10^{-9} m.

Level 'I': C-S-H phases – Gel Porosity – C-S-H matrix The solid phase of level '0' together with gel-porosity forms different types of C-S-H phases. These phases manifest themselves in units roughly larger than 16.6×10^{-9} m (Jennings, 2004), as sketched in Figure 9. We refer to this level as level I, as it represents the smallest material length scale that is presently accessible by mechanical testing, *ie.* nanoindentation (Constantinides et al., 2003; Constantinides and Ulm, 2004). At level I, the C-S-H exists in at least two different forms, a low density (LD) and a high density (HD) form (Fig. 9). The difference between the two types of C-S-H relates to the gel porosity of roughly 24% for HD–C-S-H, and 37% for LD–C-S-H (Jennings, 2000), due to the different packing density of the C-S-H solid of the two types of C-S-H; in addition to the 18% nanoporosity within the C-S-H solid phase (at level '0'); see Figure 9. In contrast to the nanoporosity, the gel

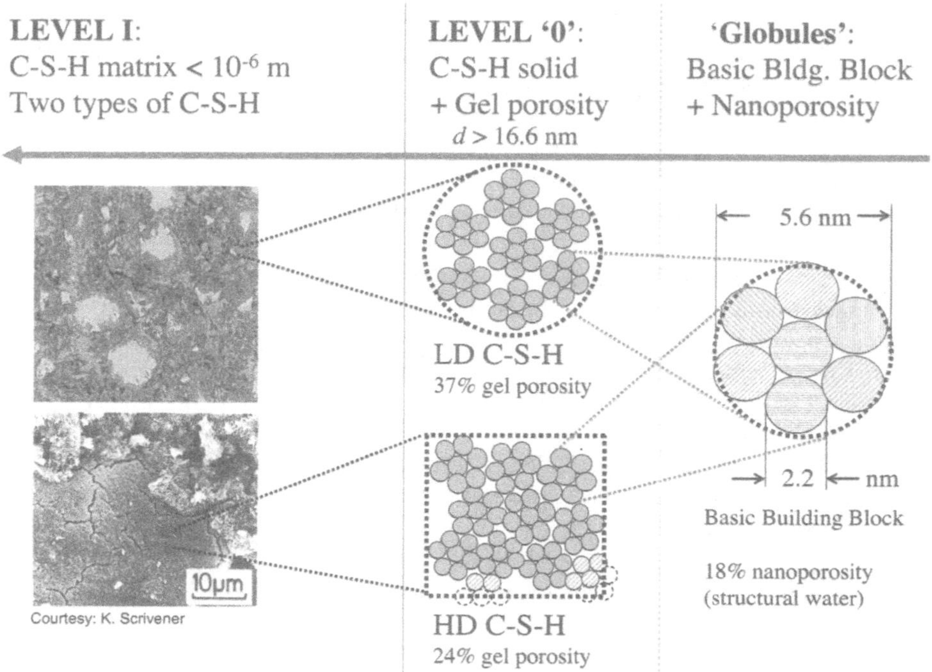

Figure 9. The Jennings colloid model of the two-types of C-S-H (Jennings, 2000; Tennis and Jennings, 2000; Jennings, 2004).

porosity has a characteristic dimension of the solid phase, *ie.* on the order of 5.6×10^{-9} m, in which the water present can be considered as a bulk water phase – in the sense of poromechanics theory. The gel porosity can be defined in a standard manner:

$$\phi_0^{LD} = \frac{V_{f,LD}}{V_{LD}} = 0.37; \; \phi_0^{HD} = \frac{V_{f,HD}}{V_{HD}} = 0.24 \tag{3.1}$$

where $V_{f,J}$ is the pore volume, and V_J the reference volume ($J = HD, LD$). The gel porosity values are also intrinsic to all cement-based materials: they are a consequence of the formation process of C-S-H in the course of hydration. That what changes from one cement paste material to the other is the volumetric proportion of HD–C-S-H and LD–C-S-H, within the C-S-H matrix.

It is intriguing to remark that the LD–C-S-H packing density, $1 - \phi_0^{LD} = 0.63$, almost coincides with the random packing density of spheres of 64%, which corresponds to the maximum packing density in the random close-packed limit (known as RCP)[5]. In return,

[5] More recent concepts refer to the RCP as the maximally random jammed state (MRJ), corresponding to the least ordered among all jammed packings, which has been shown to have

the HD–C-S-H packing density, $1 - \phi_0^{HD} = 0.76$ exceeds by little the densest possible spherical packing in three-dimensions of 74% (a problem known as the Kepler problem (Jaeger and Nagel, 1992; Sloane, 1998; Donev et al., 2004), which is the close-packed hexagonal or cubic structure. This is significant as it would confirm the underlying conjecture of the Jennings colloid model, that the two types of C-S-H are manifestations of the same solid, the only difference being an unstructured (random) order of the LD–C-S-H vs. a highly structured order of the HD–C-S-H.

Level 'II': Capillary Porosity – n Solid phases Level II refers to the cement paste, which manifests itself at a characteristic length scale of $10^{-6} - 10^{-4}$ m (Fig. 8). At this scale, the porous C–S–H matrix composed of two types of C-S-H together with the unhydrated cement products (*ie.* the four clinker phases C_3S, C_2S, C_3A, C_4AF), large Portlandite crystals ($CH = Ca(OH)_2$), aluminates and the macro-porosity ϕ_0 (which is often referred to as capillary porosity, and which is generally present only in high w/c–materials) form the cement paste. The total porosity p of a cement paste is the sum of the nanoporosity, the gel-porosity and the capillary porosity (Ulm et al., 2004):

$$p^{II} = [\varphi_0 (1 - \phi_{LD}) + \phi_{LD}] f_{LD} + [\varphi_0 (1 - \phi_{HD}) + \phi_{HD}] f_{HD} + \phi_0 \tag{3.2}$$

where f_{LD} and f_{HD} represent the volume fractions of the LD–C-S-H phase and the HD–C-S-H phase present in the cement paste. The total porosity is the relative change of mass that is measured by means of weighting experiments on a fully saturated and a dried cement paste oven dried at 105°C. By contrast, other techniques such as poromercury intrusion (PMI), do not allow to assess the total porosity, as the technique typically fails for pore throat radii smaller than 7×10^{-9} m. From a poromechanics point of view, the porosity which is filled by a bulk water phase is the total porosity minus the nanoporosity (filled by structural water):

$$\phi_0^{II} = \phi_{LD} f_{LD} + \phi_{HD} f_{HD} + \phi_0 \tag{3.3}$$

The dominating phase, at this scale, is the porous C-S-H matrix, which can occupy up to 90% of the volume of a cement paste.

Level 'III': Mortar and Concrete Level III of a characteristic length scale greater than 10^{-3}m refers to mortar and concrete; that is a composite material composed of a porous cement paste matrix, and sand particle inclusions in the case of mortar, and additional aggregate inclusions in the case of concrete. Some authors consider in addition the Interfacial Transition Zone (ITZ) between inclusions and matrix, which has been focus of many micromechanical modeling attempts (e.g. Garboczi, 1993; Li et al., 1999; Hashin and Monteiro, 2002; Heukamp and Ulm, 2002). From a poromechanics point of view, the cement paste matrix is a porous matrix, while –except for special applications– any porosity contained in the aggregates is rather of occluded nature. The total porosity at level III, that can be assessed by weighting experiments, is:

$$p^{III} = (1 - f_I) p^{II} + \left(\phi_{ITZ} - p^{II}\right) f_{ITZ} \tag{3.4}$$

a density of 63.7%, which is very close the traditional definition of the random close-packed limit; see Donev et al. (2004), and references cited herein.

where f_I and f_{ITZ} represent the volume fractions of the inclusions and of the ITZ, while ϕ_{ITZ} is the porosity of the ITZ, which has been argued to differ (in some cases) from the porosity of the cement paste. A lower bound of the inclusion volume fraction is obtained by letting $\phi_{ITZ} = p^{II}$.

3.2 Indentation Analysis

Indentation tests on cementitious materials have been reported by several authors: Igarashi et al. (1996) used a Vickers indenter (four-sided pyramid) with a maximum penetration depth on the order of $h_{\max} \simeq 10^{-5}$m, which gives access to the bulk properties of cement paste at a sub-millimeter to millimeter material length scale. To our knowledge, the first nanoindentation results of constituents of Portland cement were provided by Velez et al. (2001) and Acker (2001). The first reported nanoindentation results of the elastic modulus and hardness of the major clinker phases (C_2S,C_3S, C_3A, C_4AF), and the second provided values for Portlandite ($CH{=}Ca(OH)_2$) and the C-S-H gel for different C/S-ratio. These results were obtained on an ultra-high performance cementitious composite material, DUCTAL^{TM}, with a Berkovich indenter with penetration depths of about $h_{\max} \simeq 0.3 - 0.5 \times 10^{-6}$m; corresponding thus to bulk properties of the different phases at a characteristic length scale in the micrometer range. The stiffness value for CH was confirmed by Constantinides and Ulm (2004), who provided stiffness values of the two types of C-S-H of an ordinary Portlandite cement paste prepared at a water-cement ratio of $w/c = 0.5$, in a non-degraded and an asymptotically leached state. The key result obtained by this investigation was that the stiffness property of the two types of C-S-H obtained by nanoindentation at level I are independent of the mix proportions of the material (see also Constantinides et al. (2003)). Ever since, the question has been raised why?

Recently, we performed a series of tests on heat-cured cement paste samples, that provide some hints as to the origin of the invariant nature of the C-S-H properties (Constantinides and Ulm, 2004b). The general interest in the behavior of heat treated cementitious materials stems from the use of high-temperature curing in concrete technology for obtaining both a high early-strength and a significant decrease of time dependent deformation (creep and shrinkage), particularly in the precast concrete industry. However, it was quickly recognized that long-term properties are often negatively influenced by elevated curing temperature, as strength and other mechanical properties are often reduced and permeability increased. Beyond this particular industrial context, what we aim to show is the synergy of nanoindentation and microporomechanics analysis for understanding and assessment of the properties of complex natural composites, such as cementitious materials.

The cement paste samples were prepared using a white portland cement with a low aluminates content (US Gypsum Co., Chicago, IL) at a water/cement ratio of $w/c = 0.5$, to form bars measuring $0.25\,\mathrm{m} \times 0.025\,\mathrm{m} \times 0.025\,\mathrm{m}$. The paste was hydrated under limewater at room temperature (Control specimen, labeled 'C'), and at a temperature of $60\,^\circ$C for 28 days (Heat-cured specimen, labeled 'HC-28'). The heat curing was performed using the procedure described by Thomas and Jennings (2002): specimen were placed into a programmable water bath of limewater, and the temperature was controlled

Specimen	Control (C)	HC-28
w/c [1]	0.5	0.5
Heat-curing duration [d]	0	28
Age [mth]	5	5
No. of Micro-Tests	10 (MM*)	30 (MM*)
No. of Nano-Tests	200 (HYS**)	509 (HYS**)

Table 2. Specimen description and experimental program. Control specimen was cured at 20° C. Heat curing temperature was 60° C. After heat-curing specimens were kept at 20° C. Micro-Tests refer to microindentation tests operated with a maximum indentation depth of 10×10^{-6} m, Nano-Tests refer to nanoindentation tests operated with a maximum indentation depth of 300 nm. Machines: (*) MM = MicroMaterials, (**) HYS = Hysitron.

using a thermometer placed next to the specimen. The maximum temperature of 60 °C was chosen to avoid secondary high-temperature reactions, such as the decomposition of ettringite into monosulfate, which would hamper interpretation of the results. After the heat curing, the specimen were cooled to room temperature under sealed conditions, and kept in limewater until testing. The tested specimen, w/c ratio and age at testing are reported in Table 2. For the indentation testing, the square plate specimens were cut into slices of approximate thickness $5 - 10$ mm. The surfaces were ground and polished with silicon carbide papers and diamond particles to obtain a very flat and smooth surface finish. This was achieved in 6 stages of decreasing fineness with the last one being in the range of 0.25×10^{-6} m. Such a smooth surface is of critical importance for nanoindentation tests, so to avoid introducing another length scale in the similarity analysis: the surface roughness. Furthermore, special attention was paid to keep the specimens flat and parallel on both sides, since this could influence the angle of indentation and thus the result of the measurements. After polishing, the samples were placed in an ultrasonic bath to remove the dust and diamond particles left on the surface or in the pore structure.

Nanoindentation: Frequency Plots The first investigation focusses on level I and II, *ie.* on the material properties of the two-types of C-S-H. Given that the gel porosity has a characteristic dimension $\leq 10^{-8}$ m (see Fig. 9) an indentation test that captures the composite behavior of the two types of C-S-H must be on the order of $h \geq 10^{-7}$ m. Following previous experience with nanoindentation on cementitious materials (Acker, 2001; Constantinides et al., 2003; Constantinides and Ulm, 2004), a maximum penetration depth of 300 nm was chosen in the nanoindentation campaign carried out with a Berkovich indenter. 100 to 500 nanoindentation tests (see Table 2) were programmed with a grid-size of 10^{-5} m so to avoid interference in between single indents (see Fig. 6). The results are analyzed in terms of histograms of the mechanical properties (frequency plots).

Frequency plots of the nanoindentation stiffness are displayed in Figures 10 for the control specimen ('C') and the 28 day heat cured specimen ('HC-28'). The histograms display the frequency with which a specific stiffness is encountered in the experiment. The figures display a clear multi-modal distribution of the mechanical properties. Each peak corresponds to the mechanical manifestation of a chemical-morphological unit present in the microstructure. The intensity of each mechanical manifestation is representative of the intrinsic property of the phase. These are the macro-porosity, the LD–C-S-H, the HD–C-S-H, and Portlandite (CH) and residual clinker phases. The clear multi-modal distribution of these phases in Figure 10 makes it easy to graphically determine from a combination of the frequency plot and the cumulative distribution (*ie.* integral of the frequency plot), the intrinsic properties of the different phases and the volume fractions. These values are summarized in Table 3.

The indentation modulus of the LD–C-S-H and HD–C-S-H is in perfect agreement with previous reported results obtained on other cementitious materials: Acker (2001) reported two nanoindentation stiffness values for DUCTAL^{TM}, a thermally treated Ultra-High-Performance Cementitious Composite Material with $w/c < 0.20$, for which $M_{LD} = 20 \pm 2$ GPa and $M_{HD} = 31 \pm 4$ GPa. In a previous nanoindentation test campaign on another $w/c = 0.5$ hardened cement paste (Constantinides et al., 2003; Constantinides and Ulm, 2004), we found $M_{LD} = 23 \pm 2$ GPa and $M_{HD} = 31 \pm 2$ GPa. These indentation moduli are almost not affected by the thermal curing: the LD–C-S-H remains at a value of $M_{LD} = 20$ GPa, and the one of HD–C-S-H shifts only slightly from $M_{HD} = 29$ GPa for the control specimen to $M_{HD} = 31$ GPa for the thermally cured specimens, which is surely smaller than the standard deviation; and hence statistically insignificant. The almost perfect agreement with previous reported values clearly confirms that the stiffness values of the two C-S-H phases are independent not only of the mix proportion, but as well of the thermal curing in the considered temperature range and of thermal treatment (DUCTAL^{TM}). Instead, these properties are intrinsic to the formation process of the C-S-H phases.

That what changes due to thermal curing are the volumetric proportions of the different phases. The heat curing appears to favor the formation of high-density C-S-H. The relative proportion of LD-to-HD C-S-H shifts from $55/30 = 1.8$ in the control specimen to $33/45 = 0.73$ in the 28-day heat cured specimen ('HC-28'). Heat treatment, therefore, appears to be associated with a densification of the C-S-H matrix at Level I. In return, this densification of the matrix appears to occur at the expense of an increase in porosity at the considered scale. Given the indentation depth of 300 nm, this porosity has a characteristic size equal or greater than 300 nm. If we attribute an indentation stiffness ≤ 5 MPa to a nanoindentation into a bulk phase dominated by this macroporosity, the volume fraction obtained from the cumulative distribution in Figure 10 increases from 2% for the control specimen, to 8% for HC-28. What is interesting to note is that the total porosity as defined (3.2) does not change. Indeed, using the volume fractions from Table 3 in (3.2) yields a total porosity for the control specimen of $p^{II}(C) = 40\%$ and for the heat-cured specimen of $p^{II}(C) = 41\%$. The almost perfect agreement of the total porosity of control specimen and heat-cured specimen confirms that the creation of the macroporosity due to heat curing is a consequence of the conversion of LD–C-S-H (having a gelporosity of 37%) into HD–C-S-H (having a gelporosity of 24%). This conversion reduces the mesoporosity, *ie.* the total gelporosity in both LD–C-S-H and HD–C-S-H, but is compensated by the creation of macroporosity, so that the total porosity remains

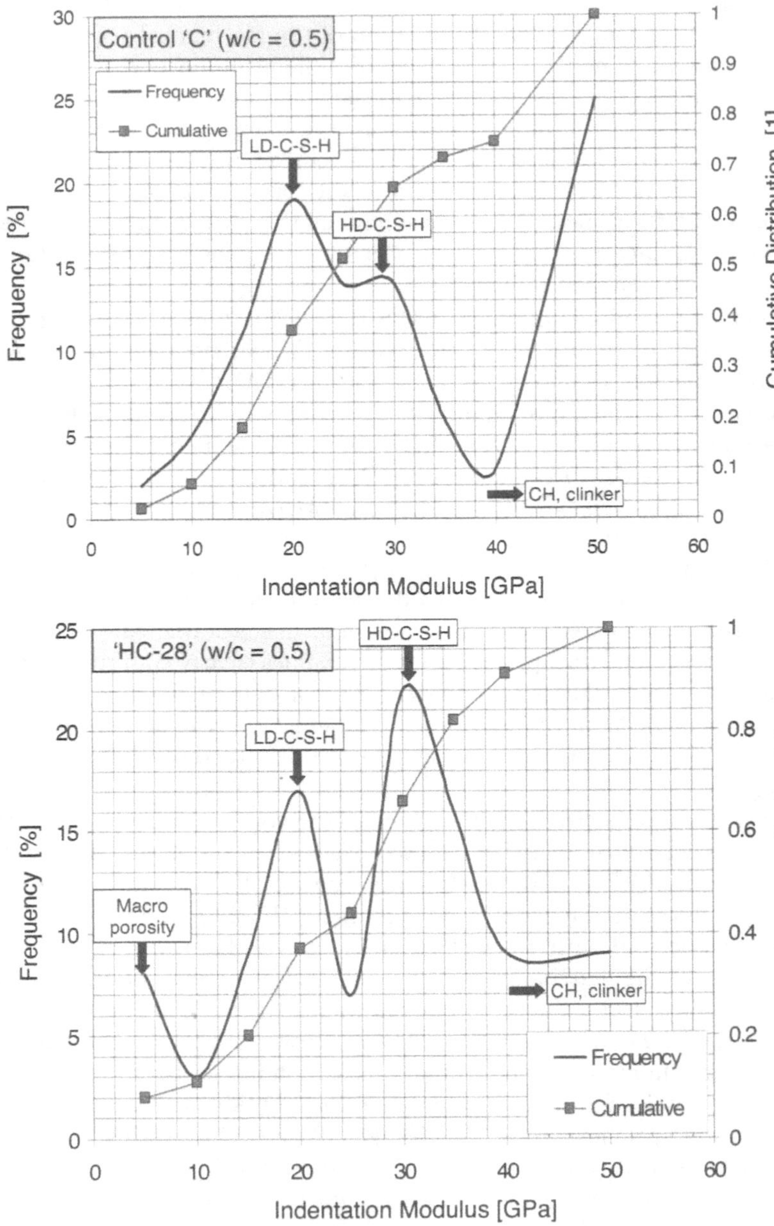

Figure 10. Frequency plot and cumulative distribution of indentation modulus determined by grid-nanoindentation: (top) control specimen (w/c=0.5) (No. of tests = 200); (bottom) heat cured specimen (No. of tests = 508).

		Control ('C')		HC-28	
		M [GPa]	Vol. Fr. [%]	M [GPa]	Vol. Fr. [%]
Level I	Macroporosity	≤ 5	2	≤ 5	8
	LD–C-S-H	20	55	20	33
	HD–C-S-H	29	30	31	45
	CH + Clinker	≥ 40	13	≥ 40	14
		μ [GPa]	σ/μ [%]	μ [GPa]	σ/μ [%]
Level I	LD–C-S-H, H	0.52	68	0.45	47
	HD–C-S-H, H	1.07	63	0.87	40
Level II	M^{II}_{exp}	19.1	9	19.5	25
	H^{II}_{exp}	0.50	2	0.41	4

Table 3. Summary of indentation results. The maximum *nano*indentation depth (level I) was 300nm, and the maximum *micro*indentation depth was 10μm.

almost constant.

Mechanical Mapping of Nanoindentation Results A second way of analyzing the nanoindentation test results consists in mapping the mechanical properties obtained on a grid of indentation tests. Each indentation result is assigned to a point on the grid that corresponds to the center of the indent. The discrete data points are linearly interpolated in between grid points to obtain continuous fields of mechanical properties, with a resolution defined by the grid-size of 10^{-5} m. The results are displayed on the $x-y$ plane in form of contour plots that capture ranges of mechanical properties. Based on the frequency plots and the values summarized in Table 3, we map the nanoindentation stiffness values into four domains:

1. Macroporosity domain $0 - 13$ GPa: Values situated in this range are associated with regions for which the mechanical response is dominated by high porosity. For purpose of comparison, these areas are indicative of the macroporosity.
2. Low-Density C-S-H domain $13 - 26$ GPa : Values situated in this range are situated around the first peak in the frequency plots 10, and are associated with regions in which the mechanical response is dominated by the LD–C-S-H (having a characteristic gel porosity at a scale below of 37%).
3. High-Density C-S-H domain $26 - 39$ GPa: Values situated in this range correspond to the second peak in the frequency plots 10, and are attributed to the dominant mechanical effect of the HD–C-S-H (having a characteristic gel porosity at a scale below of 24%).
4. Unhydrated clinker and CH domain $\geq$ 40 GPa: The higher stiffness values are indicative of unhydrated clinker phases and Portlandite. While the latter tends to grow in regions of high w/c ratios adjacent to macropores and in between LD–C-S-H, the residual clinker phases are generally rimmed by HD–C-S-H.

Figure 11 displays maps of 100 nanoindentation results for the control specimen and the 28-day heat cured specimen. The contour plots provide evidence of the creation of a new pore class and of a substantial change of percolation of the C-S-H phases as a consequence of the heat curing. Indeed, while almost absent in the control specimen 'C' (a single point in Fig. 11 represents rather a statistical event than a morphological pattern), the macroporosity domain becomes an identifiable morphological pattern in the map of the 28-days heat cured specimen , in which the macropores occupy regions several times the grid size of 10^{-5} m. Furthermore, in the control specimen, the continuous (percolated) solid phase appears to be the LD–C-S-H, which forms a matrix that accommodates residual clinker phase encapsulated into a rim of HD–C-S-H. By contrast, in the heat cured specimen, there is a larger proportion of HD–C-S-H present which tends to percolate throughout the microstructure, while the LD–C-S-H appears to be disconnect and situated around the large macropores. It almost appears as if the HD–C-S-H forms a continuous matrix that accommodates the macroporosity encapsulated into a rim of LD–C-S-H.

Nanohardness Measurements Figure 12 shows a log-log plot of the hardness vs. the stiffness. Besides showing a large scattering of the hardness values for a given indentation stiffness, the general trend is that the hardness increases as the stiffness increases. This correlation, however, should not be seen as a physical link between hardness and stiffness, but should be rather attributed to the fact that both stiffness and strength properties are affected by the microstructure[6]. The high scatter of the $H - M$ relation in Figure 12 highlights the sensitivity of the hardness w.r.t. the microstructure. The figure also displays fitted power functions for the tested specimen. The exponent of the power function is found to decrease as a result of heat curing: it is $H \propto M^{1.49}$ for the control specimen 'C', and $H \propto M^{1.28}$ for the 28-day heat cured specimen 'HC-28'. This power relation cannot be explained by the determination process, since the hardness scales with $H \sim A^{-1}$ and the indentation modulus with $M \sim A^{-1/2}$, so that any bias induced by application of relations (2.1) and (2.11) should be visible in a quadratic power relation, $H \sim M^2$. The hardness-stiffness scaling, therefore, should be rather of microstructural origin. In fact, since H/M is a constant for a homogeneous material (*ie.* relation (2.13)), the scaling $H \propto M^{\delta}$ is an expression of the heterogeneity of the material investigated by nanoindentation. It is higher for the control specimen ($\delta = 1.49$) than for the heat cured specimen ($\delta = 1.28$), which is an indication that heat-curing, on-average, 'homogenizes' the microstructure. This observation is consistent with the contour plots in Figure 11.

Using the stiffness values that characterize the LD–C-S-H domain ($E_{LD} \in [13, 26]$ GPa) and the HD–C-S-H domain ($E_{HD} \in [26, 39]$ GPa), we can determine the corresponding hardness values, representative of the two-types of C-S-H. Mean values and standard deviations are given in Table 3 for the LD–C-S-H phase and the HD–C-S-H

[6] We recall that the stiffness relates to the recovery (in form of deformation) of the elastic energy stored into the microstructure during loading; while the hardness relates to the irreversible dissipation of the external work during loading into heat form. In the absence of fracture processes, the only link that exists between the two quantities is the geometry and morphology of the microstructure.

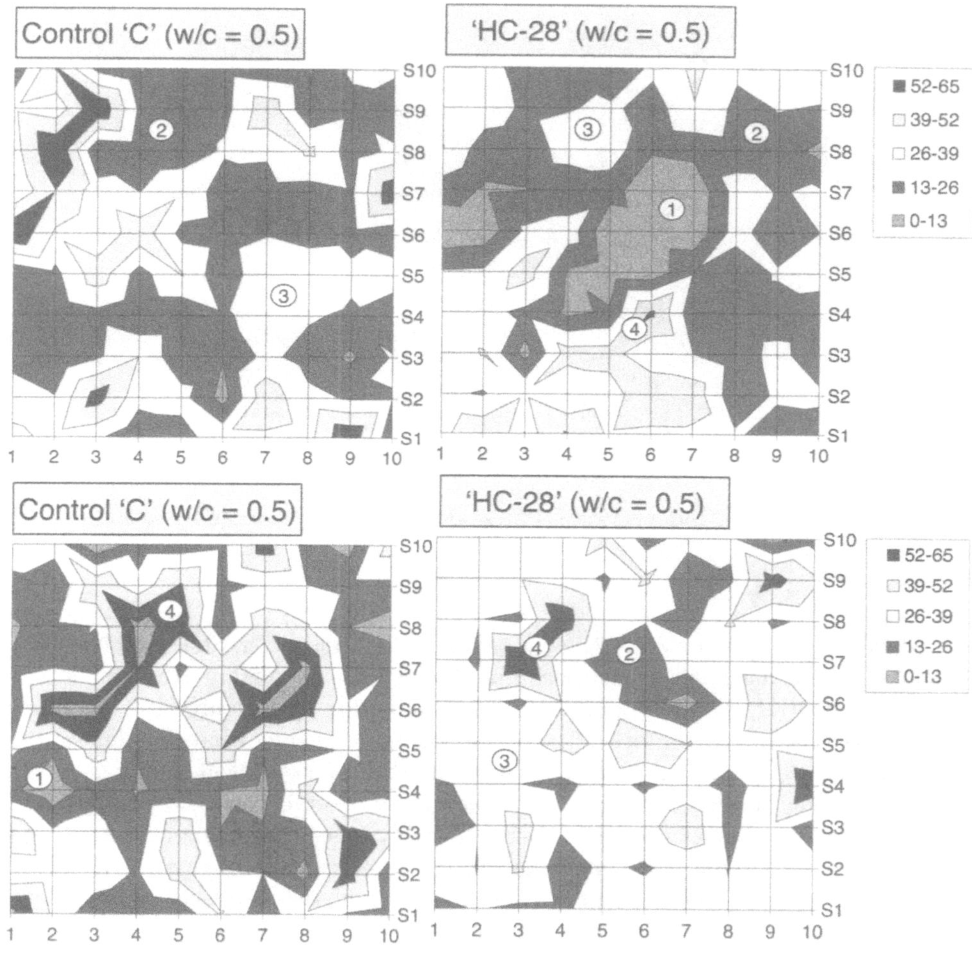

Figure 11. Contour plots of the spatial distribution of stiffness (in GPa) in the microstructure obtained by grid nanoindentation (grid-size = 10^{-5} m). Each grid point corresponds to a single nanoindentation test (unloading from a maximum indentation depth of 3×10^{-7} m). The numbers in the figure are attributed to 1 = porosity, 2 = LD-C-S-H, 3 = HD-C-S-H, 4 = CH and clinker phases.

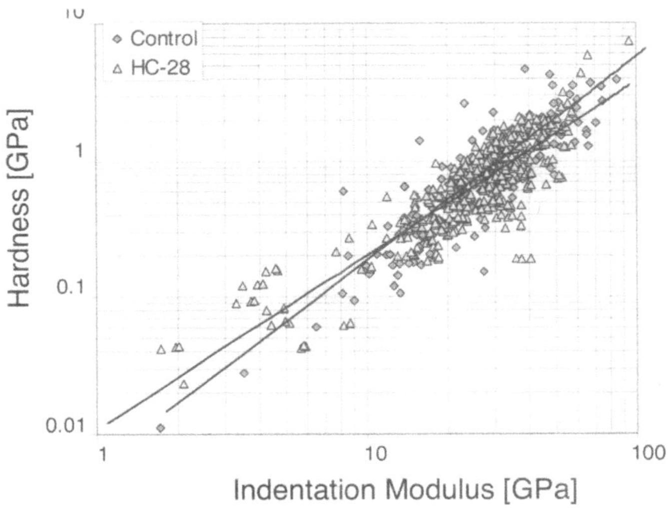

Figure 12. Hardness vs. indentation stiffness determined by nanoindentation.

phase. There is a trend that heat curing reduces the intrinsic hardness of the two C-S-H phases, as the hardness of the heat cured specimen is roughly 10% smaller than the one of the control specimen, and a hardness reduction by roughly 20% is found for the HD-C-S-H phase. On the other hand, this decrease is somewhat smaller than the standard deviation, so that it is difficult to conclude on the statistical significance of the heat-curing induced hardness drop. Two observations deserve particular attention: (1) The mean hardness values of the HD–C-S-H is roughly twice the value of the LD–C-S-H; and (2) the standard deviation, which can be seen as a measure of the microstructural disorder, reduces as a consequence of heat curing. Both observations are consistent with the disordered vs. ordered packing densities of the two types of C-S-H, or more precisely with the number of contacts that characterize the two types of packing densities. Indeed, it has been recently shown that the random packing of spheres which appears to characterize the LD–C-S-H, is characterized by on-average 6 points of contacts (Donev et al., 2004), while a high-density packing of spheres has a maximum of 12 points of contacts (Sloane, 1998), that is twice as much as the one in the random close-packed limit (RCP), which almost perfectly correlates with the measured nanohardness ratio $H_{HD}/H_{LD} \sim 2$. It is then not surprising that the overall scatter of the hardness values decreases as the amount of ordered HD–C-S-H increases in the material system. Rather than a coincidence, we see here a second evidence (in addition to the packing density) of the characteristic packing patterns of the LD–C-S-H and the HD–C-S-H. It suggests that the hardness (and thus the strength properties) of the two types of C-S-H is a mere consequence of the packing density, and more precisely of the number of contacts stabilizing each solid sphere in the system. The higher the packing density and thus the number of contact points, the greater the number of degrees of freedom along which the system can dissipate energy at the limit state associated with strength, and which translates into the measured nanohardness values.

Microindentation Analysis To complete the analysis, we performed a series of *micro*indentation tests on the control specimen ('C') and the heat cured specimen ('HC-28'), with a maximum penetration depth of roughly 10^{-5} m. Table 3 summarizes the indentation results (level II), which are average values and standard deviation of respectively 10 indentation tests for the control specimen 'C', and 30 indentation tests for the 28-days heat cured specimen 'HC-28' (see Table 2). The higher number of indentation tests for the HC-28 sample was necessary because of a higher standard deviation of the elasticity, which is due to the large macropores that interfere at the micro-indentation scale of 10^{-5} m. While the mean values obtained for the control specimen allow an interpretation as characteristic properties of the homogenized material, *ie.* cement paste, given the size of the macropores generated in heat-cured specimen, a continuous interpretation should be handled with care. Nevertheless some interesting trends become apparent.

The elastic stiffness remains practically unaffected by heat curing, as the slight increase in mean value from 19.1 GPa for the control specimen 'C' to 19.5 GPa for 'HC-28' is smaller than the standard deviation. It appears that the two competing mechanisms that characterize the mechanical effect of heat curing at a scale below, that is the densification of the C-S-H matrix and the generation of a macroporosity, cancel each other out. In return, the higher standard deviation for heat cured specimens is indicative for a higher degree of disorder at level II, induced at a scale below by the creation of the macroporosity.

The hardness which is a measure of strength properties, decreases by roughly 20% as a consequence of heat curing. This decrease in strength performance of heat cured specimen seems to be a consequence of the macroporosity generation which is not compensated by the densification of the C-S-H matrix. This 'coarsening of the macroporosity', as it is often referred to in the literature (Bentur, 1980), leads to high stress concentrations in the C-S-H matrix, leading to a lower overall material strength, particular in high w/c ratio cement-based materials. Moreover, the microhardness is roughly on the same order of magnitude of nanohardness of the LD–C-S-H phase at a scale below. This may not be a coincidence. In fact, regarding the control specimen, with almost non-detectable macroporosity, the closeness of nano- and micro-hardness values would hint towards relevance of the weakest link theory, according to which the strength behavior is governed by the weakest phase in the material system, which is the LD–C-S-H phase, and which appears to dominate the strength behavior of the cement paste. A similar argument can be made for the 28-day heat-cured material ('HC-28'): the generation of the macroporosity leads to higher stress concentrations in the C-S-H matrix; and in particular in the LD-C-S-H surrounding the macropores (see Fig. 11), which therefore becomes dominant in the strength behavior of the cement paste. Finally, it is also worth mentioning that the scatter of the microhardness values, expressed by the standard deviation, reduces significantly compared to the nanoindentation results. The smaller scatter is a clear indication of the homogenized nature of the properties obtained by indentation at the scale of the cement paste, but as well an indication that one detrimental phenomenon is at work governing the strength properties of the cement paste: the coarsening of the macroporosity.

3.3 Microporoelastic Analysis of Cement-Based Materials

Microporomechanics is a powerful framework to analyze and model indentation data, and ultimately to predict macroscopic material properties from the mere knowledge of some material independent phase properties, phase morphologies and volume fractions, determined either experimentally (e.g. nanoindentation and/or chemical analysis (Constantinides and Ulm, 2004)) or by modeling the physical chemistry processes that lead to the formation of the phases (see e.g. Bernard et al., 2003). Microporomechanics can start at the smallest scale, at which the composite material manifests itself composed of a continuous solid phase and a pore space that is large enough for water to be present as a bulk phase, characterized by a hydrostatic stress field, $\boldsymbol{\sigma} = -p\boldsymbol{\delta}$, where p is the fluid pressure. In the case of cement-based materials, this is the scale of the two-types of C-S-H (Ulm et al., 2004).

Level '0': Invariant Material Properties of C-S-H The information that is available about the two types of C-S-H are the porosities (3.1) determined by mass density measurements (see Table 1), and composite stiffness values determined independently by nanoindentation (see Table 3). The overall picture that emerges from a combination of these independent measurements is that the difference in mechanical performance of the two types of C-S-H is a consequence of their packing density, while the solid phase is the same, having the same mass density and as a consequence same mechanical properties. The poroelastic behavior of the two types of C-S-H is defined by the classical poroelastic state equations (Biot, 1941; Coussy, 1995; Coussy, 2004):

$$\Sigma_m = KE_v - bp; \; S_{ij} = 2GE_{ij}^d \tag{3.5a}$$

$$\phi - \phi_0 = bE_v + \frac{p}{N} \tag{3.5b}$$

where $\Sigma_m = \frac{1}{3}\Sigma_{ii}$ is the mean stress, $S_{ij} = \Sigma_{ii} - \Sigma_m \delta_{ij}$ is the stress deviator, $E_v = E_{ii}$ is the volume strain, and $E_{ij}^d = E_{ii} - \frac{1}{3}E_v\delta_{ij}$ is the strain deviator. Using standard relations of microporoelasticity (Chateau and Dormieux, 2002; Dormieux et al., 2002; Dormieux and Bourgeois, 2003; and the contribution of Dormieux in this book), the poroelastic properties, K, G, b, N, are estimated from:[7]

$$K = g_s \frac{4(1-\phi_0)}{3\phi_0 + 4(g_s/k_s)} \quad (3.6a)$$

$$G = g_s \frac{(1-\phi_0)(8g_s+9k_s)}{6\phi_0(2g_s+k_s)+8g_s+9k_s} \quad (3.6b)$$

$$b = \phi_0 \frac{4g_s+3k_s}{3\phi_0 k_s + 4g_s} \quad (3.6c)$$

$$\frac{1}{N} = \frac{3\phi_0(1-\phi_0)}{3\phi_0 k_s + 4g_s} \quad (3.6d)$$

where k_s and g_s are the solid bulk modulus and the solid shear modulus. These nanomechanical properties are currently still out of reach to direct mechanical measurements; and that what is measured are the homogenized properties of the two types of C-S-H. The determination of the nanomechanical properties, therefore, can only be achieved by an inverse application of the homogenization relations. As input we have values for the indentation modulus, M_{HD} and M_{LD}, and for the porosity, ϕ_0^{LD} and ϕ_0^{HD}. The unknowns are the stiffness properties of the C-S-H solid phase, k_s and g_s, and one 'homogenized property' per type of C-S-H; e.g. the Poisson's ratio, ν_{HD} and ν_{LD}. This approach (Ulm et al., 2004) leads to identifying a solid bulk and shear modulus of $k_s = 32$ GPa, $g_s = 19$ GPa, which corresponds to a Young's modulus of $E_s \simeq 48$ GPa, a Poisson's ratio of $\nu_s = 0.25$, and a Biot coefficient of the C-S-H phases situated in between:

$$b_{HD}^0 = 0.61 \leq b \leq 0.71 = b_{LD}^0 \quad (3.7)$$

Level 'I': Poroelastic Properties of C-S-H Matrix – Double Porosity Model
The C-S-H matrix is composed of two porous materials, the LD–C-S-H phase and the HD–C-S-H phase, each of which is described by poroelastic constitutive equations of the form (3.5). To derive the poroelastic constitutive equations of the C-S-H matrix, we adopt a continuous description of the stress field in the heterogeneous r.e.v.:

$$\text{in}\, V : \sigma_m(\underline{z}) = k(\underline{z})\,\epsilon(\underline{z}) + \sigma^p(\underline{z})\,;\ s_{ij}(\underline{z}) = 2g(\underline{z})\,e_{ij}(\underline{z}) \quad (3.8)$$

together with the distributions of the elastic properties $(k(\underline{z}), g(\underline{z}))$ and the eigenstress $\sigma^p(\underline{z})$:

$$(k(\underline{z}), g(\underline{z})) = \begin{cases} (k_{LD}, g_{LD}) \text{ in } V_{LD} \\ (k_{HD}, g_{HD}) \text{ in } V_{HD} \end{cases} \quad \sigma^p(\underline{z}) = \begin{cases} -b_{LD}^0\, p_1 \text{ in } V_{LD} \\ -b_{HD}^0\, p_2 \text{ in } V_{HD} \end{cases} \quad (3.9)$$

where $\epsilon = \varepsilon_{ii}$ is the microscopic volume strain; and $e_{ij} = \varepsilon_{ij} - \frac{1}{3}\epsilon\delta_{ij}$ is the microscopic deviator strain. $(k_j, g_j), (j = LD, HD)$ are the elastic properties of the LD– and HD–C-S-H phases that occupy domains V_{LD} and V_{HD} in the r.e.v. For purpose of argument, we will assume that the fluids present within the two types of C-S-H are not at the same pressure.

[7] Relations (3.6) are the poroelastic constants that can be derived from the Hollow Sphere Model, or by application of a Mori-Tanaka scheme to an isotropic solid-porosity composite.

Following the linear microporomechanics approach (Dormieux and Bourgeois, 2003), we decompose the problem in two sub-problems:

1. The first sub-problem corresponds to overall drained conditions, for which $p_1 = p_2 = 0$. We assume that the r.e.v. is subjected to a uniform displacement boundary condition, so that the solid boundary conditions to which the solid phase is subjected read:

$$\text{on}\,\partial V \quad : \quad \underline{\xi}'(\underline{z}) = \mathbf{E} \cdot \underline{z} \tag{3.10a}$$

$$\text{on}\,\mathcal{I}_{sf} \quad : \quad \boldsymbol{\sigma} \cdot \underline{n} = 0 \tag{3.10b}$$

Here, ∂V stands for the boundary of the r.e.v., and $\mathcal{I}_{sf}$ represents the solid-fluid interfaces. $\underline{\xi}'(\underline{z})$ stands for the microscopic displacement field, and $\mathbf{E}$ is the macroscopic strain tensor, which is related to the microscopic strain by an isotropic strain localization condition:

$$\boldsymbol{\varepsilon}'(\underline{z}) = \mathbf{e}'(\underline{z}) + \frac{1}{3}\epsilon'(\underline{z})\,\boldsymbol{\delta} = a^d(\underline{z})\,\mathbf{E}_d + \frac{1}{3}a^v(\underline{z})\,E_v\boldsymbol{\delta} \tag{3.11}$$

where $a^d(\underline{z})$ and $a^v(\underline{z})$ are the deviator strain and the volumetric strain localization factors. Use in (3.8) together with the elastic distribution (3.9) yields after volume averaging the macroscopic stress:

$$\Sigma'_m = \overline{\sigma_m} = K^I E_v; \quad K^I = f_{LD}k_{LD}A^v_{LD} + f_{HD}k_{HD}A^v_{HD} \tag{3.12a}$$

$$\mathbf{S} = \overline{\mathbf{s}} = 2G^I\mathbf{E}_d; \quad G^I = f_{LD}g_M A^d_M + f_{HD}g_{HD}A^d_{HD} \tag{3.12b}$$

where $A^v_J = \langle a^v(\underline{z})\rangle_{V_J}$ and $A^d_J = \langle a^d(\underline{z})\rangle_{V_J}$ stand for volume averages of the localization factors over the C-S-H subdomains, and $f_J = V_J/V$ represent the volume fractions satisfying $f_{LD} + f_{HD} = 1$. On the other hand, the change of the gelporosity in the subproblem reads ($J = LD, HD$):

$$\left[(\phi-\phi_0)^I_J\right]' = f_J\left[(\phi-\phi_0)^0_J\right]' = b^I_J E_v \tag{3.13a}$$

$$b^I_J = f_J b^0_J A^v_J \tag{3.13b}$$

where superscript $0, I$ indicates the scale ('level') at which the quantity is defined. For instance, b^0_J stands for the Biot coefficients given by (3.7).

2. The second sub-problem we consider is the zero-displacement boundary problem, while the gelporosity is pressurized. From linear microporoelasticity, it is known that the problem is conveniently solved using Levine's theorem, delivering a relation between the macroscopic and the microscopic eigenstresses (Chateau and Dormieux, 2002; Dormieux et al., 2002; Dormieux and Bourgeois, 2003):

$$\boldsymbol{\Sigma}'' = \overline{\boldsymbol{\sigma}^p(\underline{z}) : \mathbb{A}(\underline{z})} \tag{3.14}$$

where $\mathbb{A}(\underline{z})$ is the forth-order strain localization tensor. Application of the eigenstress distribution (3.9) in (3.14) yields the macroscopic mean stress in this subproblem:

$$\Sigma''_m = -b^I_{LD}p_1 - b^I_{HD}p_2 \tag{3.15}$$

where we verify that $b^I_{LD} = f_{LD} b^0_{LD} A^v_{LD}$ and $b^I_{HD} = f_{HD} b^0_{HD} A^v_{HD}$ obtained in the first sub-problem (3.13) are the Biot coefficients associated with the pressures in gelporosity of the two type of C-S-H. Because the constitutive equation for the change of the porosity in each subdomain is given by (3.5b), the change of porosity reads:

$$\left[(\phi - \phi_0)^I_J\right]'' = f_J \left(b^0_J \langle \epsilon'' \rangle_{V_J} + \frac{p_J}{N^0_J} \right) \tag{3.16}$$

We need to eliminate $\langle \epsilon'' \rangle_{V_J}$ in (3.16). To this end, we determine the stress average in this sub-problem on account of (3.8) and (3.9):

$$\begin{aligned} \Sigma''_m &= f_{LD} \langle \sigma''_m \rangle_{V_{LD}} + f_{HD} \langle \sigma''_m \rangle_{V_{HD}} \\ &= f_{LD} k_{LD} \langle \epsilon'' \rangle_{V_{LD}} + f_{HD} k_{HD} \langle \epsilon'' \rangle_{V_{HD}} - f_{LD} b^0_{LD} p_1 - f_{HD} b^0_{HD} p_2 \end{aligned} \tag{3.17}$$

Then, we use (3.15) in (3.17) to express the average strain in the porous matrix $\langle \epsilon'' \rangle_{V_J}$ as a function of p_J:

$$\left(f_{LD} b^0_{LD} - b^I_{LD} \right) p_1 + \left(f_{HD} b^0_{HD} - b^I_{HD} \right) p_2 = f_{LD} k_{LD} \langle \epsilon'' \rangle_{V_{LD}} + f_{HD} k_{HD} \langle \epsilon'' \rangle_{V_{HD}} \tag{3.18}$$

which suggests, using (3.13) or (3.15):

$$\langle \epsilon'' \rangle_{V_J} = \frac{b^0_J}{k_J} \left(1 - A^v_J\right) p_J \tag{3.19}$$

with no summation on repeated subscripts. Finally, use of (3.19) in (3.16) yields:

$$\left[(\phi - \phi_0)^I_J\right]'' = \frac{p_J}{N^I_J};\ \frac{1}{N^I_J} = \frac{f_J}{N^0_J} + f_J \frac{\left(b^0_J\right)^2}{k_J} \left(1 - A^v_J\right) \tag{3.20}$$

Last, a superposition of the two subproblems yields the macroscopic state equations of the C-S-H matrix:

$$\Sigma_m = \Sigma'_m + \Sigma_m'' = K^I E_v - b^I_{LD} p_1 - b^I_{HD} p_2 \tag{3.21a}$$

$$(\phi - \phi_0)^I_J = b^I_J E_v + \frac{p_J}{N^I_J};\ J = LD, HD \tag{3.21b}$$

The resulting state equations are recognized as the state equations of a double-porosity material without interaction of the fluid pressures on the deformation of the solid part of the porous subdomains (*ie.* relation (3.19)). Given the similar size of the porosity in the LD– and HD–C-S-H, as defined by the packing densities, it can be suggested that there should be little difference in pressure in between the two types of C-S-H. In this case, the poroelastic state equations of the C-S-H matrix reduce to:

$$\Sigma_m = K^I E_v - b^I p \tag{3.22a}$$

$$(\phi - \phi_0)^I = b^I E_v + \frac{p}{N^I} \tag{3.22b}$$

and the poroelastic constants are given by:

$$K^I = f_{LD}k_{LD}A^v_{LD} + f_{HD}k_{HD}A^v_{HD} \tag{3.23a}$$

$$b^I = b^I_{LD} + b^I_{HD} = f_{LD}b^0_{LD}A^v_{LD} + f_{HD}b^0_{HD}A^v_{HD} \tag{3.23b}$$

$$\frac{1}{N^I} = \frac{1}{N^I_{LD}} + \frac{1}{N^I_{HD}} \tag{3.23c}$$

Level 'II': Poroelastic Properties of Cement Paste – Two-Scale Double Porosity Model Let us turn now to cement paste: a composite of a porous C-S-H matrix and macroporosity, in addition to some inclusion phases (residual clinker, Portlandite). In contrast to the C-S-H matrix (level I), the main feature here at level II is that the porosity manifests itself at two different scales, a microporosity situated within the porous matrix and the macroporosity. We want to derive the poroelastic state equations of the cement paste. For purpose of clarity, we set any additional inclusion phases aside (see Ulm et al. (2004) for an account of multiple solid phases), and restrict ourselves, for a microporoelastic analysis, to a porous matrix and a macroporosity. The constitutive equations of the porous matrix are assumed to be defined by (3.22). Given the difference in size of the porosities involved, we assume a different pressure in the micro- and macroporosity.

Proceeding as before, we adopt a continuous description of the stress field in the heterogeneous r.e.v., of the form (3.8), together with the following distributions of the elastic properties $(k(\underline{z}), g(\underline{z}))$ and the eigenstress $\sigma^p(\underline{z})$:

$$(k(\underline{z}), g(\underline{z})) = \begin{cases} (K^I, G^I) \text{ in } V_M \\ (0,0) \text{ in } V_{\phi 2} \end{cases} \quad \sigma^p(\underline{z}) = \begin{cases} -b^I p_1 \text{ in } V_M \\ -p_2 \text{ in } V_{\phi 2} \end{cases} \tag{3.24}$$

where V_M stands for the volume occupied by the porous matrix in the r.e.v., and $V_{\phi 2}$ for the macroscopic pore space. We decompose the problem in (the meanwhile classical) two sub-problems:

1. The first sub-problem corresponds to overall drained conditions, for which $p_1 = p_2 = 0$. The boundary conditions are still defined by (3.10), and an isotropic strain localization condition of the form (3.11) is applied. This yields after volume averaging the macroscopic stress in the form:

$$\Sigma'_m = K^{II}E_v;\ \ K^{II} = (1-\phi_{02})\,K^I A^v_M \tag{3.25a}$$

$$\mathbf{S} = 2G^{II}\mathbf{E}_d;\ \ G^{II} = (1-\phi_{02})\,G^I A^d_M \tag{3.25b}$$

where $A^v_M = \langle a^v(\underline{z})\rangle_{V_M}$ and $A^d_M = \langle a^d(\underline{z})\rangle_{V_M}$; $1-\phi_{02} = V_M/V$ is the volume fraction of the porous matrix in the r.e.v. K^{II} and G^{II} are recognized as the macroscopic drained elastic stiffness properties of the cement paste. On the other hand, there are two associated changes of porosity; the microporosity and the macroporosity. The microporosity is defined per unit of (undeformed) matrix volume V_M so that the change of porosity at the macroscale in the considered subproblem is:

$$\left[(\phi_1 - \phi_{01})^{II}\right]' = (1-\phi_{02})\left[(\phi_1 - \phi_{01})^{I}\right]' = b_1^{II} E_v \tag{3.26a}$$

$$b_1^{II} = (1-\phi_{02})\, b^I A_M^v \tag{3.26b}$$

In return, the change in macroporosity reads:

$$\left[(\phi_2 - \phi_{02})^{II}\right]' = b_2^{II} E_v \tag{3.27a}$$

$$b_2^{II} = \phi_{02} A_{\phi 2}^v \tag{3.27b}$$

where $A_{\phi 2}^v = \langle a^v(\underline{z})\rangle_{V_{\phi 2}}$. Herein, b_1^{II} and b_2^{II} are the Biot coefficients associated with the micro- and the macroporosity, respectively.

2. The second sub-problem we consider is the zero-displacement boundary problem, for which Levine's theorem (3.14) applies. Application of the eigenstress distribution (3.24) in (3.14) yields the macroscopic stress in this subproblem:

$$\Sigma_m'' = -(1-\phi_{02})\, b^I A_M^v\, p_1 - \phi_{02} A_{\phi 2}^v p_2 = -b_1^{II} p_1 - b_2^{II} p_2 \tag{3.28}$$

There are two changes of porosity to be considered; the change in microporosity,

$$\left[(\phi_1 - \phi_{01})^{II}\right]'' = (1-\phi_{02})\left(b_M \langle \epsilon'' \rangle_{V_M} + \frac{p_1}{N^I}\right) \tag{3.29}$$

and the change in macroporosity:

$$\left[(\phi_2 - \phi_{02})^{II}\right]'' = \phi_{02} \langle \epsilon'' \rangle_{V_{\phi 2}} = -(1-\phi_{02}) \langle \epsilon'' \rangle_{V_M} \tag{3.30}$$

We need to eliminate $\langle \epsilon'' \rangle_{V_M}$ in (3.29) and (3.30). To this end, we determine the stress average in this sub-problem on account of (3.24):

$$\begin{aligned} \Sigma_m'' &= (1-\phi_{02}) \langle \sigma_m'' \rangle_{V_M} + \phi_{02} \langle \sigma_m'' \rangle_{V_{p2}} \\ &= (1-\phi_{02})\, K^I \langle \epsilon'' \rangle_{V_M} - (1-\phi_{02})\, b^I p_1 - \phi_{02} p_2 \end{aligned} \tag{3.31}$$

Then, we use (3.28) in (3.31) to express the average strain in the porous matrix $\langle \epsilon'' \rangle_{V_M}$ as a function of p_1 and p_2:

$$\begin{aligned} (1-\phi_{02}) \langle \epsilon'' \rangle_{V_M} &= \frac{1}{K^I}\left[\Sigma_m'' + (1-\phi_{02})\, b^I p_1 + \phi_{02} p_2\right] \\ &= \frac{1}{K^I}\left[\left((1-\phi_{02})\, b^I - b_1^{II}\right) p_1 + \left(\phi_{02} - b_2^{II}\right) p_2\right] \end{aligned} \tag{3.32}$$

Finally, substitution in (3.29) and (3.30) yields the change of the microporosity:

$$(\phi_1 - \phi_{01})'' = \frac{p_1}{N_{11}} + \frac{p_2}{N_{12}} \tag{3.33a}$$

$$\frac{1}{N_{11}} = (1-\phi_{02})\left(\frac{b^I}{K^I}\left(b^I - b_1^{II}\right) + \frac{1}{N^I}\right) \tag{3.33b}$$

$$\frac{1}{N_{12}} = \frac{b^I}{K^I}\left(\phi_{02} - b_1^{II}\right) \tag{3.33c}$$

and the change of the macroporosity:

$$(\phi_2 - \phi_{02})'' = \frac{p_1}{N_{21}} + \frac{p_2}{N_{22}} \tag{3.34a}$$

$$\frac{1}{N_{21}} = \frac{1}{K^I}\left(b_1^{II} - (1-\phi_{02})\, b^I\right) \tag{3.34b}$$

$$\frac{1}{N_{22}} = \frac{1}{K^I}\left(b_2^{II} - \phi_{02}\right) \tag{3.34c}$$

The symmetry of the skeleton Biot moduli $N_{12} = N_{21}$ is readily shown, by substituting (3.27b) in (3.33c) and (3.26b) in (3.33b):

$$\frac{1}{N_{12}} = \phi_{02}\frac{b^I}{K^I}\left(1 - A^v_{\phi 2}\right) \tag{3.35a}$$

$$\frac{1}{N_{21}} = (1-\phi_{02})\frac{b^I}{K^I}\left(A^v_M - 1\right) \tag{3.35b}$$

Then, use of the strain localization compatibility condition in one of the two relations yields the other one. For instance, replacing $\phi_{02}A^v_{\phi 2} = 1 - (1-\phi_{02})\, A^v_M$ in (3.35a) gives:

$$\frac{1}{N_{12}} = (1-\phi_{02})\frac{b^I}{K^I}\left(A^v_M - 1\right) = \frac{1}{N_{21}} \tag{3.36}$$

Last, a superposition of the two subproblems yields the macroscopic state equations of the isotropic double-porosity system:

$$\Sigma_m = \Sigma'_m + \Sigma_m'' = K^{II}E_v - b_1^{II}p_1 - b_2^{II}p_2 \tag{3.37a}$$

$$\phi_1 - \phi_{01} = b_1^{II}E_v + \frac{p_1}{N_{11}} + \frac{p_2}{N_{12}} \tag{3.37b}$$

$$\phi_2 - \phi_{02} = b_2^{II}E_v + \frac{p_1}{N_{21}} + \frac{p_2}{N_{22}} \tag{3.37c}$$

where all seven poroelastic constants are known here as functions of the poroelastic properties of the porous matrix, and of two strain localization factors, $A^v_{\phi 2}$ (or A^v_M) and A^d_M. An estimate of these factors provided e.g. by a Mori-Tanaka scheme reads:

$$A^v_{\phi 2} = \frac{4G^I + 3K^I}{3\phi_{02}K^I + 4G^I};\; A^v_M = \frac{1-\phi_{02}A^v_{\phi 2}}{1-\phi_{02}} \tag{3.38a}$$

$$A^d_M = \frac{1-\phi_{02}A^d_{\phi 2}}{1-\phi_{02}} = \frac{8G^I + 9K^I}{6\phi_0\left(2G^I + K^I\right) + 8G^I + 9K^I} \tag{3.38b}$$

For purpose of comparison of the poroelastic properties, it is useful to evaluate the poroelastic constants for the same pressure in both meso- and macroporosity, for which the state equations are given by (3.5), with:

$$b^{II} = b_1^{II} + b_2^{II} = b^I + \left(1 - b^I\right)\phi_{02}A^v_{\phi 2} \tag{3.39a}$$

$$\frac{1}{N^{II}} = \frac{1}{N_{11}} + \frac{2}{N_{12}} + \frac{1}{N_{22}} \tag{3.39b}$$

Summary of Microporoelastic Analysis The poroelastic analysis of cement-based materials at different scales illustrates the important role of microporoelasticity theory in the understanding and assessment of material properties at different scales of cement-based materials. Table 4 summarizes the poroelastic properties of the $w/c = 0.5$ control specimen and thermally cured specimen, determined from the presented theory. The error of the prediction of the Young's modulus is smaller than the standard deviation of the microindentation results. The input to the analysis are the indentation stiffness and the volume fractions determined by nanoindentation:

- Level '0': On very small scales of cement-based materials, a combination of nanoindentation results and microporomechanics provides a means to determine material invariant mechanical properties, that are readily employed as a starting point for the application of microporomechanics theory. From a materials science point of view, cement-based materials, at this scale, are organized in two characteristic packing features, leaving a pore space behind that is at the origin of the pronounced porous material behavior of cement-based materials at larger scales.

- Level I: The C-S-H matrix manifests itself, from a poromechanics point of view, as a double porosity material system, one representing the LD–C-S-H phase, the other the HD–C-S-H phase. At this scale, the poroelastic properties depend on the relative proportion of the LD– and HD–C-S-H phase present, which depends on the mix proportions, curing conditions, and so on. It is not surprising that thermal curing which increases the amount of HD-C-S-H on the expenses of LD–C-S-H yields a higher stiffness of the C-S-H matrix, and a (slightly) lower Biot coefficient.

- Level II: Cement paste is a double porosity material system as a consequence of the porosity at (at least) two different scales; the mesoporosity in the C-S-H matrix, and the macroporosity. Despite an increase of the stiffness of the C-S-H matrix due to heat curing, it is found that stiffness of the cement paste is almost the same for both control specimen and thermally cured specimen. Microporoelastic theory thus confirms that the stiffening of the C-S-H phase is compensated by the macroporosity generation in its overall effect on the (drained) elasticity of heat cured materials. As a result of the macroporosity generation, the Biot coefficient increases.

		Control ('C')		HC-28	
		E_{hom} [GPa]	b_{hom} [1]	E_{hom} [GPa]	b_{hom} [1]
Level 0	LD–C-S-H	18.8	0.71	18.8	0.71
	HD–C-S-H	27.3	0.61	29.2	0.61
Level I	C-S-H Matrix	21.5	0.68	24.2	0.66
Level II	Cement Paste	20.5	0.70	20.2	0.72
Error	$1 - E_{\text{hom}}/E_{\text{exp}}^{II}$	−7%	–	−3.5%	–

Table 4. Summary of microporoelastic analysis. The input are experimentally determined nanoindentation stiffness and volume fractions from Table 3. The assumed Poisson's ratio for the LD– and HD–C-S-H is 0.24.

4 Experimental Microporomechanics of Shales

Is shale anisotropy a consequence of the intrinsic anisotropy of the clay minerals or rather an effect of texture, or both?[8] –

Shales are probably one of the most complicated and intriguing natural materials present on earth. Shales make up about 75% of sedimentary basins (Jones and Wang, 1981) and cover many hydrocarbon-bearing reservoirs. Knowing and predicting the anisotropy plays a critical role in many fields of exploitation, ranging from seismic exploration (log-data interpretation), to well drilling (well bore stability) and production (Thomsen, 2001). But the properties of shales are still an enigma that has deceived many decoding attempts from experimental or theoretical sides. Two schools of thought exist:

- The shale geomechanics school of thought, which is based on observational techniques, such as Scanning Electron Microscopy (SEM), Environmental Scanning Electron Microscopy (ESEM), Transmission Electron Microscopy (TEM), etc., considers the elastic anisotropy as a result of the microfabric signature of clay that originates from physicochemical, electrochemical, thermomechanical, interface dynamics, bioorganic and, most importantly, burial diagenesis processes (Bennett et al., 1991; Mitchell, 1993).
- The basis of a second school of thought is the existence of the 'perfect' shale, a concept introduced in a 1994 paper by Hornby et al. (1994), and which has found strong support particularly in the shale acoustics community (Jacobsen et al., 2003). The origin of anisotropy, in this concept, is attributed to a perfect alignment of shale particles at very small scales, that is eventually disturbed at larger scales

[8] The investigation of the sources of anisotropy of shale materials was suggested to us by Prof. Younane Abousleiman (Oklahoma Uinversity) and was carried out in a collaborative research effort between MIT, the Poromechanics Institute at Oklahoma University at Norman, and the Chevron Texaco team led by Dr. Russ Ewy. The financial support of the study by Chevron Texaco is gratefully acknowledged. The investigated shale specimens, mineralogy, porosity and UPV measurements were provided by the Chevron Texaco team, Drs. Russ Ewy, Luca Duranti, Douglas K. McCarty, whose collaboration, comments and suggestions were critical in this multidisciplinary project.

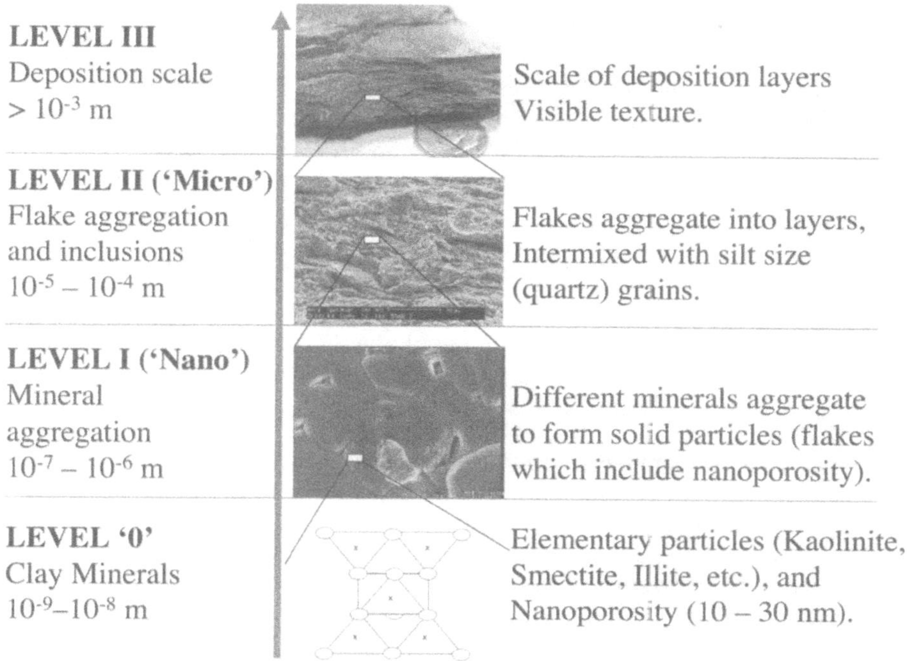

Figure 13. Four-level microstructural think-model of shale materials.

through a partially oriented distribution of clay minerals (Sayers, 1994; Johansen et al., 2004), and by the presence of (almost spherical) silt inclusions that reduce the intrinsic anisotropy of shale materials.

Shale anisotropy is a formidable play ground for an experimental microporomechanics approach, combining nanoindentation with advanced microporoelasticity theory (Delafargue and Ulm, 2004b).

4.1 Multiscale Indentation Analysis of Shale Materials

Inarguably, shale materials are highly heterogeneous multiphase natural composite materials. Figure 13 displays the four-level think model of the multiscale structure of shale materials, which forms much of the backbone of our experimental and theoretical microporomechanics approach.

Level '0': Scale of Elementary Particles The lowest level we consider is the scale of the elementary clay particles at a length scale of some nanometer. This scale has been focus of much research in clay mineralogy; and it is now well known (Velde, 1993), from both XRD-analysis and high resolution TEM-images, that the clay crystals (kaolonite, illite, smectite, etc.) form platelets with typical aspect ratios of 1/20.

However, unlike many other minerals, clay stiffness values are extremely rare in handbooks (Mavko et al., 1998). To our knowledge, the only direct measurements of the anisotropic elasticity of clay minerals were reported for large muscovite crystals, possessing transversely isotropy: $C_{11} = 178$ GPa, $C_{33} = 55$ GPa, $C_{44} = 12$ GPa, $C_{12} = 42$ GPa, $C_{13} = 15$ GPa, or equivalently expressed in terms of indentation stiffness (2.44) and (2.52),

$$M^0_{1,\mathrm{exp}} = 118 \text{ GPa}; \; M^0_{3,\mathrm{exp}} = 46 \text{ GPa}; \; \frac{M^0_{1,\mathrm{exp}}}{M^0_{3,\mathrm{exp}}} = 2.6 \tag{4.40}$$

The main difficulty of measuring the mineral elasticity stems from the fact that clay particles are too small to be tested in pure solid form. Several attempts to overcome this difficulty have been reported (Table 5):

- The stiffness of compacted clay samples has been measured and extrapolated to a zero porosity assuming that this extrapolation technique yields a 'pure clay' stiffness value (Marion et al., 1992; Hornby et al., 1994; Vanorio et al., 2003). There is quite some difference in stiffness values obtained with this technique, which makes it difficult to state with certainty whether such extrapolation techniques give access to intrinsic stiffness values or not. In particular, depending on the degree of saturation of the sample and other parameters influencing surface forces, the interactions between clay particles might exhibit different compliances which may in turn modify the extrapolated stiffness value. In order to have relevant experimental measures of clay minerals stiffness, Prasad et al. (2002) have recently made some dynamic measurement of clays using Atomic Force Acoustic Microscopy (AFAM). They reported a very low Young's modulus for dickite of 6.2 GPa.
- Ultrasonic velocities of composite mixture of individual clay particles (powder) diluted, at various concentrations, in an epoxy matrix have been measured (Wang et al., 2001). Using a backward homogenization derivation, the Young's modulus for randomly distributed clay particles was found to be on the order of $50-60$ GPa for Kaolonite, $65-80$ GPa for illite, $40-50$ GPa for montmorillonite, and greater than 100 GPa for Chlorite. These results appear to be very consistent with theoretical estimates of single crystal elastic properties by Katahara (1996) for kaolonite, illite and chlorite, based on data of Alexandrov and Ryzhova (1961; cited from Wang et al. (2001)).
- Very recently, Steered Molecular Dynamics simulations of pyrophyllite clay provided an order of magnitude of the interlayer stiffness of 130 GPa, while the clay layers are almost rigid (Katti et al., 2003), with a stiffness value of 547 GPa in the direction perpendicular to crystal planes for a simple structure of two crystal layers with an interlayer space.

The scarcity of experimental values for solid clay stiffness and the large range of reported values highlight the difficulty to assess the intrinsic elasticity of single clay crystals. This has been recognized as one current major weakness of existing micromechanical models of shales (Jacobsen et al., 2003). We performed some nanoindentation tests on a kaolin powder composed of 97% kaolonite (EPK Kaolin[9], Feldspar Corporation, Atlanta Georgia) having an average particle size of 1.36×10^{-6} m, a specific surface area of $24.25\,\mathrm{m}^2/\mathrm{g}$, and a loose (dense) mass density of $519 - 740\,\mathrm{kg}/\mathrm{m}^3$. Using indentation depths of $70 - 90$ nm, that is an order of length magnitude smaller than the grain size, we obtain indentation stiffness values of

$$\text{EPK Kaolin: } M^0_{\text{exp}} = 40.3 \pm 8.8 \text{ GPa} \tag{4.41}$$

Figure 14 displays an Atomic Force Microscopy image of a kaolin powder grain on which we indented, together with five force–indentation depth curves, from which we determine, from the unloading branch, the indentation stiffness. The order of magnitude of the indentation stiffness of kaolin powder is found to be situated in between the values reported by Marion et al. (1992) and Wang et al. (2001), respectively; and on the order of the weak axis indentation stiffness of Muscovite (see Table 5).

Level 'I': Porous Clay Composite Clay particles with nanoporosity form a porous clay composite at a characteristic length scale in the hundreds of nanometer range. ESEM images (Fig. 15) of the porous clay composite reveal that the porous clay composite in shale materials comes in a large variety of forms and shapes, ranging from highly ordered sheet bundles to wavy flake structures and highly pressed and crushed structure of clay sheets. The characteristic dimension of the clay sheets visible in these pictures is on the order of $500 - 1000$ nm, and $20 - 50$ nm thick, which confirms the aspect ratio of roughly $1/25 - 1/20$, as generally admitted in the clay literature. However, the microphotographs in figures 15 also display a high degree of disorder of the clay sheet orientation, and this both normal to the bedding direction (labeled '$x3$') and in bedding direction (labeled '$x1$' or '$x2$'). At the considered scale of the porous clay composite, it is quite frequent that clay packages are found to strongly divert from the deposition orientation.

Furthermore, the space in between the clay sheets forms almost the totality of the porosity of shales, as poromercury intrusion studies show, displaying a very low characteristic pore access radius[10] of some nanometers; thus situating the porosity of shales at a scale below the porous clay composite. This justifies considering shales at a length scale of $500 - 5{,}000$ nm as porous composites. This is level 'I' of shale materials, *ie.* the lowest scale currently accessible to mechanical testing. Prasad et al. (2002) used atomic force acoustic microscopy (AFAM) to determine the Young's modulus of a thin layer of dickite to be on the order of 6.2 GPa. The authors concluded from this and a subsequent study (Vanorio et al., 2003), in which $P-$ and $S-$ wave velocity measurements on cold-pressed clay aggregates of different porosity/grain density were extrapolated to zero porosity values, that the actual stiffness of clay crystals was on the order of $10 - 15$ GPa, which is

[9] For detailed chemical and mineralogy information on EPK Kaolin, see http://www.feldspar.com/minerals/epk.html.

[10] The poromercury intrusion model is based on a pore cylinder model, and it is generally admitted that MPI allows invasion of (cylindrical) pores of pore radius 3.6×10^{-9} m to 60×10^{-6} m. Given the sheet shape of the clay particles, it is most unlikely that the pores in shales are cylindrical, as the porosity is the space in between the sheets, so that the pore geometry should be similar to the sheet geometry.

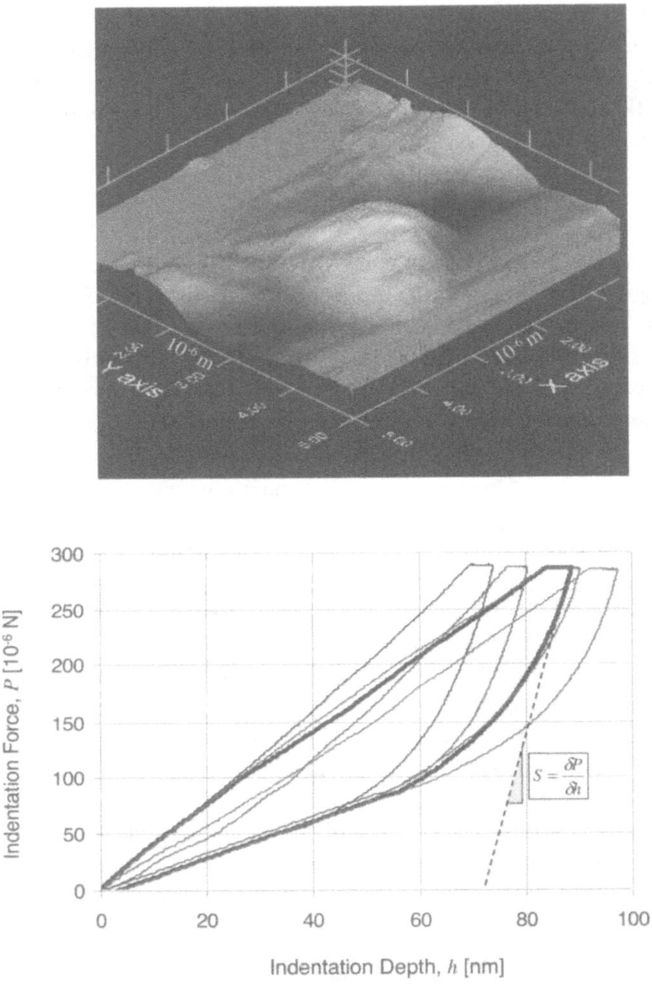

Figure 14. Nanoindentation on kaolin powder: (top) AFM image; (bottom) load-displacement curves.

Clay Mineral	Sample	Experim. Techn.	Stiffness [GPa] $M^0_{1,\text{exp}}$	Stiffness [GPa] $M^0_{3,\text{exp}}$	Source
Muscovite	nat. crystal		118.1	46.2	Mavko et al., 1998
Kaolinite	clay mixture	acoustic (MicroMech)	30.3		Marion et al., 1992 Hornby et al., 1994
Dicktite	clay mixture	AFAM	$E = 6.2$		Prasad et al., 2002
Kaolinite Smectite Montmorillonite	clay-mixture suspensions	acoustic (Extrapol.)	$10.4 - 16.8$		Vanorio et al, 2003
Kaolinite	powder in epoxy	acoustic (MicroMech)	$57.9 - 85.9$		Wang et al., 2001
Illite			$73.9 - 84.3$		Wang et al., 2001
Smectite/Illite			51.5		Wang et al., 2001
Montmorillonite			$44.7 - 54.5$		Wang et al., 2001
Chlorite			$82.2 - 214.0$		Wang et al., 2001
Pyrophyllite	–	Mol. Dyn.	weak axis	547	Katti et al., 2003
		$\Sigma/(\Delta L/L)$	interlayer	130	Katti et al., 2003

Table 5. Reported elastic stiffness values of clay minerals. For purpose of comparison, values are expressed, if possible, as indentation stiffness.

much smaller than typical values reported previously. Given the inarguably small size of the porosity of clay materials, which is a consequence of the packing density, and our confirmation of the clay crystal stiffness by nanoindentation (*ie.* Eq. (4.41) and Fig. 14), it appears to us very likely that the reported values by Prasad et al. (2002) are those of the porous clay composite rather than the ones of the clay crystals themselves! Clear evidence is provided by nanoindentation on shale materials:

Table 6 reports the indentation results on three types of shale materials of different mineralogy and porosities. The shale materials are labeled 1,2 and 3, 'N' respectively 'M' stands for respectively nanoindentation and microindentation test, and 'xi (i=1,2,3)' stands for the indentation axis. Each series of tests consist of a hundred force driven indentation tests performed on a grid of 50 μm grid size. The maximum indentation depth in the nanoindentation tests is on the order of $150 - 400$ nm, and the derived properties, therefore, should be representative of a bulk material size on the order of $1,000$ nm. This is the scale of the porous clay composite at level I, that includes the inter-grain clay porosity. We remark that the order of magnitude of the average nanoindentation stiffness is in deed on the same order of magnitude as the values reported by Vanorio et al. (2003):

$$M^I_{\text{exp}} = 8 - 14 \text{ GPa} \tag{4.42}$$

Moreover, depending on the specific shale material, there are some anisotropy effects present. Quantified in terms of the indentation stiffness ratio, we find that the anisotropy ratio of the three materials at level I is roughly half the value of the clay minerals:

$$\frac{M^I_{1,\text{exp}}}{M^I_{3,\text{exp}}} = 1.06 - 1.46 - 1.16 \tag{4.43}$$

Three particular experimental observation deserve attention:

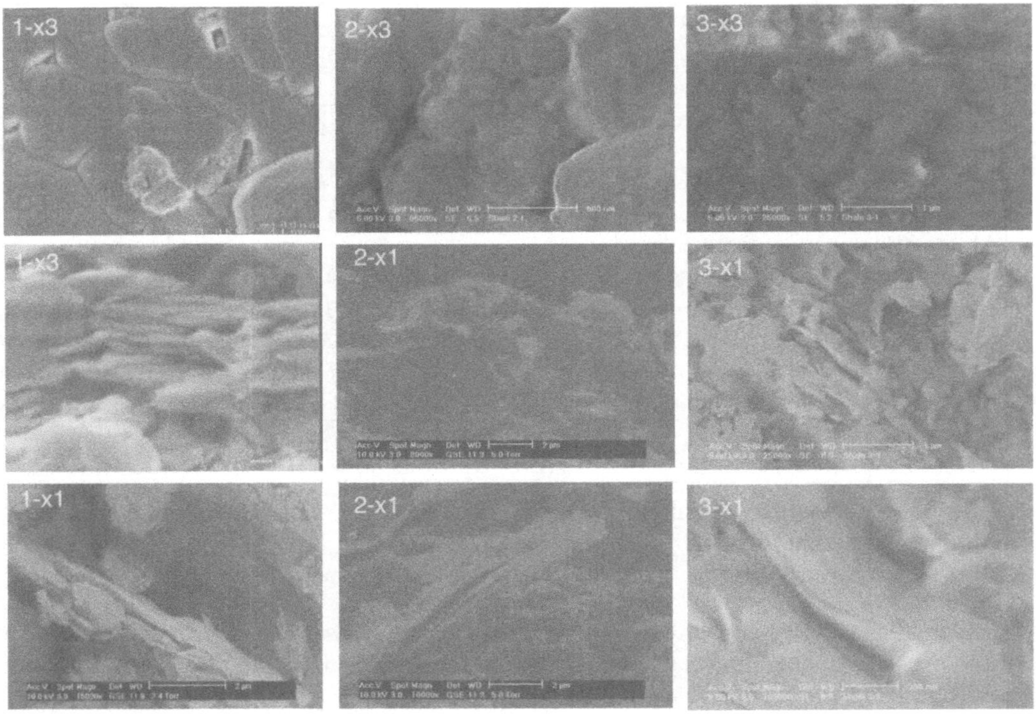

Figure 15. SEM and ESEM images of three shale materials, viewed normal to the deposition direction (x3) and in the deposition plane (x1 and x2).

1. What is sensed in a force driven indentation test on natural composites, is the highly heterogeneous nature of such materials, which translates into a stiffness scaling relation as a result of the BASh equation (2.11), as shown in Figure 16 for the three shale materials. While the mean value of the indentation stiffness shows some anisotropy feature, the continuity of the scaling relation for different indentation directions is an indication that the same material is present –on-average– in different indentation directions.
2. The maps of the nanoindentation stiffness values for different indentation directions (Figure 17) display a highly heterogeneous spatial distribution. There is no visible morphology at the grid size scale of 50 μm that could explain the (small but existing) anisotropy of the average stiffness values obtained at level I.
3. There is some difference in frequency distribution of the stiffness values obtained by grid indentation on different material planes, as shown in Figure 18. Recalling that the same distribution in different indentation directions is a condition for isotropy, we recognize that it is the frequency distribution which is at the origin of the observed anisotropy of shale materials at a nanoscale.

Shale #	Scale (Level)	Name #–N/M–dir	$P_{\max}$ [μN]		$h_{\max}$ [nm]		M [GPa]	
			σ	μ	σ	μ	σ	μ
1	Nano (I)	1-N-x3	262	21	230	61	9.5	3.1
	$\phi^I = 0.31$	1-N-x1	256	18	275	109	10.1	5.4
	Micro (II)	1-M-x3	12,348	111	2,183	618	6.9	2.5
	$\phi^{II} = 0.31$	1-M-x1	12,363	98	2,252	533	9.5	2.7
		1-M-x2	12,332	96	2,310	593	9.4	3.0
	Macro (III)	1-UPV-x3	–	–	–	–	10.9	NaN
	$\phi^{III} = 0.26$	1-UPV-x1	–	–	–	–	14.7	NaN
2	Nano (I)	2-N-x3(1)	257	17	270	100	9.7	4.3
	$\phi^I = 0.18$	2-N-x3(2)	251	18	302	107	8.0	4.3
		2-N-x1	265	12	219	70	14.2	5.5
	Micro (II)	2-M-x3	12,305	70	1,947	396	8.0	1.9
	$\phi^{II} = 0.18$	2-M-x1	12,206	31	1,414	194	15.0	2.8
	Macro (III)	2-UPV-x3	–	–	–	–	13.8	NaN
	$\phi^{III} = 0.13$	2-UPV-x1	–	–	–	–	22.9	NaN
3	Nano (I)	3-N-x3	251	18	306	107	10.2	5.1
	$\phi^I = 0.10$	3-N-x1	260	12	254	73	11.8	3.9
	Micro (II)	3-M-x3	12,255	80	1,634	452	10.1	2.5
	$\phi^{II} = 0.10$	3-M-x1	12,268	107	1,735	607	14.9	4.0
	Macro (III)	3-UPV-x3	–	–	–	–	21.0	NaN
	$\phi^{III} = 0.075$	3-UPV-x1	–	–	–	–	31.7	NaN

Table 6. Overview of experimental results.

Bringing these three observations together provides evidence for the origin of the anisotropy at level I: Given the continuity of the stiffness scaling relation and the absence of any specific morphology in the stiffness maps, the origin of this anisotropy cannot be attributed neither to the anisotropy of different clay minerals, nor to the difference of porosity (see Table 6); but rather to some privileged orientation of clay minerals at level I, as a consequence of the deposition history of each shale material. Indeed, while almost absence in shale 1 ($\phi^I = 0.31$), and rather weak in shale 3 ($\phi^I = 0.10$), the nanoscale shale anisotropy is quite strong for shale 2 ($\phi^I = 0.18$).

Level 'II': Textured Clay Layer Composite The porous clay composite from level I forms a layered textured matrix at level II that is representative of shales at a scale of tens of micrometers. This microfabric signature of shale materials, which is most likely a consequence of the deposition history is clearly visible in SEM images of 100 to 1000 times magnification, as shown in Figure 19 for the three shale materials. Some trendlines have been superimposed on the microphotographs to highlight the layered texture at the considered scale, as well as the perturbation of the layered structure through the presence of silt inclusions of micrometer size. The smaller the magnification the more regular and homogeneous the visible layered structure, which suggests that the shale fabric becomes

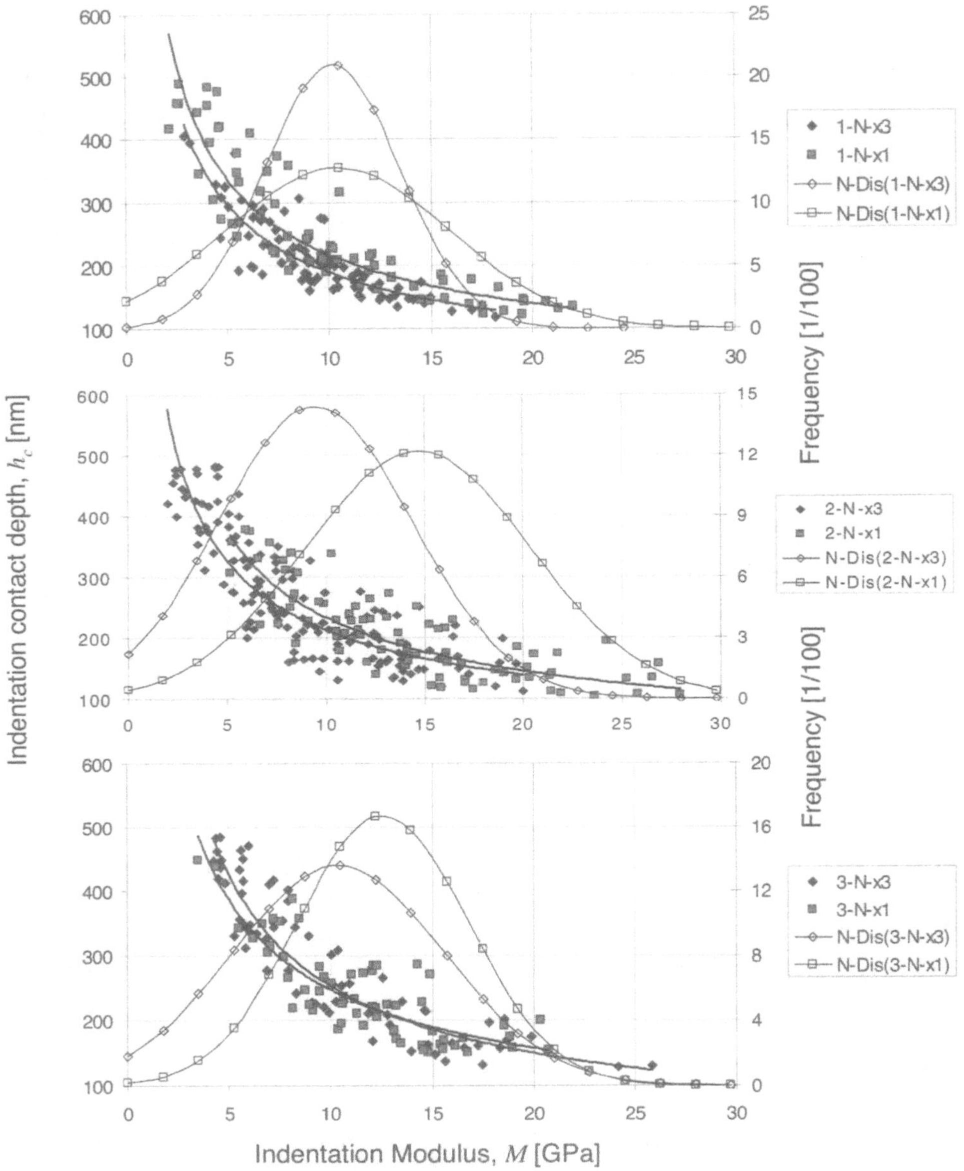

Figure 16. Nanoindentation stiffness scaling relation.

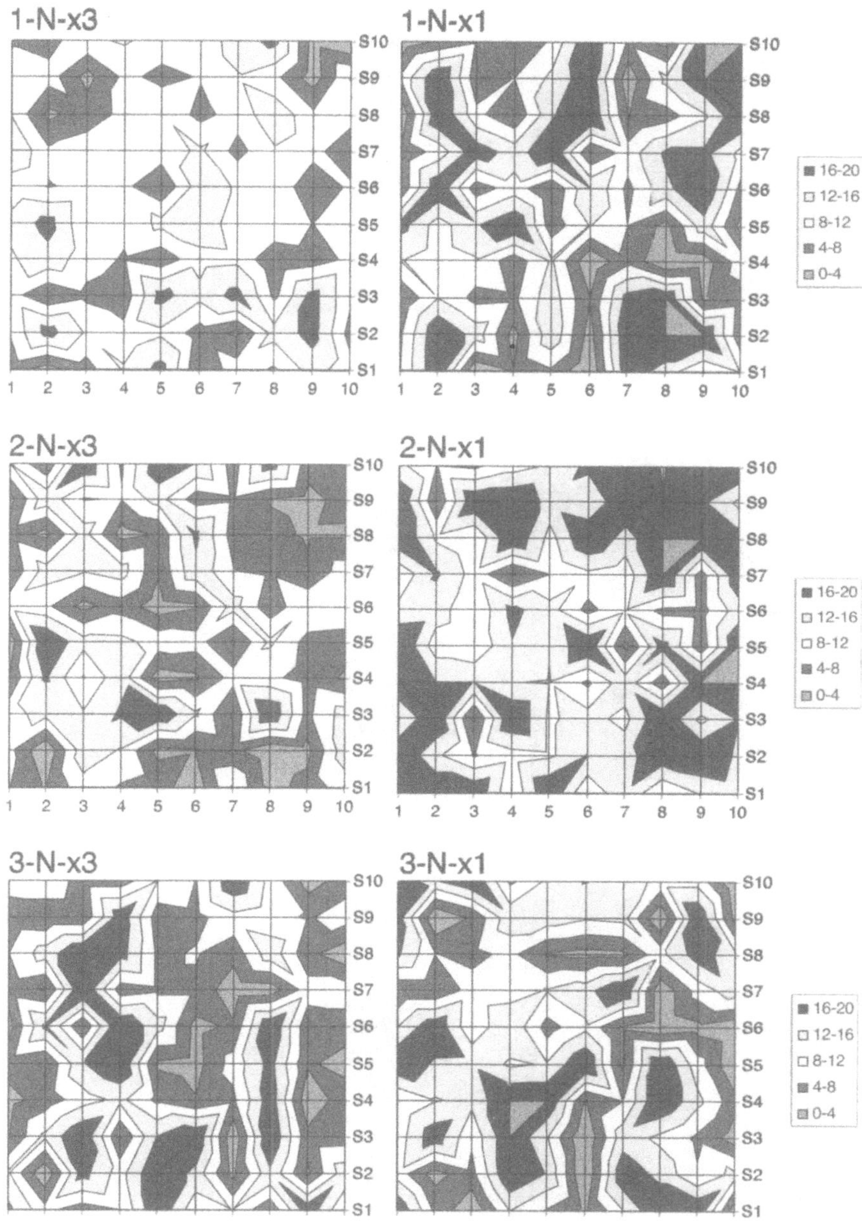

Figure 17. Nanoindentation stiffness maps.

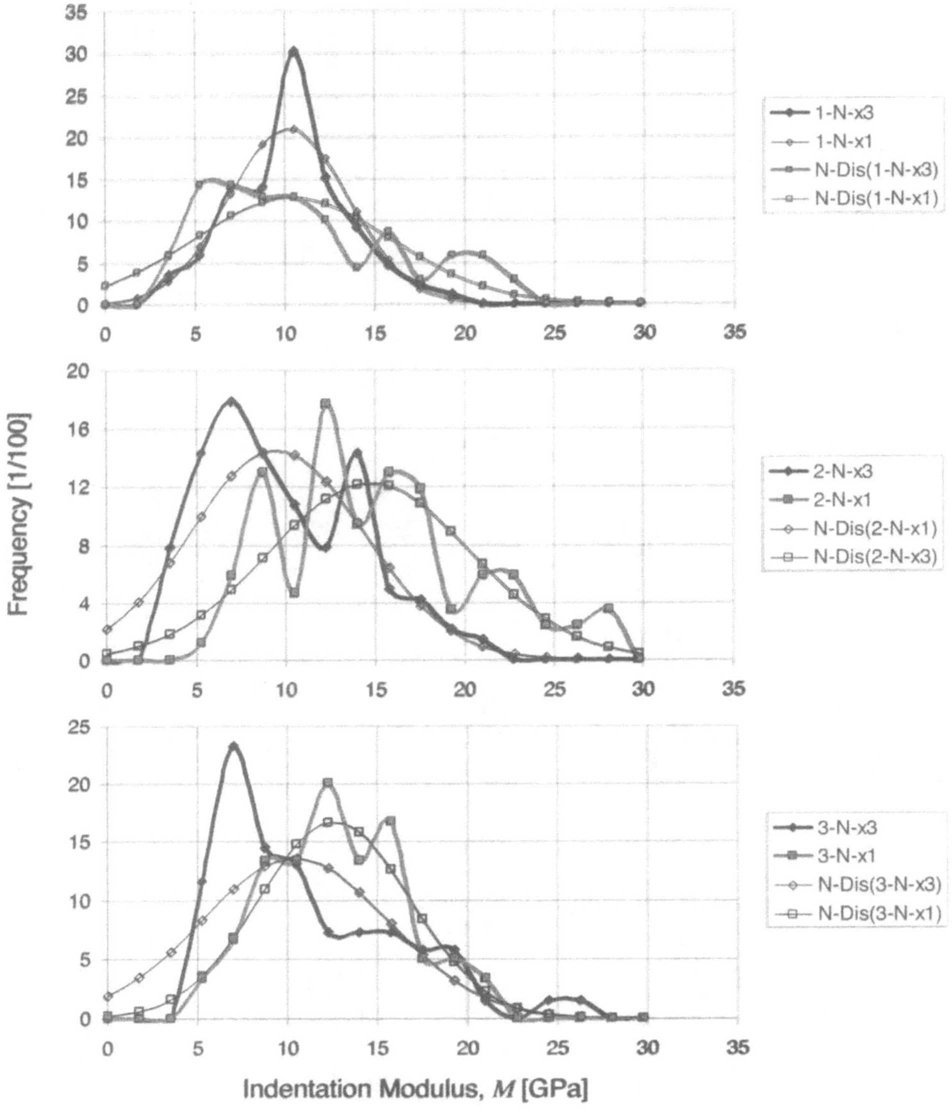

Figure 18. Nanoindentation stiffness frequency distribution.

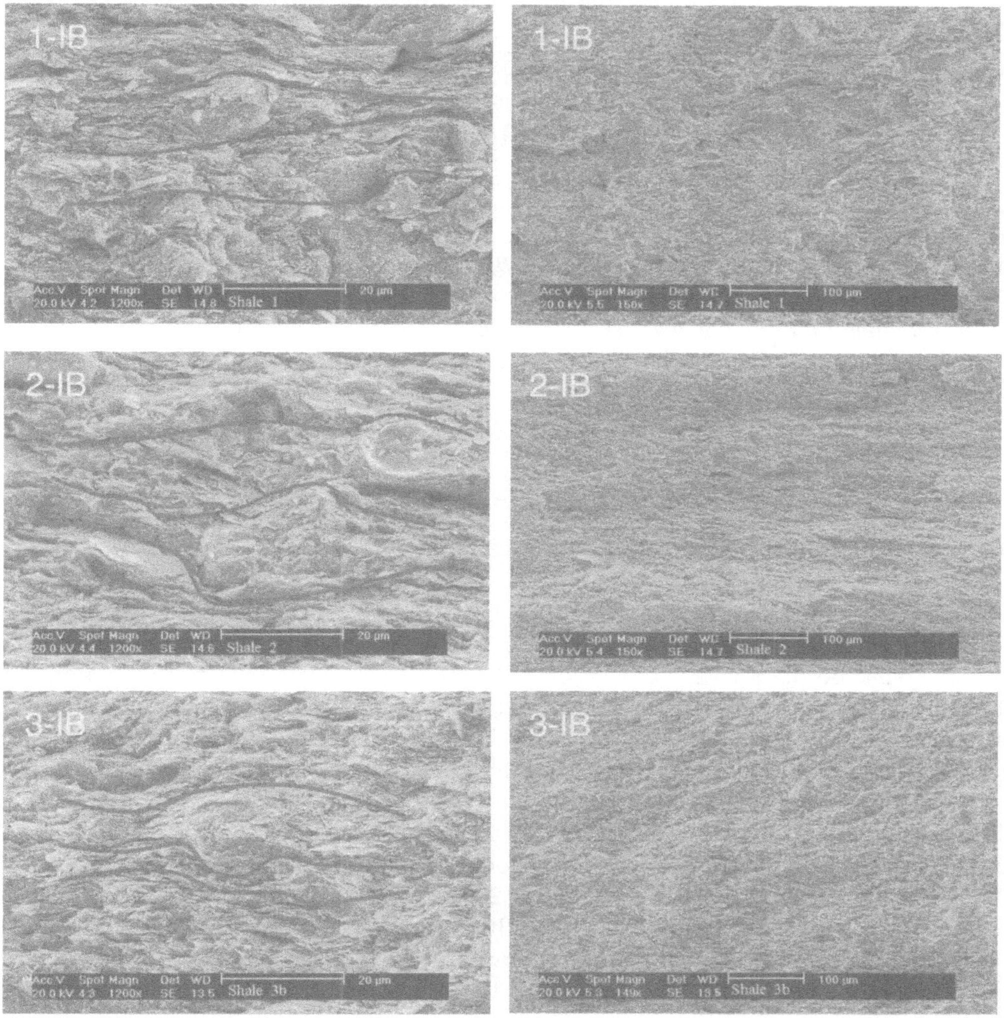

Figure 19. SEM Images of shales at level II (from Chevron Texaco).

more and more dominant at larger than at smaller scales. That what changes from Level I to Level II is the texture.

A clear indication of the effect of texture are *micro*indentation results obtained with a maximum force that is roughly 50 times greater than the one employed in the series of *nano*indentation tests (see Table 6). The resulting maximum indentation depth in the nanoindentation tests is on the order of $1,000 - 5,000\,\text{nm}$, *ie.* roughly 10 times greater than nanoindentation depths. The derived properties, therefore, should be representative of shale materials on the order of tens of micrometer. The results which are summarized in Table 6, show a clear trend that the stiffness increases as a function of the porosity, and this both normal to bedding and in bedding direction:

$$M^{II}_{3,\text{exp}} = 6.9\,(6.4) - 8.0\,(8.0) - 10.1\,(9.9)\ \text{GPa} \quad (4.44a)$$

$$M^{II}_{1,\text{exp}} = 9.5\,(9.6) - 14.2\,(15.3) - 14.9\,(16.2)\ \text{GPa} \quad (4.44b)$$

In contrast to the nanoindentation results, there is a clear separation of the microindentation stiffness scaling relation (Figure 20), and a clear one-peak distribution of the stiffness values (Figure 21), that are both indicative of the (micro)mechanical anisotropic nature of shale materials at level II. The peak values (which are given in parenthesis in (4.44)) are used to calculate the anisotropy ratio:

$$\frac{M^{II}_{1,\text{exp}}}{M^{II}_{3,\text{exp}}} = 1.5 - 1.9 - 1.6 \quad (4.45)$$

A comparison of (4.43) and (4.45) shows the added anisotropy due to the layered structure, exceeding the anisotropy present at smaller scales.

Level 'III': Macroscopic Clay–Quartz Inclusion Composite The textured clay composite of level II together with quartz inclusions forms a macroscopic composite material, that becomes visible at the sub-millimeter to millimeter scale. The material at this scale is composed of a textured clay matrix with an in-general abundant population of poorly sorted detrital grains (quartz inclusions), that are either concentrated into laminations located between thinner, clay-rich (or quartz starved), lens shaped lams, or homogeneously distributed throughout. This scale has been extensively researched, in the shale acoustics and exploration geophysics community, by means of compressional and shear-wave velocity measurements (Thomsen, 1986; Brittan et al., 1995; Tsvankin, 1996; Hornby, 1998; Sayers, 1999; Wang, 2002; Pratson et al., 2003), and it is now generally agreed that shales behave elastically as transverse isotropic media. The typical wave length employed in such studies operated at frequencies in the MHz range, is in the millimeter range, which captures well the composite stiffness of the macroscopic clay-quartz inclusion composite.

Table 6 provides the values of the equivalent indentation moduli derived from UPV measurements of the three shale materials. It should be noted that UPV-measurements carried out on saturated shale materials are undrained stiffness values, which may explain some hydraulic stiffening effects, in addition to stiffening induced by the presence of quartz inclusions. For pure purpose of comparison, we calculate the anisotropy ratio:

$$\frac{M^{III}_{1,\text{exp}}}{M^{III}_{3,\text{exp}}} = 1.4 - 1.7 - 1.5 \quad (4.46)$$

The fact that the anisotropy of the macroscopic composite is slightly smaller than the one of the textured clay composite (level II; (4.45)) is not surprising: mixing a (more-or-less) isotropic inclusions into an anisotropic material system reduces the anisotropy.

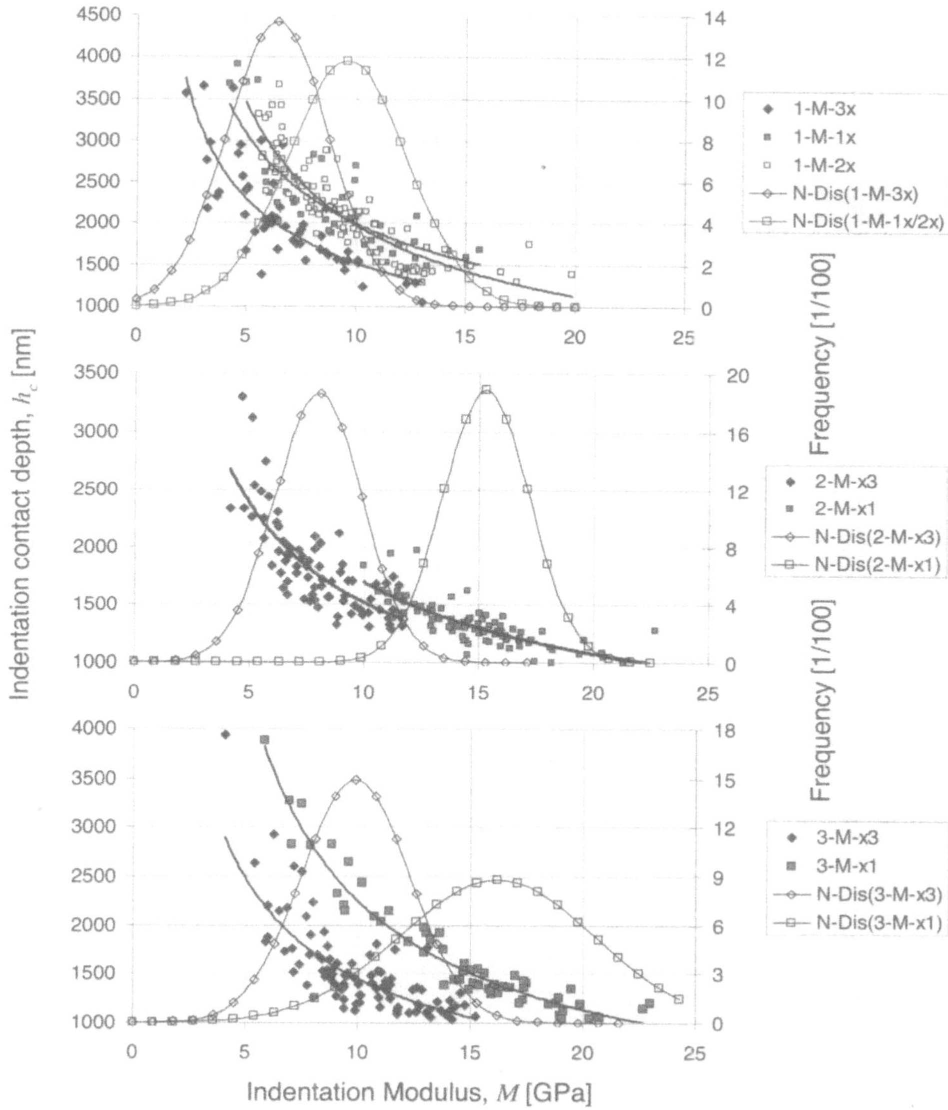

Figure 20. Microindentation scaling relation.

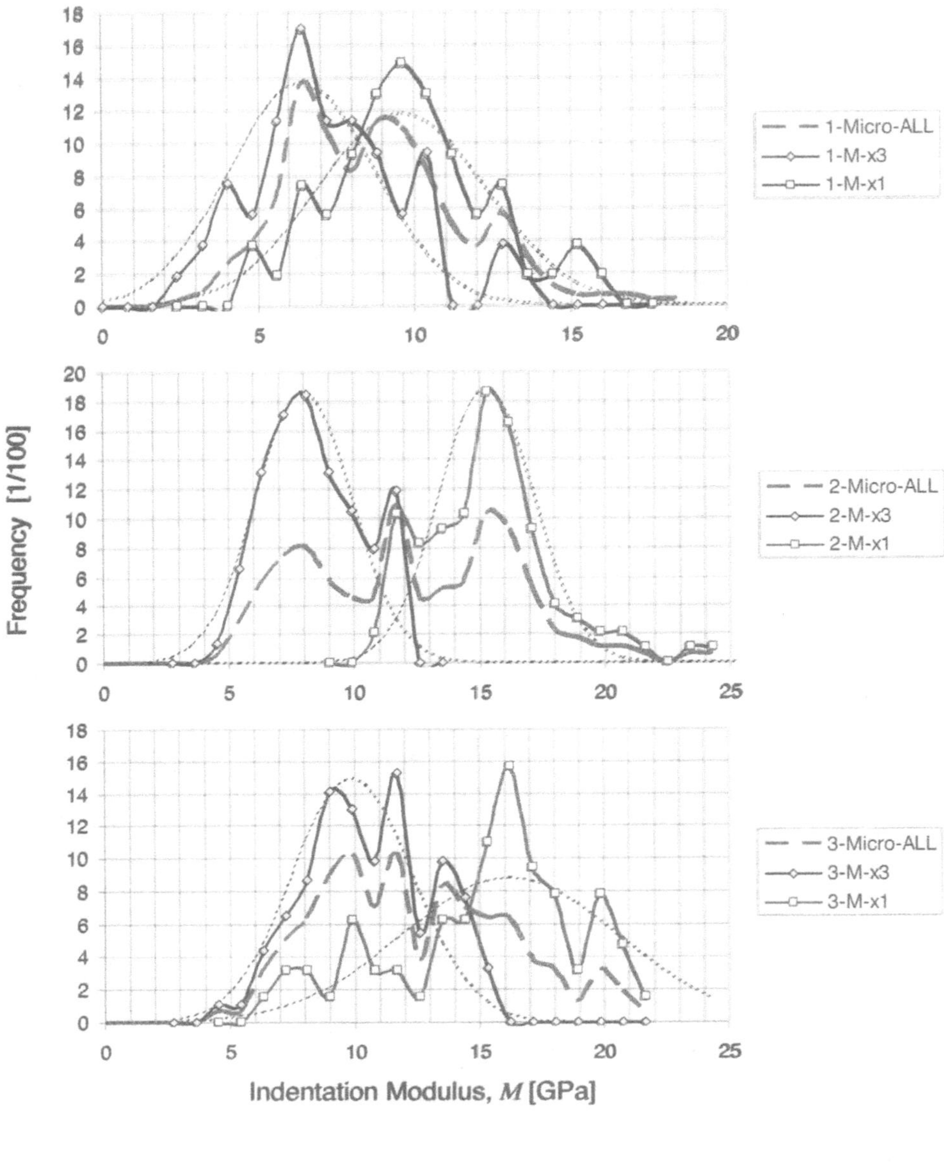

Figure 21. Microindentation frequency plot.

4.2 Anisotropic Microporoelasticity of Shale Materials

The overall trend that emerges from the multiscale indentation analysis is that the anisotropy of shale materials increases from smaller to larger scales.

Sources of Anisotropy Representation Our first objective is to analyze indentation results by means of a microporomechanics modeling approach. As an experimental base, we have six experimental indentation stiffness values for each shale material: $M^{I}_{1,\exp}, M^{I}_{3,\exp}, M^{II}_{1,\exp}, M^{II}_{3,\exp}, M^{III}_{1,\exp}, M^{III}_{3,\exp}$ (see Table 6). Using (2.44) and (2.52), each of these indentation stiffness values provides a snapshot of the elasticity of the stiffness values $C^J_{1111}, C^J_{3333}, C^J_{1122}, C^J_{1133}, C^J_{2323}$ (or $C^J_{11}, C^J_{33}, C^J_{12}, C^J_{13}, C^J_{44}$) at each considered scale $J = I, II, III$. Since there is *a priori* no reason that the five elasticity constants at different scales should coincide, there is a total of 6 knowns for a total of 15 unknowns. This imbalance highlights the necessity for a micromechanics approach, requiring at each scale three further information to complete the stiffness characterization, that cannot rely on an empirical choice only. The micromechanics model aims at answering the following questions:

1. Level '0': *In the light of the decreasing anisotropy at smaller scales, is it possible that the elementary building block of shale materials were isotropic?* – Given the anisotropy of the clay minerals it is unlikely that shales at the nanoscale are mechanical isotropic materials. However, the stiffness values of clay minerals (Table 5) are (at least) one order of magnitude greater than the nanoindentation stiffness values (Table 6). This observation suggests that it is not the mineral elasticity itself, but rather the contact regions in between minerals, *ie.* the inter-mineral porosity, that governs the elasticity (See Figure 22, top right). In fact, deformation in the crystalline structure vis-à-vis deformation at the interface between clay particles involve much higher forces, energies and stiffness. The simplest continuous representation of such a discrete material system comes to replace a discrete assembly of grains by a continuous solid phase intermixed with some porosity (Fig. 22, top left). This assumption (and it is one!) provides a convenient way to localize the total elastic deformation in a conceptual continuous pure clay phase, whereas it physically occurs both in very stiff clay flakes and in the contact regions between them. The two possible sources of anisotropy are then the anisotropy of the solid phase and the pore morphology. Given the trend that clay minerals appear in packages, forming the surrounding solid of the porosity, it is convenient to attribute the entire anisotropy at the scale of the elementary building block of shale materials, to the pore morphology by assuming the pure clay phase isotropic. This reduces the number of unknowns to three: two elastic constants and the pore aspect ratio. More specifically, since the shape of clay particles can be approached by the one of flat ellipsoids, we will assume that the pores have the same morphology: an oblate spheroid with a varying aspect ratio $\rho = h/D$, where h is the thickness and D the diameter of the pore (Figure 22, bottom left). Hornby et al. (1994) choose a unique value for the pore aspect ratio in his model, 1/20, equivalent to the one of clay particles. The pore aspect ratio depends more generally on other parameters as well, like pore pressure and depth (Jacobsen et al., 2003). Moreover, if we con-

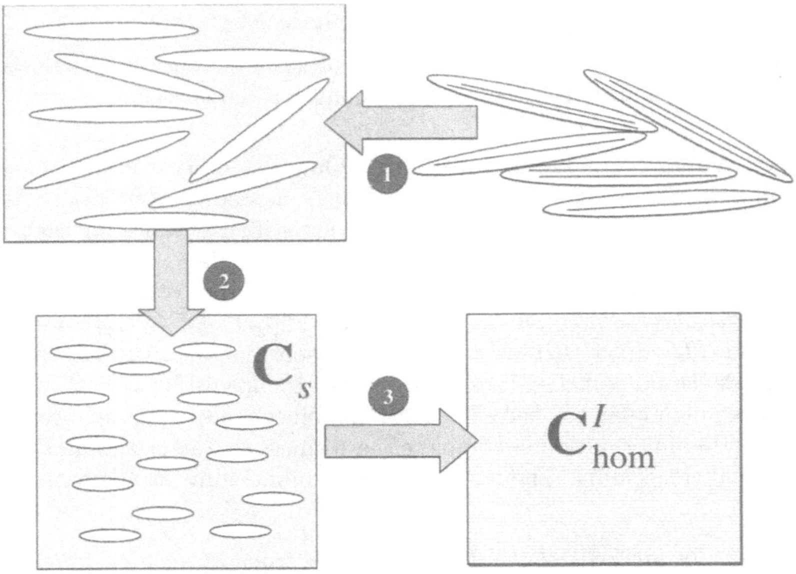

Figure 22. Three steps to model the shale elementary building block from level 0 (top right) to level 1 (bottom right): The first and second ones show the continuous and isotropic representation of clay particles (1) and simplified manifestation of porosity (2); the third one regards it as a (micro)homogeneous phase.

sider that the pores are the sole source of the shale anisotropy at level '0' defined by a unique value of ρ, one should expect a significant decrease in macroscopic anisotropy with decreasing porosity, other parameters being equal.

2. Level 'I': *If the shale building block is clearly anisotropic, why did nanoindentation operated to nanoindentation depths of 100 and 400 nm which activate approximately a bulk comparable to the shale building block, give very close average values when performed in bedding or normal to bedding planes?* – The nanoindentation results show a fair amount of randomness of the distribution of the stiffness in the microstructure. However, since the stiffness distributions in different indentation directions do not coincide (see Figure 16), the randomness of the extracted properties should not be attributed to randomness in a mechanical sense, *ie.* isotropy. Instead, it appears as a consequence of the randomness of the indentation locus defined by the grid size that is much larger than both the indentation depth and the distance in between clay particles. In other words, the elementary building blocks –and the associated anisotropy– manifest themselves at a scale that is much smaller than the indentation grid size. As a consequence the statistical averages of nanoindentation stiffness values are not averages in the sense of a mechanical homogenization; but reflect the orientations of shale building blocks encountered in both in-bedding and normal-to-bedding tests as shown on Figure 23. For instance, if the

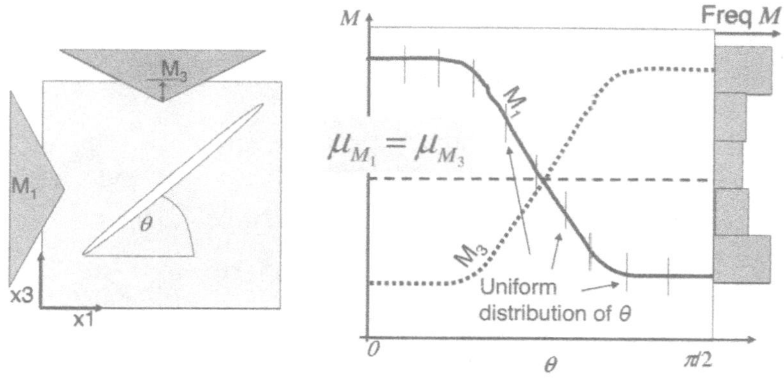

Figure 23. Nanoindentation on a shale elementary building block, level I (left). The indentation stiffness depends on the orientation of the pores, ie of the building block. In the remarkable case where all orientations are uniformly distributed in the shale (right), the average indentation modulus of a grid of tests does not depend on the indented surface orientation. The stiffness frequency exhibits an interesting two-peak distribution.

pore orientation of building blocks has a uniform orientation distribution, then the statistical average of individual stiffness values would be the same for in-bedding and the normal-to-bedding indentations, although the stiffness distribution in each indentation direction contains a large range of values (Fig. 23). The highest (resp. lowest) expected stiffness is obtained when the pore normal direction is perpendicular (resp. parallel) to the axis of indentation. Any difference in between the two averages can be explained by a non-uniform orientation distribution function (ODF) of pores, flakes or shale building blocks. This non-uniform orientation distribution is a third model parameter capturing the disorder of shale materials at level I.

3. Level 'II': *How does the pore morphology from the building block translates into the pronounced anisotropy of the layered shale structure?* – The microindentation results display a (more-or-less) clear one-peak distribution, so that values obtained at this scale can be associated with averages in a mechanical homogenization sense, averaging stiffness values of shale building blocks over a large volume. At this scale, the anisotropy increases for all shales which suggests that an additional morphological feature appears. Porosity, pore shape and orientation distribution functions are already determining the anisotropy in the statistical analysis of nanoindentation. The only possible reason that remains is a particular spatial distribution of shale building blocks. From SEM pictures at that scale, the orientation of the flakes is locally variable, but some wavy separation lines in the bedding directions indicate a clear privileged direction (Figure 19). This can be translated in a schematic laminate of shale building blocks oriented in the bedding directions, where each layer corresponds to one particular orientation of building blocks, as Figure

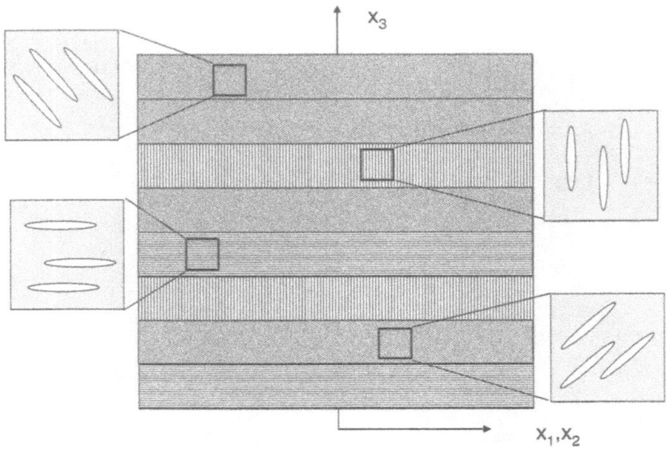

Figure 24. The schematic layered model for the shale microfabric. Each layer is defined by a unique orientation of pores, ie shale building blocks. The resulting laminate is aligned along the shale bedding planes.

24 shows. Similar to a brick wall composed of bricks and joints, such a representation suggests that horizontal pores (joints) are connected whereas vertical ones are interrupted, so that the pronounced anisotropy of shales at larger scales appears as a consequence of the connectivity of pores, with lower stiffness values in the direction normal to bedding (axis $x3$) than in-bedding (axis $x1$ and $x2$).

The suggested microporomechanics representation of shale materials is no-doubt reductionist, reducing the complexity of shales to an isotropic pure clay phase, an oriented pore space and an orientation distribution function. The model has a total of four model parameters, if we exclude the two volume fractions, porosity and inclusion, which are generally known from independent measurements. The porosity at level I and II is the same, while the porosity at level III is related to the former by:

$$\phi_0^{III} = (1 - f_I)\, \phi_0^{II} \tag{4.47}$$

Given this reduced number of model input parameters, we use the four indentation stiffness values $M^I_{1,\text{exp}}, M^I_{3,\text{exp}}, M^{II}_{1,\text{exp}}, M^{II}_{3,\text{exp}}$ as a first data set for a calibration of the model. Using the calibrated model parameters, we then predict the macroscopic stiffness, and compare these predictions with the stiffness values from a second independent data set obtained by UPV-measurements.

Level '0': Shale Building Block We assume that the shale building bock is composed of a solid phase and a porosity, for which the classical relations of anisotropic poroelasticity apply (Coussy, 2004):

$$\Sigma = \mathbb{C}^{\text{hom}} : \mathbf{E} - \mathbf{b}p \tag{4.48a}$$

$$\phi - \phi_0 = \mathbf{b} : \mathbf{E} + \frac{p}{N} \tag{4.48b}$$

where $\mathbb{C}_{\text{hom}}$ is the stiffness tensor, $\mathbf{b}$ the 2nd order tensor of Biot coefficients, and N the solid Biot modulus. The homogenization rules for this classical poroelastic material system are well established by now (Chateau and Dormieux, 2002; Dormieux and Bourgeois, 2003):

$$\mathbb{C}^I_{\text{hom}} = (1 - \phi_0)\, \mathbb{C}_s : \langle \mathbb{A}\,(\underline{z}) \rangle_{V_s} \tag{4.49a}$$

$$\mathbf{b}^I = \boldsymbol{\delta} : \phi_0 \langle \mathbb{A}\,(\underline{z}) \rangle_{V_\phi} = \boldsymbol{\delta} : \left(\mathbb{I} - \mathbb{C}_s^{-1} : \mathbb{C}^I_{\text{hom}}\right) \tag{4.49b}$$

$$\frac{1}{N^I} = \boldsymbol{\delta} : \mathbb{C}_s^{-1} : (\mathbf{b} - \phi_0 \boldsymbol{\delta}) \tag{4.49c}$$

where $\mathbb{C}_s$ is the stiffness tensor of the (assumed) isotropic clay solid (bulk modulus k_s, shear modulus g_s):

$$\mathbb{C}_s = 3k_s\mathbb{J} + 2g_s\mathbb{K} \tag{4.50}$$

where we recall that $J_{ijkl}=1/3\ (\delta_{ij}\delta_{kl})$ and $\mathbb{K} = \mathbb{I} - \mathbb{J}$; $\mathbb{I}$ is the fourth order identity tensor $I_{ijkl} = \frac{1}{2}(\delta_{ik}\delta_{jl} + \delta_{il}\delta_{jk})$, and δ_{ij} is the Kronecker delta. Furthermore, $\mathbb{A}\,(\underline{z})$ is a forth-order strain localization tensor, which concentrates the macroscopic strain tensor prescribed at the boundary into the microscopic solid phase. For the solid-pore morphology, a good estimate of the volume averages $\langle \mathbb{A}\,(\underline{z}) \rangle_{V_s}$ and $\langle \mathbb{A}\,(\underline{z}) \rangle_{V_\phi}$ is provided by the Mori-Tanaka scheme (see e.g. Zaoui (2002)):

$$\langle \mathbb{A}\,(\underline{z}) \rangle_{V_s} \simeq \mathbb{A}_s^{est} = \left[(1-\phi_0)\mathbb{I} + \phi_0 \left(\mathbb{I} - \mathbb{S}^s_\rho\right)^{-1}\right]^{-1} \tag{4.51}$$

$$\langle \mathbb{A}\,(\underline{z}) \rangle_{V_\phi} \simeq \mathbb{A}_\phi^{est} = \left(\mathbb{I} - \mathbb{S}^s_\rho\right)^{-1} : \left[(1-\phi_0)\mathbb{I} + \phi_0 \left(\mathbb{I} - \mathbb{S}^s_\rho\right)^{-1}\right]^{-1} \tag{4.52}$$

where $\mathbb{S}^s_\rho$ is the Eshelby tensor (Eshelby, 1957) that depends on the aspect ratio $\rho = h/D$ of the pores and the Poisson's ratio of the matrix (see Appendix). Using the estimate (4.51) in (4.49a) yields the homogenized stiffness of a transversely isotropic material in the principle material axis, defined by the pore plane that extends in the $x1' \times x2'$ plane, and whose normal is oriented in direction $x3'$. In this base, the components of the stiffness tensor are of the form (2.40), and the tensor of Biot coefficients is a diagonal tensor having as components:

$$b_{ij} = \begin{bmatrix} b_{11} & 0 & 0 \\ & b_{11} & 0 \\ & & b_{33} \end{bmatrix} \tag{4.53}$$

where b_{11} and b_{33} represent the Biot coefficients in the horizontal and vertical direction in the shale building block:

$$b_{11} = \phi_0^I \left[A^{est}_{\phi,1111} + A^{est}_{\phi,1122} + A^{est}_{\phi,1133}\right] \tag{4.54a}$$

$$b_{33} = \phi_0^I \left[2A^{est}_{\phi,1133} + A^{est}_{\phi,3333}\right] \tag{4.54b}$$

It is convenient to express the homogenized stiffness of the shale building block in a coordinate system $(x1, x2, x3)$, in which the pore plane's normal orientation is defined by the Euler angles θ and φ (See Figure 25):

$$\underline{n} = \sin\theta\cos\varphi\,\underline{e}_1 + \sin\theta\sin\varphi\,\underline{e}_2 + \cos\theta\,\underline{e}_3 \tag{4.55}$$

$\theta = \varphi = 0$ corresponds to a pore plane oriented by the normal pointing in direction $x3$; and $\theta = \pi/2, \varphi = 0$ and $\theta = 0, \varphi = 0$ corresponds to pore planes oriented by a unit normal in direction $x1$ and $x2$, respectively. The relations between the second order strain and stress tensors in the reference and rotated configurations are given by:

$$\mathbf{E}_{\theta,\varphi} = \mathbf{R}(\theta,\varphi)\cdot\mathbf{E}_0\cdot\mathbf{R}(\theta,\varphi)^T \tag{4.56a}$$

$$\mathbf{\Sigma}_{\theta,\varphi} = \mathbf{R}(\theta,\varphi)\cdot\mathbf{\Sigma}_0\cdot\mathbf{R}(\theta,\varphi)^T \tag{4.56b}$$

where $\mathbf{R}(\theta,\varphi)$ is the rotation matrix:

$$R_{ij}(\theta,\varphi) = \begin{bmatrix} 1 & 0 & 0 \\ 0 & \cos(\theta) & -\sin(\theta) \\ 0 & \sin(\theta) & \cos(\theta) \end{bmatrix} \begin{bmatrix} \cos(\varphi) & -\sin(\varphi) & 0 \\ \sin(\varphi) & \cos(\varphi) & 0 \\ 0 & 0 & 1 \end{bmatrix} \tag{4.57}$$

Using (4.56) in (4.48), the poroelastic constants read in the rotated configuration:

$$\mathbb{C}^I_{\text{hom}}(\theta,\varphi) = \mathbf{R}(\theta,\varphi)\cdot\mathbb{C}^I_{\text{hom}}(0,0)\cdot\mathbf{R}(\theta,\varphi)^T \tag{4.58a}$$

$$\mathbf{b}^I(\theta,\varphi) = \mathbf{R}(\theta,\varphi)\cdot\mathbf{b}^I(0,0)\cdot\mathbf{R}(\theta,\varphi)^T \tag{4.58b}$$

Level 'I': Statistical Averages We consider that the mean nanoindentation results represent statistical averages of a representative sample of indentation tests on elementary building blocks present on a specific surface (Fig. 17). The statistical average of the indentation tests on this surface oriented in the xJ–direction is:

$$\mu_{M_J} = \frac{1}{2\pi}\int_{\theta=0}^{\pi}\int_{\varphi=0}^{2\pi} M_J(\theta,\varphi)\sin\theta\,d\varphi\,d\theta \tag{4.59}$$

where $M_J(\theta,\varphi)$ is the indentation modulus of a single indentation test on an elementary building block rotated by Euler angles θ and φ w.r.t. the indentation axis xJ. Assuming the function to be continuous, the indentation modulus has two extrema: a minimum for indentation normal to the pore plane, *ie.* $\underline{n}\cdot\underline{e}_J = 1$, and a maximum for indentation in the pore plane, *ie.* $\underline{n}\cdot\underline{e}_J = 0$:

$$\forall\underline{n}\cdot\underline{e}_J = 1 \Rightarrow M_J(\theta,\varphi) = \min M \tag{4.60a}$$

$$\forall\underline{n}\cdot\underline{e}_J = 0 \Rightarrow M_J(\theta,\varphi) = \max M \tag{4.60b}$$

Combining (4.55) with (4.60), we recognize for indentation in bedding directions (axis $x1$ and $x2$) and normal to it (axis $x3$):

$$\begin{array}{llll} \min M & = M_1(\pi/2,0) & = M_2(\pi/2,\pi/2) & = M_3(0,\varphi) \\ \max M & = M_1(0,\varphi) & = M_2(0,\varphi) & = M_3(\pi/2,\varphi) \\ & = M_1(\theta,\pi/2) & = M_2(\theta,0) & \end{array} \tag{4.61}$$

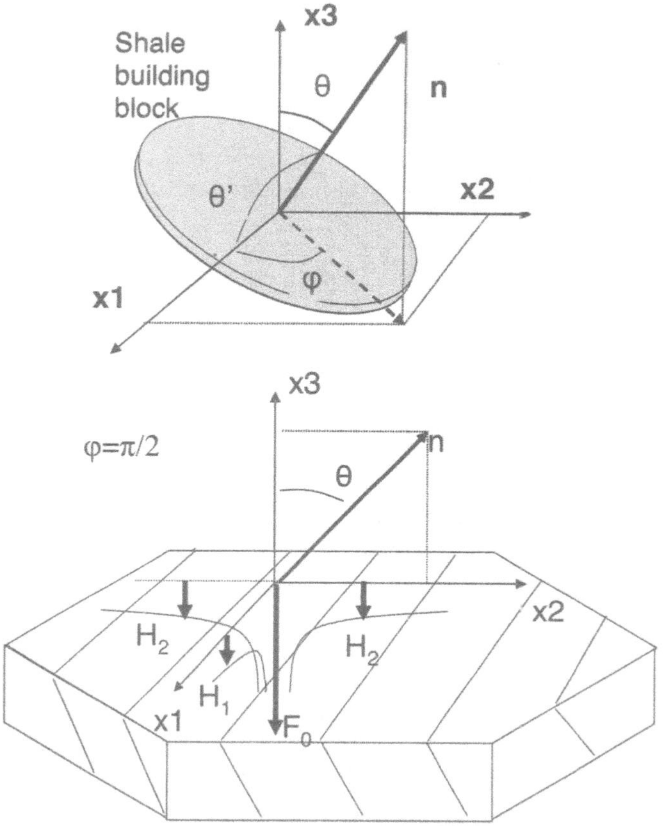

Figure 25. (top) The orientation of a transverse isotropic shale elementary building block is defined by the direction of its axis of symmetry, with the two Euler angles θ and φ in the cartesian system $Ox_1x_2x_3$; (bottom) Schematic representation of the surface Green function extreme profiles, when the axes of material symmetry and indentation (x_3) form any angle θ. Without loss of generality we consider here $\varphi = \pi/2$, and the extremal values of $\mathcal{H}$, H_1 and H_2, are obtained in the $x_1$and x_2 directions.

The minimum value $\min M$ is strictly the indentation modulus for indentation in the symmetry axis of a transversely isotropic material, which is related to the elastic constants $\mathbb{C}_{\text{hom}}(0,0)$ through (2.44). Similarly, the maximum value $\max M$ is the indentation modulus for indentation in the plane of symmetry of a transversely isotropic material, which is related to $\mathbb{C}_{\text{hom}}(0,0)$ through (2.52). For any other orientation of the pore plane defined by $\cos\theta'(\theta,\varphi) = \underline{n}\cdot\underline{e}_J$, the link between the indentation modulus $M_J(\theta')$ and the elasticity of the building block $\mathbb{C}_{\text{hom}}(0,0)$ is estimated from a generalization of (2.52):

$$M_J(\theta') \simeq \frac{1}{\pi\sqrt{\mathcal{H}_K\mathcal{H}_L}} \tag{4.62}$$

satisfying:

$$M_J(\theta'=0) = \min M \tag{4.63a}$$

$$M_J(\theta'=\pi/2) = \max M \tag{4.63b}$$

where $\mathcal{H}_K = \mathcal{H}(\theta'=0)$ and $\mathcal{H}_L = (\theta'=\pi/2)$ are the extreme values of the Green's function angular part $\eta(\theta')$ given by (2.35)[11] and are represented schematically on Figure 25. However, in contrast to the closed form expressions (2.44) and (2.52), these extreme values cannot be expressed in closed form for $\theta' =]0,\pi/2[$, and need to be numerically evaluated.

Relation (4.62) contains not only the elastic stiffness information of the elementary building block, but as well the distribution of indentation moduli as a function of the pore plane orientation $\cos\theta' = \underline{n}\cdot\underline{e}_J$ situated in between the two extrema, $\min M$ and $\max M$. This distribution corresponds to a uniform orientation distribution of building blocks (*ie.* pore orientations) on a specific surface. It implies that shale building blocks have their axis of symmetry pointing in all directions with uniform probability density. A refinement of this theory can be achieved by considering an orientation distribution function $W(\theta,\varphi)$ in the evaluation of the statistical average (4.59):

$$\mu_{M_J} = \frac{1}{2\pi}\int_{\theta=0}^{\pi}\int_{\varphi=0}^{2\pi} M_J(\theta,\varphi)\,W(\theta,\varphi)\sin\theta\,d\varphi\,d\theta \tag{4.64}$$

This orientation distribution function needs to satisfy:

$$\frac{1}{2\pi}\int_{\theta=0}^{\pi}\int_{\varphi=0}^{2\pi} W(\theta,\varphi)\sin\theta\,d\varphi d\theta = 1 \tag{4.65}$$

For shale materials, orientation distribution functions were first introduced by Sayers (1994), and have gained some popularity as an important tool to account for a non-uniform distribution of the orientation of clay particles, that can be determined experimentally (see e.g. Johansen et al. (2004)). We employ this concept here to account for a

[11] This interpolation of the extreme values of the Green's function angular part $\eta(\theta')$ is justified by the fact that the half-space Green function, π-periodic by definition, conserves its eveness, because of a material plane symmetry.

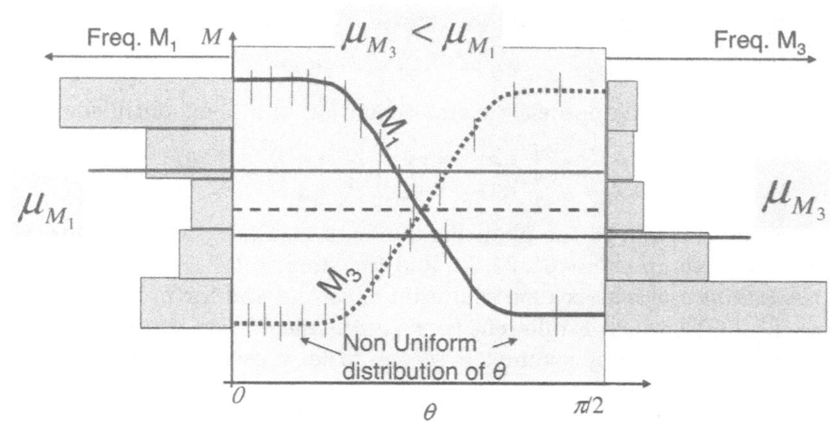

Figure 26. Average and distribution of indentation moduli in the two indented direction at the nano-scale. The anisotropy depends on the anisotropy of the shale building blocks, and on the orientation distribution function and the k factor.

non-uniform orientation distribution of building blocks; and chose a π-periodic Gaussian like function:

$$W(\theta, k) = w_k \cosh(k \cos\theta) \tag{4.66}$$

where $w_k = 2/k \cosh k$ is a normalization constant, and k a parameter referring to the degree of preferred alignment: $k = 0$ means an absence of any preferred alignment, and $k > 0$ induces a second source of anisotropy in the shale micromechanics model.

The resulting statistical average (4.64) illustrated on Figure 26 depends on the elastic constants of the isotropic solid clay phase –bulk modulus k_s, shear modulus g_s (see (4.50))–, the pore aspect ratio $\rho = h/D$ and the $k-$ factor; in addition to the clay porosity ϕ^I; and provides a direct link between the 4 + 1 model parameters and the experimental mean nanoindentation stiffness values $M^I_{1,\text{exp}} \equiv \mu_{M_1}$ and $M^I_{3,\text{exp}} \equiv \mu_{M_3}$, respectively the anisotropy ratio $M^I_{1,\text{exp}}/M^I_{3,\text{exp}}$.

Level 'II': Microporoelasticity of Porous Laminates We consider that the textured clay layer composite has an infinite number of layers, each being characterized by the orientation of the pores (θ, φ) (Fig. 24). The microindentation results are averages in the sense of micromechanics theory of a layered composite. The inherent assumption in layered composites mechanics is that the strain parameters ε_{11}, ε_{22}, ε_{12} are uniform, as well as the stress components σ_{13}, σ_{23}, σ_{33}. This is true for laminates with vanishing layer thicknesses, which is a special case of the scale separability condition (2.53). There are many different formulations for stress-strain averages of laminates. The one we retain here is the method proposed by El Omri et al. (2000), which provides an explicit formulation of the laminate stiffness tensor as a function of each layer stiffness, using stress-strain hybrid averages. We extend this method to the porous laminate, for which the layer behavior is defined by the poroelastic state equation:

$$\sigma_{ij} = C_{ijkl}\,\varepsilon_{kl} - b_{ij}\,p \tag{4.67}$$

The idea is to rewrite the poroelastic state equation in a plane-antiplane basis:

$$\begin{pmatrix} \sigma_P \\ \sigma_A \end{pmatrix} = \begin{bmatrix} c_{PP} & c_{PA} \\ c_{AP} & c_{AA} \end{bmatrix} \begin{pmatrix} \varepsilon_P \\ \varepsilon_A \end{pmatrix} - \begin{pmatrix} b_P \\ b_A \end{pmatrix} p \tag{4.68}$$

where subscript P refers to the plane components $11, 22, 12$, whereas subscript A refers to the anti-plane components $31, 23, 33$, and the subscript A, P to the permuted configuration. Adopting the matrix representation of second and forth order tensors proposed by Helnwein (2001), which is different from the classical Voigt representation, the plane-aniplane components of any symmetric second order tensor a_{ij} are obtained from:

$$\mathbf{a} = \mathbb{P} : \begin{pmatrix} a_P & a_A \end{pmatrix}^T = \begin{pmatrix} a_{11} & a_{22} & a_{33} & \sqrt{2}\,a_{23} & \sqrt{2}\,a_{31} & \sqrt{2}\,a_{12} \end{pmatrix}^T \tag{4.69}$$

yielding:

$$\begin{pmatrix} a_P & a_A \end{pmatrix}^T = \mathbb{P}^T : \mathbf{a} = \begin{pmatrix} a_{11} & a_{22} & \sqrt{2}\,a_{12} & \sqrt{2}\,a_{31} & \sqrt{2}\,a_{23} & a_{33} \end{pmatrix}^T \tag{4.70}$$

where the matrix operator $\mathbb{P}$ reads:

$$\mathbb{P} = \begin{bmatrix} 1 & 0 & 0 & 0 & 0 & 0 \\ 0 & 1 & 0 & 0 & 0 & 0 \\ 0 & 0 & 0 & 1 & 0 & 0 \\ 0 & 0 & 0 & 0 & 0 & 1 \\ 0 & 0 & 0 & 0 & 1 & 0 \\ 0 & 0 & 1 & 0 & 0 & 0 \end{bmatrix} ; \mathbb{P}^{-1} = \mathbb{P}^T = \begin{bmatrix} 1 & 0 & 0 & 0 & 0 & 0 \\ 0 & 1 & 0 & 0 & 0 & 0 \\ 0 & 0 & 0 & 0 & 0 & 1 \\ 0 & 0 & 1 & 0 & 0 & 0 \\ 0 & 0 & 0 & 0 & 1 & 0 \\ 0 & 0 & 0 & 1 & 0 & 0 \end{bmatrix} \tag{4.71}$$

The plane-antiplane components in (4.68) thus read:

$$\begin{pmatrix} \sigma_P & \sigma_A \end{pmatrix}^T = \mathbb{P}^T : \boldsymbol{\sigma} \tag{4.72a}$$

$$\begin{pmatrix} \varepsilon_P & \varepsilon_A \end{pmatrix}^T = \mathbb{P}^T : \boldsymbol{\varepsilon} \tag{4.72b}$$

$$\begin{pmatrix} b_P & b_A \end{pmatrix}^T = \mathbb{P}^T : \mathbf{b} \tag{4.72c}$$

Similarly, the stiffness components are obtained from:

$$\mathbb{C}_{A,P} = \begin{bmatrix} c_{PP} & c_{PA} \\ c_{AP} & c_{AA} \end{bmatrix} = \mathbb{P}^T : \mathbb{C} : \mathbb{P} \tag{4.73}$$

where $\mathbb{C} = \mathbb{C}^I_{\text{hom}}(\theta, \varphi)$ is the stiffness tensor of each layer, given by (4.58a) in function of the elastic constants of the building block and the orientation of the pore plane (θ, φ). Implemented in matrix form, its components read (Helnwein, 2001):

$$C_{ijkl} = \begin{bmatrix} C_{1111} & C_{1122} & C_{1133} & \sqrt{2}C_{1123} & \sqrt{2}C_{1131} & \sqrt{2}C_{1112} \\ & C_{2222} & C_{2233} & \sqrt{2}C_{2223} & \sqrt{2}C_{2231} & \sqrt{2}C_{2212} \\ & & C_{3333} & \sqrt{2}C_{3323} & \sqrt{2}C_{3331} & \sqrt{2}C_{3312} \\ & & & 2C_{2323} & 2C_{2331} & 2C_{2312} \\ & & & & 2C_{3131} & 2C_{3112} \\ sym & & & & & 2C_{1212} \end{bmatrix} \tag{4.74}$$

The plane-antiplane representation allows for the following remarkable averaging relations:

$$\varepsilon_P \equiv \overline{\varepsilon_P} = E_P = \begin{pmatrix} E_{11} & E_{22} & \sqrt{2}\,E_{12} \end{pmatrix}^T \tag{4.75a}$$

$$\sigma_A \equiv \overline{\sigma_A} = \Sigma_A = \begin{pmatrix} \sqrt{2}\,\Sigma_{31} & \sqrt{2}\,\Sigma_{23} & \Sigma_{33} \end{pmatrix}^T \tag{4.75b}$$

The remaining stress and strain components, not included in laminate averaging relations (4.75) are obtained from a partial inversion of (4.68):

$$\begin{pmatrix} \sigma_P \\ \varepsilon_A \end{pmatrix} = \begin{bmatrix} k_{PP} & k_{PA} \\ k_{AP} & k_{AA} \end{bmatrix} \begin{pmatrix} E_P \\ \Sigma_A \end{pmatrix} - p \begin{pmatrix} l_P \\ l_A \end{pmatrix} \tag{4.76}$$

where

$$\begin{array}{ll} k_{PP} = c_{PP} - c_{PA}\,c_{AA}^{-1}\,c_{AP} & l_P = b_P + c_{PA}\,c_{AA}^{-1}\;b_A \\ k_{AA} = c_{AA}^{-1} & \\ k_{PA} = -k_{AP}^T = c_{PA}\,c_{AA}^{-1} & l_A = -c_{AA}^{-1}\;b_A \end{array} \tag{4.77}$$

Averaging (4.76) yields:

$$\begin{pmatrix} \Sigma_P \\ E_A \end{pmatrix} = \begin{pmatrix} \overline{\sigma_P} \\ \overline{\varepsilon_A} \end{pmatrix} = \begin{bmatrix} \overline{k_{PP}} & \overline{k_{PA}} \\ \overline{k_{AP}} & \overline{k_{AA}} \end{bmatrix} \begin{pmatrix} E_P \\ \Sigma_A \end{pmatrix} - p \begin{pmatrix} \overline{l_P} \\ \overline{l_A} \end{pmatrix} \tag{4.78}$$

Finally, after partial inversion, we obtain:

$$\begin{pmatrix} \Sigma_P \\ \Sigma_A \end{pmatrix} = \begin{bmatrix} C_{PP}^{\text{hom}} & C_{PA}^{\text{hom}} \\ C_{AP}^{\text{hom}} & C_{AA}^{\text{hom}} \end{bmatrix} \begin{pmatrix} E_P \\ E_A \end{pmatrix} - p \begin{pmatrix} b_P^{\text{hom}} \\ b_A^{\text{hom}} \end{pmatrix} \tag{4.79}$$

where

$$\begin{array}{ll} C_{PP}^{\text{hom}} = \overline{k_{PP}} - \overline{k_{PA}}\left(\overline{k_{AA}}\right)^{-1}\overline{k_{AP}} & b_P^{\text{hom}} = \overline{l_P} + \overline{k_{PA}}\left(\overline{k_{AA}}\right)^{-1}\overline{l_A} \\ C_{AA}^{\text{hom}} = \left(\overline{k_{AA}}\right)^{-1} & \\ C_{PA}^{\text{hom}} = \left(C_{AP}^{\text{hom}}\right)^T = \overline{k_{PA}}\cdot\left(\overline{k_{AA}}\right)^{-1} & b_A^{\text{hom}} = -\left(\overline{k_{AA}}\right)^{-1}\overline{l_A} \end{array} \tag{4.80}$$

The homogenized stiffness of the porous laminate in the original configuration is obtained by inverse application of (4.73):

$$\mathbb{C}_{\text{hom}}^{II} = \mathbb{P} : \mathbb{C}_{A,P}^{\text{hom}} : \mathbb{P}^T \tag{4.81}$$

Similarly, application of (4.72c) yields the tensor of Biot coefficients in the original basis:

$$\mathbf{b}^{II} = \mathbb{P} : \mathbf{b}_{A,P}^{\text{hom}} \tag{4.82}$$

The last poroelastic constant, the Biot modulus N, is a scalar which is not affected by the rotation of the porosity. As a consequence, the Biot modulus is uniform throughout the layered system, and coincides with the solid Biot modulus of the layered composite, $N^{II} = N^{I}$, given by (4.49c).

Last, the volume average operation in (4.78) of the stiffness matrices $\overline{k_{IJ}}$ reads for a uniform orientation distribution of the pore planes in the laminate:

$$\overline{k_{IJ}} = \frac{1}{2\pi} \int_{\theta=0}^{\pi} \int_{\varphi=0}^{2\pi} k_{IJ}(\theta, \varphi) \sin(\theta) d\varphi d\theta \tag{4.83}$$

Similarly to (4.64) it is appropriate to introduce a non-uniform orientation distribution to account for some degree of preferred alignment:

$$\overline{k_{IJ}} = \frac{1}{2\pi} \int_{\theta=0}^{\pi} \int_{\varphi=0}^{2\pi} k_{IJ}(\theta, \varphi) W(\theta, k) \sin(\theta) d\varphi d\theta \tag{4.84}$$

where $W(\theta, k)$ is the orientation distribution function (4.66). The same non-uniform orientation average operation is applied to all averages in (4.80); including the one involved in the determination of the Biot coefficients.

The resulting homogenized stiffness (4.81) of the porous clay laminate, therefore, depends on the elastic constants of the isotropic solid clay phase –bulk modulus k_s, shear modulus g_s (see (4.50))–, the pore aspect ratio $\rho = h/D$ and the $k-$ factor; in addition to the clay porosity $\phi^I = \phi^{II}$. Using $\mathbb{C}_{\text{hom}}^{II}$ from (4.81) in (2.44) and (2.52) provides the link between these $4+1$ model parameters, and the microindentation stiffness values $M_{1,\text{exp}}^{II}$ and $M_{3,\text{exp}}^{II}$, respectively the anisotropy ratio $M_{1,\text{exp}}^{II}/M_{3,\text{exp}}^{II}$.

Level III: Addition of Inclusions Into an Anisotropic Porous Matrix We consider shale materials at the macroscopic scale to be composed of a textured porous matrix and (mainly) quartz inclusions of approximately spherical shape that are randomly distributed throughout the anisotropic porous matrix V_M (Fig. 27). Except for the anisotropy of the matrix, the microporomechanics problem is very similar to the one we encountered for cement-based materials. Proceeding as before, we adopt a continuous description of the stress field in the heterogeneous r.e.v.:

$$\text{in } V_{III} : \boldsymbol{\sigma}(\underline{z}) = \mathbb{C}(\underline{z}) : \boldsymbol{\varepsilon}(\underline{z}) + \boldsymbol{\sigma}^p(\underline{z}) \tag{4.85}$$

together with the following distributions of the elastic properties $\mathbb{C}(\underline{z})$ and the eigenstress $\boldsymbol{\sigma}^p(\underline{z})$:

$$\mathbb{C}(\underline{z}) = \left\{ \begin{array}{l} \mathbb{C}_M \text{ in } V_M \\ \mathbb{C}_i \text{ in } V_i \end{array} \right. \quad \boldsymbol{\sigma}^p(\underline{z}) = \left\{ \begin{array}{l} -\mathbf{b}_M p \text{ in } V_M \\ 0 \text{ in } V_\phi \end{array} \right. \tag{4.86}$$

where $\mathbb{C}_M = \mathbb{C}_{\text{hom}}^{II}$ is the stiffness tensor of the porous clay laminate (4.81), $\mathbb{C}_i = 3k_i\mathbb{J}+2g_i\mathbb{K}$ is the stiffness tensor of the (assumed isotropic) quartz inclusion phase; and $\mathbf{b}_M = \mathbf{b}^{II}$ is the second order Biot coefficient tensor of the porous clay laminate (4.82). Following the linear microporomechanics approach, we decompose the problem in (the meanwhile classical) two sub-problems:

1. The first sub-problem corresponds to drained conditions, for which $p = 0$. We assume that the r.e.v. is subjected to a uniform displacement boundary condition, so that the solid boundary conditions to which the solid phase is subjected read

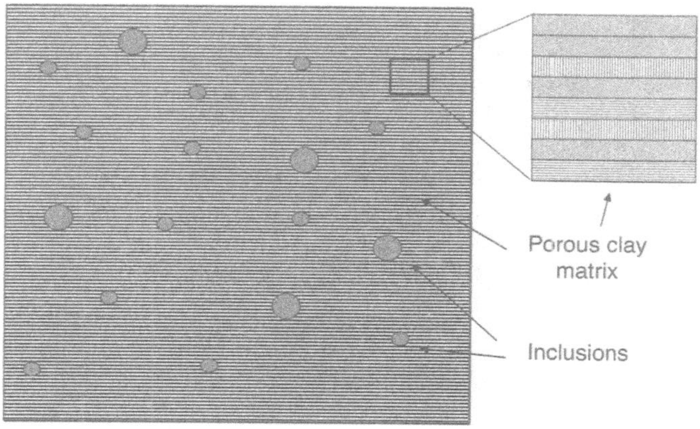

Figure 27. Macroscopic shale model: an anisotropic porous matrix of clay containing randomly distributed spherical inclusions.

(see relation (3.10)):

$$\text{on}\,\partial V_{III} \;:\; \underline{\xi}'(\underline{z}) = \mathbf{E}\cdot\underline{z} \quad (4.87a)$$

$$\text{on}\,\mathcal{I}_{sf} \;:\; \boldsymbol{\sigma}\cdot\underline{n} = 0 \quad (4.87b)$$

The macroscopic strain tensor $\mathbf{E}$ is related to the microscopic strain by the strain localization condition:

$$\boldsymbol{\varepsilon}'(\underline{z}) = \mathbb{A}(\underline{z}) : \mathbf{E} \quad (4.88)$$

$\mathbb{A}(\underline{z})$ satisfies the compatibility condition:

$$\mathbf{E} = \overline{\boldsymbol{\varepsilon}'(\underline{z})} \Leftrightarrow \mathbb{I} = (1-f_i)\langle\mathbb{A}\rangle_{V_M} + f_i\langle\mathbb{A}\rangle_{V_i} \quad (4.89)$$

where $f_i = V_i/V_{III} = 1 - f_M$ is the inclusion volume fraction. Use of (4.88) in (4.85) together with the elastic distribution (4.86) yields after volume averaging, the macroscopic stress, and the homogenized stiffness of the composite:

$$\boldsymbol{\Sigma}' = \overline{\boldsymbol{\sigma}'(\underline{z})} = \mathbb{C}^{III}_{\text{hom}} : \mathbf{E};\; \mathbb{C}^{III}_{\text{hom}} = \mathbb{C}_M + f_I(\mathbb{C}_i - \mathbb{C}_M) : \langle\mathbb{A}\rangle_{V_i} \quad (4.90)$$

The change of the porosity in the subproblem reads:

$$\left[(\phi-\phi_0)^{III}\right]' = \mathbf{b}^{III} : \mathbf{E} \quad (4.91)$$

where $\mathbf{b}^{III}$ is the second order tensor of homogenized Biot coefficients.

2. The second sub-problem we consider is the zero-displacement boundary problem, whereas the porosity is pressurized. Using Levine's theorem (3.14) for the eigenstress distribution (4.86), we obtain the macroscopic stress and the homogenized Biot coefficients:

$$\boldsymbol{\Sigma}'' = \overline{\boldsymbol{\sigma}^p(\underline{z}) : \mathbb{A}(\underline{z})} = -\mathbf{b}^{III}p;\; \mathbf{b}^{III} = \mathbf{b}_M : (\mathbb{I} - f_i\langle\mathbb{A}\rangle_{V_i}) \quad (4.92)$$

The change of the porosity in the subproblem reads:

$$\left[(\phi-\phi_0)^{III}\right]'' = (1-f_i)\left(\mathbf{b}_M : \langle\varepsilon''\rangle_{V_M} + \frac{p}{N_M}\right) \tag{4.93}$$

where $N_M = N^{II}$. We eliminate $\langle\varepsilon''\rangle_{V_M}$ in (4.93) to the benefit of pressure p, by considering the stress average in this sub-problem, in which $\mathbf{E}'' = \overline{\varepsilon''(\underline{z})} = 0$:

$$\begin{aligned}\Sigma'' &= (1-f_i)\langle\sigma''\rangle_{V_M} + f_i\langle\sigma''\rangle_{V_I} \\ &= (\mathbb{C}_M - \mathbb{C}_i) : (1-f_i)\langle\varepsilon''\rangle_{V_M} - (1-f_i)\,\mathbf{b}_M p\end{aligned} \tag{4.94}$$

Thus, from the equality of (4.92) and (4.94):

$$(1-f_i)\langle\varepsilon''\rangle_{V_M} = -(\mathbb{C}_M - \mathbb{C}_i)^{-1} : \left(\mathbf{b}^{III} - (1-f_i)\,\mathbf{b}_M\right)p \tag{4.95}$$

Finally, use of (4.95) in (4.93) yields:

$$\left[(\phi-\phi_0)^{III}\right]'' = \frac{p}{N^{III}};$$

$$\frac{1}{N^{III}} = \frac{1-f_i}{N_M} - f_i\mathbf{b}_M : (\mathbb{C}_M - \mathbb{C}_i)^{-1} : \mathbf{b}_M : \left(\mathbb{I} - \langle\mathbb{A}\rangle_{V_i}\right) \tag{4.96}$$

In summary, a superposition of the two subproblems yields the macroscopic state equations of the shale composite material:

$$\Sigma = \mathbb{C}^{III}_{\text{hom}} : \mathbf{E} - \mathbf{b}^{III}p \tag{4.97a}$$

$$(\phi-\phi_0)^{III} = \mathbf{b}^{III} : \mathbf{E} + \frac{p}{N^{III}} \tag{4.97b}$$

together with the poroelastic properties:

$$\mathbb{C}^{III}_{\text{hom}} = \mathbb{C}_M + f_I\,(\mathbb{C}_i - \mathbb{C}_M) : \langle\mathbb{A}\rangle_{V_i} \tag{4.98a}$$

$$\mathbf{b}^{III} = \mathbf{b}_M : \left(\mathbb{I} - f_i\langle\mathbb{A}\rangle_{V_i}\right) \tag{4.98b}$$

$$\frac{1}{N^{III}} = \frac{1-f_i}{N_M} + f_i\mathbf{b}_M : (\mathbb{C}_i - \mathbb{C}_M)^{-1} : \mathbf{b}_M : \left(\mathbb{I} - \langle\mathbb{A}\rangle_{V_i}\right) \tag{4.98c}$$

The input to this homogenization step are the poroelastic properties of the porous clay laminate, $\mathbb{C}_M = \mathbb{C}^{\text{hom},II}$, $\mathbf{b}_M = \mathbf{b}^{II}$, $N_M = N^{II}$, the (quartz) inclusion volume fraction f_i, and an estimate of the volume average of the inclusion strain concentration tensor $\langle\mathbb{A}\rangle_{V_i}$. Given the matrix–inclusion morphology, such an estimate is suitably provided by the Mori-Tanaka scheme:

$$\langle\mathbb{A}\rangle_{V_M} \simeq \mathbb{A}^{est}_M = \left[(1-f_i)\mathbb{I} + f_i\left(\mathbb{I} + \mathbb{P}^M : (\mathbb{C}_i - \mathbb{C}_M)\right)^{-1}\right]^{-1} \tag{4.99}$$

where $\mathbb{P}^M$ is the so-called P-tensor that can be obtained from the matrix Green's function $G^0_{ik}(\underline{x} - \underline{x}')$ which expresses the displacement at point $\underline{x}$ in direction i in a linear elastic solid of stiffness $\mathbb{C}^0$ ($= \mathbb{C}_M$ for the Mori-Tanaka scheme (Zaoui, 2002)) subjected to a unit force $\delta(\underline{x} - \underline{x}')$ applied in direction k at the point $\underline{x}'$:

$$\mathbb{P}^0_{ijkl} = -\left(\partial^2 \left(\int_V G^0_{ik}(\underline{x} - \underline{x}')dV\right) / \partial x_j \partial x_l\right)_{(ij)(kl)} \tag{4.100}$$

where $(ij)(kl)$ stands for the symmetrization with respect to (ij) and (kl). The expression of the P-tensor for spherical inclusions in a transversely isotropic matrix can be found in Laws (1985), and is given in the Appendix.

4.3 Calibration–Validation

The load bearing phase in shales is essentially a colloidal system of clay particles in which repulsive contact forces are in equilibrium with cohesive bonds. Like other natural composites, it behaves macroscopically as an elastic solid but with a particular microstructure. One could (and it has been!) argue(d) that continuum-based micromechanics is not the best theory suited for shales multiscale mechanical modeling. However, it may be the best currently available as far as (poro)elastic properties are concerned. Elastic homogenization theories for particulate systems are still very limited, and could not be used anyway with the continuum analysis of indentation tests.

The present model relies on two main assumptions: (1) The porous clay phase of shale is a continuous solid with pores; and (2) the pure clay stiffness is isotropic. By means of these two assumptions, we arrive at reducing the total numbers of model degrees of freedom to four: the elastic constants of the isotropic solid clay phase –bulk modulus k_s, shear modulus g_s (see (4.50)) or plane stress modulus M_s and Poisson's ratio ν_s–, the pore aspect ratio $\rho = h/D$ and the $k-$ factor. To complete the model we need to calibrate and validate the model. This will be achieved in two steps:

- For the calibration, we will only use the four results of the nano- and microindentation analysis from level I and II, and search for each shale material (porosity and inclusion volume fraction given) for the optimum set of model parameters (M_s, ν_s, ρ, k) that best fit the experimental constraints:

$$\begin{array}{l} \left|M^I_{1,\exp} - \mu_{M_1}(M_s, \nu_s, \rho, k)\right| \leq \epsilon^I_1 \\ \left|M^I_{3,\exp} - \mu_{M_3}(M_s, \nu_s, \rho, k)\right| \leq \epsilon^I_3 \\ \left|M^{II}_{1,\exp} - M^{II}_1(M_s, \nu_s, \rho, k)\right| \leq \epsilon^{II}_1 \\ \left|M^{II}_{3,\exp} - M^{II}_3(M_s, \nu_s, \rho, k)\right| \leq \epsilon^{II}_3 \end{array} \tag{4.101}$$

  where μ_{M_J} is the statistical model average (4.64), $M^{II}_J\ (J = 1, 3)$ is the model indentation modulus given by an application of the model stiffness $\mathbb{C}^{II}_{\text{hom}}$ from (4.81) in (2.44) and (2.52), and ϵ^J_j are some prescribed (absolute) error values. Ideally, $\epsilon^J_j = 0$.
- For validation purposes, we use a second experimental data set for each shale material: the dynamic elastic stiffness values determined by UPV measurements, $\mathbb{C}^{III,dyn}_{\exp}$. We compare these dynamic values to the undrained model stiffness values, which are obtained from the poroelastic model properties $\mathbb{C}^{III}_{\text{hom}}, \mathbf{b}^{III}, N^{III}$ (relation

Shale #		1		2		3	
ν_s	[1]	0.4	0.48	0.4	0.48	0.315	0.415
M_s	[GPa]	32.0 – 36.7	34.2 – 38.9	36.4 – 38.0	38.5 – 41.0	36.3	38.7
ρ	[1]	0.054	0.0515	0.036	0.034	0.0155	0.0148
k	[1]	0.9		3.4		1.4	

Table 7. Summary of model calibration procedure.

(4.98)), through (see e.g. Coussy (1995)):

$$\mathbb{C}_{\text{hom}}^{III,un} = \mathbb{C}_{\text{hom}}^{III} + \mathcal{M}^{III}\mathbf{b}^{III} \otimes \mathbf{b}^{III} \tag{4.102}$$

where $\mathcal{M}^{III}$ is the overall Biot modulus:

$$\frac{1}{\mathcal{M}^{III}} = \frac{1}{N^{III}} + \frac{\phi^{III}}{K^{fl}} \tag{4.103}$$

with K^{fl} the fluid bulk modulus. The validation then reads:

$$\left\|\mathbb{C}_{\text{exp}}^{III,dyn} - \mathbb{C}_{\text{hom}}^{III,un}\right\| \leq \epsilon^{III} \tag{4.104}$$

Calibration: A Unique Clay Solid Stiffness? Table 7 provides a summary of the calibration procedure. An interesting observation is that the plane-stress modulus of the pure clay phase, $M_s = E_s/\left(1-\nu_s^2\right)$, is almost the same for all three shale materials (and even more so for the Young's modulus E_s!), which are quite different in mineralogy, burial depth, porosity, etc. It is somewhat smaller than the indentation modulus of pure mineral crystals (compare with Table 5), and clearly greater than average indentation stiffness values obtained by nanoindentation at larger scales. The main difference in between the different shale materials is the pore aspect ratio ρ and the alignment factor k. The first scales almost linearly with the clay porosity, $\rho = c\phi^I$ (where $c = 0.1745$), whereas the second appears to be unrelated to the porosity and the pore aspect ratio, as one should expect for an independent (random) model variable.

Validation: Drained and Undrained Poroelastic Properties Table 8 provides a summary of the predicted (drained) poroelastic properties at different scales for the three shale materials considered in this study. The input parameters are the values from Table 7. It is worth noting that the difference in between the Biot coefficients strongly reflects the anisotropy induced by the pore aspect ratio. Use of the level III values in (4.102) yields the undrained stiffness values to be compared with the experimental dynamic stiffness values. The comparison of experimental vs. predicted undrained stiffness properties is displayed in Figure 28. Since this data set has not been used in the calibration of the model parameters, the good agreement is a good indication of the model's capability to capture the essential poroelastic behavior of shale materials, by means of four model parameters. Two of the model parameters, the elastic constants of the pure clay phase, seem to be (almost) independent of the specific shale material, and in particular of the

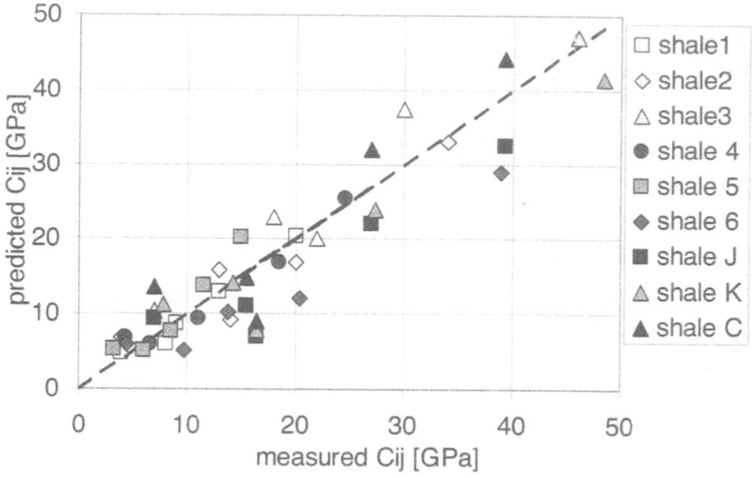

Figure 28. Predicted vs. measured C_{ij}, for the three indented shales using the calibrated input parameters, and for six other shales (J = Jurassic shale; K = Kimmeridge shale; C = Cretaceous shale).

mineralogy. A third model parameter, the pore aspect ratio, appears to be related to the porosity; most probably as a result of the burial diagenesis: the greater the burial depth, the higher the packing density of the clay particles. Such high packing densities are indicative of a more ordered structure of the shale building blocks (clay particle packages) at level 0. For the same porosity, this increases the likelihood of pore connectivity at the nanoscale, which we capture through the pore aspect ratio. The anisotropy induced by the pore aspect ratio can be considered as the base source of anisotropy of shales. A second source of anisotropy of shales stems from some preferential orientation of building blocks at larger scales (level I and II), which is captured in our model through the $k-$factor. The $k-$factor which is the only remaining (true) degree of freedom of the model, increases the base anisotropy. Motivated from nanoindentation results and SEM and ESEM images of shale materials, it is very likely that a preferred alignment of building blocks relates to silt-size inclusions that may disturb the deposition alignment of clay particles in the course of the deposition history. It should therefore be related to size and distribution of inclusions within the porous clay laminate. However, based on the multiscale investigation of three shale materials only, it is still to early to be conclusive.

Finally, Figure 28 also displays a comparison of predicted vs. experimental values for some other shale materials found in the literature. What is given for these shale materials is the porosity, the inclusion volume fraction and the clay mineralogy. The input for the predictions are the elastic constants of the pure clay phase ($M_s = 35$ GPa,

	Shale #		1		2		3	
Level 0	ν_s	[1]	0.4	0.48	0.4	0.48	0.315	0.415
	M_s	[GPa]	32.0	34.2	38.0	41.0	36.3	38.7
Level I	C_{11}	[GPa]	26.2	25.3	32.9	34.9	35.8	39.3
$\theta = 0$	C_{12}	[GPa]	11.3	13.3	14.4	18.7	11.3	17
$\varphi = 0$	C_{13}	[GPa]	2.3	2.8	3.1	4.2	1.9	3.0
	C_{33}	[GPa]	3.3	3.0	4.5	4.5	4.2	4.4
	C_{44}	[GPa]	2	1.7	2.7	2.5	3.0	2.8
	b_{11}	[1]	0.74	0.94	0.68	0.92	0.39	0.56
	b_{33}	[1]	0.95	0.99	0.93	0.98	0.78	0.78
	N	[GPa]	5.13	4.46	5.99	5.63	15.8	25.7
Level II	C_{11}	[GPa]	13.2	11.9	19.2	19.1	19.0	19.7
	C_{12}	[GPa]	4.4	4.6	6.6	7.7	5.1	6.4
	C_{13}	[GPa]	2.4	2.7	2.8	3.4	2.1	3.1
	C_{33}	[GPa]	6.8	6.2	6.6	6.5	8.4	8.7
	C_{44}	[GPa]	3.3	2.8	4.2	3.8	5.0	4.8
	b_{11}	[1]	0.87	0.94	0.82	0.96	0.58	0.67
	b_{33}	[1]	0.92	0.99	0.92	0.98	0.70	0.73
	N	[GPa]	5.13	4.46	5.99	5.63	15.8	25.7
Level III	C_{11}	[GPa]	18.1	16.5	30.2	29.8	36.6	42.8
	C_{12}	[GPa]	6.4	6.5	11.2	12.5	11.5	18.5
	C_{13}	[GPa]	3.5	3.7	4.8	5.5	8.1	15.4
	C_{33}	[GPa]	10.3	9.3	13.1	12.8	25.1	32.8
	C_{44}	[GPa]	4.8	4.8	7.3	6.8	10.5	10.4
	b_{11}	[1]	0.82	0.92	0.73	0.85	0.50	0.56
	b_{22}	[1]	0.89	0.95	0.87	0.92	0.62	0.58
	N	[GPa]	6.1	5.3	7.7	7.2	20.3	32.3

Table 8. Predicted poroelastic properties at different scales.

$\nu_s = 0.4$), and the pore aspect ratio determined from the clay matrix porosity. The $k-$factor was interpolated from the values of our three shale materials. The comparison of predicted vs. experimental dynamic measurements shows generally a good agreement for all shale materials, in which the clay phase is the load bearing phase. Some limitations of the predictive capabilities of our model relate to either high silt inclusion materials or extremely low or high porosity shale materials. Both limitations are probably due to limitations of the homogenization scheme we employ. Indeed, the schemes employed assume that the load bearing phase is the clay phase (and not the inclusion phase), so that predictions will certainly be inaccurate for high quartz inclusion shale materials. Furthermore, it is quite unlikely that the schemes can capture extremely high packing density beyond 90% (*ie.* ultra-low porosity materials) or extremely low packing density, *ie.* much below the random close-packed limit (RCP). From packing theory, it is known that such high packing densities can only be achieved by different sized solid building blocks, and would probably require considering several pore families.

5 Conclusions

We raised the question whether it were possible to break down natural porous material systems down to a scale where materials no longer change from one material to another, and to upscale ('nanoengineer') the behavior from the nanoscale to the macroscale of engineering applications. Based on a combination of indentation analysis and microporomechanics analysis of cement-based materials and shale materials, the tentatively answer is yes!

1. Level '0': At a nanoscale, most geomaterials are colloidal material systems, in which repulsive contact forces are in equilibrium with cohesive bonds. The material response at very fine scales is dominated by these surface properties, rather than by mineral properties. It is most likely that the scale of material invariant properties of natural composites is related to the predominant effect of such surface properties, which no-doubt depend on the type of mineral bonding. We come to this conclusion from the backanalysis of measured nanoindentation results of cement-based materials and shales, that seem to converge towards a single set of material properties for each class of materials: the C-S-H solid stiffness and the pure clay solid phase properties.
2. Level 'I': At a scale of tens of nanometers, most geomaterials possess characteristic packing densities, as a consequence of the material genesis, *ie.* hydration in the case of cement-based materials; deposition and burial diagenesis in the scale of shales. This packing density defines a particular solid-pore space microstructure, unique for each class of materials. In the case of cement-based materials, these are the LD– and HD–C-S-H packing; and in the case of shales, it is the shale building block, *ie.* the packages of clay particles. The material invariant properties (from level '0') and the packing mode (from level 'I') translate into a base mechanics behavior. The porosity ('one minus packing density') which manifests itself at this scale, defines much of the poromechanics behavior of natural composites at larger scales.
3. Level 'II': At a micrometer scale, the behavior of natural composites is a consequence of texture; *ie.* the LD–/HD–C-S-H volume proportions of cement-based materials, and the porous clay laminate structure of shales. Texture can explain much of the large variety of geomaterials. For instance, the lower strength of heat cured cement-based materials appears as a consequence of the conversion of LD–C-S-H into HD–C-S-H, creating a net macroporosity. In the case of shales, it is the preferred alignment of building blocks increasing the elastic anisotropy.
4. Level 'III': All things come together at the macroscale, where the textured load bearing porous material and inclusions (sand and aggregates in the case of mortar and concrete; quartz, feldspar, pyrite, etc. in the case of shales) determine the mechanical behavior of natural composites.

Our experimental microporomechanics analysis focussed primarily on the poroelastic behavior of natural composites, for which a combination of advanced experimental indentation techniques and linear microporoelasticity theory provides a unique opportunity to understand and assess nanoproperties and microstructure, as a new basis for the engineering prediction of macroscopic poroelastic properties of natural composite materials. Things will still take some time regarding strength properties, requiring advances in the interpretation of strength properties from indentation tests and in non-linear microporomechanics, to ultimately arrive at identifying material invariant nanostrength properties. This is a formidable challenge!

Bibliography

P. Acker. Micromechanical analysis of creep and shrinkage mechanisms, In F.-J. Ulm, Z.P. Bažant, and F.H.Wittmann, editors, *Creep, Shrinkage and Durability Mechanics of Concrete and other quasi-brittle Materials.* Elsevier, Oxford UK, 15–25, 2001.

D.M. Barnett and J. Lothe. Line force loadings on anisotropic half-spaces and wedges. *J. Appl. Phys.*, 23:470–474, 1975.

Z.P. Bažant. Thermodynamics of interacting continua with surfaces and creep analysis of concrete structures. *Nuclear Engrg. and Design*, 20:477–505, 1972.

R.H. Bennett, N.R. O'Brien, and M.H. Hulbert. Determinants of clay and shale microfabric signatures: Processes and mechanisms. In R. Bennett, W. Bryant, M. Hulbert, editors, *Microstructure of fine grained sediments: from mud to shale*, Springer-Verlag, New York, Chapter 2, 5–32, 1991.

A. Bentur. Effect of curing temperature on the pore structure of tricalcium silicate pastes. *Journal of Colloid and Interface Science*, 74(2):549–560, 1980.

O. Bernard, F.-J. Ulm, and E. Lemarchand. A multiscale micromechanics–hydration model for the early-age elastic properties of cement-based materials. *Cem. Concr. Res.*, 33(8):1155–1173, 2003.

M.A. Biot. General theory of three dimensional consolidation. *Journal of Applied Physics*, 12:155–164, 1941.

F.M. Borodich, L.M. Keer, and C.S. Korach. Analytical study of fundamental nanoindentation test relations for indenters of non-ideal Shapes. *Nanotechnology*, 14:803–808, 2003.

F.M. Borodich and L.M. Keer. Contact problems and depth-sensing nanoindentation for frictionless and frictional boundary conditions. *Int. J. Solids Struct.*, 41: 2479–2499, 2004.

J. Brittan, M. Warner, and G. Pratt. Short Note: Anisotropic parameters of layered media in terms of composite elastic properties. *Geophysics*, 60(4):1243–1248, 1995.

X. Chateau and L. Dormieux. Micromechanics of saturated and unsaturated porous media. *Int. J. Numer. Anal. Meth, Geomech.*, 26:830–844, 2002.

Y.-T. Cheng and C.-M. Cheng. Scaling relationships in conical indentation of elastic-perfectly plastic solids. *Int. J. Solids Struct.*, 36:1231–1243, 1999.

Y.-T. Cheng and C.-M. Cheng. Scaling, dimensional analysis, and indentation measurements. *Materials Science and Engineering*, R44:91–149, 2004.

G. Constantinides, F.-J. Ulm, and K.J. van Vliet. On the use of nanoindentation for cementitious materials. *Materials and Structures* (Special issue of Concrete Science and Engineering) RILEM, 205:191–196, 2003.

G. Constantinides and F.-J. Ulm. The effect of two types of C–S–H on the elasticity of cement-based materials: Results from nanoindentation and micromechanical modeling. *Cem. Concr. Res.*, 34(1):67–80, 2004.

G. Constantinides and F.-J. Ulm. *Material invariant properties of cement-based materials: an experimental microporomechanics approach.* MIT-CEE Res. Rep. R04-03 (Ph.D.-Dissertation), Cambridge, MA, In Preparation.

O. Coussy. *Mechanics of porous continua.* J. Wiley & Sons, Chichester, UK, 1995.

O. Coussy. *Poromechanics.* J. Wiley & Sons, Chichester, UK, 2004.

A. Delafargue and F.-J. Ulm. Explicit approximations of the indentation modulus of elastically orthotropic solids for conical indenters. Int. J. Solids Struct., 41: 7351–7360, 2004.

A. Delafargue and F.-J. Ulm. *Material invariant properties of shales: nanoindentation and microporoelastic analysis.* MIT-CEE Res. Rep. R04-02 (SM-Dissertation), Cambridge, MA, 2004.

M.F. Doerner and W.D. Nix. A method for interpreting the data from depth-sensing indentation instruments. *J. Mater. Res.*, 1:601–609, 1986.

H.A. Elliot. Axial symmetric stress distributions in aelotropic hexagonal crystals: The problem of the plane and related problems. *Proc. Camb. Phil. Soc.*, 45:621–630, 1949.

A. Donev, I. Cisse, D. Sachs, E.A. Variano, F.H. Stillinger, R. Connely, S. Torquato, P.M. Chaikin. Improving the density of jammed disordered packings using ellipsoids. *Science*, 303:990–993, 2004.

L. Dormieux, A. Molinari, and D. Kondo. Micromechanical approach to the behaviour of poroelastic materials, *J. Mech. Phys. Solids*, 50:2203–2231, 2002.

L. Dormieux and E. Bourgeois. *Introduction à la micromécanique des milieux poreux.* Presses de l'Ecole nationale des ponts et chaussées, Paris, France, 2003.

A. El Omri, A. Fennan, F. Sidoroff, and A. Hihi A. Elastic-Plastic homogenization for layered composites. *European Journal of Mechanics*, A/Solids 19:585–601, 2000.

J. Eshelby. The determination of the elastic field of an ellipsoidal inclusion, and related problems. *Proceedings of the Royal Society London*, Series A, 241:376–396, 1957.

R.F. Feldman and P.J. Sereda. A new model of hydrated cement and its practical implications. *Eng. J. Can.*, 53:53–59, 1970.

F.P. Ganneau and F.-J. Ulm. *From nanohardness to strength properties of cohesive-frictional Materials — Application to shale materials.* MIT-CEE Res. Rep. R04-01 (SM-Dissertation), Cambridge, MA, 2004.

F.P. Ganneau, G. Constantinides, and F.-J. Ulm. Dual indentation technique for the assessment of strength properties of cohesive-frictional materials. *Int. J. Solids Struct.*, (in review), 2004.

E.J. Garboczi. Computational materials science of cement-based materials. *Materials and Structures*, RILEM, 26(2):191–195, 1993.

M.T. Hanson. The elastic field for conical indentation including sliding friction for transverse isotropy. *J. Appl. Mech.*, 59:S123–S130, 1992.

Z. Hashin and P.J.M. Monteiro. An inverse method to determine the elastic properties of the interphase between the aggregate and the cement paste. *Cem. Concr. Res.*, 32(8):1291–1300, 2002.

P. Helnwein. Some remarks on the compressed matrix representation of symmetric second-order and fourth-order tensors. *Comput. Methods Appl. Mech. Engrg.*, 190:2753–2770, 2001.

F.H. Heukamp and F.-J. Ulm. *Chemomechanics of calcium leaching of cement-based materials at different scales: The role of CH-dissolution and C-S-H degradation on strength and durability performance of materials and structures.* MIT-CEE Report R02-03 (D.Sc.-Dissertation), Cambridge, MA, 2002.

H.G. Hopkins, A.D. Cox, and G. Eason. Axially symmetric plastic deformation in soils. *Philosophical Transactions, Royal Society*, A264:1–45, 1961.

B. Hornby, L. Schwartz, and J. Hudson. Anisotropic effective medium modeling of the elastic properties of shales. *Geophysics*, 59(10):1570–1583, 1994.

B. Hornby. Experimental laboratory determination of the dynamic elastic properties of wet, drained shales. *Journal of Geophysical Research*, 103(B12), 29,945–29,964, 1998.

S. Igarashi, A. Bentur, and S. Mindess. Characterization of the microstructure and strength of cement paste by microhardness testing. *Advances in Cement Research*, 8(30):877–890, 1996.

H.M. Jaeger and S.R. Nagel. 'Physics of granular state. *Science*, 255(5051):1523–1531, 1992.

M. Jakobsen, J.A. Hudson, and T.A. Johansen. T-matrix approach to shale acoustics. *Geophys. J. Int.*, 154:533–558, 2003.

H.M. Jennings. A model for the microstructure of calcium silicate hydrate in cement paste. *Cem. Concr. Res.*, 30(1):101–116, 2000.

H.M. Jennings. Colloid model of C-S-H and implications to the problem of creep and shrinkage. *Materials and Structures* (Special issue of Concrete Science and Engineering) RILEM, 265:59–70, 2004.

T.A. Johansen, B.O. Ruud, and M. Jakobsen. Effect of grain scale alignment on seismic anisotropy and reflectivity of shales. *Geophysical Prospecting*, 52:133–149, 2004.

L.E.A. Jones and H.F. Wang. Ultrasonic velocities in Cretaceous shales from the Williston basin. *Geophysics*, 46:288–297, 1994.

D. Katti, S. Schmidt, D. Gosh, and K. Katti. Modeling response of pyrophyllite clay interlayer to applied stress using steered moelcular dynamics. *personal communication*, 2003.

N. Laws. A note on penny-shaped cracks in transversely isotropic materials. *Mechanics of Materials*, 4:209–212, 1985.

G.Q. Li, Y. Zhao, S.S. Pang, and Y.Q. Li. Effective Young's modulus estimation of concrete. *Cem. Concr. Res.*, 29(9):1455–1462, 1999.

D. Marion, A. Nur, H. Yin, and D. Han. Compressional velocity and porosity in sand-clay mixtures. *Geophysics*, 57(4):554–563, 1992.

M. Matar and J. Salençon. Capacité portante des fondations superficielles circulaires. *Journal de Mechanique Théorique et Appliquée*, 1(2):237—267, 1982. Available in English In *Foundation Engineering, Volume 1, Soil properties-Foundation design and construction*, Presse de l'Ecole Nationale des Ponts et Chaussées, France.

G. Mavko, T. Mukerji, and J. Dvorkin. *The Rock Physics Handbook.* Cambridge University Press, UK, 1998.

J. Mitchell. *Fundamentals of soil behavior.* J. Wiley & Sons, New York, 1993.

W.C. Oliver and G.M. Pharr. An improved technique for determining hardness and elastic modulus using load and displacement sensing indentation experiments. *J. Mater. Res.*, 7(6):1564–1583, 1992.

G.M. Pharr. Understanding nanoindentation unloading curves. *J. Mater. Res.*, 17(10):2660–2671, 2002.

T.C. Powers and T.L. Brownyard. *Studies of the physical properties of hardened portland cement paste.* PCA Bulletin 22, 1948.

M. Prasad, M. Kopycinska, U. Rabe, and W. Arnold, W. Measurement of Young's modulus of clay minerals using atomic force acoustic microscopy. *Geophysical Research Letters*, 29(8), 1172, 13-1–13-4, 2002.

L.F. Pratson, A. Stroujkova, D. Herrick, F. Boadu, and P. Malin. Predicting seismic velocity and other rock properties from clay content only. *Geophysics*, 68(6):1847–1856, 2003.

J. Salençon. *Calcul à la rupture et analyse limite.* Presses de l'Ecole nationale des ponts et chaussées, Paris, France, 1983.

C.M. Sayers. The elastic anisotrophy of shales, *Journal of Geophysical Research*, 99(B1):767–774, 1994.

C. Sayers. Stress-dependent seismic anisotropy of shales. *Geophysics*, 64(1):93–98, 1999.

N.J.A. Sloane. Kepler's conjecture confirmed. *Nature*, 395:435–436, 1998.

I. Sneddon, editor. *Application of integral transforms in the theory of elasticity.* Springer Verlag, Wien-New York, 1977.

P. Suquet, editor. *Continuum micromechanics*, CISM Courses and Lectures no. 377, 1997. Springer Verlag, Wien-New York.

J.G. Swadener and G.M. Pharr. Indentation modulus of elastically anisotropic half-spaces by cones and parabolae of revolution. *Philos. Mag.*, A81:447–466, 2001.

P.D. Tennis and H.M. Jennings. A model for two types of calcium silicate hydrate in the microstructure of portland cement pastes. *Cem. Concr. Res.*, 30(6):855–863, 2000.

J.J. Thomas and H.M. Jennings. Effect of heat treatment on the pore structure and drying shrinkage behavior of hydrated cement paste. *Journal of the American Ceramic Society*, 85(9):2293–2298, 2002.

L. Thomsen. Weak elastic anisotropy. *Geophysics*, 52(10):1954–1966, 1986.

L. Thomsen. Seismic anisotropy. *Geophysics*, 66:40–41, 2001.

I. Tsvankin. P-wave signatures and notation for transversely isotropic media: An overview. *Geophysics*, 61(2):467–483, 1996.

F.-J. Ulm and O. Coussy. *Mechanics and durability of solids. Vol. I: Solid Mechanics.* Prentice Hall, Upper Saddle River, New Jersey, 2003.

F.-J. Ulm, G. Constantinides, and F.H. Heukamp. Is concrete a poromechanics material? – A multiscale investigation of poroelastic properties, *Materials and Structures* (Special issue of Concrete Science and Engineering), 265:43–58, 2004.

T. Vanorio, M. Prasad, and A. Nur. Elastic properties of dry clay mineral aggregates, suspensions and sandstones. *Geophys. J. Int.*, 155:319–326, 2003.

K. Velez, S. Maximilien, D. Damidot, G. Fantozzi, and F. Sorrentino. Determination by nanoindentation of elastic modulus and hardness of pure constituents of portland cement clinker. *Cem. Concr. Res.*, 31(4):555–561, 2001.

B. Velde. *Origin and mineralogy of clays.* Springer-Verlag, New York, 1995.

J.J. Vlassak and W.D. Nix. Indentation modulus of elastically anisotropic half spaces. *Philos. Mag.*, A67(5):1045–1056, 1993.

J.J. Vlassak and W.D. Nix. Measuring the elastic properties of anisotropic materials by means of indentation experiments. *J. Mech. Phys. Solids*, 42(8):1223–1245, 1994.

J.J. Vlassak, M. Ciavarella, J.R. Barber, and X. Wang. The indentation modulus of elastically anisotropic materials for indenters of arbitrary shape. *J. Mech. Phys. Solids*, 51:1701–1721, 2003.

Z. Wang, H. Wang, and M.E. Cates. Effective elastic properties of solid clays. *Geophysics*, 66(2):428–440, 2001.

Z. Wang, Z. Seismic anisotropy in sedimentary rocks. Part 1: A single plug laboratory method; Part 2: Laboratory data. *Geophysics*, 67(5):1415–1422 (part I) and 67(5):1423–1440 (part II).

J.R. Willis. Herztian contact of anisotropic bodies. *J. Mech. Phys. Solids*, 14:163–176, 1966.

F.H. Wittmann. Bestimmung physikalischer Eigenschaften des Zementsteins. *Deutscher Ausschuss fuer Stahlbeton*, Heft 232:1–63, 1974, W. Ernst & Sohn, Berlin, Germany.

F.H. Wittmann. Creep and shrinkage mechanisms in concrete. In Z.P. Bažant and F.H.Wittmann, editors, *Creep & Shrinkage of Concrete*, J. Wiley & Sons, New York, 1988.

A. Zaoui. Continuum micromechanics: Survey. *Journal of Engineering Mechanics* (ASCE), 128(8):808–816, 2002.

6 Appendix: Eshelby and P tensors

6.1 Oblate spheroidal inclusion in an isotropic matrix

The non-zero components of the Eshelby tensor $\mathbb{S}$ for oblate spheroidal inclusions of aspect ratio $\rho = h/D < 1$ in a matrix of Poisson's ratio ν read:

$$S_{ijkl} = S_{jikl} = S_{ijlk}$$

$$
\begin{aligned}
S_{1111} &= \frac{3}{8\pi(1-\nu)} I_{11} + \frac{1-2\nu}{8\pi(1-\nu)} I_1 = S_{2222} \\
S_{1122} &= \frac{1}{8\pi(1-\nu)} I_{12} + \frac{1-2\nu}{8\pi(1-\nu)} I_1 = S_{2211} \\
S_{1133} &= \frac{1}{8\pi(1-\nu)} \rho^2 I_{13} + \frac{1-2\nu}{8\pi(1-\nu)} I_1 \\
S_{3311} &= \frac{1}{8\pi(1-\nu)} I_{13} + \frac{1-2\nu}{8\pi(1-\nu)} I_3 = S_{3322} \\
S_{1212} &= \frac{1}{8\pi(1-\nu)} I_{12} + \frac{1-2\nu}{16\pi(1-\nu)} (I_1 + I_2) \\
S_{2323} &= \frac{1}{16\pi(1-\nu)} (1+\rho^2) I_{13} + \frac{1-2\nu}{16\pi(1-\nu)} (I_1 + I_3)
\end{aligned}
$$

where:

$$
\begin{aligned}
I_1 &= I_2 = \frac{2\pi\rho}{(1-\rho)^{3/2}} \left\{ \cos^{-1}\rho - \rho\left(1-\rho^2\right)^{1/2} \right\},\ I_3 = 4\pi - 2I_1 \\
I_{13} &= I_{23} = \frac{(I_1 - I_3)}{(\rho^2 - 1)},\ 3I_{33} = 4\pi/\rho^2 - 2I_{13} \\
I_{11} &= I_{22} = I_{12} = \pi - I_{13}/4 = \pi - \frac{(I_1 - I_3)}{4(\rho^2 - 1)}
\end{aligned}
$$

Note that the Eshelby tensor in matrix notationis is not symmetric ($S_{iijj} \neq S_{jjii}$).

6.2 Spherical Inclusion in a Transversely Isotropic Matrix

The $\mathbb{P}$ tensor has the form of the stiffness tensor of a transverse isotropic material:

$$P_{2222} = P_{1111},\ P_{2211} = P_{1122},\ P_{3311} = P_{1133},\ P_{2323} = P_{1313},\ P_{1212} = 1/2(P_{1111} - P_{1122})$$

where:

$$
\begin{aligned}
P_{1111} &= \frac{3}{16} I_1(C_{44}, C_{33} - 2C_{44}, C_{44} - C_{33}) + \frac{1}{16} I_2(1, -1) \\
P_{1122} &= \frac{1}{16} I_1(C_{44}, C_{33} - 2C_{44}, C_{44} - C_{33}) - \frac{1}{16} I_2(1, -1) \\
P_{1133} &= \frac{1}{4} I_1(0, -C_{13} - C_{44}, C_{13} + C_{44}) \\
P_{3333} &= \frac{1}{2} I_1(0, C_{11}, C_{44} - C_{11}) \\
P_{1313} &= \frac{1}{16} I_1(C_{11}, -2(C_{11} + C_{13}), C_{11} + C_{33} + 2C_{13}) - \frac{1}{16} I_2(0, -1)
\end{aligned}
$$

and

$$I_1(l,m,n) = \frac{1}{C_{33}C_{44}}\int_{-1}^{1}\frac{\rho}{\left\{1-(1-\rho^2)\xi^2\right\}^{3/2}}\cdot\frac{l+m\xi^2+n\xi^4}{\left\{\gamma_1^2+(1-\gamma_1^2)\xi^2\right\}\left\{\gamma_2^2+(1-\gamma_2^2)\xi^2\right\}}d\xi$$

$$I_2(l,m) = \int_{-1}^{1}\frac{\rho}{\left\{1-(1-\rho^2)\xi^2\right\}^{3/2}}\cdot\frac{l+m\xi^2}{\left\{\frac{1}{2}(C_{11}-C_{12})+(C_{44}-\frac{1}{2}(C_{11}-C_{12}))\xi^2\right\}}d\xi$$

where γ_1^2, γ_2^2 are the roots of:

$$C_{33}C_{44}x^2-(C_{11}C_{33}-2C_{13}C_{44}-C_{13}^2)x+C_{11}C_{44}=0$$

Microelasticity of Bone

Christian Hellmich [*‡]

[*] Vienna University of Technology (TU Wien), Institute for Mechanics of Materials and Structures, Karlsplatz 13/202, A-1040 Wien (Vienna)

Abstract Despite the complex hierarchical organization and the astonishing variety of bones, it was recently possible to identify a few elementary mechanical components at the micro and nanolevel of the material: hydroxyapatite crystals, collagen molecules, and water. The mechanical properties (i.e. elasticity) of these elementary components are (up to experimental scattering) the same for all bones; they are 'universal', i.e., independent of tissue-type, species, and anatomical location. The mechanical interaction between these elementary components (mechanical morphology) and the dosages of these components in different tissues determine the macroscopic material properties. Whereas the mechanical interaction is limited to certain patterns, it is mainly the tremendous number of different dosages of the elementary components by which nature produces the admirable diversity of different bones.

This contribution covers the description and validation of a microelastic model for bone, based on a four-step homogenization scheme. At a length scale of several hundred nanometers, oriented, highly organized collagen molecules, the minority of the hydroxyapatite crystals present in bone tissues, and water build up (mineralized) fibrils. At the same length scale, but in the extrafibrillar space, the majority of (largely disordered) hydroxyapatite crystals build up a mineral foam (polycrystal), with water filling the inter-crystalline (nano)space. At a length scale of several micrometers, the fibrils and the extrafibrillar space build up solid bone matrix or ultrastructure. Finally, at a length scale of several milimeters, macroscopic bone material (cortical or trabecular bone) comprises solid bone matrix and the microporous space (Haversian and Volkmann canals, as well as the intertrabecular space).

Model validation rests on statistically and physically independent experiments: The ultrastructural or macroscopic (=microstructural) stiffness values predicted by the micromechanical model on the basis of tissue-independent phase stiffness properties of hydroxyapatite, collagen, and water (experimental set I) for tissue-specific composition data (experimental set IIa) are compared to corresponding experimentally determined tissue-specific stiffness values (experimental set IIb). Experinental set I comprises ultrasonic and Brillouin light scattering tests; experimental set IIa comprises dehydration-demineralization tests, neutron diffraction tests, mass density tests based on Archimedes' principle, and microradiography; experimental set IIb comprises ultrasonic tests at different frequency ranges, and quasi-static tests.

Having in mind that the aforementioned dosages are dependent on complex biochemical control cycles (defining the metabolism of the organism), the purely mechanical theory can be linked to biology, biochemistry, and, on the applied side, to clinical practice.

1 Introduction – Hierarchical organization of bone

Quoting a recent book on skeletal tissue mechanics (Martin et al., 2001), p.177, 'no one has developed a theory of bone that can satisfactorily explain its mechanical behavior on the basis of its internal structure and composition'. In more recent years, however, continuum micromechanics turned out to be a powerful tool for developing such a theory (Hellmich and Ulm, 2002b; Hellmich et al., 2004b,a). The present contribution summarizes and extends results of the aforementioned references.

The hierarchical organization of bone and its mechanical implications have attracted researchers for hundreds of years. In the line of pertinent contributions in the field (Katz et al., 1984; Weiner and Wagner, 1998), we here distinguish five levels of hierarchical organization:

- The macrostructure at an observation scale of several mm to cm, where cortical (or compact) bone and trabecular (or spongy) bone can be distinguished [Fig. 1(a) and (b)];
- The microstructure at an observation scale of several 100 μm to several mm, where cylindrical units called osteons build up cortical bone, and where the single trabecular struts or plates can be distinguished [Fig.1(c) and (d)];
- The ultrastructure (or extracellular solid bone matrix) at an observation scale of 5 to 10 μm, comprising the material building up both trabecular struts and osteons [Fig.1(e)].
- Within the ultrastructure, collagen-rich domains [light areas in Fig.1(e)] and collagen-free domains [dark areas in Fig.1(e)] can be distinguished at an observation scale of several hundred nanometers. Commonly, these domains are referred to as fibrils and extrafibrillar space.
- Finally, at an observation scale of several ten nanometers, the so-called elementary components of mineralized tissues can be distinguished. These are
  - Plate or needle-shaped mineral crystals consisting of impure hydroxyapatite (HA; $Ca_{10}[PO_4]_6[OH]_2$) with typical 1 to 5 nm thickness, and 25 to 50 nm length (Weiner and Wagner, 1998) [Fig.1(f)];
  - Long cylindrically shaped collagen molecules with a diameter of about 1.2 nm and a length of about 300 nm (Lees, 1987), which are self-assembled in staggered organizational schemes (fibrils) with characteristic diameters of 50 to 500 nm (Cusack and Miller, 1979; Miller, 1984; Lees et al., 1990, 1994b; Weiner et al., 1997; Weiner and Wagner, 1998; Rho et al., 1998; Prostak and Lees, 1996), [Fig.1(g)]; several covalently bonded fibrils are sometimes referred to as fibers;
  - Different non-collagenous organic molecules, predominantly lipids and proteins (Urist et al., 1983; Hunter et al., 1996); and
  - Water.

The organization of the elementary components within the ultrastructure of mineralized tissues (bone and mineralized tendons) has provoked some controversy. This controversy concerns first the distribution of mineral and collagen in the ultrastructure, i.e. the question how much mineral is present inside and outside the fibrils. In fact, the ratio of

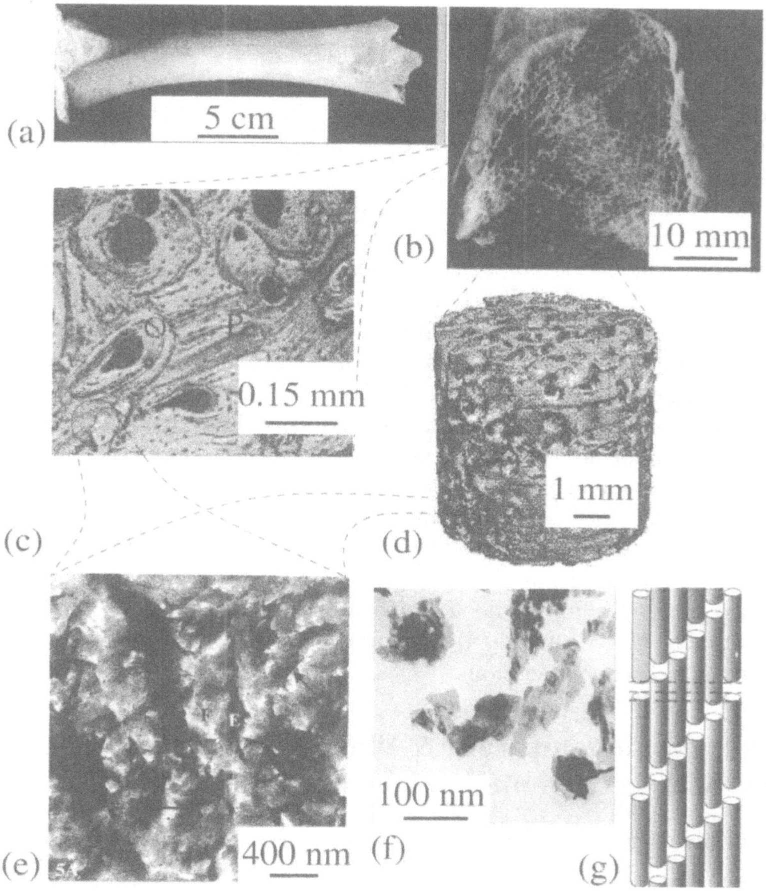

Figure 1: Hierarchical organization of bone: (a) whole long bone (macrostructure)(+); (b) section through long bone (macrostructure)(+); (c) osteonal cortical bone (microstructure)(o); (d) trabecular spaceframe (microstructure)(□); (e) ultrastructure(×); (f) hydroxyapatite crystals (elementary components)(+); (g) collagen molecules (elementary components)(+); (+)... from (Weiner and Wagner, 1998), permission for reproduction granted by Annual Reviews, www.AnnualReviews.org; (o)... from (Hoffler et al., 2000), permission for reproduction granted by Elsevier Science; (□)... from (Ding and Hvid, 2000), permission for reproduction granted by Elsevier Science; (×) ... from (Prostak and Lees, 1996), permission for reproduction granted by Springer Verlag, p.478, Fig. 5a

extrafibrillar to intrafibrillar mineral varies from tissue to tissue. However, the average mineral concentration in the non-collageneous space, i.e. in the space not occupied by collagen molecules, is – probably in all tissues – the same inside and outside the fibrils, as Hellmich and Ulm (2001, 2003) have shown recently, based on experimental work of Bonar et al. (1985); Lees (1987); Prostak and Lees (1996); Zylberberg et al. (1998). Secondly, the controversy concerns the impact of the organization of the elementary components on the mechanical interaction of collagen and hydroxyapatite in the ultrastructure. While the proposition of Currey (1969), a 'mineral-reinforced collagen matrix', attributing the anisotropy of bone ultrastructure to the mineral, has inspired a lot of scientific work, e.g. (Katz, 1980, 1981; Sasaki, 1991; Mammone and Hudson, 1993; Pidaparti et al., 1996; Ascenzi, 1999; Jäger and Fratzl, 2000), also the inverse proposition of a 'collagen-reinforced mineral matrix' has been made. In this vision, the collagen is thought to be the anisotropy-inducing element, as Lees and Rollins (1972); Lees et al. (1979); Cusack and Miller (1979); Lees et al. (1981, 1990) have stated as early as 1979. In recent years, this vision has been corroborated independently by material science (Benezra Rosen et al., 2002), by basic mechanical energy considerations close to those of cellular solid mechanics (Hellmich and Ulm, 2002a), by computational homogenization techniques for elasticity and fracture of cortical bone (Crolet et al., 1993; Aoubiza et al., 1996; Arnoux et al., 2002) and from continuum micromechanics approaches for the elasticity of all types of (cortical and trabecular) bone (Hellmich and Ulm, 2002b; Hellmich et al., 2004b,a). Especially the latter approaches have proven to be powerful predictive tools. They constitute a solid basis for the subsequent sections.

2 Multi-step homogenization of elastic properties – Fundamentals of continuum micromechanics

2.1 Representative volume element – averaging

The following statements follow closely the reviews of Zaoui (1997b,a, 2002).

A central concept of continuum micromechanics is that for the study of deformations of a solid, the observer must physically label or mark constitutive material elements (*representative volume elements – RVE*), (Zaoui, 1997b; Salencon, 2001), p.8. The characteristic length of such an RVE, ℓ, must be both considerably larger than the dimensions of heterogeneities within the RVE, d, and significantly smaller than the characteristic lengths of geometry or loading[1] of a structure built up by the material defined on the RVE, $\mathcal{L}$; in mathematical terms, this separation-of-scales requirement reads as: $d \ll \ell \ll \mathcal{L}$.

In general, the microstructure within each RVE is so complicated that it cannot be described in complete detail. Rather, only some properties of the constituents (phases) of the RVE are known. Such properties can be specified in a quantitative or in a qualitative way. Quantitative properties are e.g. volume fractions of phases, or (average) elastic properties of phases. Qualitative information concerns the 'morphological' description of the inhomogeneous material, as, e.g., the isotropy or the symmetries of anisotropy of the

[1]In case of dynamic excitations, a wavelength λ is typically such a characteristic length of loading.

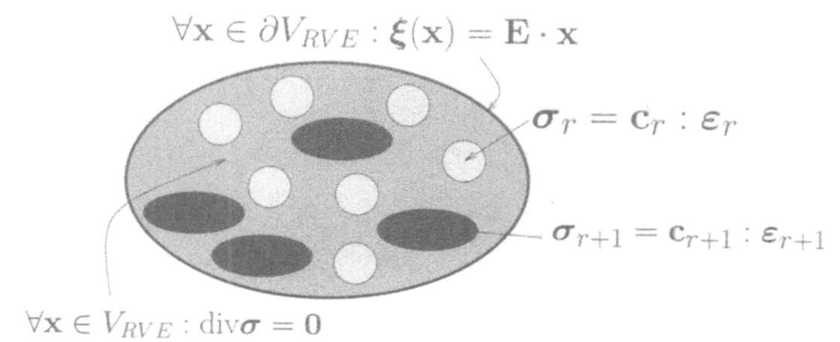

Figure 2: Loading of a representative volume element, built up by phases r with stiffness $\mathbf{C}_r$, according to continuum micromechanics (in the linear elastic case): Displacements $\boldsymbol{\xi}$, related to a constant (homogenized) strain $\boldsymbol{E}$, are imposed at the boundary of the RVE

spatial distribution of the phases, the existence of one connected 'matrix phase' in which one or several 'inclusion phases' with different shapes are embedded (as in a reinforced composite material), or the disordered arrangement of all phases (as in a polycrystal).

One goal of continuum micromechanics is to estimate the elastic properties $\mathbf{C}$ of the material defined on the RVE (the macrohomogeneous heterogeneous medium) from the phase properties (volume fractions f_r and elastic stiffness $\mathbf{c}_r$) as well as from the morphological information. This procedure is referred to as homogenization or one homogenization step. Thereby, the phase stiffness $\mathbf{c}_r$ relates the (average 'microscopic') strain in phase r, $\boldsymbol{\varepsilon}_r$, to the (average 'microscopic') stress in phase r, $\boldsymbol{\sigma}_r$;

$$\boldsymbol{\sigma}_r = \mathbf{c}_r : \boldsymbol{\varepsilon}_r \tag{2.1}$$

On the other hand, the homogenized ('macroscopic') stress

$$\boldsymbol{\Sigma} = \langle \boldsymbol{\sigma} \rangle = \frac{1}{V_{RVE}} \int_{V_{RVE}} \boldsymbol{\sigma} dV = \sum_r f_r \boldsymbol{\sigma}_r \tag{2.2}$$

and the homogenized ('macroscopic') strain

$$\boldsymbol{E} = \langle \boldsymbol{\varepsilon} \rangle = \frac{1}{V_{RVE}} \int_{V_{RVE}} \boldsymbol{\varepsilon} dV = \sum_r f_r \boldsymbol{\varepsilon}_r \tag{2.3}$$

are related by the homogenized ('macroscopic') stiffness $\mathbf{C}$,

$$\boldsymbol{\Sigma} = \mathbf{C} : \boldsymbol{E} \tag{2.4}$$

whereby V_{RVE} is the volume of the RVE. Imposition of displacements $\boldsymbol{\xi}(\boldsymbol{x}) = \boldsymbol{E} \cdot \boldsymbol{x}$ at the boundary of the RVE ensures validity of Eqn.(2.3); with $\boldsymbol{x}$ as the position vector for locations within or at the boundary of the RVE (Figure 2).

Inside the RVE, the ('microscopic') stresses fulfill the equilibrium condition

$$\text{div}\boldsymbol{\sigma} = \mathbf{0} \tag{2.5}$$

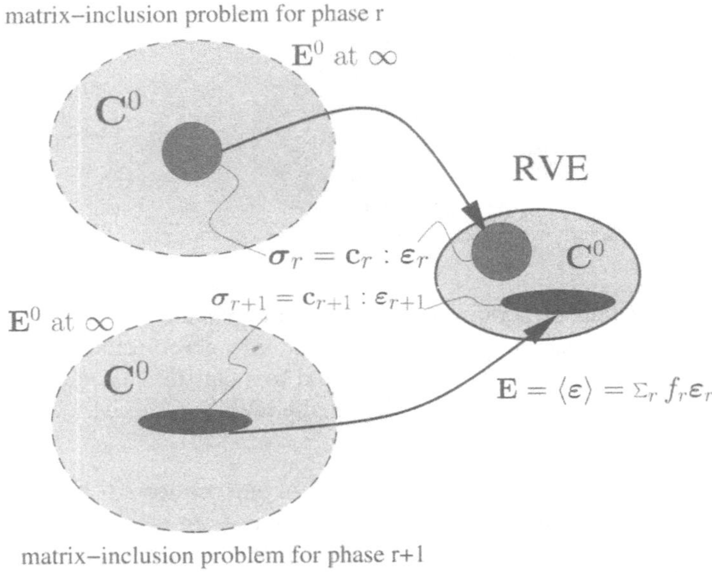

Figure 3: Estimation of relation between phase strain $\boldsymbol{\varepsilon}_r$ and overall homogenized strain $\boldsymbol{E}$ ('strain concentration') from matrix inclusion problems, according to e.g. (Zaoui, 1997b)

The estimation of homogenized stiffness $\mathbf{C}$ from the stiffness properties of the phases, $\mathbf{c}_r$, is based on the estimates for the fourth-order concentration tensors $\mathbf{A}_r$ of each of the phases r. The concentration tensor of phase r, $\mathbf{A}_r$, establishes a link between the homogenized ('macroscopic') strain $\boldsymbol{E}$ and the (average 'microscopic') strain $\boldsymbol{\varepsilon}_r$ in phase r,

$$\boldsymbol{\varepsilon}_r = \mathbf{A}_r : \boldsymbol{E} \tag{2.6}$$

Insertion of Eqn.(2.6) into Eqn.(2.1) and averaging over all phases according to Eqn.(2.2) leads to

$$\boldsymbol{\Sigma} = \sum_r f_r \mathbf{c}_r : \mathbf{A}_r : \boldsymbol{E} \tag{2.7}$$

From Eqn.(2.7) and Eqn.(2.4) we can identify a relation between the phase stiffness tensors $\mathbf{c}_r$ and the overall homogenized stiffness $\mathbf{C}$ of the RVE,

$$\mathbf{C} = \sum_r f_r \mathbf{c}_r : \mathbf{A}_r \tag{2.8}$$

2.2 Matrix-inclusion problems

The concentration tensors $\mathbf{A}_r$ may be estimated from analytical solutions for the problem of an ellipsoidal inhomogeneity H with stiffness $\mathbf{c}_H$, perfectly bonded to an infinite surrounding matrix with stiffness $\mathbf{C}^0$ (Figure 3). This matrix is submitted to homogeneous strains $\boldsymbol{E}^0$ at infinity. The strains resulting from $\boldsymbol{E}^0$ in the ellipsoidal inclusion, $\boldsymbol{\varepsilon}_H$, are uniform and given by

$$\boldsymbol{\varepsilon}_H = \left[\mathbf{I} + \mathbf{P}_H^0 : (\mathbf{c}_H - \mathbf{C}^0)\right]^{-1} : \boldsymbol{E}^0 \tag{2.9}$$

where $\mathbf{I}$, $I_{ijkl} = 1/2(\delta_{ik}\delta_{jl} + \delta_{il}\delta_{kj})$, is the fourth-order unity tensor, δ_{ij} is the Kronecker delta, and $\mathbf{P}_H^0$ is a fourth-order tensor characterizing the interaction between the inclusion and the matrix,

$$P_{H,ijkl}^0 = -\left(\partial^2 \left(\int_H G_{ik}^0(\boldsymbol{x} - \boldsymbol{x}')dV'\right) / \partial x_j \partial x_l\right)_{(ij)(kl)} \tag{2.10}$$

In Eqn.(2.10), $(ij)(kl)$ refers to symmetrization with respect to (ij) and (kj). Green's function $G_{ik}^0(\boldsymbol{x} - \boldsymbol{x}')$ gives the displacement at point $\boldsymbol{x}$ in a linear elastic medium with stiffness $\mathbf{C}^0$, resulting from a unit force $\delta(\boldsymbol{x} - \boldsymbol{x}')$ at point $\boldsymbol{x}'$ of the medium; $\delta(\mathbf{0}) = \infty$, $\delta(\boldsymbol{x} \neq \boldsymbol{x}') = 0$, $\int_V \delta(\boldsymbol{x} - \boldsymbol{x}')dV(\boldsymbol{x}) = 1$. $\mathbf{P}_H^0$ depends on the stiffness properties of the matrix, and the shape of the inclusion. Analytical forms of $\mathbf{S}_H^{Esh,0} = \mathbf{P}_H^0 : \mathbf{C}^{0,-1}$, the so-called Eshelby tensor, were given for isotropic matrices and special limit cases of ellipsoidal inclusison (prolate spheroid, oblate spheroid, cylinder, sphere) by Eshelby (1957). The determination of Green's functions occuring in Eqn. (2.10) for *anisotropic matrices* is a delicate task. As a remedy, Laws (1977, 1985) has used the intimate relation between the inclusion problem and the interface problem in order to derive an integral expression for the **P**-tensor related to ellipsoidal inclusions in anisotropic media, which does not require knowledge about any Green's function,

$$P_{H,ijkl}^0 = \frac{1}{16\pi\alpha^{1/2}} \int_\Omega \frac{1}{t^3} \{\hat{g}_{il}w_j w_k + \hat{g}_{ik}w_j w_l + \hat{g}_{jl}w_i w_k + \hat{g}_{jk}w_i w_l\} dS(\boldsymbol{w}) \tag{2.11}$$

In Eqn.(2.11), the shape of the ellipsoidal inhomogeneity H is considered by $\alpha = \det\alpha_{ij}$, being related to the equation of the ellipsoid,

$$\alpha_{ij}x_i x_j = 1 \tag{2.12}$$

$dS(\boldsymbol{w})$ is a surface element of the unit sphere (with surface Ω); w_1, w_2, and w_3 are the components of the unit length vector $\boldsymbol{w}$ pointing from the origin of the sphere to the surface element $dS(\boldsymbol{w})$, and $t^2 = \alpha_{ij}^{-1}w_i w_j$. Finally, $\hat{g}_{ik}$ are the components of the inverse of the second order tensor $C_{ijkl}^0 w_j w_l$, with C_{ijkl}^0 denoting the stiffness of the anisotropic matrix.

2.3 Stiffness estimates

For estimation of $\mathbf{A}_r$, the strains $\boldsymbol{\varepsilon}_r$ in each phase r are considered to be equal to those of an ellipsoidal inhomogeneity H_r with (phase) stiffness $\mathbf{c}_r$, embedded in a fictitious homogeneous medium with stiffness $\mathbf{C}^0$, submitted to some homogeneous strain $\boldsymbol{E}^0$ at infinity (Figure 3),

$$\boldsymbol{\varepsilon}_r = \left[\mathbf{I} + \mathbf{P}_r^0 : (\mathbf{c}_r - \mathbf{C}^0)\right]^{-1} : \boldsymbol{E}^0 \tag{2.13}$$

Insertion of Eqn.(2.13) into Eqn.(2.3) provides a relation between the homogenized ('macroscopic') strain $\boldsymbol{E}$ and the homogeneous strain at the infinity of the fictitious matrix, $\boldsymbol{E}^0$,

$$\boldsymbol{E}^0 = \left\{ \sum_r f_r \left[\mathbf{I} + \mathbf{P}_r^0 : (\mathbf{c}_r - \mathbf{C}^0)\right]^{-1} \right\}^{-1} : \boldsymbol{E} \tag{2.14}$$

Backsubstitution of Eqn.(2.14) into Eqn.(2.13), and comparison of the result with Eqn.(2.6) delivers the sought estimate for the concentration tensor, $\mathbf{A}_r^{est}$, as

$$\mathbf{A}_r^{est} = \left[\mathbf{I} + \mathbf{P}_r^0 : (\mathbf{c}_r - \mathbf{C}^0)\right]^{-1} : \left\{ \sum_s f_s \left[\mathbf{I} + \mathbf{P}_s^0 : (\mathbf{c}_s - \mathbf{C}^0)\right]^{-1} \right\}^{-1} \tag{2.15}$$

Backsubstitution of Eqn.(2.15) into Eqn.(2.8) delivers the sought estimate for the homogenized ('macroscopic') stiffness tensor, $\mathbf{C}^{est}$, as

$$\mathbf{C}^{est} = \sum_r f_r \mathbf{c}_r : \left[\mathbf{I} + \mathbf{P}_r^0 : (\mathbf{c}_r - \mathbf{C}^0)\right]^{-1} : \left\{ \sum_s f_s \left[\mathbf{I} + \mathbf{P}_s^0 : (\mathbf{c}_s - \mathbf{C}^0)\right]^{-1} \right\}^{-1} \tag{2.16}$$

Choice of matrix stiffness $\mathbf{C}^0$ describes the interactions between the phases: For $\mathbf{C}^0$ coinciding with one of the phase stiffnesses (Mori-Tanaka scheme), a composite material is represented (contiguous matrix with inclusions); for $\mathbf{C}^0 = \mathbf{C}^{est}$ (self-consistent scheme), a dispersed arrangement of the phases is considered (typical for polycrystals).

2.4 Multistep homogenization

The mechanical behavior of a single phase can be estimated by introduction of RVEs within this phase, with dimensions $\ell_2 \ll d$, comprising again smaller phases with characteristic lengths $d_2 \ll \ell_2$, and so on, see Figure 4. This leads to a multistep homogenization scheme. Such a procedure should, in the end, provide access to 'universal' properties inherent to all bone tissues, at a sufficiently low observation scale. A respective strategy will be shown next.

3 Micromechanical representation of bone

Bone elasticity is predicted on the basis a four-step homogenization scheme. The extrafibrillar space [Figure 5(a)] and the fibrils [Figure 5(b)] build up the solid bone matrix or ultrastructure [Figure 5(c)]. Finally, macroscopic bone material (cortical or trabecular bone) comprises solid bone matrix and the microporous space (Haversian and Volkmann canals, as well as the intertrabecular space).

3.1 Extrafibrillar space: Hydroxyapatite foam

Within an RVE $\check{V}_{ef}$ of hydroxyapatite foam with ℓ_{ef} =100-500 nm characteristic length [see Fig. 5(a)], the phases 'hydroxyapatite (HA)' and 'ultrastructural water with non-collageneous organics (uw)' occupy volume fractions $\check{f}_{HA}$ and $\check{f}_{uw}$, $\check{f}_{HA} + \check{f}_{uw} = 1$.

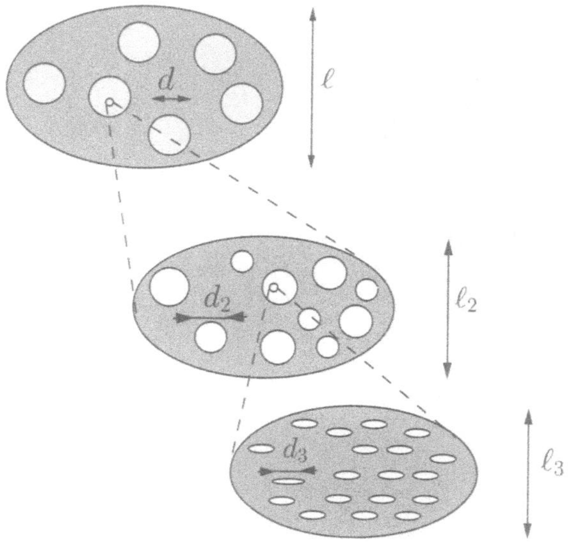

Figure 4: Multistep homogenization: Definition of several representative volume elements

The considerable disorder of the hydroxyapatite crystals in the foam, shown by Lees et al. (1994b); Fratzl et al. (1996); Benezra Rosen et al. (2002); Hellmich and Ulm (2002a), motivates the use of a self-consistent scheme with inclusions of spherical shape. Accordingly, in order to estimate the stiffness of the extrafibrillar mineral foam, $\mathbf{C}_{ef}^{SCS}$, (2.16) is specified for two (spherical) inclusion phases, i.e. for $r \in [HA, uw]$, for $\mathbf{c}_{HA}$ and $\mathbf{c}_{uw}$ according to Table 1, for $\mathbf{C}^0 = \mathbf{C}_{ef}^{SCS}$, and for $\mathbf{P}_{HA}^0 = \mathbf{P}_{uw}^0 = \mathbf{P}_{sph}^{ef}$, resulting in

$$\begin{aligned}\mathbf{C}_{ef}^{SCS} = &\left\{\textstyle\sum_r \check{f}_r \mathbf{c}_r : \left[\mathbf{I} + \mathbf{P}_{sph}^{ef} : (\mathbf{c}_r - \mathbf{C}_{ef}^{SCS})\right]^{-1}\right\} : \\ &\left\{\textstyle\sum_s \check{f}_s \left[\mathbf{I} + \mathbf{P}_{sph}^{ef} : (\mathbf{c}_s - \mathbf{C}_{ef}^{SCS})\right]^{-1}\right\}^{-1}, \quad r, s \in [HA, uw].\end{aligned} \tag{3.1}$$

The elasticity tensors $\mathbf{c}_r$ of the isotropic phases are of the form

$$\mathbf{c}_r = 3\,k_r\,\mathbf{J} + 2\,\mu_r\,\mathbf{K}, \quad r \in [HA, uw], \tag{3.2}$$

where k_r and μ_r are the bulk and shear moduli of the phases, see Table 1, $\mathbf{J}$, $J_{ijkl} = 1/3\,\delta_{ij}\,\delta_{kl}$, is the volumetric part of the fourth-order unity tensor $\mathbf{I}$, and $\mathbf{K} = \mathbf{I} - \mathbf{J}$ is the deviatoric part. $\mathbf{P}_{sph}^{ef}$ follows from specification of (2.10) for a spherical inclusion in an isotropic matrix of stiffness $\mathbf{C}_{ef}^{SCS}$. It is of the form (Zaoui, 1997b; Eshelby, 1957)

$$\mathbf{P}_{sph}^{ef} = \mathbf{S}_{sph}^{Esh,ef} : \mathbf{C}_{ef}^{SCS,-1}, \tag{3.3}$$

$$\mathbf{S}_{sph}^{Esh,ef} = \alpha_{ef}^{SCS}\,\mathbf{J} + \beta_{ef}^{SCS}\,\mathbf{K} \tag{3.4}$$

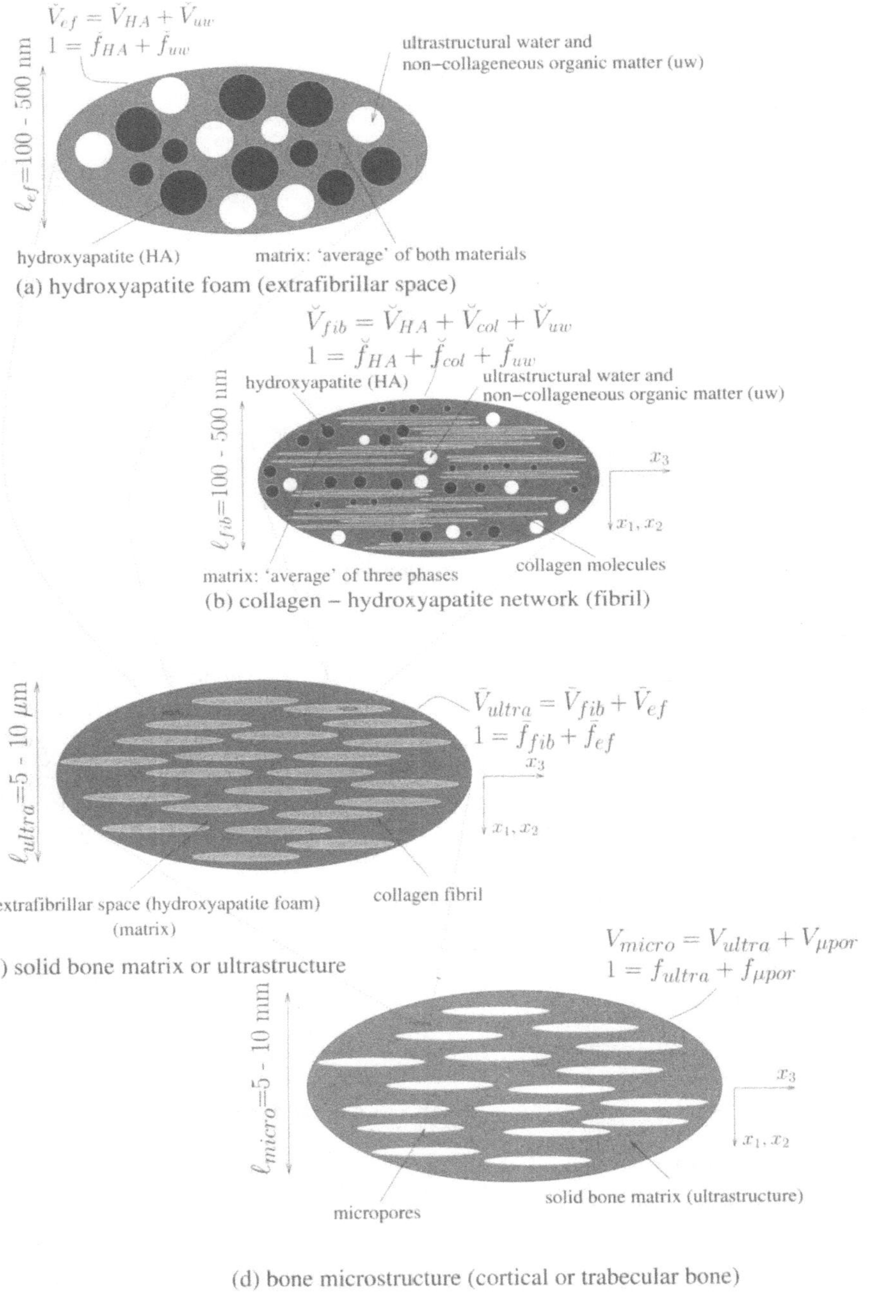

Figure 5: Micromechanical representation of bone by means of a four-step homogenization procedure

with

$$\alpha_{ef}^{SCS} = \frac{3\,k_{ef}^{SCS}}{3\,k_{ef}^{SCS} + 4\,\mu_{ef}^{SCS}}\,, \qquad \beta_{ef}^{SCS} = \frac{6\,(k_{ef}^{SCS} + 2\,\mu_{ef}^{SCS})}{5\,(3\,k_{ef}^{SCS} + 4\,\mu_{ef}^{SCS})}\,. \tag{3.5}$$

For implicit solution of (3.1), together with (3.2)-(3.5), with respect to $\mathbf{C}_{ef}^{SCS}$, we refer to (Hellmich et al., 2004a; Hellmich and Ulm, 2002b).

3.2 Mineralized collagen fibril

The mineralized collagen fibril is envisioned as spaceframe of disordered hydroxyapatite crystals with water, non-collageneous organic matter, and a predominant amount of oriented collagen molecule 'ropes' inbetween. Hence, within an RVE $\breve{V}_{fib}$ of fibrillar material with ℓ_{fib} =100-500 nm characteristic length (see Fig. 5(b)), the phases 'hydroxyapatite (HA)', 'ultrastructural water with non-collageneous organics (uw)', and 'collagen' occupy volume fractions $\breve{f}_{HA}$, $\breve{f}_{uw}$, and $\breve{f}_{col}$, $\breve{f}_{HA} + \breve{f}_{uw} + \breve{f}_{col} = 1$. The aforementioned spaceframe, where all phases are in contact with each other, is suitably represented by a self-consistent scheme with spherical inclusions for the mineral and the water phase, and cylindrical inclusions for the collagen phase. Accordingly, in order to estimate the stiffness of the mineralized collagen fibril, $\mathbf{C}_{fib}^{SCSII}$, (2.16) is specified for two spherical inclusion phases (water and mineral) and one cylindrical inclusion phase (collagen), i.e. for $r \in [HA, uw, col]$, for $\mathbf{c}_{HA}$, $\mathbf{c}_{uw}$, and $\mathbf{c}_{col}$ according to Table 1, for $\mathbf{C}^0 = \mathbf{C}_{fib}^{SCSII}$, as well as for $\mathbf{P}_{HA}^0 = \mathbf{P}_{uw}^0 = \mathbf{P}_{sph}^{fib}$ and $\mathbf{P}_{col}^0 = \mathbf{P}_{cyl}^{fib}$, resulting in

$$\begin{aligned}
\mathbf{C}_{fib}^{SCSII} = {} & \Big\{ \breve{f}_{col}\,\mathbf{c}_{col} : \Big[\mathbf{I} + \mathbf{P}_{cyl}^{fib} : \big(\mathbf{c}_{col} - \mathbf{C}_{fib}^{SCSII}\big)\Big]^{-1} \\
& + \textstyle\sum_s \breve{f}_s\,\mathbf{c}_s : \Big[\mathbf{I} + \mathbf{P}_{sph}^{fib} : \big(\mathbf{c}_s - \mathbf{C}_{fib}^{SCSII}\big)\Big]^{-1} \Big\} : \\
& \Big\{ \breve{f}_{col} \Big[\mathbf{I} + \mathbf{P}_{cyl}^{fib} : \big(\mathbf{c}_{col} - \mathbf{C}_{fib}^{SCSII}\big)\Big]^{-1} \\
& + \textstyle\sum_s \breve{f}_s \Big[\mathbf{I} + \mathbf{P}_{sph}^{fib} : \big(\mathbf{c}_s - \mathbf{C}_{fib}^{SCSII}\big)\Big]^{-1} \Big\}^{-1} ; \\
& s \in [HA, uw].
\end{aligned} \tag{3.6}$$

$\mathbf{P}_{cyl}^{fib}$ follows from specification of (2.11) for a cylindrical inclusion in a transversely isotropic matrix with stiffness $\mathbf{C}_{fib}^{SCSII}$, where the plane of isotropy is oriented perpendicular to the long axis of the cylinder, see (Hellmich et al., 2004a) for details. The non-zero components of the symmetric tensor $\mathbf{P}_{cyl}^{fib}$ read as follows:

$$\begin{aligned}
& P_{cyl,1111}^{fib} = P_{cyl,2222}^{fib} = \\
& = 1/8\,(5\,C_{fib,1111}^{SCSII} - 3\,C_{fib,1122}^{SCSII})/C_{fib,1111}^{SCSII}/\mathcal{D}_2\,,
\end{aligned} \tag{3.7}$$

$$\begin{aligned}
& P_{cyl,1122}^{fib} = P_{cyl,2211}^{fib} = \\
& = -1/8\,(C_{fib,1111}^{SCSII} + C_{fib,1122}^{SCSII})/C_{fib,1111}^{SCSII}/\mathcal{D}_2\,,
\end{aligned} \tag{3.8}$$

$$P_{cyl,2323}^{fib} = P_{cyl,1313}^{fib} = 1/(8\,C_{fib,2323}^{SCSII})\,, \tag{3.9}$$

$$P_{cyl,1212}^{fib} = 1/8\,(3\,C_{fib,1111}^{SCSII} - C_{fib,1122}^{SCSII})/C_{fib,1111}^{SCSII}/\mathcal{D}_2\,, \tag{3.10}$$

whereby

$$\mathcal{D}_2 = C^{SCSII}_{fib,1111} - C^{SCSII}_{fib,1122} \tag{3.11}$$

For implicit solution of (3.6), together with (3.7)-(3.10), with respect to $\mathbf{C}^{SCSII}_{fib}$, we refer to (Hellmich et al., 2004a).

$\mathbf{P}^{fib}_{sph}$ follows from specification of (2.11) for a spherical inclusion in a transversely isotropic matrix with stiffness $\mathbf{C}^{SCSII}_{fib}$. The non-zero components of the symmetric tensor $\mathbf{P}^{sph}_{cyl}$ can be expressed as single integrals of an auxiliary variable x, see (Hellmich et al., 2004a) for details.

3.3 Solid bone matrix or ultrastructure

Within an RVE $\bar{V}_{ultra}$ of bone ultrastructure with ℓ_{ultra}=5-10 μm characteristic length [see Figure 5(c)], cylindrical fibrillar inclusions with characteristic diameters d_{ultra} of several hundred microns are embedded in a contiguous matrix built up by the mineral foam material of homogenization step I (3.1). The volume fractions occupied by these two phases are $\bar{f}_{fib}$ and $\bar{f}_{ef}$, $\bar{f}_{fib} + \bar{f}_{ef} = 1$. Using a Mori-Tanaka scheme for the estimation of the stiffness of the composite material 'solid bone matrix or ultrastructure', $\mathbf{C}^{MT}_{ultra}$, we specify (2.16) for one (cylindrical) inclusion phase (mineralized collagen fibrils), and one extrafibrillar matrix, i.e. for $r \in [fib, ef]$, for $\mathbf{c}_{fib} = \mathbf{C}^{SCSII}_{fib}$ from (3.6), for $\mathbf{C}^0 = \mathbf{c}_{ef} = \mathbf{C}^{SCS}_{ef}$ from (3.1), and for $\mathbf{P}^0_{fib} = \mathbf{P}^{ef}_{cyl}$, resulting in

$$\begin{aligned}\mathbf{C}^{MT}_{ultra} = \; & \left\{(1-\bar{f}_{fib})\mathbf{C}^{SCS}_{ef} + \bar{f}_{fib}\mathbf{C}^{SCSII}_{fib} : \left[\mathbf{I} + \mathbf{P}^{ef}_{cyl} : (\mathbf{C}^{SCSII}_{fib} - \mathbf{C}^{SCS}_{ef})\right]^{-1}\right\} : \\ & \left\{(1-\bar{f}_{fib})\mathbf{I} + \bar{f}_{fib}\left[\mathbf{I} + \mathbf{P}^{ef}_{cyl} : (\mathbf{C}^{SCSII}_{fib} - \mathbf{C}^{SCS}_{ef})\right]^{-1}\right\}^{-1}\end{aligned} \tag{3.12}$$

The non-zero components of $\mathbf{P}^{ef}_{cyl}$ follow from substitution of 'fib' by 'ef', and of '$SCSII$' by 'SCS' in (3.7)-(3.11).

3.4 Bone microstructure (cortical or trabecular bone)

Within an RVE V_{micro} of bone microstructure (cortical or trabecular bone) with ℓ_{micro}=5-10 mm characteristic length [see Figure 5(d)], cylindrical micropores (representing Haversian and Volkmann canals as well as the intertrabecular space) with characteristic diameters of 50-500 μm, oriented in the working direction of the biological cells[2],

[2] It is interesting to relate the chosen cylindrical inclusion morphology of the micropore phase, which is obvious for Haversian canals [Fig. 6(a-c)], but less obvious for the intertrabecular space [Fig. 6(d-i)], to bone formation and remodeling activity of biological cells. In the intertrabecular porous network of trabecular bone, the same cells are at work as in the Haversian canals of cortical bone (Parfitt, 1984; Buckwalter et al., 1995a). In the course of bone remodelling, osteoblasts and osteoclasts are active on the surfaces of the solid bone matrix, at an observation scale of several tens of microns. At this scale, the working direction of cells is perpendicular to these surfaces. However, as concerns the micropore morphology, we are interested in the working direction of cells at the microstructural scale, i.e. in RVE V_{micro}

are embedded in a contiguous matrix built up by solid bone matrix material of homogenization step III (3.12). The volume fractions occupied by these two phases are f_{ultra} and $f_{\mu por}$, $f_{ultra} + f_{\mu por} = 1$. Using again the Mori-Tanaka scheme for the estimation of the stiffness of the composite material 'cortical or trabecular bone', $\mathbf{C}^{MTII}_{micro}$, we specify (2.16) for one (cylindrical) inclusion phase (micropores), and one solid bone matrix, i.e. for $r \in [\mu por, ultra]$, for $\mathbf{c}_{\mu por} = \mathbf{0}$ in case of empty micropores and $\mathbf{c}_{\mu por} = \mathbf{c}_{uw}$ in physiological conditions[3], for $\mathbf{C}^0 = \mathbf{c}_{ultra} = \mathbf{C}^{MT}_{ultra}$ from (3.12), and for $\mathbf{P}^0_{\mu por} = \mathbf{P}^{ultra}_{cyl}$, resulting in

$$\begin{aligned}\mathbf{C}^{MTII}_{micro} = \quad & \left\{(1-f_{\mu por})\mathbf{C}^{MT}_{ultra} + f_{\mu por}\mathbf{c}_{\mu por} : \left[\mathbf{I} + \mathbf{P}^{ultra}_{cyl} : (\mathbf{c}_{\mu por} - \mathbf{C}^{MT}_{ultra})\right]^{-1}\right\} : \\ & \left\{(1-f_{\mu por})\mathbf{I} + f_{\mu por}\left[\mathbf{I} + \mathbf{P}^{ultra}_{\mu por} : (\mathbf{c}_{\mu por} - \mathbf{C}^{MT}_{ultra})\right]^{-1}\right\}^{-1} \qquad (3.13)\end{aligned}$$

The non-zero components of $\mathbf{P}^{ultra}_{cyl}$ follow from substitution of 'fib' by '$ultra$', and of '$SCSII$' by 'MT' in (3.7)-(3.11).

4 Experimental validation of micromechanics model for bone

4.1 Strategy

In the line of Popper, who stated that a theory – as long as it has not been falsified – will be 'the more satisfactory the greater the severity of independent tests it survives', cited from (Mayr, 1997), p.49, the verification of the micromechanical representation of the ultrastructure and the microstructure of bone will rest on two independent experimental sets: The stiffness values $\mathbf{C}^{SCS}_{ef}$, $\mathbf{C}^{MT}_{ultra}$, and $\mathbf{C}^{MTII}_{micro}$, predicted by the micromechanical model on the basis of tissue-independent phase stiffness properties $\mathbf{c}_{HA}$, $\mathbf{c}_{col}$, $\mathbf{c}_{uw}$ (experimental set I, Table 1, see Section 4.2 for details) for tissue-specific composition data $\bar{f}_{col}$, $\bar{f}_{HA}$, d_w, and $f_{\mu por}$ (experimental set IIa, Subsection 4.3) are compared to corresponding experimentally determined tissue-specific stiffness values $\mathbf{C}^{exp}_{ef}$, $\mathbf{C}^{exp}_{ultra}$, and $\mathbf{C}^{exp}_{micro}$ (experimental set IIb, Subsection 4.4).

with $\ell_{micro} = 5-10$ mm. Since cells follow – at a smaller scale – the already present surfaces, the working direction at the scale of V_{micro} is related to the initial formation of trabecular (or cortical) bone, see e.g. (Buckwalter et al., 1995b; Farquharson and Houston, 1998; Byers et al., 2000) and references therein: Starting from a growth plate, a cartilage matrix is formed by chondrocytes which, after a series of maturation states, induce the calcification of the cartilage extracellular matrix. This matrix is then penetrated in the longitudinal direction of the prospective bone (perpendicular to the growth plate) by blood vessels, followed by osteoclastic/osteoblastic activity in the canals provided by the blood vessels, transforming the mineralized cartilage into cortical or trabecular bone tissue. Therefore, it seems reasonable to suggest that the initial direction of blood vessels is significant for the average orientation of the microporosity (Haversian canals and inter-trabecular space).

[3]The bone fluid (marrow) in the micropores acts as a phase with negligible shear stiffness and bulk modulus k_{uw}, see Table 1.

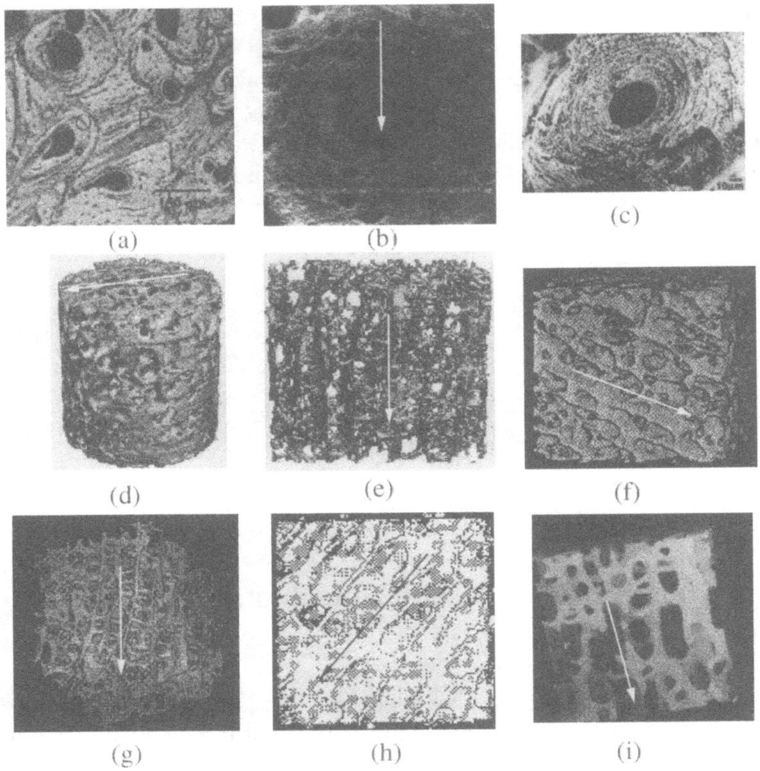

Figure 6: Microarchitecture of cortical (a-c) and trabecular bone (d-i): The cylindrical average pore shape reflects the microstructural working direction of the biological bone cells, indicated by arrows or perpendicular to the plane of the page; (a) human distal radius (Hoffler et al., 2000), (b) not specified sample (Buckwalter et al., 1995a), (c) human sample (Weiner and Wagner, 1998), (d) human proximal tibia (26 year old male) (Ding and Hvid, 2000), (e) human proximal tibia (85 year old female)(Ding and Hvid, 2000), (f) human sample (Goulet et al., 1994), (g) human proximal tibia (Odgaard et al., 1999), (h) vertebra of sperm whale (Odgaard et al., 1997), (i) human lumbar vertebra (Uchiyama et al., 1999); permission for reproduction forthcoming from Elsevier Science for (a,d,e,f,h,i), from The Journal of Bone and Joint Surgery for (b), from Annual Reviews for (c), from Kluwer Academic Publishers for (g)

phase	bulk modulus k [GPa]	shear modulus μ [GPa]	experimental source
hydroxyapatite	$k_{HA} = 82.6$	$\mu_{HA} = 44.9$	(Katz and Ukraincik, 1971)
ultrastructural water and non-collageneous organics	$k_{uw} = 2.3$	$\mu_{uw} = 0$	
	c_{ijkl} [GPa]	c_{ijkl} [GPa]	
collagen	$c_{col,3333} = 17.9$ $c_{col,1111} = 11.7$	$c_{col,1133} = 7.1$ $c_{col,1122} = 5.1$ $c_{col,1313} = 3.3$	(Cusack and Miller, 1979)

Table 1: 'Universal' (tissue and location-independent) isotropic (or transversely isotropic) phase stiffness values

4.2 'Universal' stiffnesses of elementary components – Experimental set I

Concerning the tissue-independent ('universal') phase properties, being the same for all tissues and all micromechanical representations discussed herein, we consider the following experiments: Tests with an ultrasonic[4] inferometer coupled with a solid media pressure apparatus (Katz and Ukraincik, 1971; Gilmore and Katz, 1982) reveal the isotropic elastic properties of hydroxyapatite powder, which, in view of the largely disordered arrangement of minerals (Fratzl et al., 1996; Hellmich and Ulm, 2002a), are sufficient for the characterization of the mineral phase (Hellmich and Ulm, 2002b; Hellmich et al., 2004b). Given the absence of direct measurements of (molecular) collagen, the elastic properties of (molecular) collagen are approximated by those of dry rat tail tendon, a tissue consisting almost exclusively of collagen. By means of Brillouin light scattering, Cusack and Miller (1979) have determined the respective five independent elastic constants of a transvesely isotropic material (Table 1). We assign the standard bulk modulus of water to the phase comprising ultrastructural water and non-collageneous organic matter. In case of dry tissues, we adopt $\mathbf{c}_{uw} = \mathbf{0}$.

4.3 Determination of tissue-specific volume fractions from experiments – Experimental set IIa

Dehydration - demineralization tests of Lees et al. (1979); Lees (1987); Lees et al. (1994a, 1995), as basis for determination of mineral and collagen content, $\bar{f}_{HA}$ and $\bar{f}_{col}$ Typically milimeter-sized cortical bone specimens were obtained from a wide range of vertebrates. Fully hydrated specimens were cleaned from adhering tissue and marrow. Their masses M_{wet} were determined by weighing. Their volumes V were measured by Archimedes' principle. M_{wet} and V allow for determination of the

[4]The principle of ultrasonic measurements is more discussed in Section 4.4.

specific gravity or microstructural mass density[5]

$$\rho_{micro,wet} \approx \rho_{ultra,wet} = \frac{M_{wet}}{V} \tag{4.2}$$

Afterwards, the specimens were dehydrated by vacuum dessication at 23 °C, until a constant weight M_{dry} was attained. M_{wet} and M_{dry} give access to the specimen-specific weight fraction of water,

$$WF_{H_2O} = \frac{M_{wet} - M_{dry}}{M_{wet}} \tag{4.3}$$

documented by Lees et al. (1979); Lees (1987); Lees et al. (1995) for a multitude of vertebrate (cortical) bone specimens, shown in Figure 7. Next, the specimens were immersed into a 0.5M EDTA[6] solution at a pH-value of 7.5. EDTA leached (chemically) the calcium ions out of the hydroxyapatite crystals, a process which is commonly referred to as decalcification. After the EDTA bath, the specimens were washed out or rinsed (physically), so that the dissolved substances, comprising not only the calcium ions, but the entire mineral crystals, were continuously removed out of the specimens. This sequence of leaching and rinsing was repeated until no calcium ions were detectable any more, by means of atomic absorption spectometry, in the EDTA-bath. These EDTA-washed specimens were again vacuum dessicated (dried) and subsequently, their mass $M_{EDTA,dry}$ was measured. $M_{EDTA,dry}$ and M_{wet} give access to the organic weight fraction

$$WF_{org} = \frac{M_{org}}{M_{wet}} = \frac{M_{EDTA,dry}}{M_{wet}} \tag{4.4}$$

documented by Lees et al. (1979); Lees (1987); Lees et al. (1995) for a multitude of vertebrate (cortical) bone specimens, shown in Figure 7. Based on (4.3) and (4.4), the weight fraction of mineral (hydroxyapatite) follows to be

$$WF_{HA} = 1 - WF_{H_2O} - WF_{org} \tag{4.5}$$

Considering the fact that about 90% by weight of the organic matter in bone is made up by collagen (Robinson and Elliott, 1957; Biltz and Pellegrino, 1969; Urist et al., 1983; Lees, 1987; Weiner and Wagner, 1998), yields the collagen weight fraction as

$$WF_{col} = 0.9 \times WF_{org} \tag{4.6}$$

In addition, Lees et al. (1979); Lees (1987) measured the volume of the (dense cortical) specimens after vacuum dessication (drying), $V_{dry} < V$, giving access to the mass density

[5]For cortical bone, with a microporosity $f_{\mu por}$ of typically less than 5% (Sietsema, 1995), the difference between the microstructural mass density $\rho_{micro,wet}$ and the ultrastructural mass density

$$\rho_{ultra,wet} = (\rho_{micro,wet} - \rho_{H_2O} f_{\mu por})/(1 - f_{\mu por}), \tag{4.1}$$

expressed in terms of the mass density of water ρ_{H_2O}=1 kg/dm^3, is less than 2%; therefore, we will, in the following, not distinguish between the microstructural and the ultrastructural mass density of the (dense) cortical bone tissues tested by Lees et al. (1979); Lees (1987); Lees et al. (1995).

[6]Ethylene Diamine Tetra Acetate

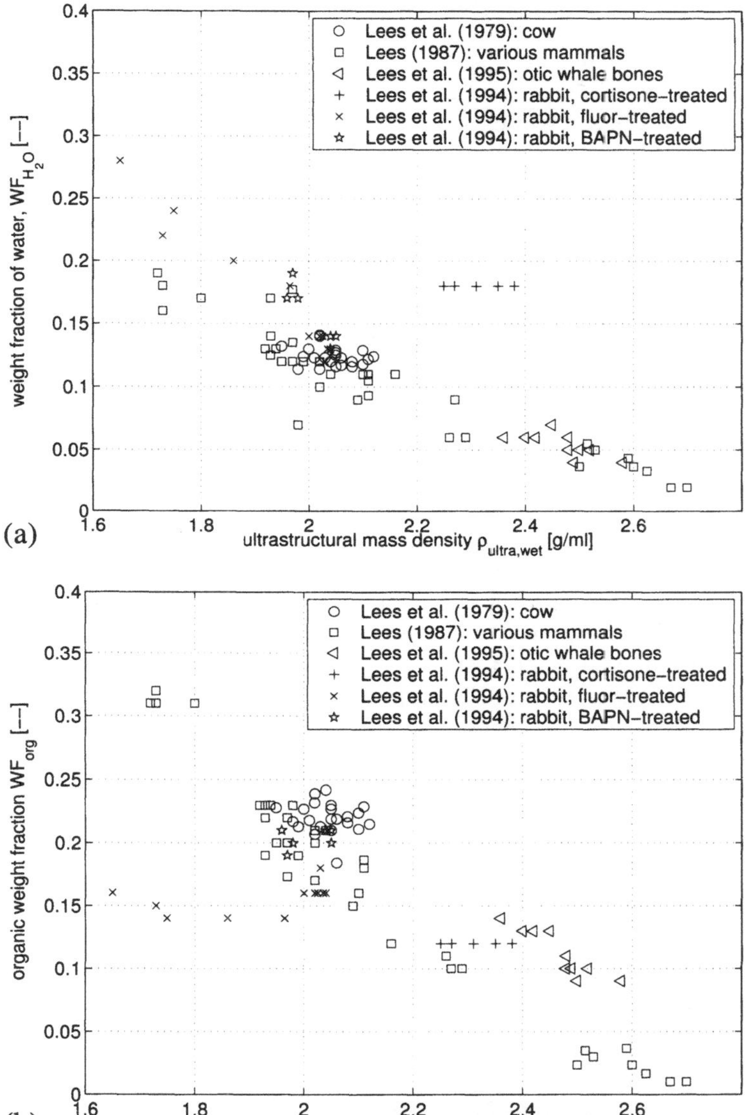

Figure 7: Dehydration-demineralization tests of Lees et al. (1979); Lees (1987); Lees et al. (1994b, 1995): Mass density versus weight fractions of (a) water and (b) organics, according to Eqn.(4.2), (4.3), and (4.4)

of the dry specimens,

$$\rho_{ultra,dry} \approx \rho_{micro,dry} = \frac{M_{dry}}{V_{dry}} \tag{4.7}$$

Together with the real mass densities of organic matter[7] and water, $\rho_{org} \approx \rho_{col}$=1.41 g/ml and ρ_{H_2O}=1.00 g/ml, the mass densities (4.2) as well as the weight fractions (4.3) and (4.4) allow for computation of the (ultrastructural) volume fractions in wet tissue, reading as

$$\bar{f}_{H_2O,wet} = \frac{\rho_{ultra,wet}}{\rho_{H_2O}} WF_{H_2O} \tag{4.8}$$

$$\bar{f}_{org,wet} = \frac{\rho_{ultra,wet}}{\rho_{org}} WF_{org} \tag{4.9}$$

$$\bar{f}_{col,wet} = \frac{\rho_{ultra,wet}}{\rho_{col}} WF_{col} \tag{4.10}$$

$$\bar{f}_{HA,wet} = 1 - \bar{f}_{org} - \bar{f}_{H_2O} \tag{4.11}$$

whereby, according to Eqn.(4.6), $WF_{col} = 0.9 \times WF_{org}$.

In contrast to ρ_{H_2O} and $\rho_{org} \approx \rho_{col}$, the real mass density of non-stoichiometric hydroxyapatite, ρ_{HA}, cannot be directly measured. It is probable to lie between the mass density ρ_{OCP}=2.61 g/ml of octacalcium phosphate, a chemical phase (precursor) preceding biological hydroxyapatite in the complex biomineralization process (Dorozhkin and Epple, 2002), and the mass density $\rho_{HA,stoi}$=3.14 g/ml (Lees, 1987), p.294, of pure stoichiometric hydroxyapatite. Lees (1987) has given an estimate of $\rho_{HA} = 3.00$ g/ml. The relevance of this estimate can be checked based on the relationship between the ultrastructural mass density $\rho_{ultra,wet}$ of a wet bone tissue and the mass densities of its individual components, ρ_{H_2O}, ρ_{org}, and ρ_{HA}, reading as

$$\begin{aligned} \rho_{ultra,wet} &= \bar{f}_{H_2O,wet}\rho_{H_2O} + \bar{f}_{org,wet}\rho_{org} + \bar{f}_{HA,wet}\rho_{HA} \\ \rightarrow \rho_{HA} &= \frac{\rho_{ultra,wet} - \bar{f}_{H_2O,wet}\rho_{H_2O} - \bar{f}_{org,wet}\rho_{org}}{\bar{f}_{HA,wet}} \end{aligned} \tag{4.12}$$

Inserting the specimen-specific values for $\rho_{ultra,wet}$, determined by of Lees et al. (1979); Lees (1987); Lees et al. (1995) through (4.2), assembled in Figure 7, as well as those for $\bar{f}_{H_2O}$, $\bar{f}_{org}$, and $\bar{f}_{HA}$ (determined from the same experimental data through (4.8), (4.9),

[7] The real mass density of (molecular) type I collagen, ρ_{col}, can be determined from different independent tests. On the one hand, (almost) all components except collagen can be removed from bone, tendon, or skin tissue. Robinson (1960) and Bear (1956) determined the mass density of dehydrated, decalcified cortical bone, Pomoroy and Mitton (1951) determined the mass density of dried calf skin. On the other hand, Chothia (1975) provided a model for the organization of proteins in general, based on structural invariants like the volumes occupied by different residues. Harley et al. (1977) applied this model to rat tail tendon for different hydration states. The state of complete dehydration refers to molecular type I collagen. All these different authors report very similar values of $\rho_{col} = 1.41 \ldots 1.45$ g/cm^3. This indicates the existence of a general value for ρ_{col}, independent of the tissue type. Following the pioneering calculations of Katz and Li (1973), p.9., and Lees (1987), p.290, we here adopt a value of ρ_{col}=1.41 g/cm^3.

and (4.11)) into (4.12) yields a large number of values for ρ_{HA}, expressed by a mean value and a standard deviation of $\rho_{HA} = 2.98 \pm 0.19$ g/ml ≈ 3 g/ml. This obviously underlines the relevance of Lees's estimate, $\rho_{HA} = 3.0$ g/ml, and allows for determination of $\bar{f}_{HA,wet}$ alternative to (4.11), namely through (Hellmich et al., 2004a),

$$\bar{f}_{HA,wet} = WF_{HA} \times \frac{\rho_{ultra,wet}}{\rho_{HA}} \tag{4.13}$$

More importantly, ρ_{HA} allows for determination of the (ultrastructural) volume fractions in dry tissues,

$$\bar{f}_{HA,dry} = \frac{WF_{HA}}{WF_{HA} + WF_{org}} \times \frac{\rho_{ultra,dry}}{\rho_{HA}}, \tag{4.14}$$

$$\bar{f}_{col,dry} = \frac{0.9WF_{org}}{WF_{HA} + WF_{org}} \times \frac{\rho_{ultra,dry}}{\rho_{col}} \tag{4.15}$$

It is interesting to note that the dehydration-demineralization tests of Lees et al. (1979); Lees (1987); Lees et al. (1995) reveal seemingly universal composition laws valid throughout the entire vertebrate animal kingdom. In fact, the ultrastructural mineral volume fraction $\bar{f}_{HA}$, obtained from the experiments of Lees et al. (1979); Lees (1987); Lees et al. (1995) by means of Eqn.(4.2) to (4.5) and (4.13), is linearly dependent on the ultrastructural tissue mass density $\rho_{ultra,wet}$ (diamond-markers in Figure 8), which can be expressed by a regression function (solid line in Figure 8) of the form,

$$\mathcal{F}_{\bar{f}_{HA,wet}} = \mathcal{A} \times \rho_{ultra,wet} + \mathcal{B} \tag{4.16}$$

with $\mathcal{A} = 0.58$ ml/g and $\mathcal{B} = -0.72$, see also (Hellmich and Ulm, 2002a). The statistical relevance of this relationship $\mathcal{F}_{\bar{f}_{HA,wet}}(\rho_{ultra,wet})$ is underlined by a correlation coefficient of $r^2 = 97$ %.

Once we accept such a relationship, see also (Torabia, 2004), we can predict the dependence of the ultrastructural organic volume fraction on the ultrastructural tissue mass density, by re-arranging of Eqn.(4.12) while considering (4.11),

$$\mathcal{F}_{\bar{f}_{org,wet}}(\rho_{ultra,wet}) = \frac{1}{\rho_{H_2O} - \rho_{org}} \times \\ \times \left\{\mathcal{F}_{\bar{f}_{HA,wet}}(\rho_{ultra,wet}) \times [\rho_{HA} - \rho_{H_2O}] - \rho_{ultra,wet} + \rho_{H_2O}\right\} \tag{4.17}$$

This prediction $\mathcal{F}_{\bar{f}_{org,wet}}(\rho_{ultra,wet})$ (dashed line in Figure 8) compares well with the experimental data of Lees et al. (1979); Lees (1987); Lees et al. (1995), evaluated by means of Eqn.(4.2), (4.4), and (4.9), indicated by circles in Figure 8. The same is true for the ultrastructural water volume fraction,

$$\mathcal{F}_{\bar{f}_{H_2O,wet}}(\rho_{ultra,wet}) = 1 - \mathcal{F}_{\bar{f}_{org,wet}}(\rho_{ultra,wet}) - \mathcal{F}_{\bar{f}_{HA,wet}}(\rho_{ultra,wet}) \tag{4.18}$$

see dash-dotted line and triangle markers in Figure 8. The relationships shown in Figure 8 provoke the seemingly reasonable concept of a constant ratio between mineral increase and water/collagen decrease during biomineralization. Although the experiments of Lees

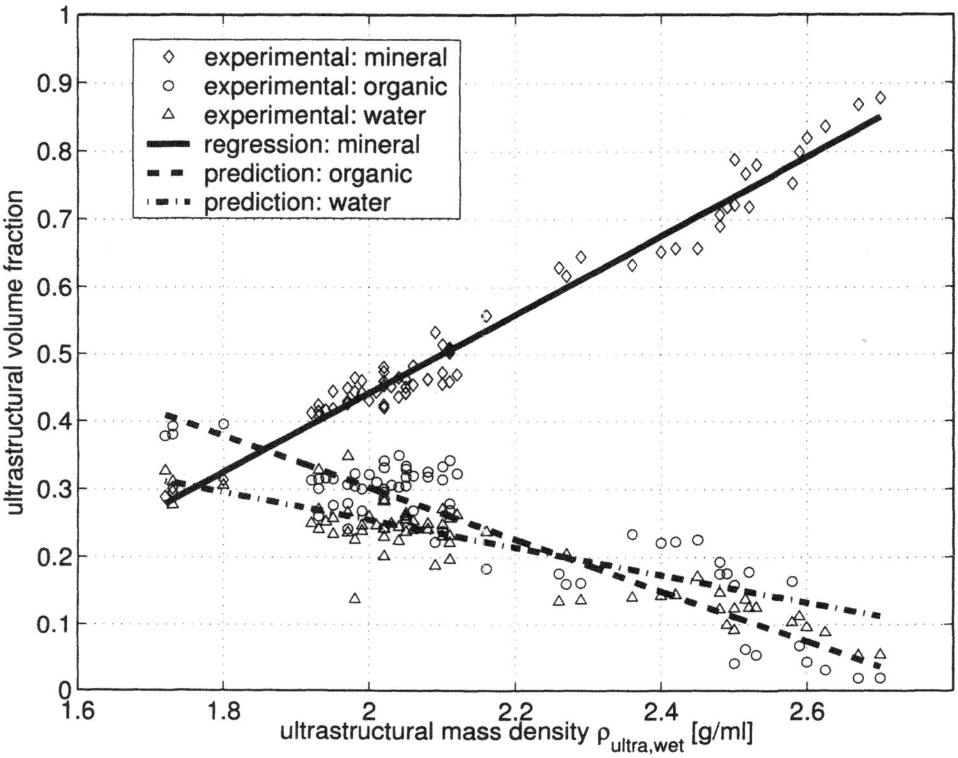

Figure 8: Composition of bone ultrastructure; testing linear relationship between ultrastructural mineral volume fraction and tissue mass density (see diamond-marked experimental data of Lees et al. (1979); Lees (1987); Lees et al. (1995), evaluated by means of Eqn.(4.2) to (4.5) and (4.13), and solid-line regression function Eqn.(4.16)): Predictions of ultrastructural organic and water volume fraction (dashed and dash-dotted lines representing Eqn.(4.17) and (4.18)) agree well with circle-marked and triangle-marked experimental data of Lees et al. (1979); Lees (1987); Lees et al. (1995), evaluated by means of Eqn.(4.2) to (4.4), Eqn.(4.8), Eqn.(4.9); see also (Torabia, 2004)

et al. (1979); Lees (1987); Lees et al. (1995) were all performed on cortical bone, the strong similarity of the ultrastructure in cortical bone and of that in trabecular bone (Gong et al., 1964) renders the relationships of Figure 8 very probable to be valid also for trabecular bone.

Neutron diffraction tests as basis for determination of fibrillar volume fraction $\bar{f}_{fib}$ Determination of the fibrillar volume fraction $\bar{f}_{fib}$ is possible thanks to the generalized packing model of Lees et al. (1984a), providing access to the average crosslink length between collagen molecules. In average, collagen molecules obey to certain packing patterns (Hulmes and Miller, 1979), which stem from crosslinks between the macromolecules, which are situated presumably at the ends of these molecules (Eyre et al., 1984; Bailey et al., 1998), Figure 10(b).[8] The packing of collagen in different mineralized tissues can be characterized by the equatorial diffraction spacing d_w, obtained from neutron diffraction experiments (Lees et al., 1984b; Bonar et al., 1985), Figure 10(a). Most remarkably, Lees (1987) provides evidence that there exists a unique relation between d_w and the wet bone density $\rho_{micro,wet} \approx \rho_{ultra,wet}$. Following dimensional analysis (Buckingham, 1914; Barenblatt, 1996), it appears that if a physically meaningful relation exists between only these two parameters, it must be of the form (Carneiro, 2000):

$$\frac{d_w}{m_0^{1/3}\rho_{ultra,wet}^{-1/3}} = \text{const.} \tag{4.19}$$

For const$\times m_0^{1/3} = 1.57\times 10^{-10}$ kg$^{1/3}$, we get the excellent fit of the diffraction data of Lees (1987), Figure 9(b,c) ($r^2 = 0.98$), with m_0 of dimension mass. d_w enters the generalized packing model of Lees et al. (1984b) for the average equatorial organization of collagen (Figure 10(a)). Figure 10(b) shows the meridional organization: In a staggered scheme with unit length (or axial macroperiod) $D \approx 64$ nm, $4.4 \times D$ long macromolecules are separated by $0.6 \times D$ hole zones (Hodge and Petruska, 1963). From Figure 10, the volume of one rhomboidal fibrillar organizational unit is seen to be:

$$v_{fib} = b d_w 5D, \tag{4.20}$$

with $b = 1.47$nm as one average (rigid) crosslink length, see Figure 10(a). If we denote by v_{col}=355.6 nm^3 (Lees, 1987) the volume of one collagen macromolecule, the ultrastructural volume fraction of fibrils, $\bar{f}_{fib}$, is obtained from the collagen volume fraction multiplied by the volume ratio v_{fib}/v_{col}:

$$\bar{f}_{fib} = \bar{f}_{col} \times \frac{v_{fib}}{v_{col}} \tag{4.21}$$

Derivation of mineral and collagen volume fractions in fibrils and extrafibrillar space, $\hat{f}_{HA}$, $\check{f}_{HA}$, $\hat{f}_{col}$ So far, we have described the determination of the ultrastructural volume fractions $\bar{f}_{col}$, $\bar{f}_{HA}$, and and $\bar{f}_{fib}$ from various experiments. As input for the

[8]Between the ends of the macromolecules, however, the single molecules are free to undergo all different kinds of deformation. The resulting large degree of lateral disorder can be reasonably simulated by a fluid-like lateral arrangement of collagen molecules (Fratzl et al., 1993).

tissue	ρ_w[g/cm^3]	d_w[nm]
cow bone matrix	1.13	1.53
fish clythrum	1.43	1.38
turkey tendon	1.59	1.33
turkey tendon	1.66	1.34
deer antler	1.78	1.29
cow tibia	2.04	1.24

(a)

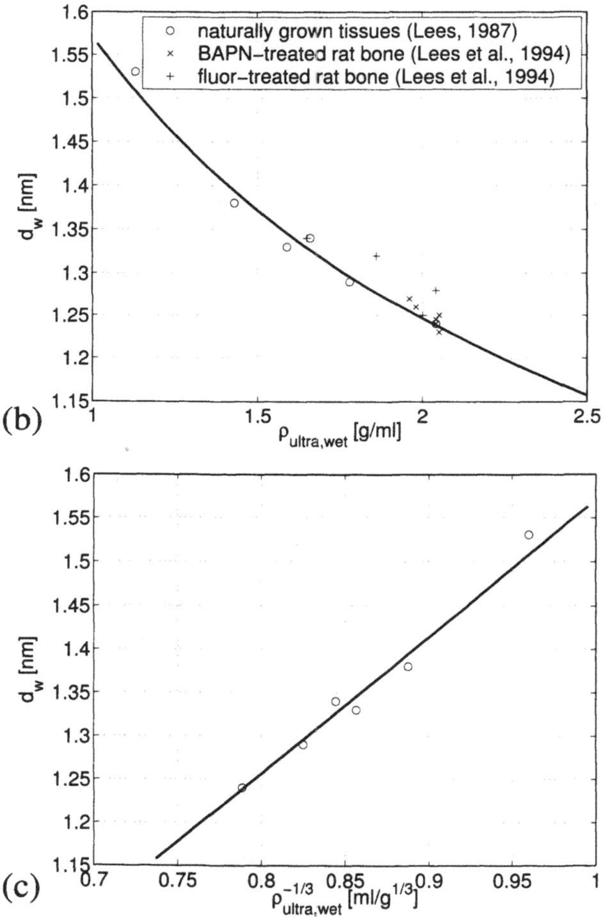

(b) (c)

Figure 9: Equatorial diffraction spacing d_w as a function wet tissue density $\rho_{ultra,wet}$: (a) experimental data compiled by Lees (1987), (b) d_w-$\rho_{ultra,wet}$-plot, (c) d_w-$\rho^{-1/3}_{ultra,wet}$-plot illustrates the existence of a dimensionless constant; $d_w \propto \rho^{-1/3}_{ultra,wet}$ (r^2=98%)

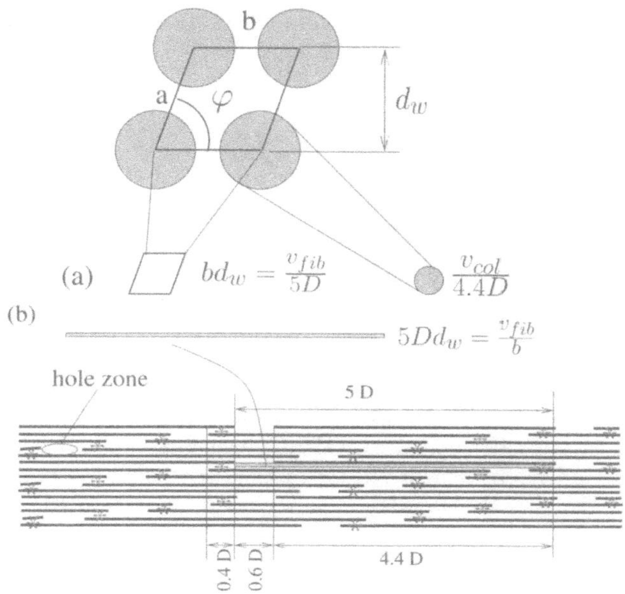

Figure 10: Organization of type I collagen molecules in mineralized tissues, determining the fibrillar space $\bar{f}_{fib}$: (a) Average equatorial organization according to Lees et al. (1984b); Lees (1986, 1987): Two constant average crosslink lengths, a=1.52 nm and b=1.47 nm, are the sides of a rhomboidal unit cell with non-constant angle φ and volume v_{fib}; the diffraction spacing is $d_w = a \times \sin\varphi$; (b) Meridional organization according to the Hodge-Petruska-scheme (Hodge and Petruska, 1963; Bailey et al., 1998); D is the axial macroperiod; * ... crosslink sites; one cylindrical collagen molecule has volume v_{col} and length $4.4D$

micromechanical model depicted in Figure 5, we need the mineral and collagen volume fractions in the fibrillar and the extrafibrillar space, $\hat{f}_{HA}$, $\check{f}_{HA}$, $\check{f}_{col}$. They are determined as follows: $\bar{f}_{fib}$ gives direct access to the volume fraction of collagen within the fibrils, through

$$\check{f}_{col} = \frac{\bar{f}_{col}}{\bar{f}_{fib}} \tag{4.22}$$

Hellmich and Ulm (2001, 2003) have provided evidence that the average hydroxyapatite concentration in the extra-collageneous space in wet mineralized tissues is the same inside and outside the fibrils. Based on this finding, the relative amount of hydroxyapatite in the extrafibrillar space reads as (Hellmich and Ulm, 2001, 2003)

$$\phi_{HA,ef} = \frac{1 - \bar{f}_{fib,wet}}{1 - \bar{f}_{col,wet}} \tag{4.23}$$

With this value at hand, the remaining volume fractions (for both wet and dry tissues) are

$$\check{f}_{HA} = \frac{\bar{f}_{HA}(1-\phi_{HA,ef})}{\bar{f}_{fib}} \tag{4.24}$$

$$\check{f}_{HA} = \frac{\phi_{HA,ef}\bar{f}_{HA}}{\bar{f}_{ef}} \tag{4.25}$$

Determination of microporosity $f_{\mu por}$ from measurements of microstructural mass density of trabecular bone For marrow-cleared trabecular bone with empty micropores, as tested by Turner et al. (1990); Keaveny et al. (1994); Rho et al. (1995), $f_{\mu por}$ can be derived from measurements of the structural (or apparent) mass density, $\rho_{micro,cleared} = (M_{H_2O} + M_{HA} + M_{col} + M_{nonc})/V_{micro}$ (compiled in Figure 14), through

$$f_{\mu por} = 1 - \frac{\rho_{micro,cleared}}{\rho_{ultra,wet}} \tag{4.26}$$

with the ultrastructural mass density $\rho_{ultra,wet} = (M_{H_2O}+M_{HA}+M_{col}+M_{nonc})/V_{ultra}$, see Figure 5(d) for definition of V_{micro} and V_{ultra}.

Determination of microporosity $f_{\mu por}$ from microradiographs McCarthy et al. (1990) tested cortical bone specimens with a comparably high microporosity, i.e. larger than the typical 5% or less (Sietsema, 1995). They estimated bone microporosity by viewing the microradiographs under a low power microscope. The area of the canals within the field of view was measured on the digitizer as was the entire area of the field of view. The microporosity was calculated as

$$f_{\mu por} = \frac{\text{area of canals in field of view}}{\text{total area of field of view}} \tag{4.27}$$

and was averaged over five fields of view for each specimen. Respective data are illustrated in Figure 11, showing also the microstructural mass density of the specimens, $\rho_{micro,wet}$.

4.4 Elasticity experiments on bone – Experimental set IIb

Ultrasonic acoustical tests – principle Ultrasonic experiments are based on the measurements of velocities of plane waves. If the waves travel along the principal material directions of an orthotropic material, through a 3D medium, the wave propagation velocities allow for determination of the material stiffness constants, according to the

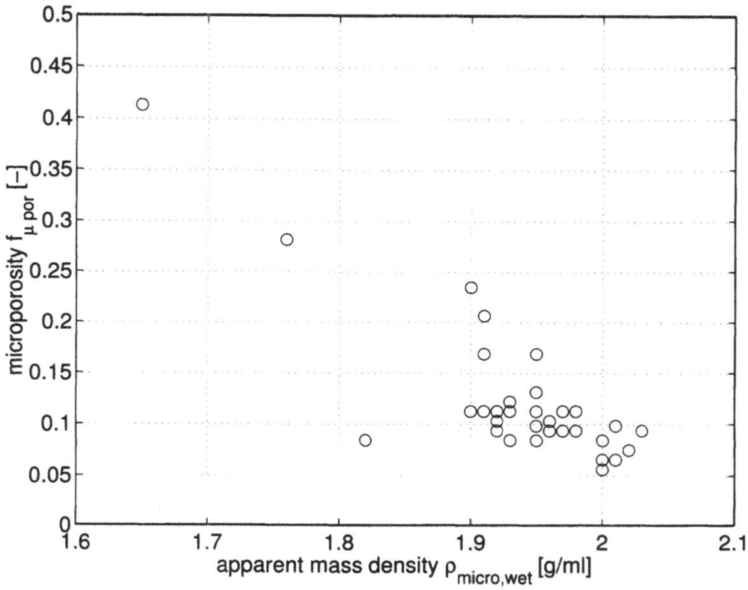

Figure 11: Determination of microporosity $f_{\mu por}$ of dorsal cortex of equine third metacarpus from radiographs by McCarthy et al. (1990): $f_{\mu por}$ as a function of microstructural (=apparent=macroscopic) mass density $\rho_{micro,wet}$

theory of elastic waves (Fedorov, 1968)

$$C_{1111} = \rho v_1^2, \quad (4.28)$$
$$C_{2222} = \rho v_2^2, \quad (4.29)$$
$$C_{3333} = \rho v_3^2, \quad (4.30)$$
$$C_{2323} = G_{23} = \rho v_{23}^2 \quad (4.31)$$
$$C_{1313} = G_{13} = \rho v_{13}^2 \quad (4.32)$$
$$C_{1212} = G_{12} = \rho v_{12}^2 \quad (4.33)$$
$$C_{1122} = (\rho v_1^2 + \rho v_{1/2}^2 - 2\rho \times v_{12/12}^2)(\rho v_2^2 + \rho v_{1/2}^2 - 2\rho \times v_{12/12}^2) - \rho v_{1/2}^2 \quad (4.34)$$
$$C_{1133} = (\rho v_1^2 + \rho v_{1/3}^2 - 2\rho \times v_{13/13}^2)(\rho v_3^2 + \rho v_{1/3}^2 - 2\rho \times v_{13/13}^2) - \rho v_{1/3}^2 \quad (4.35)$$
$$C_{2233} = (\rho v_2^2 + \rho v_{2/3}^2 - 2\rho \times v_{23/23}^2)(\rho v_3^2 + \rho v_{2/3}^2 - 2\rho \times v_{23/23}^2) - \rho v_{2/3}^2 \quad (4.36)$$

Hereby, ρ refers to the mass density of the material (at the ultrastructural or microstructural scale, depending on the wave length), v_i refers to a longitudinal wave travelling in the i-direction, and v_{ij} denotes the velocity of transverse waves traveling in the i-direction with particle motion in the j-direction. $v_{ij/ij}$ denotes the velocity of a longitudinal or a transverse wave traveling in the $(i+j)/\sqrt{2}$-direction. Indices 1 and 2 refer to the radial (transverse) direction, and 3 refers to the axial (longitudinal) direction.

If the waves travel through a one dimensional (rod-like) structure with one material direction i coinciding with the structural direction, the wave propagation velocities allow for determination of Young's modulus in this direction, (Kolsky, 1953)

$$E_i = \rho v_i^2 \tag{4.37}$$

whereby $E_i = 1/D_{iiii}$, with the compliance tensor $\mathbf{D} = \mathbf{C}^{-1}$. The shear moduli G_{ij} still obey Eq. (4.31) to (4.33).

A typical device for the transmission of an elastic wave through a bone sample can be sketched as follows: A transmitter excites a specimen surface by a short impulse[9] (signal) with an excitation frequency f; and after the transition time t, a receiver records the arrival of the signal at the opposite surface of the specimen. Specimen length l and transition time t allow for computation of propagation velocity $v = l/t$. Excitation frequency f and propagation velocity v govern the wave length λ, through:

$$\lambda = \frac{v}{f} \tag{4.38}$$

In terms of micromechanics (see Subsection 2.1), the wavelength λ (a measure for $\mathcal{L}$) is significantly larger than the characteristic length of a representative volume element RVE of the experimentally characterized material. This allows for interpretation of different ultrasonic tests:

- In the ultrasonic tests of Lees and Rollins (1972); Lees et al. (1979, 1983, 1994a, 1995), excitation frequencies of 10 MHz were used. Given a typical wave propagation velocity of $\sim$ 3 km/s in bone specimens, the characteristic corresponding wave lengths $\lambda = \mathcal{L}$ were of the order of 300 μm. The characteristic length of the experimentally characterized material must obey $\ell \ll 300$ μm, which is the case for the ultrastructure with $\ell \approx 10$ μm (Figures 1(e) and 5(c)). Given, in addition, the scale separation between $d \approx 200$ nm (diameter of a fibril, Figure 1(a)) and $\ell \approx 10$ μm (Table 2), the wave velocities measured in the *10 MHz - frequency regime* refer to the *elastic stiffness of bone ultrastructure.* Given the fact that all different excitation frequencies from 1 to 100 MHz deliver virtually the same results in terms of sonic velocity (Lakes et al., 1986; Katz et al., 1986; Lees and Klopholz, 1992), it seems fair to say that the MHz frequency range in general well characterizes the elastic stiffness of bone ultrastructure. This has been explicitly underlined by Ashman and Rho (1988), when testing trabecular bone with 2.25 MHz frequency.
- In the ultrasonic tests performed by Turner et al. (1990); Rho et al. (1995)[10], excitation frequencies of 50 kHz were used. Given a typical velocity of $\sim$ 1.5 km/s in trabecular bone specimens, corresponding wavelengths were, according to Eqn.(4.38), of the order of 3 cm. The characteristic length ℓ of experimentally characterized material must be significantly smaller than 3 cm. This is the case for the microstructure of trabecular bone with ℓ of several milimeters, see Figure 1(d). Given, in addition, the scale separation between $d \approx 500$ μm (thickness of individual trabecular strut or intertrabecular spacing, Figure 1(d)) and $\ell \approx 5$ mm (Table 2), the wave velocities measured in the *50 kHz - frequency regime* refer to the *stiffness of (trabecular) bone microstructure.*

[9]The excitation can be also done by a continuous wave (Ashman and Rho, 1988).
[10]Details on the experimental set-up can be found in (Ashman et al., 1987).

experiments	f	v	λ	ℓ	d	RVE
	[MHz]	[km/s]	[μm]	[μm]	[μm]	
Lees et al.	10	~ 3.00	~ 300	≈ 10	≈ 0.20	ultra
Turner et al.						
Rho et al.	0.05	~ 1.50	~ 30000	≈ 5000	≈ 500	micro (trab.)

Table 2: Experimental characterization of different RVEs (Figure 5) by different excitation frequencies f used in ultrasonic testing: f determines, via Eqn.(4.38), wave length λ of propagating elastic wave, indicating the characteristic length $\mathcal{L}$ of structural loading; separation of scales requires (i) material length ℓ of the RVE to fulfill $\ell \ll \lambda$ and (ii) length d of inhomogeneities within the RVE to fulfill $d \ll \ell$

Ultrasonic tests with 10 MHz frequency – characterization of ultrastructure Lees and Rollins (1972); Lees et al. (1979, 1983, 1994a, 1995) reported velocities of longitudinal waves traveling in the radial (transverse) direction of mammalian cortical and hyperpycnotic bone specimens, $v_1 \approx v_2$, and velocities of longitudinal waves traveling in the axial (longitudinal) direction of mammalian cortical and hyperpycnotic bone, v_3; along with mass densities $\rho_{micro} \approx \rho_{ultra}$. These values allow for determination of $C^{exp}_{ultra,1111}$ and $C^{exp}_{ultra,3333}$, by means of Eqn.(4.28) and Eqn.(4.30) with $\rho = \rho_{ultra}$, Figure 12.

Ultrasonic tests with 2.25 MHz frequency – characterization of ultrastructure McCarthy et al. (1990) reported velocities of longitudinal waves traveling in the radial (transverse) direction of specimens from the dorsal cortex of equine third metacarpus, $v_1 \approx v_2$, and velocities of longitudinal waves traveling in the axial (longitudinal) direction of the aforementioned equine bone specimens, v_3; along with microstructural (=apparent=macroscopic) mass densities $\rho_{micro,wet}$ and microporosity $f_{\mu por}$ (Figure 11). The latter give access to ultrastructural mass densities $\rho_{ultra,wet}$ via Eqn.(4.1). Insertion of these values for v_1, v_3, and $\rho = \rho_{ultra,wet}$ into Eqn.(4.28) and Eqn.(4.30) yields specimen-specific experimental stiffness values $C^{exp}_{ultra,1111}$ and $C^{exp}_{ultra,3333}$ (Figure 13).

Ashman and Rho (1988) reported velocities of longitudinal waves traveling in the axial (longitudinal) direction of bovine and human trabecular bone specimens, v_3; along with ultrastructural mass densities $\rho_{ultra,wet}$. Insertion of these values for v_3 and $\rho = \rho_{ultra,wet}$ into Eqn.(4.30) yields specimen-specific experimental stiffness values $C^{exp}_{ultra,3333}$ (Figure 13).

Ashman et al. (1984) give the average of the full experimental ultrastructural elasticity tensor of 60 human femoral cortical specimens, based on Eqn.(4.28) to Eqn.(4.36), reading

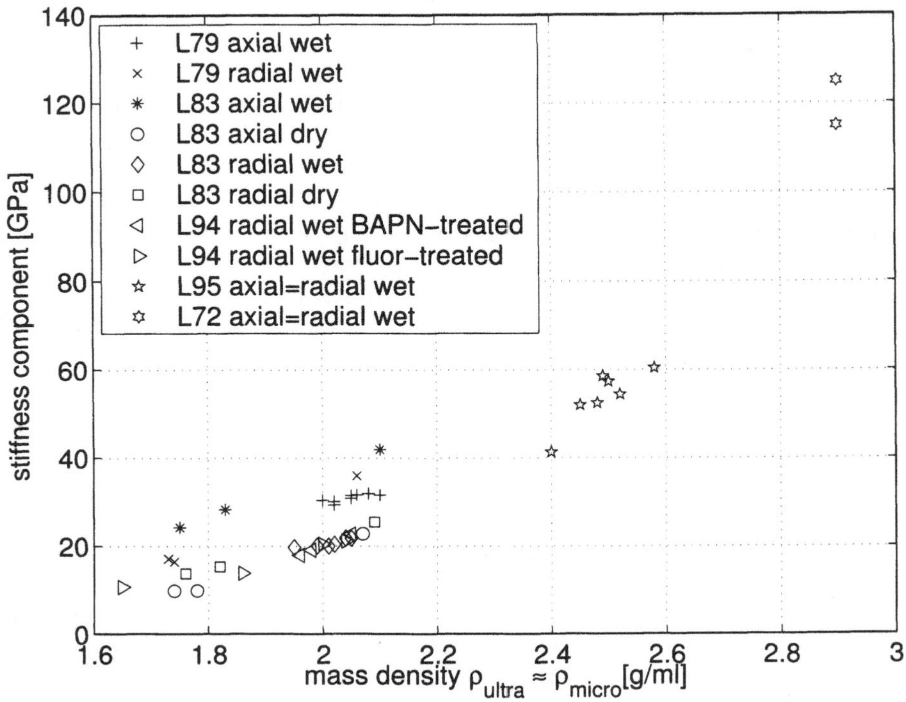

Figure 12: Experimental data used for model validation: Ultrastructural elastic stiffness constants C_{3333} (axial) and C_{1111} (radial) of cortical bone, determined from ultrasonic tests with 10 MHz frequency, as functions of mass density $\rho_{micro} \approx \rho_{ultra}$; experiments of L79... (Lees et al., 1979), L83... (Lees et al., 1983), L94... (Lees et al., 1994a), L72... (Lees and Rollins, 1972), L95... (Lees et al., 1995),

as

$$\mathbf{C}_{ultra}^{exp} = \begin{pmatrix} 18.0 & 9.98 & 10.1 & 0 & 0 & 0 \\ 9.98 & 20.2 & 10.7 & 0 & 0 & 0 \\ 10.1 & 10.7 & 27.6 & 0 & 0 & 0 \\ 0 & 0 & 0 & 12.5 & 0 & 0 \\ 0 & 0 & 0 & 0 & 11.2 & 0 \\ 0 & 0 & 0 & 0 & 0 & 9.04 \end{pmatrix} \tag{4.39}$$

Ultrasonic tests with 50 kHz frequency – characterization of microstructure As for the marrow-cleared trabecular bone specimens with a wet solid bone matrix tested by Turner et al. (1990); Rho et al. (1995), the longitudinal wavelengths, $\lambda = 2 \ldots 5$ cm, are *larger* than the characteristic cross-sectional dimensions of the bone specimens, 10x10x20 mm. In this case, the one-dimensional theory Eq.(4.37) and Eqn.(4.31) to (4.33) is appropriate. Insertion of experimental values for velocities of longitudinal and

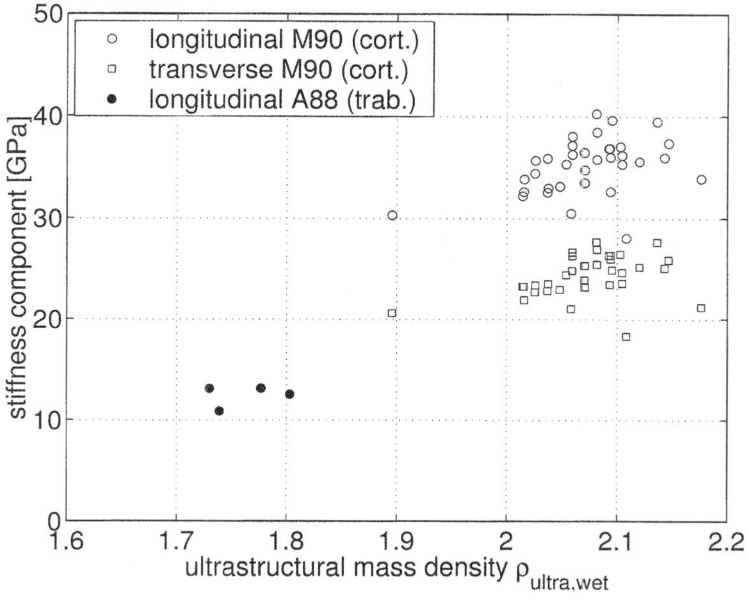

Figure 13: Experimental data used for model validation: Ultrastructural elastic stiffness constants C_{3333} (axial=longitudinal) and C_{1111} (radial=transverse) of cortical and trabecular bone, determined from ultrasonic tests with 2.25 MHz frequency, as functions of ultrastructural mass density $\rho_{ultra,wet}$; experiments of M90 ... (McCarthy et al., 1990), A88 ... (Ashman and Rho, 1988)

transverse waves, traveling in the axial and radial material directions, and for the microstructural (=apparent=macroscopic) mass density $\rho_{micro,cleared}$ into these equations yields corresponding experimental stiffness values, see Figure 14. It is noteworthy that the bovine specimens of Turner et al. (1990) were not cut in the anatomical directions, but in the principal direction of the *a priori* determined fabric tensor. In this sense, they are – strictly speaking – not similar to the other specimens. However, we include them here for comparative purposes.

Quasi-static testing For transverse isotropy, quasi-static mechanical tests (Keaveny et al., 1994) reveal the longitudinal (i.e. superior-inferior) Young's modulus as

$$E_3 = \frac{1}{D_{3333}}, \tag{4.40}$$

with the compliance tensor $\mathbf{D} = \mathbf{C}^{exp,-1}_{micro}$, and the experimental stiffness tensor $\mathbf{C}^{exp}_{micro}$. Experimental data obtained by Keaveny et al. (1994) are compiled in Figure 14.

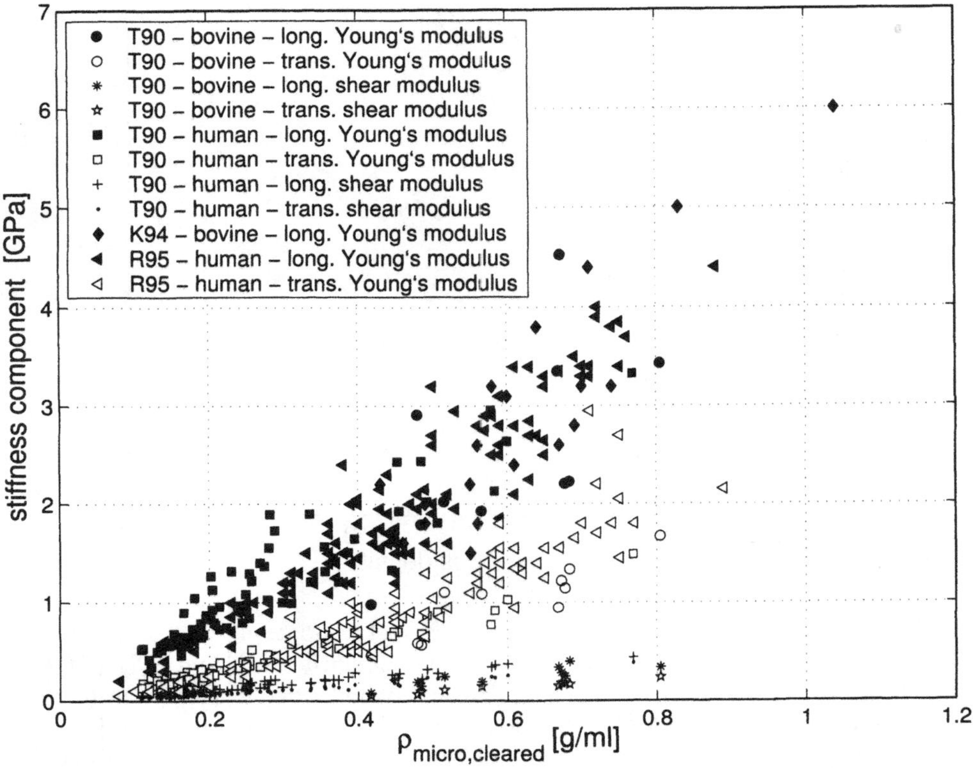

Figure 14: Experimental data used for model verification: Microstructural (=apparent=macroscopic) elastic stiffness constants of trabecular bone, determined from ultrasonic tests with 50 kHz frequency (T90, R95) or from quasi-static tests (K94), as function of apparent (=microstructural=macroscopic) mass density $\rho_{app} = \rho_{micro,cleared}$; T90... (Turner et al., 1990); K94... (Keaveny et al., 1994); R95... (Rho et al., 1995)

4.5 Comparison between stiffness predictions and experiments

The stiffness values predicted by the four-step homogenization procedure (model output or dependent variables; Figure 5) on the basis of the tissue-independent ('universal') phase stiffness properties of Table 1 for the tissue-specific volume fractions (model input or independent variables; derived from Eq. (4.2) to Eq. (4.27)) compare well to the corresponding experimentally determined tissue-specific stiffness values from experimental set IIb (Subsection 4.4). This is true for different observation scales. In fact, it holds for the hydroxyapatite foam level referred to in Figure 5(a), see Figure 15, as well as for the ultrastructural level referred to in Figure 5(c), see Figures 16 and 17, and for the microstructural level referred to in Figure 5(d), see Figure 18. To quantify the model's predictive capabilities, we consider the mean and the standard deviation of the relative error between stiffness predictions and experiments,

$$\bar{e} = \frac{1}{n}\sum e_i = \frac{1}{n}\sum \frac{C^{pred}_{ijkl} - C^{exp}_{ijkl}}{C^{exp}_{ijkl}} \tag{4.41}$$

$$e_S = \left[\frac{1}{n-1}\sum (e_i - \bar{e})^2\right]^{\frac{1}{2}} \tag{4.42}$$

with summation over n values C^{exp}_{ijkl} derived from the literature and compiled in experimental set IIb (Subsection 4.4).

Hydroxyapatite foam level Isotropic hyperpycnotic tissues like the otic bones of whales or enamel, tested by Lees and Rollins (1972); Lees et al. (1995), have a very small collagen volume fraction, $\bar{f}_{col} \approx 0$, and virtually no microporosity, $f_{\mu por} \approx 0$. Therefore, homogenization step I alone (Figure 5(a)) is relevant for these tissues at all observation scales, from several hundred nanometers to several milimeters or centimeters.

Evaluation of data for otic whale bones shown in Figure 7, and for enamel specimens of Lees and Rollins (1972); Cuy et al. (2002) with $\rho_{ultra} = 2.90$ g/ml and $WF_{HA} = 0.96$, through Eqn.(4.5) and (4.13) yields composition values $\check{f}_{HA} = \bar{f}_{HA}$, corresponding to experimental stiffness values labelled L95 and L72 in Figure 12. Insertion of these composition values $\check{f}_{HA}$ into Eqn.(3.1) to Eqn.(3.5) delivers tissue-specific stiffness estimates $\mathbf{C}^{SCS}_{ef} = \mathbf{C}_{ultra}$. Stiffness predictions $C^{pred}_{1111} = C^{SCS}_{ef,1111} = C^{pred}_{3333} = C^{SCS}_{ef,3333}$ are compared to corresponding experimental stiffness values (Figure 15). The satisfactory agreement between model predictions and experiments is quantified by prediction errors of 4±9 % (mean value ± standard deviation according to Eqn.(4.41) and (4.42)).

Ultrastructural level Evaluation of data for a typical human femur tissue, characterized by $\rho_{ultra,wet} = 1.93$ g/ml in Figures 7 and 9, through Eqn.(4.5), Eqn.(4.6), Eqn.(4.8), Eqn.(4.10), Eqn.(4.13) delivers composition values of $\bar{f}_{HA} = 0.45$, $\bar{f}_{col} = 0.28$, $d_w = 1.25$ nm, while the remainder of volume fractions required for the micromechanics model at scales below the microstructure follows from Eqn.(4.20) to Eqn.(4.25). Insertion of these composition values $\check{f}_{HA}$, $\check{f}_{HA}$, $\check{f}_{col}$, and $\bar{f}_{fib}$, into Eqn.(3.1) to Eqn.(3.12)

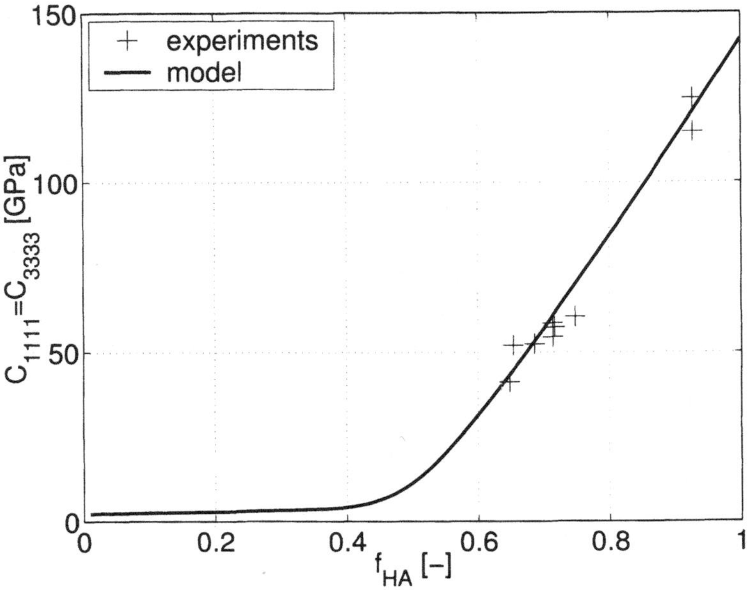

Figure 15: Comparison between model predictions and experiments at the hydroxyapatite foam scale (10 MHz-frequency tests of Lees and Rollins (1972); Lees et al. (1995))

delivers a stiffness estimate $\mathbf{C}^{MT}_{ultra}$ (model prediction) of

$$\mathbf{C}^{MT}_{ultra} = \begin{pmatrix} 23.6 & 9.1 & 9.9 & 0 & 0 & 0 \\ 9.1 & 23.6 & 9.9 & 0 & 0 & 0 \\ 9.9 & 9.9 & 28.7 & 0 & 0 & 0 \\ 0 & 0 & 0 & 15.3 & 0 & 0 \\ 0 & 0 & 0 & 0 & 15.3 & 0 \\ 0 & 0 & 0 & 0 & 0 & 14.4 \end{pmatrix} \tag{4.43}$$

This model prediction agrees quite well with the corresponding experimentally determined stiffness tensor of human femur cortical ultrastructure determined by Ashman et al. (1984), see Eqn.(4.39): A particularly good agreement is seen for the off-diagonal terms, while the agreement of the normal and shear stiffness constants is somewhat less satisfying.

Evaluation of data for cow, rabbit, and various mammals, shown in Figures 7 and 9, through Eqn.(4.5), Eqn.(4.6), Eqn.(4.8), Eqn.(4.10), Eqn.(4.13), as well as Eqn.(4.20) to Eqn.(4.25) delivers composition values $\check{f}_{HA}$, $\check{f}_{HA}$, $\check{f}_{col}$, and $\bar{f}_{fib}$, see also Figure 8 and Tables 2 to 6 in (Hellmich et al., 2004a), corresponding to the experimental stiffness values labelled L79, L83, and L94 in Figure 12. Insertion of these composition values $\check{f}_{HA}$, $\check{f}_{HA}$, $\check{f}_{col}$, and $\bar{f}_{fib}$ into Eqn.(3.1) to Eqn.(3.12) delivers tissue-specific stiffness estimates $\mathbf{C}^{MT}_{ultra}$. Stiffness predictions $C^{pred}_{1111} = C^{MT}_{ultra,1111}$ and $C^{pred}_{3333} = C^{MT}_{ultra,3333}$ are compared to corresponding experimental stiffness values (Figure 16). The satisfactory agreement between model predictions and experiments is quantified by prediction errors of -10±13 % (mean value ± standard deviation according to Eqn.(4.41) and (4.42)) for the axial (longitudinal) normal stiffness and of 5±19 % for the radial (transverse) normal stiffness.

Experimental values for $f_{\mu por}$ and $\rho_{micro,wet}$, documented by McCarthy et al. (1990), allow for determination of the ultrastructural mass density $\rho_{ultra,wet}$, through Eqn.(4.1). To account for the ultrastructural composition of the aforementioned trabecular specimens, we employ Eqn.(4.19), Eqn.(4.16), Eqn.(4.17), and Eqn.(4.18). Via Eqn.(4.20), Eqn.(4.21), and Eqn.(4.23), we obtain the mineral and collagen volume fractions within the fibrils and in the extrafibrillar space, given by Eqn.(4.22), Eqn.(4.24), and Eqn.(4.25), corresponding to the experimental stiffness values labelled M90 in Figure 13. Insertion of these composition values $\check{f}_{HA}$, $\check{f}_{HA}$, $\check{f}_{col}$, and $\bar{f}_{fib}$ into Eqn.(3.1) to Eqn.(3.12) delivers tissue-specific stiffness estimates $\mathbf{C}^{MT}_{ultra}$. Stiffness predictions $C^{pred}_{1111} = C^{MT}_{ultra,1111}$ and $C^{pred}_{3333} = C^{MT}_{ultra,3333}$ are compared to corresponding experimental stiffness values (Figure 17). The satisfactory agreement between model predictions and experiments is quantified by prediction errors of -10±12 % (mean value ± standard deviation according to Eqn.(4.41) and (4.42)) for the axial (longitudinal) normal stiffness and of 9±16 % for the radial (transverse) normal stiffness.

Despite of their moderate magnitude, the aforementioned prediction errors might still deserve an explanation: In the ultrastructure of cortical bone, where the extrafibrillar stiffness exceeds the stiffness of the fibrils, neglection of the (quite moderate) deviation of the collagen orientation from the anatomical axis may well lead to an overestimation of the longitudinal stiffness and to an underestimation of the transverse stiffness component (Figures 16 and 17).

Ashman and Rho (1988) report experimental values for the ultrastructural mass density $\rho_{ultra,wet}$ of human and bovine trabecular bone material. To account for the ultrastructural composition of the aforementioned trabecular specimens, we employ Eqn.(4.19), Eqn.(4.16), Eqn.(4.17), and Eqn.(4.18). Via Eqn.(4.20), Eqn.(4.21), and Eqn.(4.23), we obtain the mineral and collagen volume fractions within the fibrils and in the extrafibrillar space, given by Eqn.(4.22), Eqn.(4.24), and Eqn.(4.25), corresponding to the experimental stiffness values labelled A88 in Figure 13. Insertion of these composition values $\check{f}_{HA}$, $\check{f}_{HA}$, $\check{f}_{col}$, and $\bar{f}_{fib}$ into Eqn.(3.1) to Eqn.(3.12) delivers tissue-specific stiffness estimates $\mathbf{C}^{MT}_{ultra}$. Stiffness predictions $C^{pred}_{3333} = C^{MT}_{ultra,3333}$ are compared to corresponding experimental stiffness values (Figure 17). The very satisfactory agreement between model predictions and experiments is quantified by prediction errors of -3±10 % (mean value ± standard deviation according to Eqn.(4.41) and (4.42)).

Microstructural level Evaluation of apparent mass density data of Turner et al. (1990); Rho et al. (1995); Keaveny et al. (1994) (Figure 14) through Eqn.(4.26) yields specimen-specific values for the microporosity $f_{\mu por}$. As regards respective tests on bovine trabecular bone, performed by Turner et al. (1990); Keaveny et al. (1994), we use the value of $\rho_{ultra,wet}$ = 1.74 g/ml, reported by Ashman and Rho (1988). As regards respective tests on human trabecular bone, performed by Turner et al. (1990); Rho et al.

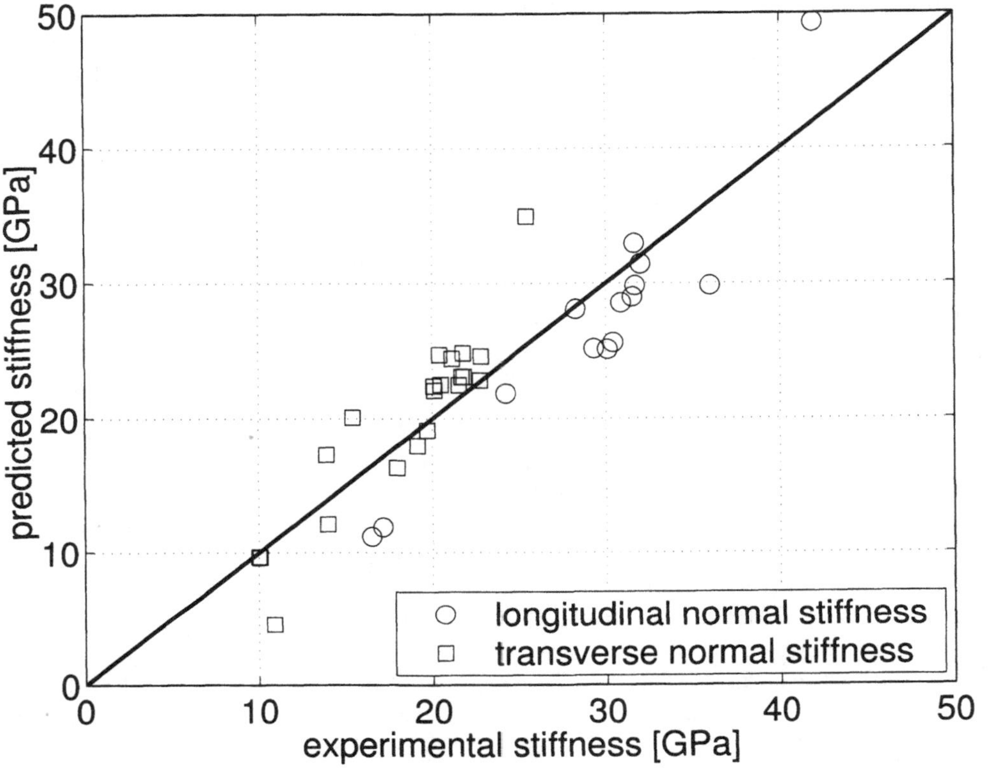

Figure 16: Comparison between model predictions and experiments at the ultrastructural scale (10 MHz-frequency tests of Lees et al. (1979, 1983, 1994a))

(1995), we use the value of $\rho_{ultra,wet} = 1.76$ g/ml, the mean value of the data set reported by Ashman and Rho (1988). To account for the ultrastructural composition of the aforementioned trabecular specimens, we employ Eqn.(4.19), Eqn.(4.16), Eqn.(4.17), and Eqn.(4.18). Via Eqn.(4.20), Eqn.(4.21), and Eqn.(4.23), we obtain the mineral and collagen volume fractions within the fibrils and in the extrafibrillar space, given by Eqn.(4.22), Eqn.(4.24), and Eqn.(4.25). Insertion of these composition values $\check{f}_{HA}$, $\check{f}_{HA}$, $\check{f}_{col}$, $\bar{f}_{fib}$, and $f_{\mu por}$, corresponding to the experimental stiffness values depicted in Figure 14, into Eqn.(3.1) to Eqn.(3.13) yields tissue-specific stiffness estimates $\mathbf{C}^{MT}_{micro}$. Stiffness predictions $E_1^{pred} = 1/[\mathbf{C}^{MTII,-1}_{micro}]_{1111}$, $E_3^{pred} = 1/[\mathbf{C}^{MTII,-1}_{micro}]_{3333}$, $G_{23}^{pred} = C^{MTII}_{micro,2323}$, and $G_{12}^{pred} = C^{MTII}_{micro,1212}$ are compared to corresponding experimental stiffness values (Figure 18). The satisfactory agreement between model predictions and experiments is quantified by prediction errors of 8±24 % (mean value ± standard deviation according to Eqn.(4.41) and (4.42)) for the axial (longitudinal) Young's modulus and of -10±25 % for the radial (transverse) Young's modulus. The predictions of the shear moduli are

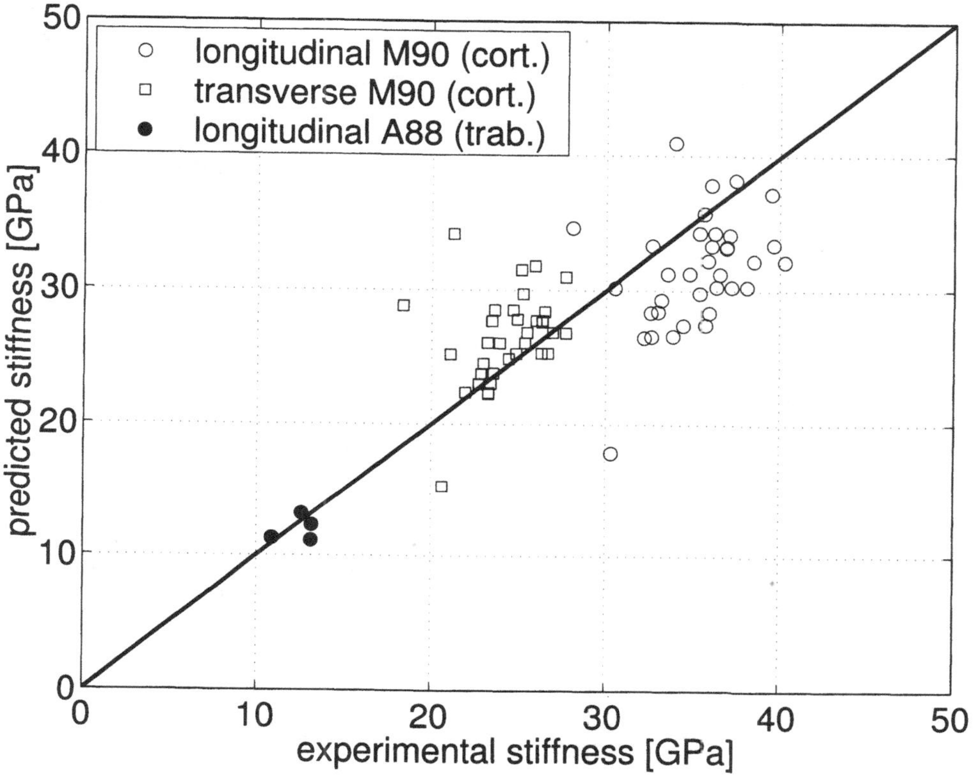

Figure 17: Comparison between model predictions and experiments at the ultrastructural scale (2.25 MHz-frequency tests of A88 ... (Ashman and Rho, 1988), M90 ... (McCarthy et al., 1990))

remarkably less satisfactory; they are quantified by errors of 35±34 % for shear in a plane parallel to the fibrils (longitudinal shear modulus), and of 64±42 % in a plane perpendicular to the fibrils (transverse shear modulus).

This deserves further discussion: The deviation of both collagen and micropore orientation from the anatomical axis is presumably more pronounced in trabecular bone than in cortical bone (Rubin et al., 2003; Prostak and Lees, 1996). This higher disorder in the microstructure is reflected in higher standard deviations of prediction errors. In the ultrastructure of trabecular bone, where the extrafibrillar stiffness exceeds the stiffness of the fibrils, neglection of the collagen misalignment in the current micromechaniacal model may well result in an overestimation of the longitudinal Young's modulus (circle markers in Figure 18), and in an underestimation of the transverse Young's modulus (square markers in Figure 18). While these overestimations and underestimations are not more severe than for the cortical bone specimens investigated before, the effects of the neglection of the micropore orientation in trabecular bone on the prediction error of

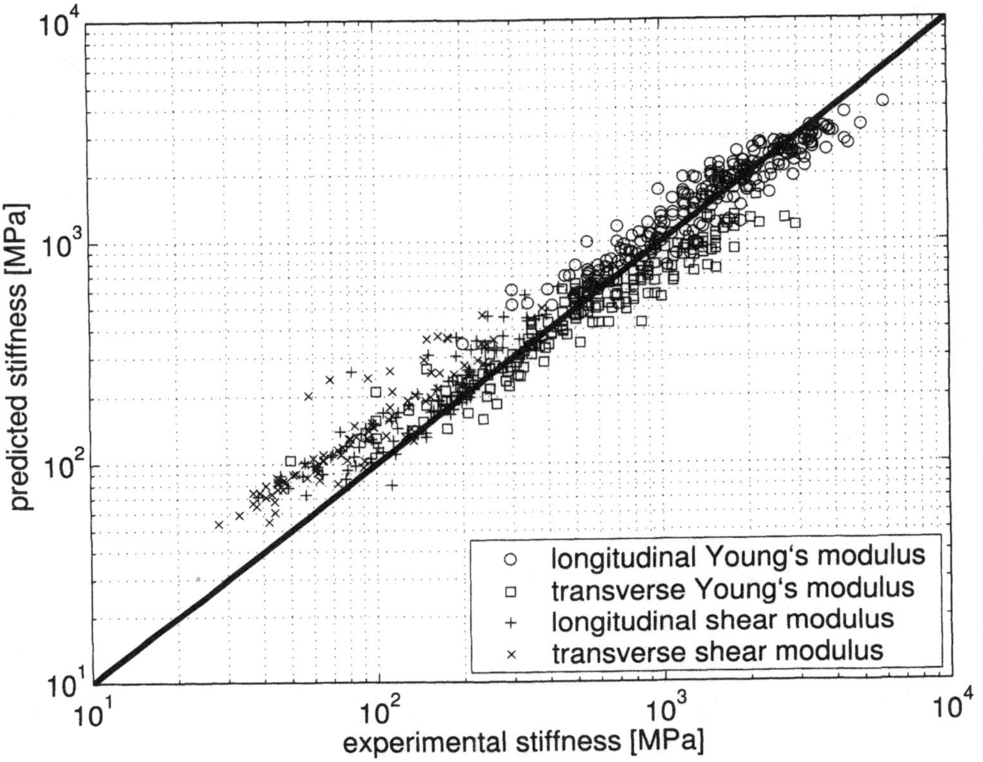

Figure 18: Comparison between model predictions and experiments at the microstructural scale (50-kHz frequency tests of Turner et al. (1990); Rho et al. (1995) and quasi-static tests of Keaveny et al. (1994))

the shear moduli are actually more severe: The shear moduli are highly overestimated, up to a factor of more than two (cross markers in Figure 18). While these errors are still lower than those frequently encountered with other models in the bone biomechanics literature, further model improvement is obviously needed here. Based on the mean intercept length concept (Whitehouse, 1974; Harrigan and Mann, 1984; Turner et al., 1990), we currently try to incorporate information on collagen/micropore misalignment into the microelastic model.

5 Conclusion and Outlook

Given the predictive capabilities of the micromechanics model, exceeding significantly those of other models described in the open literature, the developments collected in this contribution may well be regarded as substantial contribution to modern bone biomechanics. Linking (Torabia, 2004) the micromechanics model to standard clinical tools such as Computer Tomography (Kalender, 2000) will probably improve bone fracture risk assessment in a significant way, once micromechanical developments beyond elasticity will have been achieved on the basis of the current model; be they related to strength, crack propagation, or poroelastic couplings. As for the latter, some success has already been made (Hellmich and Ulm, 2004a,b).

6 Acknowledgments

The author is grateful to Franz-Josef Ulm and Emilio Silva, Massachusetts Institute of Technology (MIT), Cambridge, USA, to Sidney Lees, The Forsyth Institute, Boston, USA, to Eric Lemarchand, Lille University of Science and Technology, France, to Luc Dormieux and Jean-Francois Barthélémy, National School of Civil Engineering (ENPC), Marne-la-Vallée, France, as well as to Karin Hofstetter, Sherin Torabia, and Andreas Fritsch, Vienna University of Technology (TU Wien), Vienna, Austria, for numerous enlightening discussions and helpful support in the domains of continuum micromechanics and bone biophysics.

Bibliography

B. Aoubiza, J.M. Crolet, and A. Meunier. On the mechanical characterization of compact bone structure using homogenization theory. *Journal of Biomechanics*, 29(6):1539 – 1547, 1996.

P.J. Arnoux, J. Bonnoit, P. Chabrand, M. Jean, and M. Pithioux. Numerical damage models using a structural approach: application to bones and ligaments. *The European Physical Journal Applied Physics*, 17:65 – 73, 2002.

M.-G. Ascenzi. A first estimation of prestress in so-called circularly fibered osteonic lamellae. *Journal of Biomechanics*, 32:935 – 942, 1999.

R.B. Ashman, J.D. Corin, and C.H. Turner. Elastic properties of cancellous bone: measurement by an ultrasonic technique. *Journal of Biomechanics*, 20(10):979 – 986, 1987.

R.B. Ashman, S.C. Cowin, W.C. van Buskirk, and J.C. Rice. A continuous wave technique for the measurement of the elastic properties of cortical bone. *Journal of Biomechanics*, 17(5):349 – 361, 1984.

R.B. Ashman and J.Y. Rho. Elastic modulus of trabecular bone material. *Journal of Biomechanics*, 21(3):177 – 181, 1988.

A.J. Bailey, R.G. Paul, and L. Knott. Mechanisms of maturation and ageing of collagen. *Mechanisms of Aging and Development*, 106:1 – 56, 1998.

G.I. Barenblatt. *Scaling, Self-Similarity, and Intermediate Asymptotics.* Cambridge University Press, Cambridge, United Kingdom, 1 edition, 1996.

R.S. Bear. The structure of collagen molecules and fibrils. *Journal of Biophysical and Biochemical Cytology*, 2:363 – 368, 1956.

V. Benezra Rosen, L.W. Hobbs, and M. Spector. The ultrastructure of anorganic bovine bone and selected synthetic hydroxyapatites used as bone graft substitute material. *Biomaterials*, 23:921 – 928, 2002.

R.M. Biltz and E.D. Pellegrino. The chemical anatomy of bone. *Journal of Bone and Joint Surgery*, 51-A(3):456 – 466, 1969.

L.C. Bonar, S. Lees, and H.A. Mook. Neutron diffraction studies of collagen in fully mineralized bone. *Journal of Molecular Biology*, 181:265 – 270, 1985.

E. Buckingham. On physically similar systems. illustrations of the use of dimensional analysis. *Physical Review*, 4:345 – 376, 1914.

J.A. Buckwalter, M.J. Glimcher, R.R. Cooper, and R. Recker. Bone biology, Part I: Structure, blood supply, cells, matrix, and mineralization. *The Journal of Bone and Joint Surgery*, 77-A(8):1256 – 1275, 1995a.

J.A. Buckwalter, M.J. Glimcher, R.R. Cooper, and R. Recker. Bone biology, Part II: Formation, form, modeling, remodeling, and regulation of cell functions. *The Journal of Bone and Joint Surgery*, 77-A(8):1276 – 1288, 1995b.

S. Byers, A.J. Moore, R.W. Byard, and N.L. Fazzalari. Quantitative histomorphometric analysis of the human growth plate from birth to adulescence. *Bone*, 27(4):495 – 501, 2000.

F.L.L.B. Carneiro. On the use, by Einstein, of the principles of dimensional homogeneity in three problems of the physics of solids. *Annuals of the Brazilian Academy of Sciences*, 72(4):591 – 596, 2000.

C. Chothia. Structural invariants in protein folding. *Nature*, 254:304 – 308, 1975.

J.M. Crolet, B. Aoubiza, and A. Meunier. Compact bone: numerical simulation of mechanical characteristics. *Journal of Biomechanics*, 26(6):677 – 687, 1993.

J.D. Currey. The relationship between the stiffness and the mineral content of bone. *Journal of Biomechanics*, 2:477 – 480, 1969.

S. Cusack and A. Miller. Determination of the elastic constants of collagen by Brillouin light scattering. *Journal of Molecular Biology*, 135:39 – 51, 1979.

J.L. Cuy, A.B. Mann, K.J. Livi, M.F. Teaford, and T.P. Weihs. Nanoindentation mapping of the mechanical properties of human molar tooth enamel. *Archives of Oral Biology*, 47:281 – 291, 2002.

M. Ding and I. Hvid. Quantification of age-related changes in the structure model type and trabecular thickness of human tibial cancellous bone. *Bone*, 26:291 – 295, 2000.

S.V Dorozhkin and M. Epple. Biological and medical significance of calcium phosphates. *Angew. Chem. Int. Ed.*, (41):3130– 3146, November 2002.

J.D. Eshelby. The determination of the elastic field of an ellipsoidal inclusion, and related problems. *Proceedings of the Royal Society London, Series A*, 241:376 – 396, 1957.

D.R. Eyre, M.A. Paz, and P.M. Gallop. Cross-linking in collagen and elastin. *Annual Review of Biochemistry*, 53:717 – 748, 1984.

C. Farquharson and B. Houston. Growth-plate chondrocytes in normal and abnormal growth of bone. In *1997/98 Annual Report of the Roslin Institute, Edinburgh, Scotland, UK*, 1998. published at www.roslin.ac.uk.

F.I. Fedorov. *Theory of elastic waves in crystals*. Plenum Press, New York, 1968.

P. Fratzl, N. Fratzl-Zelman, and K. Klaushofer. Collagen packing and mineralization – an X-ray scattering investigation of turkey leg tendon. *Biophysical Journal*, 64:260 – 266, 1993.

P. Fratzl, S. Schreiber, and K. Klaushofer. Bone mineralization as studied by small-angle X-ray scattering. *Connective Tissue Research*, 34(4):247 – 254, 1996.

R.S. Gilmore and J.L. Katz. Elastic properties of apatites. *Journal of Materials Science*, 17:1131 – 1141, 1982.

J.K. Gong, J.S. Arnold, and H. Cohn, S. Composition of trabecular and cortical bone. *The Anatomical Record*, 149:325 – 332, 1964.

R.W. Goulet, S.A. Goldstein, M.J. Ciarelli, J.L. Kuhn, M.B. Brown, and L.A. Feldkamp. The relationship between the structural and orthogonal compressive properties of trabecular bone. *Journal of Biomechanics*, 27:375 – 389, 1994.

R. Harley, D. James, A. Miller, and J.W. White. Phonons and the elastic moduli of collagen and muscle. *Nature*, 267:285 – 287, 1977.

T.P. Harrigan and R.W. Mann. Characterization of microstructural anisotropy in orthotropic materials using a second rank tensor. *Journal of Materials Science*, 19:761 – 767, 1984.

Ch. Hellmich, J.-F. Barthélémy, and L. Dormieux. Mineral-collagen interactions in elasticity of bone ultrastructure – a continuum micromechanics approach. *European Journal of Mechanics A-Solids*, 23:783 – 810, 2004a.

Ch. Hellmich and F.-J. Ulm. Hydroxyapatite is uniformly concentrated in the extra-collagenous ultrastructure of mineralized tissue. In J. Middleton, N.G. Shrive, and M.L. Jones, editors, *Proceedings of the Fifth International Symposium on Computer Methods in Biomechanics and Biomedical Engineering*, Rome, Italy, 2001.

Ch. Hellmich and F.-J. Ulm. Are mineralized tissues open crystal foams reinforced by crosslinked collagen? – some energy arguments. *Journal of Biomechanics*, 35:1199 – 1212, 2002a.

Ch. Hellmich and F.-J. Ulm. A micromechanical model for the ultrastructural stiffness of mineralized tissues. *Journal of Engineering Mechanics (ASCE)*, 128(8):898 – 908, 2002b.

Ch. Hellmich and F.-J. Ulm. Average hydroxyapatite concentration is uniform in extracollageneous ultrastructure of mineralized tissue. *Biomechanics and Modeling in Mechanobiology*, 2:21 – 36, 2003.

Ch. Hellmich and F.-J. Ulm. Drained and undrained poroelastic properties of healthy and pathological bone: a poro-micromechanical investigation. *Transport in Porous Media*, 2004a. available online at www.kluweronline.com.

Ch. Hellmich and F.-J. Ulm. Micro-porodynamics of bones: prediction of the 'Frenkel-Biot' slow compressional wave. *Journal of Engineering Mechanics (ASCE)*, 2004b. in print.

Ch. Hellmich, F.-J. Ulm, and L. Dormieux. Can the diverse elastic properties of trabecular and cortical bone be attributed to only a few tissue-independent phase properties and their interactions? – arguments from a multiscale approach. *Biomechanics and Modeling in Mechanobiology*, 2:219 – 238, 2004b.

A.J. Hodge and J.A. Petruska. Recent studies with the electron microscope on ordered aggregates of the tropocollagen molecule. In G.N. Ramachandran, editor, *Aspects of Protein Structure – Proceedings of a Symposium held in Madras 14 - 18 January 1963 and organized by the University of Madras, India*, pages 289 – 300. Academic Press, London and New York, 1963.

C.E. Hoffler, K.E. Moore, K. Kozloff, P.K. Zysset, M.B. Brown, and S.A. Goldstein. Heterogeneity of bone lamellar-level elastic moduli. *Bone*, 26:603 – 609, 2000.

D.J.S. Hulmes and A. Miller. Quasi-hexagonal molecular packing in collagen fibrils. *Nature*, 282:878 – 880, 1979.

G.K. Hunter, P.V. Hauschka, A.R. Poole, L.C. Rosenberg, and H.A. Goldberg. Nucleation and inhibition of hydroxyapatite formation by mineralized tissue proteins. *Biochemical Journal*, 317:59 – 64, 1996.

I. Jäger and P. Fratzl. Mineralized collagen fibrils: a mechanical model with a staggered arrangement of mineral particles. *Biophysical Journal*, 79:1737 – 1746, 2000.

W.A. Kalender. *Computed Tomography*. Publicis MCD Verlag, Munich, Germany, 2000.

E.P. Katz and S.-T. Li. Structure and function of bone collagen fibrils. *Journal of Molecular Biology*, 80:1 – 15, 1973.

J.K. Katz and K. Ukraincik. On the anisotropic elastic properties of hydroxyapatite. *Journal of Biomechanics*, 4:221 – 227, 1971.

J.L. Katz. Anisotropy of Young's modulus of bone. *Nature*, 283:106 – 107, 1980.

J.L. Katz. *Composite material models for cortical bone*, pages 171 – 184. American Society of Mechanical Engineers, New York, NY, USA, 1981.

J.L. Katz, A.L. Maharidge, and H.S. Yoon. *The frequency and temperature dependence of longitudinal ultrasonical velocities in bovine bone*. Elsevier, 1986.

J.L. Katz, H.S. Yoon, S. Lipson, R Maharidge, A. Meunier, and P. Christel. The effects of remodelling on the elastic properties of bone. *Calcified Tissue International*, 36: S31 – S36, 1984.

T.M. Keaveny, E.F. Wachtel, C.M. Ford, and W.C. Hayes. Differences between the tensile and compressive strengths of bovine tibial trabecular bone depend on modulus. *Journal of Biomechanics*, 27(9):1137 – 1146, 1994.

H. Kolsky. *Stress waves in solids*. Clarendon Press, Oxford, UK, Cambridge, United Kingdom, 1953.

R. Lakes, H.S. Yoon, and J.L. Katz. Ultrasonic wave propagation and attenuation in wet bone. *Journal of Biomedical Engineering*, 8:143 – 148, 1986.

N. Laws. The determination of stress and strain concentrations at an ellipsoidal inclusion in an anisotropic material. *Journal of Elasticity*, 7(1):91 – 97, 1977.

N. Laws. A note on penny-shaped cracks in transversely isotropic materials. *Mechanics of Materials*, 4:209 – 212, 1985.

S. Lees. Water content in type I collagen tissues calculated from the generalized packing model. *International Journal of Biological Macromolecules*, 8:66 – 72, 1986.

S. Lees. Considerations regarding the structure of the mammalian mineralized osteoid from viewpoint of the generalized packing model. *Connective Tissue Research*, 16:281 – 303, 1987.

S. Lees, J.M. Ahern, and M. Leonard. Parameters influencing the sonic velocity in compact calcified tissues of various species. *Journal of the Acoustical Society of America*, 74(1):28 – 33, 1983.

S. Lees, L.C. Bonar, and H.A. Mook. A study of dense mineralized tissue by neutron diffraction. *International Journal of Biological Macromolecules*, 6:321 – 326, 1984a.

S. Lees, D. Hanson, and E.A. Page. Some acoustical properties of the otic bones of a fin whale. *Journal of the Acoustical Society of America*, 99(4):2421 – 2427, 1995.

S. Lees, D. Hanson, E.A. Page, and H.A. Mook. Comparison of dosage-dependent effects of beta-aminopropionitrile, sodium fluoride, and hydrocortisone on selected physical properties of cortical bone. *Journal of Bone and Mineral Research*, 9(9):1377 – 1389, 1994a.

S. Lees, J.D. Heeley, and P.F. Cleary. A study of some properties of a sample of bovine cortical bone using ultrasound. *Calcified Tissue International*, 29:107 – 117, 1979.

S. Lees, J.D. Heeley, and P.F. Cleary. Some properties of the organic matrix of a bovine cortical sample in various media. *Calcified Tissue International*, 33:81 – 86, 1981.

S. Lees and D.Z. Klopholz. Sonic velocity and attenuation in wet cow femur for the frequency range 5 to 100 MHz. *Ultrasound in Medicine and Biology*, 18(3):303 – 308, 1992.

S. Lees, M. Pineri, and M. Escoubes. A generalized packing model for type I collagen. *International Journal of Biological Macromolecules*, 6:133 – 136, 1984b.

S. Lees, K.S. Prostak, V.K. Ingle, and K. Kjoller. The loci of mineral in turkey leg tendon as seen by atomic force microscope and electron microscopy. *Calcified Tissue International*, 55:180 – 189, 1994b.

S. Lees and F.R. Rollins. Anisotropy in hard dental tissues. *Journal of Biomechanics*, 5: 557 – 566, 1972.

S. Lees, N.-J. Tao, and M. Lindsay. Studies of compact hard tissues and collagen by means of Brillouin light scattering. *Connective Tissue Research*, 24:187 – 205, 1990.

J.F. Mammone and S.M. Hudson. Micromechanics of bone strength and failure. *Journal of Biomechanics*, 26:439 – 446, 1993.

R.B. Martin, D.B. Burr, and N.A. Sharkey. *Skeletal Tissue Mechanics*. Springer, New York, 2001.

E. Mayr. *This is biology – the science of the living world*. Harvard University Press, Cambridge, MA, USA, 1 edition, 1997.

R.N. McCarthy, L.B. Jeffcott, and R.N. McCartney. Ultrasound speed in equine cortical bone: effects of orientation, density, porosity and temperature. *Journal of Biomechanics*, 23(11):1139 – 1143, 1990.

A. Miller. Collagen: the organic matrix of bone. *Philosophical Transactions of the Royal Society in London Series B*, 304:455 – 477, 1984.

A. Odgaard, J. Kabel, B. van Rietbergen, M. Dalstra, and R. Huiskes. Fabric and elastic principal directions of cancellous bone are closely related. *Journal of Biomechanics*, 30(5):487 – 495, 1997.

A. Odgaard, J. Kabel, B. van Rietbergen, and R. Huiskes. Architectural 3-D parameters and anisotropic elastic properties of cancellous bone. In P. Pederson and M.P. Bendsoe, editors, *IUTAM Symposium on Synthesis in Bio Solid Mechanics*, pages 33 – 42. Kluwer Academic Publishers, The Netherlands, 1999.

A.M. Parfitt. The cellular basis of bone remodeling: the quantum concept reexamined in light of recent advances in cell biology of bone. *Calcified Tissue International*, 36: S37 – S45, 1984.

R.M.V. Pidaparti, A. Chandran, Y. Takano, and C.H. Turner. Bone mineral lies mainly outside the collagen fibrils: Predictions of a composite model for osteonal bone. *Journal of Biomechanics*, 29(7):909 – 916, 1996.

C.D. Pomoroy and R.J. Mitton. The real densities of chrome and vegetable tanned leather. *Journal of the Society of the Leather Trades Chemists*, 35:360 – 375, 1951.

K.S. Prostak and S. Lees. Visualization of crystal-matrix structure. In situ demineralization of mineralized turkey leg tendon and bone. *Calcified Tissue International*, 59: 474 – 479, 1996.

J.-Y. Rho, M.C. Hobatho, and R.B. Ashman. Relations of mechanical properties to density and CT numbers in human bone. *Medical Engineering & Physics*, 17(5):347 – 355, 1995.

J.-Y. Rho, L. Kuhn-Spearing, and P. Zioupos. Mechanical properties and the hierarchical structure of bone. *Medical Engineering & Physics*, 20:92 – 102, 1998.

R.A. Robinson. Chemical analysis and electron microscopy of bone. In K. Rodale, J. Nicholson, and E.M. Brown, editors, *Bone as Tissue*, pages 186 – 250. McGraw-Hill, New York, 1960.

R.A. Robinson and S.R. Elliott. The water content in bone: I. The mass of water, inorganic crystals, organic matrix, and co$_2$ space components in a unit volume of dog bone. *The Journal of Bone and Joint Surgery*, 39-A:167 – 188, 1957.

M.A. Rubin, I. Jasiuk, J. Taylor, J. Rubin, T. Ganey, and R.P. Apkarian. TEM analysis of the nanostructure of normal and osteoporotic bone. *Bone*, 33:270 – 282, 2003.

J. Salencon. *Handbook of Continuum Mechanics – General Concepts.Thermoelasticity.* Springer, Berlin, Germany, 1 edition, 2001.

N. Sasaki. Orientation of mineral in bovine bone and the anisotropic mechanical properties of plexiform bone. *Journal of Biomechanics*, 24:57 – 61, 1991.

W.K. Sietsema. Animal models of cortical porosity. *Bone*, 17(4):297S – 305S, 1995.

S. Torabia. Identification of a relationship between the chemical constituents of mineralized tissues: bridging computed tomography and micromechanical modeling, for noninvasive determination of elastic properties of bone. Master's thesis, Vienna University of Technology, 2004.

C.H. Turner, S.C. Cowin, J.Y. Rho, R.B. Ashman, and J.C. Rice. The fabric dependence of the orthotropic elastic constants of cancellous bone. *Journal of Biomechanics*, 23 (6):549 – 561, 1990.

T. Uchiyama, T. Tanizawa, H. Muramatsu, N. Endo, H.E. Takahashi, and T. Hara. Three-dimensional microstructural analysis of human trabecular bonein relation to its mechanical properties. *Bone*, 25(4):487 – 491, 1999.

M.R. Urist, R.J. DeLange, and G.A.M. Finerman. Bone cell differentiation and growth factors. *Science*, 220:680 – 686, 1983.

S. Weiner, T. Arad, I. Sabanay, and W. Traub. Rotated plywood structure of primary lamellar bone in the rat: Orientation of the collagen fibril arrays. *Bone*, 20:509 – 514, 1997.

S. Weiner and H.D. Wagner. The material bone: structure - mechanical function relations. *Annual Review of Materials Science*, 28:271 – 298, 1998.

W.J. Whitehouse. The quantitative morphology of anisotropic trabecular bone. *Journal of Microscopy*, 101:153 – 168, 1974.

A. Zaoui. Matériaux hétérogène et composites [Heterogeneous materials and composites], 1997a. Lecture Notes from École Polytechnique, Palaiseau, France, in French.

A. Zaoui. *Structural morphology and constitutive behavior of microheterogeneous materials*, pages 291 – 347. Springer, Wien – New York, 1997b.

A. Zaoui. Continuum micromechanics: Survey. *Journal of Engineering Mechanics (ASCE)*, 128(8):808 – 816, 2002.

L. Zylberberg, W. Traub, V. de Buffrenil, F. Allizard, T. Arad, and S. Weiner. Rostrum of a toothed whale: Ultrastructural study of a very dense bone. *Bone*, 23:241 – 247, 1998.

FSC
www.fsc.org
MIX
Papier | Fördert gute Waldnutzung
FSC® C083411

Kolibri 360 GmbH
Ferdinand-Jühlke-Straße 7
99095 Erfurt, Deutschland
produktsicherheit@kolibri360.de